ADVANCES IN CHEMICAL ENGINEERING
Volume 28

Molecular Modeling and Theory in Chemical Engineering

ADVANCES IN
CHEMICAL ENGINEERING

Editor-in-Chief

JAMES WEI
School of Engineering and Applied Science
Princeton University
Princeton, New Jersey

Editors

MORTON M. DENN
College of Chemistry
University of California at Berkeley
Berkeley, California

JOHN H. SEINFELD
Department of Chemical Engineering
California Institute of Technology
Pasadena, California

GEORGE STEPHANOPOULOS
Department of Chemical Engineering
Massachusetts Institute of Technology
Cambridge, Massachusetts

ARUP CHAKRABORTY
Department of Chemical Engineering
University of California at Berkeley
Berkeley, California

JACKIE YING
Department of Chemical Engineering
Massachusetts Institute of Technology
Cambridge, Massachusetts

NICHOLAS PEPPAS
Purdue University
School of Chemical Engineering
West Lafayette, Indiana

ADVANCES IN
CHEMICAL
ENGINEERING

Volume 28

Molecular Modeling and Theory in Chemical Engineering

Edited by
ARUP CHAKRABORTY

Department of Chemical Engineering
University of California at Berkeley
Berkeley, California

ACADEMIC PRESS

A Harcourt Science and Technology Company

San Diego San Francisco New York Boston London Sydney Tokyo

Cover image (paperback only): Illustrations courtesy of Dimitrios Maroudas (see Chapter 7, Figure 12).

This book is printed on acid-free paper. ∞

Copyright © 2001 by ACADEMIC PRESS

All Rights Reserved.
No part of this publication may be reproduced or transmitted in any form or by any means, electronic or mechanical, including photocopy, recording, or any information storage and retrieval system, without permission in writing from the Publisher.

The appearance of the code at the bottom of the first page of a chapter in this book indicates the Publisher's consent that copies of the chapter may be made for personal or internal use of specific clients. This consent is given on the condition, however, that the copier pay the stated per copy fee through the Copyright Clearance Center, Inc. (222 Rosewood Drive, Danvers, Massachusetts 01923), for copying beyond that permitted by Sections 107 or 108 of the U.S. Copyright Law. This consent does not extend to other kinds of copying, such as copying for general distribution, for advertising or promotional purposes, for creating new collective works, or for resale. Copy fees for pre-2001 chapters are as shown on the title pages. If no fee code appears on the title page, the copy fee is the same as for current chapters.
0065-2377/2001 $35.00

Explicit permission from Academic Press is not required to reproduce a maximum of two figures or tables from an Academic Press chapter in another scientific or research publication provided that the material has not been credited to another source and that full credit to the Academic Press chapter is given.

Academic Press
A Harcourt Science and Technology Company
525 B Street, Suite 1900, San Diego, California 92101-4495, USA
http://www.academicpress.com

Academic Press
Harcourt Place, 32 Jamestown Road, London NW1 7BY, UK
http://www.academicpress.com

International Standard Book Number: 0-12-008528-3 (case)
International Standard Book Number: 0-12-743274-4 (pb)

PRINTED IN THE UNITED STATES OF AMERICA
01 02 03 04 05 06 QW 9 8 7 6 5 4 3 2 1

CONTENTS

CONTRIBUTORS xi
PREFACE xiii

Hyperparallel Tempering Monte Carlo and Its Applications

QILIANG YAN AND JUAN J. DE PABLO

I. Introduction 1
II. Methodology 3
III. Applications 5
   A. Lennard–Jones Fluid 5
   B. Primitive Model Electrolyte Solutions 7
   C. Homopolymer Solutions and Blends 11
   D. Semiflexible Polymers and Their Blends with Flexible Polymers 15
   E. Block Copolymers and Random Copolymers 17
IV. Discussion and Conclusion 18
   References 20

Theory of Supercooled Liquids and Glasses: Energy Landscape and Statistical Geometry Perspectives

PABLO G. DEBENEDETTI, FRANK H. STILLINGER, THOMAS M. TRUSKETT, AND CATHERINE P. LEWIS

I. Introduction 22
   A. Phenomenology of Vitrification by Supercooling 23
   B. Open Questions 29
   C. Structure of This Article 32
II. The Energy Landscape 33
III. Statistical Geometry and Structure 39
   A. Void Geometry and Connections to the Energy Landscape 40
   B. Quantifying Molecular Disorder in Equilibrium and Glassy Systems . . 45
IV. Landscape Dynamics and Relaxation Phenomena 50
V. Thermodynamics 60

VI. Conclusion 70
   References 72

A Statistical Mechanical Approach to Combinatorial Chemistry

MICHAEL W. DEEM

I. Introduction 81
II. Materials Discovery 83
   A. The Space of Variables 84
   B. Library Design and Redesign 85
   C. Searching the Variable Space by Monte Carlo 87
   D. The Simplex of Allowed Compositions 89
   E. Significance of Sampling 91
   F. The Random Phase Volume Model 92
   G. Several Monte Carlo Protocols 94
   H. Effectiveness of the Monte Carlo Strategies 95
   I. Aspects of Further Development 96
III. Protein Molecular Evolution 97
   A. What Is Protein Molecular Evolution? 98
   B. Background on Experimental Molecular Evolution 100
   C. The Generalized NK Model 102
   D. Experimental Conditions and Constraints 104
   E. Several Hierarchical Evolution Protocols 105
   F. Possible Experimental Implementations 109
   G. Life Has Evolved to Evolve 111
   H. Natural Analogs of These Protocols 113
   I. Concluding Remarks on Molecular Evolution 115
IV. Summary 117
   References 118

Fluctuation Effects in Microemulsion Reaction Media

VENKAT GANESAN AND GLENN H. FREDRICKSON

I. Introduction 123
II. Reactions in the Bicontinuous Phase 127
   A. Diffusion Equations 127
   B. Objectives 128
   C. Mean-Field Analysis 129
   D. Renormalization Group Theory 132
   E. Discussion 134
   F. Summary 135
III. Reactions in the Droplet Phase 136
   A. Outline 136

B. Fluctuations of the Droplet Phase	137
C. Diffusion Equation and Perturbation Expansion	139
D. Consideration of Temporal Regimes	141
E. Intermediate Times	143
F. Short Time Regime	143
G. Effect of the Péclet Number	144
H. Discussion	145
I. Other Effects	146
J. Summary	146
References	147

Molecular Dynamics Simulations of Ion–Surface Interactions with Applications to Plasma Processing

DAVID B. GRAVES AND CAMERON F. ABRAMS

I. Introduction	149
A. Plasma Processing	149
B. Length Scales in Plasma Processing	152
C. The Nature of Plasma–Surface Interactions	153
D. Ion–Surface Interactions in Plasma Processing	155
II. Use of Molecular Dynamics to Study Ion–Surface Interactions	156
A. Simulation Procedure	156
III. Mechanisms of Ion-Assisted Etching	161
A. Experimental Studies of Ion-Assisted Etching Mechanisms	161
B. Molecular Dynamics Studies of Ion-Assisted Etching Mechanisms	164
C. Ion–Surface Scattering Dynamics	172
D. Ion–Surface Interactions with both Deposition and Etching: CF_3^+/Si	180
IV. Concluding Remarks	198
References	199

Characterization of Porous Materials Using Molecular Theory and Simulation

CHRISTIAN M. LASTOSKIE AND KEITH E. GUBBINS

I. Introduction	203
II. Disordered Structure Models	206
A. Porous Glasses	206
B. Microporous Carbons	209
C. Xerogels	213
D. Templated Porous Materials	216
III. Simple Geometric Pore Structure Models	218
A. Molecular Simulation Adsorption Models	222
B. Density Functional Theory Adsorption Models	225

C. Semiempirical Adsorption Models 231
D. Classical Adsorption Models 239
IV. Conclusions 244
References 246

Modeling of Radical–Surface Interactions in the Plasma-Enhanced Chemical Vapor Deposition of Silicon Thin Films

DIMITRIOS MAROUDAS

I. Introduction 252
II. Computational Methodology 254
   A. The Hierarchical Approach 255
   B. Density-Functional Theory 257
   C. Empirical Description of Interatomic Interactions 258
   D. Methods of Surface Preparation 260
   E. Methods of Surface Characterization and Reaction Analysis . . . 263
III. Surface Chemical Reactivity with SiH_x Radicals 264
   A. Structure of Crystalline and Amorphous Silicon Surfaces . . . 265
   B. Interactions of SiH_x Radicals with Crystalline Silicon Surfaces . . . 266
   C. Interactions of SiH_x Radicals with Surfaces of Amorphous Silicon Films 270
IV. Plasma–Surface Interactions during Silicon Film Growth 273
   A. Surface Chemical Reactions during Film Growth 274
   B. Mechanism of Amorphous Silicon Film Growth 280
   C. Surface Evolution and Film Structural Characterization . . . 281
   D. Film Surface Composition and Comparison with Experiment . . . 283
   E. The Role of the Dominant Deposition Precursor 284
   F. The Role of Chemically Reactive Minority Species 286
V. Summary 290
References 291

Nanostructure Formation and Phase Separation in Surfactant Solutions

SANAT K. KUMAR, M. ANTONIO FLORIANO,
AND ATHANASSIOS Z. PANAGIOTOPOULOS

I. Introduction 298
II. Simulation Details 300
   A. Models and Methods 300
   B. Some Methodological Issues 301
III. Results 302
   A. Homopolymer Chains 302
   B. Role of Different Interaction Sets 304
IV. Discussion 308

V. Conclusions 310
References 310

Some Chemical Engineering Applications of Quantum Chemical Calculations

STANLEY I. SANDLER, AMADEU K. SUM, AND SHIANG-TAI LIN

I. Introduction 314
II. *Ab Initio* Interaction Potentials and Molecular Simulations 315
III. Infinite Dilution Activity Coefficients and Partition Coefficients from Quantum Mechanical Continuum Solvation Models 325
IV. Use of Computational Quantum Mechanics to Improve Thermodynamic Property Predictions from Group Contribution Methods 335
V. Use of *ab Initio* Energy Calculations for Phase Equilibrium Predictions . . . 341
VI. Conclusions 347
References 348

Car–Parrinello Methods in Chemical Engineering: Their Scope and Potential

BERNHARDT L. TROUT

I. Introduction 353
II. Objectives and Description of This Article 355
III. Objectives of Car–Parrinello Methods and Classes of Problems to Which They Are Best Applicable 356
IV. Methodology 357
   A. Classical Molecular Dynamics 357
   B. Density-Functional Theory 358
   C. Choice of Model and Solution of the Equations Using Plane-Wave Basis Sets and the Pseudopotential Method 362
   D. Car–Parrinello Molecular Dynamics 368
V. Applications 370
   A. Gas-Phase Processes 372
   B. Processes in Bulk Materials 376
   C. Properties of Liquids, Solvation, and Reactions in Liquids 378
   D. Heterogeneous Reactions and Processes on Surfaces 382
   E. Phase Transitions 386
   F. Processes in Biological Systems 389
VI. Advances in Methodology 392
VII. Concluding Remarks 393
Appendix A: Further Reading 393
Appendix B: Codes with Capabilities to Perform Car–Parrinello Molecular Dynamics 394
References 394

Theory of Zeolite Catalysis

R. A. van Santen and X. Rozanska

I. Introduction	400
II. The Rate of a Catalytic Reaction	401
III. Zeolites as Solid Acid Catalysts	403
IV. Theoretical Approaches Applied to Zeolite Catalysis	407
A. Simulation of Alkane Adsorption and Diffusion	407
B. Hydrocarbon Activation by Zeolitic Protons	414
C. Kinetics	427
V. Concluding Remarks	432
References	433

Morphology, Fluctuation, Metastability, and Kinetics in Ordered Block Copolymers

Zhen-Gang Wang

I. Introduction	439
II. Anisotropic Fluctuations in Ordered Phases	441
III. Kinetic Pathways of Order–Order and Order–Disorder Transitions	445
IV. The Nature and Stability of Some Nonclassical Phases	450
V. Long-Wavelength Fluctuations and Instabilities	452
VI. Morphology and Metastability in ABC Triblock Copolymers	456
VII. Conclusions	460
References	460

Index	465
Contents of Volumes in this Serial	487

CONTRIBUTORS

Numbers in parentheses indicate the pages on which the authors' contributions begin.

CAMERON F. ABRAMS, *Department of Chemical Engineering, University of California–Berkeley, Berkeley, California 94720* (149)

PABLO G. DEBENEDETTI, *Department of Chemical Engineering, Princeton University, Princeton, New Jersey 08544* (21)

JUAN J. DE PABLO, *Department of Chemical Engineering, University of Wisconsin–Madison, Madison, Wisconsin 53706* (1)

MICHAEL W. DEEM, *Chemical Engineering Department, University of California–Los Angeles, Los Angeles, California 90095–1592* (81)

M. ANTONIO FLORIANO, *Department of Chemistry, Università della Calabria, 87036 Arcavacata di Rende (CS), Italy* (297)

GLENN H. FREDRICKSON, *Department of Chemical Engineering, University of California–Santa Barbara, Santa Barbara, California 93106* (123)

VENKAT GANESAN, *Department of Chemical Engineering, University of California–Santa Barbara, Santa Barbara, California 93106* (123)

DAVID B. GRAVES, *Department of Chemical Engineering, University of California–Berkeley, Berkeley, California 94720* (149)

KEITH E. GUBBINS, *Department of Chemical Engineering, North Carolina State University, Raleigh, North Carolina 27695* (203)

SANAT K. KUMAR, *Department of Materials Science and Engineering, Pennsylvania State University, University Park, Pennsylvania 16802* (297)

CHRISTIAN M. LASTOSKIE, *Department of Chemical Engineering, Michigan State University, East Lansing, Michigan 48824* (203)

CATHERINE P. LEWIS, *Department of Chemical Engineering, Princeton University, Princeton, New Jersey 08544* (21)

SHIANG-TAI LIN, *Center for Molecular and Engineering Thermodynamics, Department of Chemical Engineering, University of Delaware, Newark, Delaware 19716* (313)

DIMITRIOS MAROUDAS, *Department of Chemical Engineering, University of California–Santa Barbara, Santa Barbara, California 93106–5080* (251)

ATHANASSIOS Z. PANAGIOTOPOULOS, *Department of Chemical Engineering, Princeton University, Princeton, New Jersey 08544* (297)

X. ROZANSKA, *Schuit Institute of Catalysis, Eindhoven University of Technology, Eindhoven 5600 MB, The Netherlands* (399)

STANLEY I. SANDLER, *Center for Molecular and Engineering Thermodynamics, Department of Chemical Engineering, University of Delaware, Newark, Delaware 19716* (313)

FRANK H. STILLINGER, *Bell Laboratories, Lucent Technologies, Murray Hill, New Jersey 07974; and Princeton Materials Institute, Princeton, New Jersey 08544* (21)

AMADEU K. SUM, *Center for Molecular and Engineering Thermodynamics, Department of Chemical Engineering, University of Delaware, Newark, Delaware 19716* (313)

BERNHARDT L. TROUT, *Department of Chemical Engineering, Massachusetts Institute of Technology, Cambridge, Massachusetts 02139* (353)

THOMAS M. TRUSKETT, *Department of Chemical Engineering, Princeton University, Princeton, New Jersey 08544* (21)

R. A. VAN SANTEN, *Schuit Institute of Catalysis, Eindhoven University of Technology, Eindhoven 5600 MB, The Netherlands* (399)

ZHEN-GANG WANG, *Division of Chemistry and Chemical Engineering, California Institute of Technology, Pasadena, California 91125* (439)

QILIANG YAN, *Department of Chemical Engineering, University of Wisconsin–Madison, Madison, Wisconsin 53706* (1)

PREFACE

The ultimate objective of the profession of chemical engineering is the manufacture of chemicals and products that improve societal and economic conditions. Achieving this goal often requires that chemical engineers understand and exploit many physical, chemical, and biological phenomena. In recent years, chemical engineers have increasingly been involved in the design, synthesis, and manufacture of high-value-added products and chemicals. These technologies often demand that product properties and processing methods be controlled with precision. Similar issues arise in research aimed toward developing more efficient processing of petroleum products and developing catalysts for synthesis of alternative fuels. In some cases, product properties must be precisely controlled at the macroscopic level, and in other cases, the properties that we seek are on much smaller scales (nanometers to micrometers). One way to confront this challenge with both classes of systems is to learn how to manipulate system characteristics at the molecular and/or mesoscopic scales so that we obtain the desired properties. Learning how molecular constitution and mesoscopic characteristics influence the properties of a system of interacting components can only be addressed by synergistic experimental and theoretical research. The pertinent experimental and theoretical methods must be able to interrogate systems on a wide range of length and time scales. Chemical engineers are playing an important role in the development and application of a number of such experimental and theoretical tools. These research efforts are taking steps toward developing the knowledge base required to relate structures to properties for both synthetic and biological systems. This volume of *Advances in Chemical Engineering* focuses on theoretical and computational efforts at the frontiers of a number of different application areas that benefit from such research.

The bedrocks of the theoretical and computational methods that allow study of relationships between molecular and mesoscopic scale events and system properties are quantum and statistical mechanics. Thus, this volume comprises chapters that describe the development and application of quantum and statistical mechanical methods to various problems of technological relevance. The application areas include catalysis and reaction engineering, processing of materials for microelectronic applications, polymer science and engineering, fluid phase equilibrium, and combinatorial methods for materials discovery. The theoretical methods that are discussed in the various chapters include electronic structure calculations, *ab initio* molecular dynamics simulations, Monte-Carlo simulation methods, field-theoretic methods, and various theories of the liquid state. The diversity of application areas represented in this volume reflects the fact that methods based on quantum and statistical mechanics now play an important role in research that is relevant to a variety of technologies. The diversity of methods discussed in this volume reflects the fact that for complex problems no single method can serve as a panacea. In other words, studying properties influenced by phenomena at different length and time scales requires a hierarchy of methods.

This collection of articles is not a comprehensive compendium of the interesting work being done to study complex systems using quantum and statistical mechanical methods. It is hoped, however, that this representative sampling of work being carried out by chemical engineers in this broad area will provide the beginning graduate student and the experienced practitioner with a sense of the current state of the art and the challenges that need to be confronted in the future. My personal opinion is that future volumes dedicated to this broad topic will witness a greater emphasis on nonequilibrium phenomena, the coupling of quantum and statistical mechanical approaches, and more applications focused on biomedical problems.

My fellow editors of *Advances in Chemical Engineering*, the staff at Academic Press, and I thank the authors for taking time out of their busy schedules to contribute to this volume. The effort involved in writing good review articles is a selfless service to the profession and is truly appreciated. A personal note of thanks is also extended to the authors for their patience during the review and production process.

ARUP K. CHAKRABORTY

HYPERPARALLEL TEMPERING MONTE CARLO AND ITS APPLICATIONS

Qiliang Yan and Juan J. de Pablo

Department of Chemical Engineering, University of Wisconsin–Madison,
Madison, Wisconsin 53706

I. Introduction 1
II. Methodology 3
III. Applications 5
  A. Lennard–Jones Fluid 5
  B. Primitive Model Electrolyte Solutions 7
  C. Homopolymer Solutions and Blends 11
  D. Semiflexible Polymers and Their Blends with
    Flexible Polymers 15
  E. Block Copolymers and Random Copolymers 17
IV. Discussion and Conclusion 18
  References 20

This review discusses a newly proposed class of tempering Monte Carlo methods and their application to the study of complex fluids. The methods are based on a combination of the expanded grand canonical ensemble formalism (or simple tempering) and the multi-dimensional parallel tempering technique. We first introduce the method in the framework of a general ensemble. We then discuss a few implementations for specific systems, including primitive models of electrolytes, vapor–liquid and liquid–liquid phase behavior for homopolymers, copolymers, and blends of flexible and semiflexible polymers. © 2001 Academic Press.

I. Introduction

Complex fluids such as electrolyte solutions, polymer solutions, and biological macromolecule solutions pose significant obstacles to molecular simulation, particularly at low temperatures and elevated densities. Conventional molecular dynamics methods are unable to generate trajectories that are long enough to cover the inherently long characteristic relaxation times that characterize polymeric fluids, and naïve Monte Carlo techniques are unable to sample their configuration space efficiently. All of these systems, however, are of engineering importance. Unfortunately, these are also systems for which our theoretical understanding is far from complete. Predictive models for the equilibrium thermodynamic and structural properties of such fluids are required to design chemical and separation processes; to formulate new models, it would be useful to have access to the results of simulations.

When only the equilibrium properties of a complex fluid are of interest, it is possible to devise "nonphysical" simulation techniques that are sometimes able to circumvent the sampling problems that are usually associated with complex fluids. Examples of such techniques include configurational bias Monte Carlo methods, multicanonical ensemble simulations, J-walking, $1/k$ sampling, simulated tempering, and parallel tempering [1–14]. In this review we discuss some of our recent experiences with parallel tempering. This method has a number of useful features, which make it attractive for the study of complex fluids. Interestingly, while the idea of parallel tempering is not new [8, 9], its application to the study of many-body fluids has been limited. We therefore present results for a variety of systems, and in each case we try to emphasize the advantages provided by tempering over more conventional techniques.

The basic idea of parallel tempering consists of simulating several copies of a system in parallel; each copy or "replica" is constructed to represent the same system in a different thermodynamic state. Conventional Monte Carlo methods are employed to sample the configuration of each distinct replica under the relevant thermodynamic conditions. In addition to the trial moves involved in such methods, however, attempts are made to interchange the configurations corresponding to any two replicas of the system. Such trial "swaps" are accepted according to probability criteria that ensure the appropriate ensembles are sampled. The benefit of swapping is that if one of the replicas relaxes much faster than the others (e.g., a replica at a high temperature), the fast-evolving configurations in that replica can be artificially "propagated" to other boxes via exchanges, thereby effectively accelerating the relaxation of all other copies of the system.

Depending on the system and the ensemble of choice, the thermodynamic state of a replica can be specified through the number of molecules of each species, the volume, the temperature, the pressure, and the chemical potential. Our experience (and that of others [15, 16]) suggests that, from the point of view of improving sampling, open ensembles offer a number of advantages over closed systems. In open ensembles, molecules can be completely removed from a system and reinserted at a later point in completely different positions and configurations, thereby circumventing diffusional bottlenecks. Furthermore, for difficult systems, such as polymers, deletions, and insertions can be facilitated significantly by resorting to expanded ensemble methods [17, 18]. Most of the implementations of hyperparallel tempering Monte Carlo (HPTMC) reported here are carried out in open ensembles, and whenever possible we also capitalize on the benefits provided by configurational bias and expanded ensemble techniques. As discussed in this review, it turns out that in some cases HPTMC can provide striking efficiency increases over traditional methods for the simulation of complex fluids with minimal changes to existing simulation algorithms and codes.

II. Methodology

Formally, we consider a generalized ensemble whose partition function is given by

$$Z(\mathbf{f}) = \sum_{x} \Omega(x) w(x, \mathbf{f}), \tag{1}$$

where $\mathbf{f}$ denotes a set of specified generalized forces or potentials, which determine the thermodynamic state of the system. In Eq. (1), x is used to denote a microscopic state or an instantaneous configuration of the system, $\Omega(x)$ is the density of states, and $w(x, \mathbf{f})$ is an arbitrary weighting function for state x, at the given set of generalized potentials $\mathbf{f}$. The grand canonical ensemble is recovered by writing

$$\mathbf{f} = \{T, \mu\}, \qquad w(x, \mathbf{f}) = \exp(-\beta U(x) + N(x)\beta\mu), \tag{2}$$

where $\beta = 1/k_B T$, T is the temperature, k_B is Boltzmann's constant, μ is the specified chemical potential, $U(x)$ is the potential energy corresponding to configuration x, and $N(x)$ is the number of particles in configuration x.

Hyperparallel tempering simulations are conducted on a composite ensemble, which consists of M, noninteracting replicas of the above-mentioned generalized ensemble. Each replica can have a different set of generalized potentials. The complete state of the composite ensemble is specified through $\mathbf{x} = (x_1, x_2, ..., x_M)^T$, where x_i denotes the state of the ith replica. The partition function Z_c of the composite ensemble is given by

$$Z_c = (\mathbf{f}_1, \mathbf{f}_2, \cdots, \mathbf{f}_M) = \prod_{i=1}^{M} Z(\mathbf{f}_i). \tag{3}$$

The unnormalized probability density of the complete state **x** is proportional to

$$p(\mathbf{x}) \prod_{i=1}^{M} \Omega(x_i) w(x_i, \mathbf{f}_i). \tag{4}$$

In expanded grand canonical ensemble simulations [18] (also called simple tempering simulations), the system can jump along a set of expanded states, in addition to the conventional (N, U) phase-space variables of a grand canonical ensemble. For the particular implementation to polymeric fluids, chain molecules are inserted or removed gradually, i.e., several segments at a time. In other words, a simulation box contains several "regular" chain molecules and a tagged chain, whose length n_y fluctuates during the simulation; n_y therefore serves as the expanded state variable. A preweighting factor $\exp(\Psi(y))$ is assigned to each expanded state y. In the language of Eq. (1), the weighting function for the expanded grand canonical ensemble is

$$\mathbf{f} = \{T, \mu, \Psi\}, \qquad w(x, \mathbf{f}) = \exp[-\beta U(x) + N(x)\beta\mu + \Psi(y)]. \tag{5}$$

If we assume that the segmental chemical potential is independent of chain length, we can set the preweighting function to be

$$\Psi(y) = \frac{n_y}{n}\beta\mu^r = \frac{n_y}{n}\left[\beta\mu - \ln\left(\frac{N_y}{V}\right)\right], \tag{6}$$

where $N_y = N + n_y/n$; n_y is the length of the tagged chain and n is the length of a full polymer chain. In Eq. (6), μ^r denotes the residual chemical potential of a polymer chain.

Figure 1 illustrates schematically the implementation of HPTMC. Each box in the figure represents a replica of the simulation system; each replica has a different value of T, μ, and n_y. To implement a hyperparallel tempering algorithm, three types of trial moves are necessary. (1) Conventional canonical Monte Carlo moves are used to sample configurations in each replica of the system. These moves include translational or rotational displacements and configurational-bias or reptation moves for polymers. (2) Trial shrinking or growing moves are proposed to change the length of a tagged chain in each replica, thereby implementing the underlying expanded grand canonical formalism. (3) Configuration swaps or exchanges are attempted between any two randomly chosen replicas. The arrows in Fig. 1 correspond to different types of moves.

The acceptance criteria corresponding to trials moves of type 1 or 2 are fairly standard and have been reported in a number of texts and articles. We therefore limit the remainder of this section to a brief discussion of the acceptance criteria for trial swap moves. Consider a swap between two replicas, i and j. After the swap, the new state of replica i will be the current

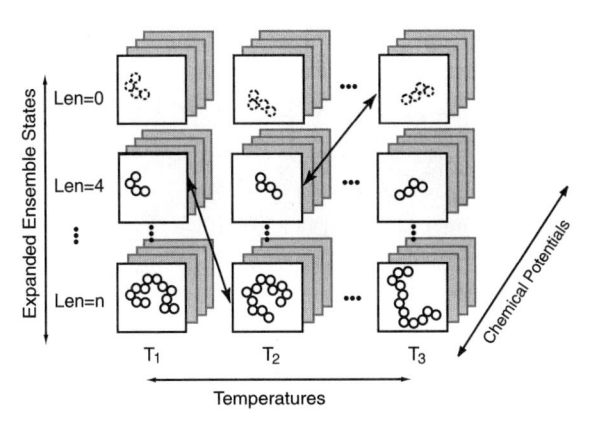

FIG. 1. Schematic illustration of the implementation of the hyperparallel Monte Carlo method. Each box represents a distinct replica of the simulated system; these replicas are simulated simultaneously in a single run. In addition to traditional Monte Carlo trial moves, these replicas can (1) change their state variables in the expanded dimension and (2) exchange configurations with each other, thereby visiting different values of T and μ.

state of replica j, and the new state of replica j will be the current one for replica i; i.e.,

$$x_i^{\text{new}} = x_j^{\text{old}}$$
$$x_j^{\text{new}} = x_i^{\text{old}}. \tag{7}$$

Metropolis acceptance criteria are applied to trial swap moves. Given that simulations are being conducted in a composite ensemble of several replicas, a trial swap move is accepted with probability

$$p_{\text{acc}}(x_i \leftrightarrow x_j) = \min\left[1, \frac{w(x_j, \mathbf{f}_i)w(x_i, \mathbf{f}_j)}{w(x_i, \mathbf{f}_i)w(x_j, \mathbf{f}_j)}\right]. \tag{8}$$

The particular swapping acceptance criteria for hyperparallel tempering are obtained by substituting Eq. (6) into Eq. (8),

$$p_{\text{acc}} = \min[1, \exp(\Delta\beta\Delta U - \Delta N_y \Delta(\beta\mu))], \tag{9}$$

where $\Delta\beta = \beta_i - \beta_j$, $\Delta U = U_i - U_j$, $\Delta N_y = N_{y,i} - N_{y,j}$, and $\Delta(\beta\mu) = \beta_i\mu_i - \beta_j\mu_j$.

III. Applications

A. LENNARD–JONES FLUID

The phase behavior of the Lennard–Jones fluid has been studied extensively in the past. It therefore provides an ideal example for examining the

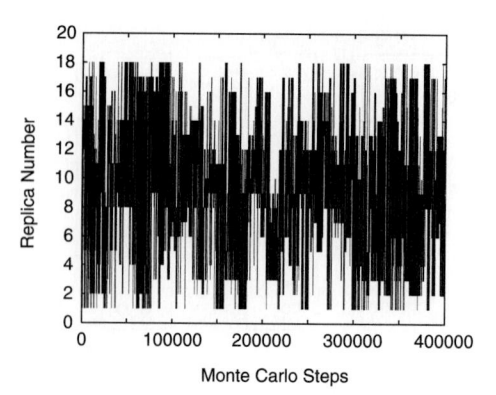

FIG. 2. Replica number as a function of Monte Carlo steps, for $T^* = 0.73$ and $\beta\mu = -5.30$.

performance and accuracy of HPTMC vis-à-vis those of other, more conventional methods. In this application, 18 replicas are simulated in parallel. Histogram reweighting techniques are employed to calculate the phase diagram of the system [22–24]. Figure 2 illustrates how replicas are swapped during a simulation. Each particular configuration or collection of particles can be labeled according to the replica number that it occupies at the beginning of a simulation. When a trial swap move is accepted, two configurations move to new replicas, and the transition is registered as a step change in the figure. Figure 2 describes the trajectory of configuration number 13 as it travels through different replicas during the course of a simulation. After a successful swap move, two replicas receive completely new configurations, thereby reducing dramatically the correlation time corresponding to a given thermodynamic state. Furthermore, through such swapping, configurations that were originally in low-temperature and high-density boxes can be passed over to high-temperature and moderate-density boxes, where they can relax more rapidly. Relaxed configurations can subsequently return to their original box, thereby accelerating the overall relaxation of the global system and facilitating sampling of phase space under adverse conditions. It is instructive to note that swapping provides an efficient way of sampling the tails of the distribution function corresponding to a given thermodynamic state; such tails often contribute significantly to thermodynamic averages, and they can be difficult to sample using conventional techniques.

Figure 3 compares the phase diagram calculated from histograms corresponding to all 18 replicas with literature data for the same fluid [26]. The two sets of data are in quantitative agreement [13]. The slight discrepancies at high temperatures are due mainly to different definitions of the equilibrium saturated density. (In this work, we regard the mean density corresponding to a peak of the distribution as the equilibrium value; Wilding *et al.* [26] define

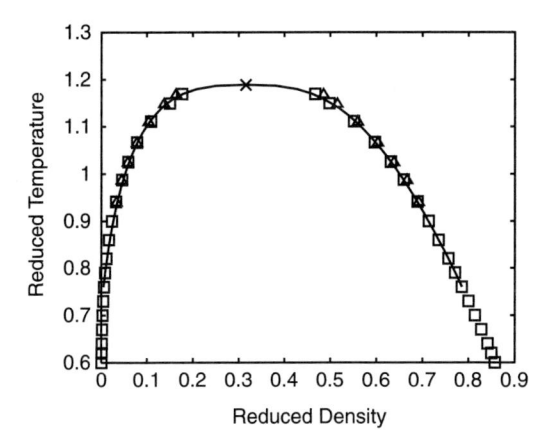

FIG. 3. Phase diagram (vapor–liquid equilibria) for a truncated Lennard–Jones fluid. The squares correspond with the results of this work, and the triangles show results reported by Wilding et al. [26]. Statistical errors are smaller than the symbol size. The solid line is an Ising form fit to the simulation data.

it as the peak value of the distribution.) As shown in Fig. 3, the proposed method is able to generate phase equilibrium data at temperatures and densities in the near vicinity of the triple point of the truncated Lennard–Jones fluid (e.g., $T^* = 0.60$ and $\rho^* = 0.86$). A simple Gibbs ensemble method or conventional grand canonical simulations would be much more demanding under such conditions.

B. PRIMITIVE MODEL ELECTROLYTE SOLUTIONS

Electrolyte solutions play a central role in chemical engineering practice. As such, they have been studied extensively (both theoretically and experimentally) in the past, but simulations have been limited. Simulations could provide much needed numerical data to verify the accuracy and validity of currently available, approximate predictions, and also yield useful insights for development of new theories. Primitive models provide one of the simplest representations of electrolyte solutions. In these models, the system is described by a binary mixture of charged hard spheres immersed in a dielectric continuum. Notwithstanding the apparent simplicity of the model, the calculation of the phase behavior of primitive electrolytes has presented a challenge to molecular simulations for several decades. The tendency of ions to associate, the long-range nature of Coulombic interactions, and the low temperatures that are often of interest in the study of electrolytes have all conspired to render such calculations particularly demanding. Primitive model electrolytes are believed to exhibit a low-temperature gas–liquid region of coexistence [13, 27–31], and only recently has some consensus begun to emerge regarding the nature of the coexistence curve and the precise location of the critical point of the *restricted* primitive model (the special case in which both cations and anions have exactly the same size and charge) [31, 13]. The phase behavior of primitive models in general (having cations and anions of different size and charge) remains unknown, and researchers have had to rely on elaborate theoretical predictions to delineate the phase behavior of such systems [27, 32, 33].

We have used parallel tempering methods to study the general case of asymmetric primitive models. We use approximately 10 to 15 replicas in our calculations, and the composite system is simulated in parallel for at least 2×10^6 Monte Carlo steps to calculate a coexistence curve. Each Monte Carlo step consists of 200 particle displacements and 100 insertion or deletion attempts. Configuration swaps are attempted every 20 Monte Carlo steps. To estimate critical properties, four or five boxes are simulated in parallel for at least 10×10^6 Monte Carlo steps. Longer simulations are required as the asymmetry of the ions increases.

Figure 4 shows simulated binodal curves for asymmetric ionic systems. Coexistence curves are shown for $\lambda = 1, 0.75, 0.50,$ and 0.25, where $\lambda = \sigma_+/\sigma_-$ is the ratio of the radius of cations and anions and therefore provides a measure of the asymmetry of the system. The results of our simulations indicate that as λ decreases from 1 (the restricted model) to 0.05 (the most asymmetric system considered in our simulations), both T_c^* and ρ_c^* decrease. For nearly symmetric electrolytes (e.g., $\lambda = 0.75$), the effect of λ is relatively

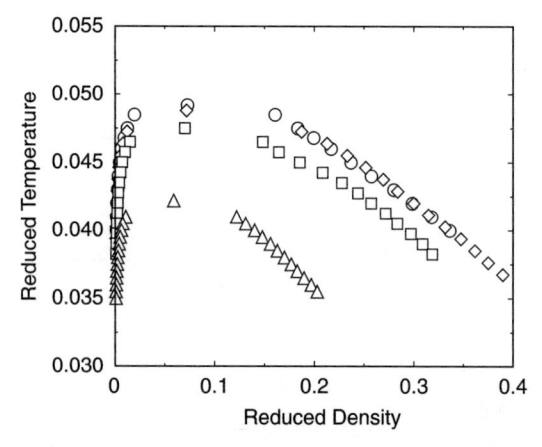

FIG. 4. Simulated binodal curves for size-asymmetric electrolyte systems with different λ values. Circles, $\lambda = 1$; diamonds, $\lambda = 0.75$; squares, $\lambda = 0.5$; triangles, $\lambda = 0.25$.

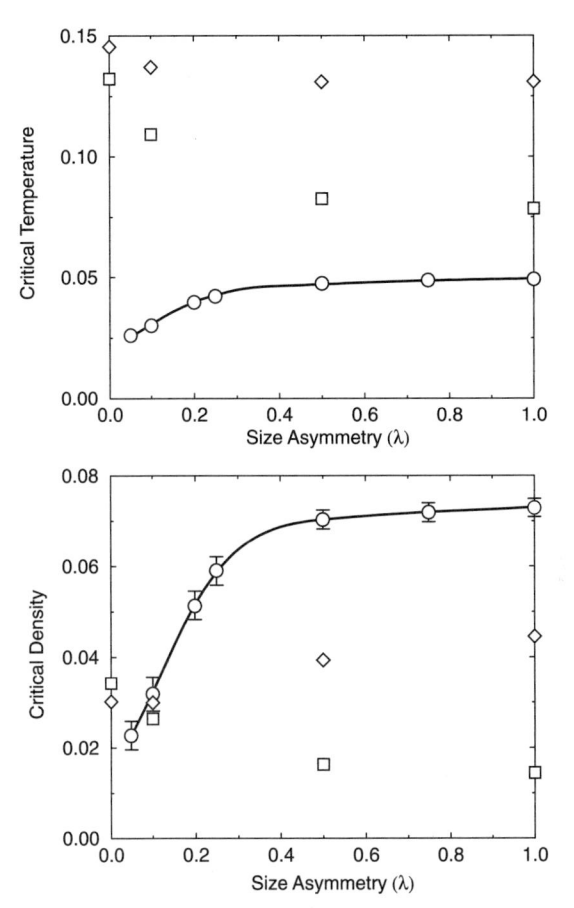

FIG. 5. Simulated binodal curves for size-asymmetric electrolyte systems with different λ values. Circles, $\lambda = 1$; diamonds, $\lambda = 0.75$; squares, $\lambda = 0.5$; triangles, $\lambda = 0.25$.

small. The binodal curves corresponding to $\lambda = 1$ and $\lambda = 0.75$ are almost identical. For highly asymmetric systems ($\lambda < 0.4$), the effect of λ is much stronger and the coexistence curves show pronounced differences.

The simulated behavior of T_c^* and ρ_c^* with size asymmetry can be compared to existing integral-equation theoretical predictions using a mean-spherical approximation (MSA). Figure 5 shows the simulated critical parameters as a function of the size asymmetry. Recently reported predictions of MSA are also shown for both the virial and the energy routes [27]. As expected from a mean-field calculation, the MSA critical predictions are not in quantitative agreement with the results of simulations. What is perhaps more surprising, however, is that the trends predicted by the theory disagree with those observed in simulations. The MSA theory predicts that both the critical temperature and the critical density *increase* as λ decreases.

We have also examined the effects of charge asymmetry on the phase behavior of primitive electrolytes. For 2–1 electrolytes (the cation has a charge of 2, while the anions have unit charge), if the sizes of cations and anions are the same, the MSA theory predicts that the critical point is identical to that of the RPM (same anion–cation charge and size), namely, $T_c^* = 0.049$ and $\rho_c^* = 0.062$. Our simulations predict that the critical temperature is reduced significantly, to $T_c^* = 0.046$, and the critical density increases to $\rho_c^* = 0.105$.

A key feature of electrolyte systems is their tendency to associate. Figure 6 shows several clusters from a simulation of the $\lambda = 0.1$ system in a box of size $L^* = 55$ at $T^* = 0.03$. The instantaneous density is $\rho^* = 0.00122$. As shown in the figure, ions form polymer-like structures whose shapes include chains, rings, and branched chains. This pronounced tendency to cluster can be rationalized by considering a simple aggregate of only four ions. Figure 7 shows the fraction of ions involved in clusters of a given size n, for $\lambda = 0.1$, at $T^* = 0.03$ and $\langle \rho^* \rangle = 0.003$ (for comparison, we also show results for the RPM model at a similar density). The cluster size distribution exhibits an

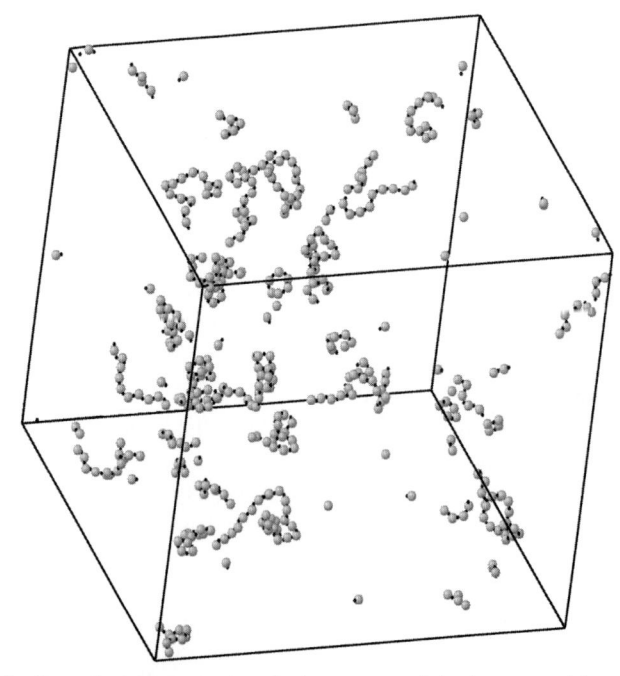

FIG. 6. Configuration representative of a size-asymmetric ionic system with $\lambda = 0.1$ and $L^* = 55$ at $T^* = 0.03$. The instantaneous density corresponding to this configuration is $\rho^* = 0.00122$.

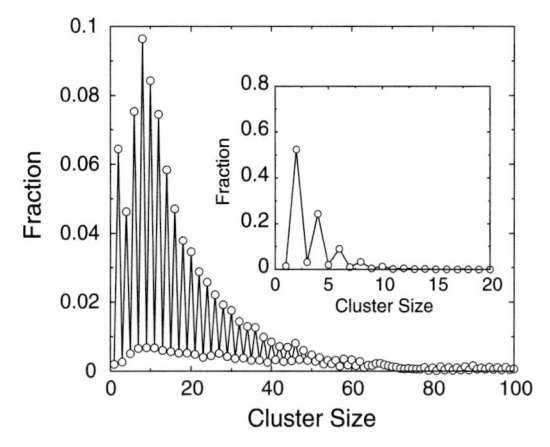

FIG. 7. Probability of finding an ion involved in a cluster of size n at $T^* = 0.03$ and $\rho^* = 0.003$ for a $\lambda = 0.1$ system of size $L^* = 91$. The inset shows results for symmetric electrolytes (RPM model) at $T^* = 0.051$ and $\rho^* = 0.002$ in a system of size $L^* = 50$.

interesting maximum at $n = 8$, indicating that at this density and temperature, ions are more likely to be part of an octamer than a dimer; furthermore, this distribution extends even beyond 100 ions. In contrast, for symmetric systems at approximately the same density, most of the ions associate into simple pairs and most of the clusters involve fewer than 10 ions. These results suggest that theories capable of describing the low-temperature phase behavior of electrolytes should take into consideration the many-body nature of clusters; theoretical efforts in that direction have appeared recently [32], and it will be interesting to compare predictions of those theories to the results of simulations.

C. HOMOPOLYMER SOLUTIONS AND BLENDS

Polymeric fluids constitute another class of systems for which theory has made considerable progress in the last decades; simulations of the phase behavior and thermodynamic properties of polymers, however, have been scarce. Sampling the configuration space of polymeric molecules is computationally demanding, and advanced simulation methods are generally necessary to generate meaningful statistical averages for the properties of interest.

We begin by reviewing applications of the proposed HPTMC method to polymer solutions and blends. For pure polymer solutions, we simulate chains consisting of up to 16,000 sites for simple-cubic lattice models and 500 sites for bond-fluctuation lattice models. For polymer blends, we simulate two highly asymmetric systems on a cubic lattice: in the first, polymers have 16 sites and 64 sites, respectively. In the second, polymer chains have 50 sites and 500 sites, respectively. Both blend systems are simulated at constant temperature and chemical potential. For all systems, each Monte Carlo step consists of 50% chain growth or shrinking moves and 50% local moves (kinkjump and crankshaft moves). Configuration swaps are attempted every 10 Monte Carlo steps. Depending on the conditions, 18 to 20 replicas are used to calculate phase diagrams.

To measure the performance of hyperparallel tempering, we perform a series of simulations using the NVT ensemble, the grand canonical ensemble, multidimensional parallel tempering in a grand canonical ensemble, an expanded grand canonical ensemble, and the HPTMC method; in all cases we examine the decay of the end-to-end vector autocorrelation function for the polymer, which provides a stringent test of efficiency for a polymer-simulation technique. Figure 8 shows some of our results. The dotted curve shows the decay of this function for the NVT ensemble; the relaxation is slow and reaches a value of approximately 0.6 after 5000 steps. The curve below that of the NVT ensemble corresponds to the grand canonical simulation; the decay is only marginally better than that for the NVT method. This is due to the extremely low molecule-insertion acceptance rate experienced in

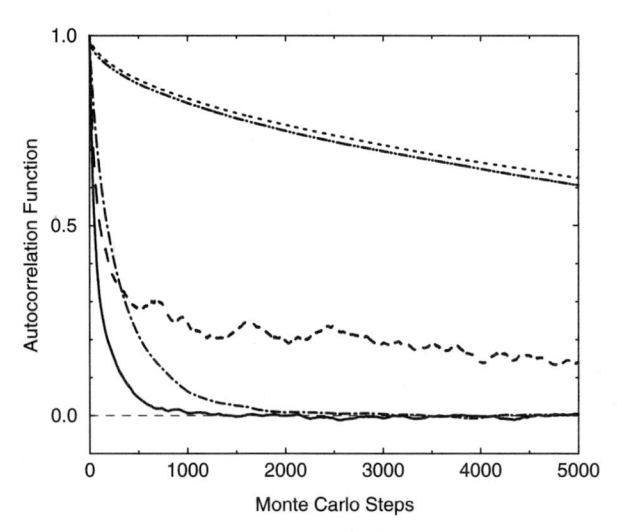

FIG. 8. End-to-end vector autocorrelation function for polymer chains obtained by different simulation methods: (1) short dashed line, canonical ensemble; (2) dash–dot–dotted line, grand canonical ensemble; (3) dashed line, multidimensional parallel tempering; (4) dash–dotted line, expanded grand canonical ensemble; (5) solid line, hyperparallel tempering.

conventional grand canonical simulations of macromolecules. The dashed line shows results from multidimensional parallel tempering simulations in the grand canonical ensemble; the performance is much better than that of naïve grand canonical or *NVT* simulations. The dash–dotted curve shows results for the *expanded* grand canonical ensemble; it decays to zero after about 3000 MC steps. The performance of the expanded grand canonical method is more than one order of magnitude better than that of a naïve grand canonical simulation. The solid curve shows the results for HPTMC; the decay to zero is even faster than for the expanded grand canonical method. The relaxation of the end-to-end autocorrelation function occurs in fewer than 1000 steps. In this context, HPTMC is several times more efficient than the expanded grand canonical technique, and more importantly, it is several orders of magnitude more efficient than some of the methods that have traditionally been used to simulate polymeric fluids.

Figure 9 shows coexistence curves for polymers of 100, 600, 1000, and 2000 sites. The lines are the results of this work, and the open symbols are simulation data from the literature [25]. For $n = 100$ and $n = 600$, our results are in good agreement with literature reports. Note, however, that with the new method, we are able to explore the phase behavior of long polymer chains down to fairly low temperatures. The computational demands of the new method are relatively modest. For example, calculation of the full phase diagram for polymer chains of length 2000 required less than 5 days on a workstation. It is important to emphasize that, for the cubic lattice model adopted here, chains of 2000 segments correspond to polystyrene solutions

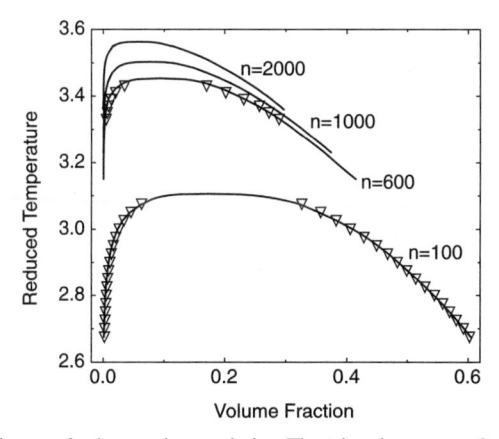

FIG. 9. Phase diagram for long polymer chains. The triangles are results reported by Panagiotopoulos and Wong [25]. The curves show results of this work. Note that with the HPTMC, we are able to calculate the phase diagram for longer polymers, down to lower temperatures, with modest computational requirements.

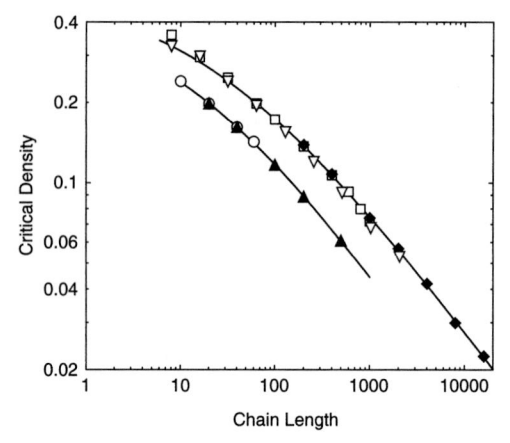

FIG. 10. Scaling of the critical density with chain length. The filled triangles show our results for the bond-fluctuation model, the circles are results reported by Wilding et al. [26], the diamonds show our results for the cubic lattice model, the open triangles are results reported by Frauenkron and Grassberger [38], and the squares are results reported by Panagiotopolous and Wong [25]. The curves are fits to our simulation results (using the functional form $\phi_c(n) = (b_1 + b_2 n^{x_2})^{-1}$). The uncertainty in the critical density is comparable to the size of the symbols.

of relatively high molecular weight (approximately 1,400,000). These are truly polymeric materials, and they therefore exhibit many of the key features (e.g., entanglement) that give rise to true polymeric behavior. The same calculation using traditional grand canonical or Gibbs ensemble techniques would require several months or years of computer time [34]. The results of simulations are also consistent with the experimental data for polystyrene–cyclohexane solutions [35, 36]. More importantly, our calculations indicate that, as shown in Fig. 10, for ultrahigh molecular weights, polymer solutions exhibit a crossover to classical scaling behavior [36]. This is contrary to the results of previous simulations for shorter polymers [25], but it is consistent with recent theoretical arguments by Grassberger and Frauenkron [38].

Another interesting application of HPTMC is encountered in the study of compressibility effects on the phase behavior of polymer blends. While much theoretical work has been devoted to describe the temperature dependence of miscibility for polymer blends, there are relatively few studies of the effect of pressure on miscibility. To a large extent, this is due to the fact that most theoretical models assume that polymer blends are incompressible. HPTMC can be used to determine the miscibility of blends as a function of pressure. Figure 11 shows the phase diagrams for two above-mentioned asymmetric polymer blends. For both of these systems, we can see that in

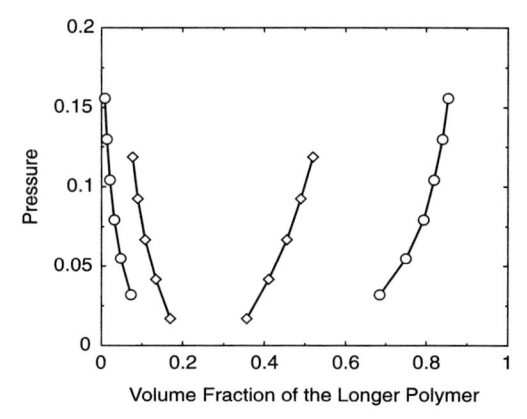

FIG. 11. Phase diagram for asymmetric polymer blends. Circles are for the system with chains of 16 and 64 sites ($kT/\varepsilon_{11} = 2.33, kT/\epsilon_{22} = 2.80, kT/\epsilon_{12} = 2.95$); diamonds are for the system with chains of 50 and 500 sites ($kT/\epsilon_{11} = 2.75, kT/\epsilon_{22} = 3.30, kT/\epsilon_{12} = 3.10$).

some circumstances the pressure can have a nonnegligible negative effect on the miscibility of the polymers. To the best of our knowledge, these results constitute the only available simulation report of the phase diagram of such highly asymmetric and compressible polymer blends; comparisons between our results and previous work are therefore not available at this time.

D. SEMIFLEXIBLE POLYMERS AND THEIR BLENDS WITH FLEXIBLE POLYMERS

We have also applied HPTMC to the simulation of phase coexistence for semiflexible polymers. As before, we use a lattice model to represent the polymers. Stiffness is modeled by introducing an energy penalty ε_B for each kink in a chain. For the particular system studied here, the chain length is $n = 100$, the energy penalty is $\varepsilon_B = 5$, the simulation box size is $L = 50$, and eight replicas are simulated in parallel.

Figure 12 shows the calculated vapor–liquid phase diagram of the semiflexible polymer system. The corresponding phase diagram for the fully flexible polymer is also shown in the figure. The stiffness of the polymer affects the phase diagram dramatically. The critical temperature is higher, and the critical density is slightly lower. The shape of the chains reveals several interesting features. As Fig. 13 shows, in the liquid phase chains adopt extended configurations to avoid energetic penalties; in the vapor phase, chains exhibit typical coiled configurations, similar to those encountered with flexible chains. At low temperatures, semiflexible chains also exhibit a transition to

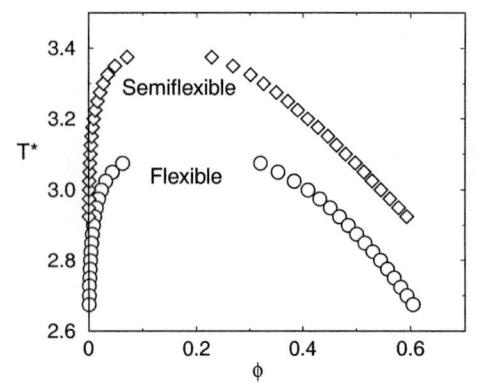

FIG. 12. Phase diagram for semiflexible polymer and fully flexible polymer. Circles are results for flexible polymers, and diamonds are results for semiflexible polymers. Both systems have a chain length $n = 100$. The bending energy penalty for the semiflexible polymer is $\varepsilon_B = 5$.

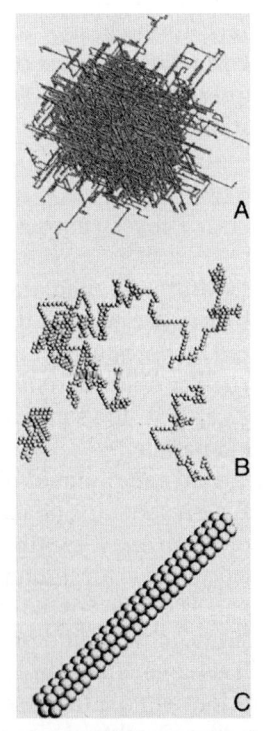

FIG. 13. Configuration snapshot of semiflexible polymer systems. (A) Saturated liquid phase at $T^* = 3.0$; (B) saturated vapor phase at $T^* = 3.0$; (C) single molecule at $T^* = 0.5$.

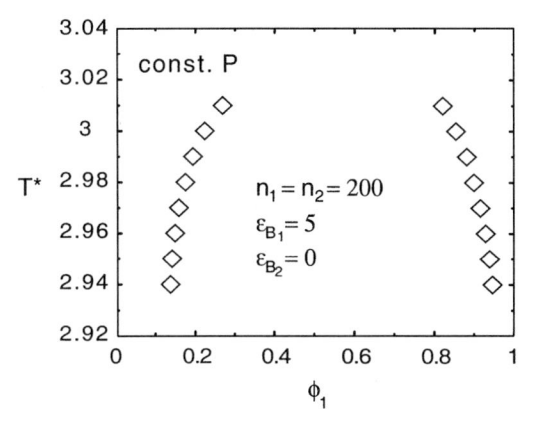

FIG. 14. Liquid–liquid phase diagram for a mixture of semiflexible and flexible polymers at constant pressure. The chain lengths of both species are $n = 200$.

more compact configurations. Figure 13 includes a configuration representative of some of the rod-like shapes that occur at $T^* = 0.5$.

Figure 14 shows the phase diagram of a flexible–semiflexible polymer blend at a constant pressure. Theoretical calculations and experimental results show that such mixtures can exhibit an isotropic–isotropic and isotropic–nematic phase separation. Our calculations are able to capture the isotropic–isotropic phase separation and serve to show that the origin of such a transition can be purely entropic.

E. Block Copolymers and Random Copolymers

As a last example of the application of HPTMC, we calculate the phase behavior of block copolymers and random copolymers. Again, lattice models are used in these calculations. For block copolymers, we study the influence of the number of blocks on the phase behavior; for random copolymers, we examine the effect of sequence length. We use a one-dimensional Ising model to represent the random copolymer. Sequence length is statistically determined by the "temperature" of the one-dimensional Ising model. When this "temperature" approaches infinity, the sequence of the copolymer is completely random; when the "temperature" approaches zero, the random copolymer becomes a diblock copolymer. For all calculations, the chain length is $n = 1000$.

Figure 15 shows the phase diagram of block copolymers and random copolymers. For comparison, the phase diagram of a homopolymer having the same molecular weight is also shown in the figure. As we can see, the number of blocks on the polymer has a dramatic effect on the phase behavior. The

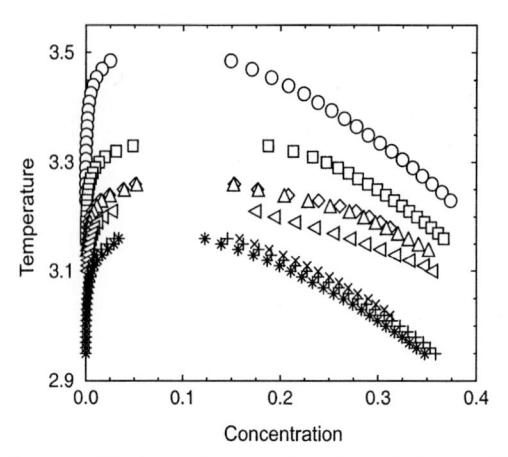

FIG. 15. Phase diagram of block copolymers and random copolymers. Circles are results for homopolymers; squares are results for diblock copolymers; diamonds are results for triblock copolymers; up triangles are results for a mixture of A–B–A and B–A–B triblock copolymers; left triangles are results for tetrablock copolymers; X's are results for random copolymers with average sequence length $l = 55$; crosses are results for random copolymers with $l = 20$; asterisks are results for a completely random copolymer.

coexistence curves for diblocks, triblocks, or tetrablocks lie well below that for the homopolymer; the critical temperature decreases as the number of blocks increases. In contrast, the coexistence curve for random copolymers is not affected considerably by the details of the "blocks" or their number. Our simulations indicate that the sequence length has only marginal effects on the phase behavior, at least as far as vapor–liquid coexistence is concerned.

IV. Discussion and Conclusion

Many of the applications described above, particularly the simulations of highly asymmetric electrolytes and long polymeric system, have been possible only through the use of the HPTMC method. The advantages of this new method have been shown to arise from the combination of biased, open ensemble simulations with replica swapping. Biased open ensemble simulations facilitate elimination of molecules from a simulation box and insertion into completely new positions and configurations; one needs not wait for particles to diffuse slowly through the system. Through replica swapping, the entire configuration residing in a simulation box can be completely replaced, thereby circumventing the slow diffusion through phase space that characterizes complex fluids.

Several factors affect the performance of HPTMC. First, factors that affect the performance of the underling expanded ensemble simulation clearly influence the performance of HPTMC. With regard to HPTMC itself, the frequency and success rate of configuration swaps are the most important factors. A simple rule-of-thumb that we have adopted in these applications is to make the frequency of successful swaps of the same order of magnitude as the frequency of successful particle insertions/removals. Note, however, that simulations of different complex fluids are likely to require some fine-tuning to arrive at optimal parallel tempering algorithms for complex fluids.

It is important to emphasize that, by construction, the acceptance rate for trial swap moves depends on the overlap between the probability distributions corresponding to the state points for two replicas; a high acceptance rate therefore requires closely spaced state points. This overlap depends on how far apart the state points of the replicas are, as well as the characteristics of the system. The width of the distribution functions decreases as the system size increases. For large systems, state points must be relatively close to achieve significant overlap. A possible solution to the problem is to combine further multicanonical sampling [1, 2] with HPTMC; by artificially widening the probability distribution, one should be able to simulate large systems and fewer boxes. Further studies in this direction are currently under way. A second shortcoming of the proposed method is the large memory required for simultaneous simulation of several boxes. This problem, however, can be alleviated by a parallel processor architecture.

In this paper, we have reviewed some recent applications of the HPTMC method. We have attempted to demonstrate its versatility and usefulness with examples for Lennard–Jones fluids, asymmetric electrolytes, homopolymer solutions and blends, block copolymer and random copolymer solutions, semiflexible polymer solutions, and mixtures. For these systems, the proposed method can be orders of magnitude more efficient than traditional grand canonical or Gibbs ensemble simulation techniques. More importantly, the new method is remarkably simple and can be incorporated into existing simulation codes with minor modifications. We expect it to find widespread use in the simulation of complex, many-molecule systems.

ACKNOWLEDGMENTS

This work was supported by the Division of Chemical Sciences, Office of Basic Energy Sciences, Office of Sciences, U.S. Department of Energy. Acknowledgment is also made to the Donors of the Petroleum Research Fund, administered by the ACS, for partial support of this research.

REFERENCES

1. Berg, B., and Neuhaus, T., *Phys. Lett. B* **267**, 249 (1991).
2. Berg, B., and Neuhaus, T., *Phys. Rev. Lett.* **68**, 9 (1992).
3. Hessebo, B., and Stinchcombe, R. B., *Phys. Rev. Lett.* **74**, 2151 (1995).
4. Lyubartsev, A. P., Martinovski, A. A., Shevkunov, S. V., and Vorontsov-Velyaminov, P. N., *J. Chem. Phys.* **96**, 1776 (1992).
5. Marinari, E., and Parisi, G., *Europhys. Lett.* **19**, 451 (1992).
6. Frantz, D. D., Freeman, D. L., and Doll, J. D., *J. Chem. Phys.* **93**, 2769 (1990).
7. Ortiz, W., Perlloni, A., and Lopez, G. E., *Chem. Phys. Lett.* **298**, 66 (1998).
8. Geyer, C. J., and Thompson, E. A., *J. Am. Stat. Assoc.* **90**, 909 (1995).
9. Marinari, E., Parisi, G., and Ruiz-Lorenzo, J., in "Directions in Condensed Matter Physics," (A. P. Young, Ed.), Vol. 12, pp. 59–98. Singapore, World Scientific (1998).
10. Tesi, M. C., Janse van Rensburg, E. J., Orlandini, E., and Whittington, S. G., *J. Stat. Phys.* **82**, 155 (1996).
11. Hansmann, U. H. E., *Chem. Phys. Lett.* **281**, 140 (1997).
12. Wu, M. G., and Deem, M. W., *Mol. Phys.* **97**, 559 (1999).
13. Yan, Q. L., and de Pablo, J. J., *J. Chem. Phys.* **111**, 9509 (1999).
14. Yan, Q. L., and de Pablo, J. J., *J. Chem. Phys.* **113**, 1276 (2000).
15. de Pablo, J. J., Yan, Q. L., and Escobedo, F. A., *Annu. Rev. Phys. Chem.* **50**, 377 (1999).
16. Wilding, N. B., and Binder, K., *Physica A* **231**, 439 (1996).
17. Escobedo, F. A., and de Pablo, J. J., *J. Chem. Phys.* **103**, 2703 (1995).
18. Escobedo, F. A., and de Pablo, J. J., *J. Chem. Phys.* **105**, 4391 (1996).
19. de Pablo, J. J., Laso, M., Siepmann, J. I., and Suter, U. W., *Mol. Phys.* **80**, 55 (1993).
20. Frenkel, D., Mooij, G. C. A. M., and Smit, B., *J. Phys. Condens. Matter* **4**, 3053 (1992).
21. Deleted at Proof.
22. Ferrenberg, A. M., and Swendsen, R. H., *Phys. Rev. Lett.* **23**, 2635 (1988).
23. Ferrenberg, A. M., and Swendsen, R. H., *Phys. Rev. Lett.* **63**, 1195 (1989).
24. Ferrenberg, A. M., and Swendsen, R. H., *Comput. Phys.* **Sept./Oct.** 101 (1989).
25. Panagiotopoulos, A. Z., and Wong, V., *Macromolecules* **31**, 912 (1998).
26. Wilding, N. B., Müller, M., and Binder, K., *J. Chem. Phys.* **105**, 802 (1996).
27. González-Tovar, E., *Mol. Phys.* **97**, 1203 (1999).
28. Panagiotopoulos, A. Z., *Fluid Phase Equil.* **76**, 97 (1992).
29. Caillol, J. M., *J. Chem. Phys.* **100**, 2161 (1994).
30. Orkoulas, G., and Panagiotopoulos, A. Z., *J. Chem. Phys.* **101**, 1452 (1994).
31. Orkoulas, G., and Panagiotopoulos, A. Z., *J. Chem. Phys.* **110**, 1581 (1999).
32. Bernard, O., and Blum, L., *J. Chem. Phys.* **112**, 7227 (2000).
33. Raineri, F. O., Routh, J. P., and Stell, G., *J. Phys. IV* **10**, 99 (2000).
34. Panagiotopoulos, A. Z., *Mol. Phys.* **61**, 813 (1987).
35. Dobashi, T., Nakata, M., and Kaneko, M., *J. Chem. Phys.* **72**, 6685 (1980).
36. Yan, Q. L., and de Pablo, J. J., Critical behavior of lattice polymers studied by Monte Carlo simulations, *J. Chem. Phys.* **113**, 5954 (2000).
37. Deleted at Proof.
38. Frauenkron, H., and Grassberger, P., *J. Chem. Phys.* **107**, 9599 (1997).
39. Dudowicz, J., and Freed, K. F., *Macromolecules* **24**, 5074 (1991).
40. Gromov, D. G., and de Pablo, J. J., *J. Chem. Phys.* **109**, 10042 (1998).

THEORY OF SUPERCOOLED LIQUIDS AND GLASSES: ENERGY LANDSCAPE AND STATISTICAL GEOMETRY PERSPECTIVES

Pablo G. Debenedetti,* Thomas M. Truskett, and Catherine P. Lewis

Department of Chemical Engineering, Princeton University, Princeton New Jersey 08544

Frank H. Stillinger

Bell Laboratories, Lucent Technologies, Murray Hill, New Jersey 07974 and Princeton Materials Institute, Princeton, New Jersey 08544

I. Introduction	22
A. Phenomenology of Vitrification by Supercooling	23
B. Open Questions	29
C. Structure of This Article	32
II. The Energy Landscape	33
III. Statistical Geometry and Structure	39
A. Void Geometry and Connections to the Energy Landscape	40
B. Quantifying Molecular Disorder in Equilibrium and Glassy Systems	45
IV. Landscape Dynamics and Relaxation Phenomena	50
V. Thermodynamics	60
VI. Conclusion	70
References	72

The glassy state is ubiquitous in nature and technology. The most common way of making a glass is by cooling a liquid sufficiently fast so that it does not have time to crystallize. The manner in which such supercooled liquids acquire amorphous rigidity is poorly understood. This lack of knowledge impacts negatively on the design, formulation, and manufacturing of important products in the pharmaceutical,

* Corresponding author (E-mail: pdebene@princeton.edu).

food, communications, energy, and engineering plastics industries. We review important recent advances in the fundamental understanding of glasses that have resulted from two complementary statistical mechanical viewpoints: the energy landscape formalism and statistical geometry. The former provides a unifying analytical framework for describing the thermodynamic and transport properties of glasses and the viscous liquids from which they are commonly formed. Statistical geometry addresses the quantitative description of a glassy material's history-dependent structure. © 2001 Academic Press.

I. Introduction

Glasses are disordered solids. At the molecular level, they have a liquid-like structure and therefore lack the periodicity of crystals. Mechanically, they behave like solids, since they exhibit proportionality between stress and deformation under moderate perturbation. Glasses have played a central role in our daily lives since ancient times. Man-made glass objects, now almost 5000 years old, have been found in Egypt (Zarzycki, 1991). Ordinary window glass, made mostly of sand (SiO_2), lime ($CaCO_3$), and soda (Na_2CO_3), is the best-known example of a manufactured amorphous solid. The superior properties of pure SiO_2 sometimes justify the substantial additional costs associated with its purification and high melting point, 1713°C (Angell and Kanno, 1976). Optical wave guides, for example, consist of pure glassy silica. Most engineering plastics are amorphous, as is the silicon used in many photovoltaic cells. In the pharmaceutical industry, glasses made of sugars and small amounts of water are commonly used for the preservation of vaccines and labile biochemicals (Franks, 1994). Metallic glasses (Chaudhari and Turnbull, 1978) command technological interest because of their soft magnetism and, in the case of some alloys, their excellent corrosion resistance (Greer, 1995). The glassy state is also important in the manufacture and processing of cookies, crackers, and other cereal-based foods (Blanshard and Lillford, 1993).

In spite of the ubiquity and technological importance of the vitreous state, the literature of our profession is quite thin on the subject. In part this is because many of the important technical problems that chemical engineers have so successfully solved in the past have been closely related to the petroleum industry and, hence, involve primarily physical and chemical transformations that take place in fluids. Glasses, moreover, being structurally liquid-like but mechanically solid, and having history-dependent properties that nevertheless can persist unchanged over geological times, fall between the clear-cut boundaries into which chemical engineering science has traditionally been divided. The properties of a glass, for example, are as much a consequence of thermodynamics as they are of kinetics. The former provides the driving force toward equilibrium, the attainment of which is thwarted by following a path- and rate-dependent process that leads to the glassy state. This interplay of kinetics and thermodynamics, then, endows a glass with its physical properties. Like any nonequilibrium material, a glass has a processing history-dependent structure, another concept the quantification of which chemical engineering, with its traditional emphasis in fluid-phase transformations, has not been much concerned with in the past.

These reasons for the comparative marginality of glass science within the chemical engineering literature are rapidly becoming obsolete. Important industries that until recently had received comparatively little attention have become central to the theory and practice of modern chemical engineering. The pharmaceutical industry is perhaps the best example of this evolution. This situation represents a useful broadening of our discipline's scope, complementary rather than antithetical to its traditional petrochemical core. For chemical engineers involved in such activities as the design and synthesis of new materials, the formulation of pharmaceutical products, or the processing of cookies and crackers, knowledge of the solid states of matter is essential. We believe that the evolving needs of our practical profession, coupled with the trend in virtually all areas of contemporary scholarship toward a lowering of barriers between traditional disciplines will bring the vitreous state closer to the core of chemical engineering. This will enrich our discipline and should in turn lead to novel insights that will improve our basic understanding of the vitreous state of matter.

In this article we review important recent advances in the fundamental understanding of glasses that have resulted from two complementary statistical mechanical viewpoints: statistical geometry and the energy landscape. The former addresses the quantitative description of structure. The latter provides a unifying framework for describing both the thermodynamic and transport properties of glasses and the viscous liquids from which they are commonly formed. There are several excellent reviews of this vast topic. Kauzmann's (1948) classic article remains timely today and is still one of the best introductions to the field. More recent reviews include the articles by Angell (1995), Ediger *et al.* (1996), and Angell *et al.* (2000).

A. PHENOMENOLOGY OF VITRIFICATION BY SUPERCOOLING

Glasses can be made by a variety of processes, such as reactive precipitation, electrolytic deposition, quenching of a vapor, ion implantation, chemical vapor deposition, and cold compression of crystals (Zarzycki, 1991; Angell, 1995; Debenedetti, 1996). The most common route to the glassy state, however, is the rapid cooling of a melt. Thus, the properties of metastable liquids cooled below their freezing point (supercooled liquids) are intimately related to those of the resulting glass.

Substances known to form glasses include elements (e.g., P, S, Se); oxides (e.g., SiO_2, GeO_2, B_2O_3, P_2O_5, As_2O_3, Sb_2O_3); chalcogenides (e.g., As_2S_3); halides (e.g., BeF_2, $ZnCl_2$); salts (e.g., $KNO_3 + Ca(NO_3)_2$, $K_2CO_3 + MgCO_3$); aqueous solutions of salts, acids, or bases [e.g., H_2SO_4 (aq.), LiCl (aq.)]; organic compounds (e.g., glycerol, methanol, ethanol, glucose, o-terphenyl, fructose); polymers [e.g., polystyrene, poly(vinyl chloride), poly(ethylene oxide)]; metal alloys (e.g., Ni + Nb, Cu + Zn); and metal–metalloid alloys (e.g., Pd + Si, Ni + P) (Zarzycki, 1991; Debenedetti, 1996). A glass can be formed provided that the starting liquid is cooled fast enough to avoid crystallization. The cooling rate needed to achieve this is substance-specific. Phenyl salicylate, for example, vitrifies when cooled at a rate of 50°C/s, whereas the vitrification of Ag requires cooling at 10^{10} °C/s (Uhlmann, 1972). In general, high entropies of fusion and high interfacial tensions favor vitrification (Debenedetti, 1996).

Figure 1 (Debenedetti, 1996) illustrates the relationship between the specific volume and the temperature of a liquid as it is rapidly cooled at a constant pressure. As the temperature is lowered below the freezing point T_m, the liquid contracts (provided that its thermal expansion coefficient is positive). Cooling causes molecular motion, and hence configurational exploration, to slow down. Eventually, a condition is reached where molecules move so slowly that the liquid cannot equilibrate in the available time imposed by the cooling rate, and its structure appears "frozen" on the laboratory time scale (e.g., minutes). This falling-out of equilibrium occurs across a narrow transformation range where the thermal expansion coefficient decreases abruptly to a value generally smaller than that corresponding to the liquid and comparable to that of a crystalline solid. The resulting material is a glass. The temperature defined by the intersection of the liquid and vitreous portions of the volume-vs-temperature curve is the glass transition temperature T_g. Other thermodynamic properties behave analogously to the volume, as illustrated in Fig. 2 for the enthalpy (Debenedetti, 1996).

In contrast to the freezing point, T_g is not a true transition temperature, because the vitrification process occurs over a narrow temperature interval. Furthermore, T_g depends on the cooling rate. The slower a liquid is cooled, the longer the time available for configurational exploration, and hence the lower the temperature down to which the liquid can remain in equilibrium. Consequently, T_g increases with cooling rate (Moynihan et al., 1976; Brüning and Samwer, 1992). This means that the properties of a glass depend on the

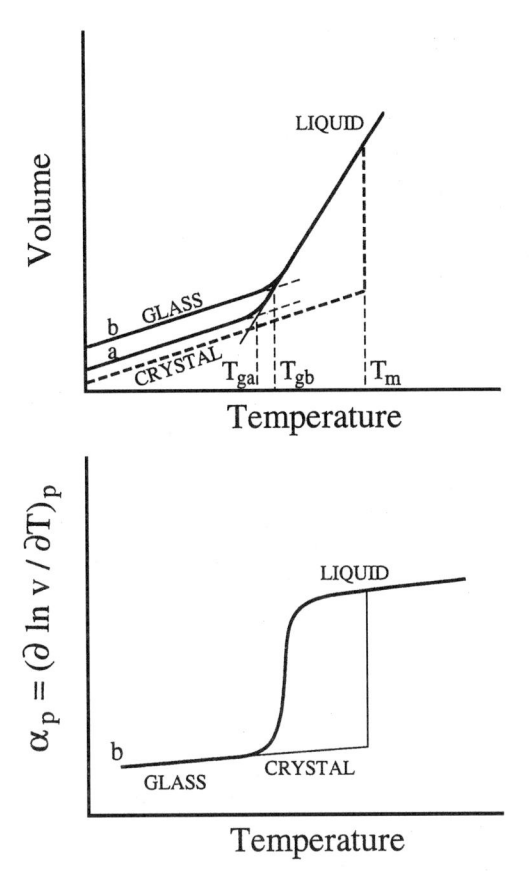

FIG. 1. Isobaric relationship between volume and temperature in the liquid, glassy, and crystalline states. T_m is the melting temperature, and T_{ga} and T_{gb} are the glass transition temperatures corresponding to slow (a) and fast (b) cooling rates. The lower diagram shows the behavior of the thermal expansion coefficient corresponding to curve b. (From Debenedetti, 1996.)

process by which it is formed. The material formed by cooling at a slower rate (Fig. 1, a), is denser and has a lower enthalpy than the faster-cooled glass, b. In practice, however, the dependence of T_g on the cooling rate is rather weak [T_g changes by only 3–5°C when the cooling rate changes by an order of magnitude (Ediger *et al.,* 1996)], and the transformation range is sufficiently narrow, so that T_g is indeed an important material characteristic (Debenedetti, 1996).

The narrow transformation range commonly referred to as *the glass transition* is the temperature interval where the characteristic molecular relaxation time becomes of the order of 100 s (the laboratory time scale). The viscosities of several glass-forming liquids are shown in Fig. 3 as a

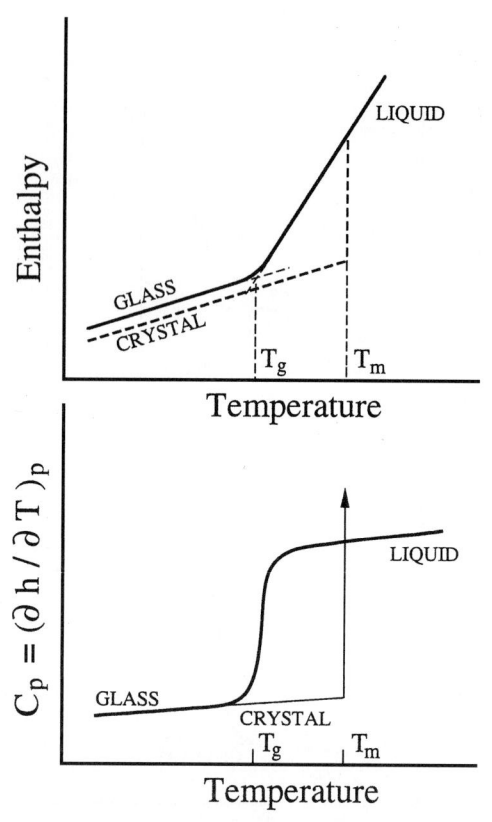

FIG. 2. Isobaric relationship between enthalpy and temperature in the liquid, glassy, and crystalline states. T_m is the melting temperature, and T_g the glass transition temperature. The lower diagram shows the behavior of the isobaric heat capacity. The arrow indicates the δ-function singularity due to latent heat at a first-order phase transition. (From Debenedetti, 1996.)

function of the reciprocal temperature. Another common definition of T_g is the temperature at which $\eta = 10^{13}$ P. Close to the glass transition the viscosity is extraordinarily sensitive to temperature. For some melts, such as silica, this dependence is well described by the Arrhenius functionality, $\eta = A \exp(E/kT)$. Other substances exhibit an even more dramatic increase in their viscosity close to the glass transition, which is often well represented by the Vogel–Tammann–Fulcher (VTF) equation (Vogel, 1921; Tammann and Hesse, 1926; Fulcher, 1925)

$$\eta = A \exp[B/(T - T_0)], \qquad (1)$$

where $T_g > T_0 > 0$. Understanding the origin of this extraordinary

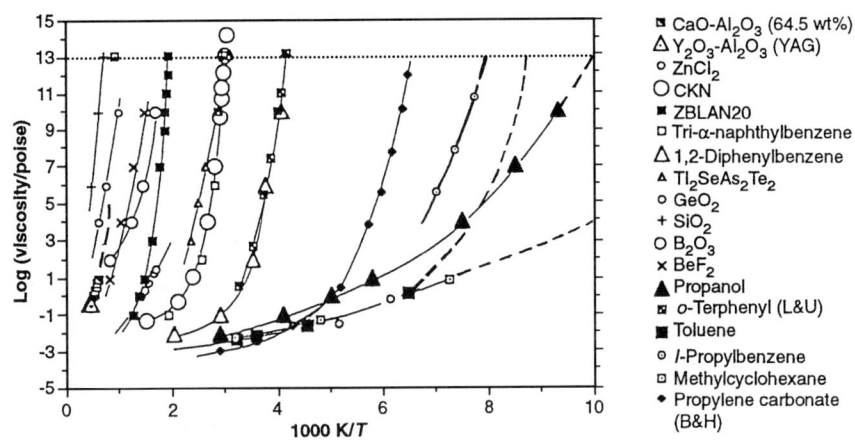

FIG. 3. Arrhenius plot of the viscosity of several supercooled liquids. The horizontal dotted line, where the viscosity reaches 10^{13} P, is commonly used as a definition of the glass transition. (Reprinted with permission from C. A. Angell. Formation of glasses from liquids and polymers. *Science* (1995); 267:1924. Copyright © (1995), American Association for the advancement of Science.)

slowing-down of molecular relaxation processes, or, equivalently, of the energy barriers that give rise to Arrhenius (or super-Arrhenius) behavior, is one of the major challenges in the physics of glasses.

Following an idea proposed by Laughlin and Uhlmann (1972), Angell (1985) plotted the viscosity of several glass-forming liquids in Arrhenius fashion but with the reciprocal temperature scaled by T_g (see Fig. 4). Since all curves coincide at T_g (where $\eta = 10^{13}$ P), and at high temperatures, where η is close to 10^{-2} P for many liquids above their melting point (e.g., water), a more orderly pattern emerges from this scaled Arrhenius representation, compared to the bare Arrhenius plot shown in Fig. 3. Angell proposed a useful classification of liquids into "strong" and "fragile" categories. The viscosity of the former behaves in nearly Arrhenius fashion, whereas fragile liquids show marked deviations from Arrhenius behavior. Silica is often mentioned as the prototypical strong liquid, whereas o-terphenyl is the canonical fragile glass-former. In general, strong liquids, such as the network oxides SiO_2 and GeO_2, have tetrahedrally coordinated structures, whereas the molecules of fragile liquids experience nondirectional, dispersive forces. The strong–fragile pattern is not limited to the viscosity. Any molecular relaxation time,[1] when plotted in scaled Arrhenius fashion, will result in a plot similar to Fig. 4 (Angell, 1995).

---

[1] The viscosity is the product of the elastic modulus, G_∞, and the shear relaxation time τ, $\eta = G_\infty \tau$.

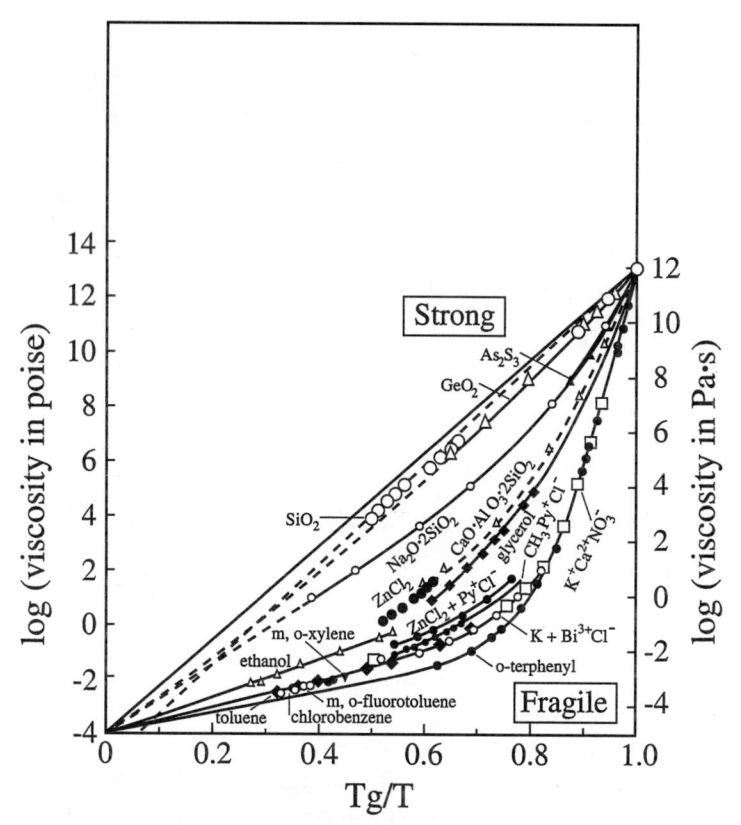

FIG. 4. The strong–fragile classification of liquids. This Arrhenius plot differs from Fig. 3 in that the temperature is scaled with T_g. Strong liquids display Arrhenius behavior; fragile liquids do not. (From Angell, 1988.)

Another important characteristic of viscous liquids close to T_g is nonexponential relaxation. Consider the response of a system to a perturbation, such as the polarization in response to an applied electric field, the strain (deformation) resulting from an applied stress, the stress in response to an imposed deformation, the volume response to applied pressure, or the temperature response to a heat flux. It is found experimentally that the temporal behavior of the response function $\Phi(t)$, following an initial "instantaneous" response, can often be described by the stretched exponential, or Kohlrausch–Williams–Watts (KWW) function (Kohlrausch, 1854; Williams and Watts, 1970),

$$\Phi(t) = \exp\{-(t/\tau)^\beta\} \qquad (\beta < 1), \tag{2}$$

where

$$\Phi(t) = [\sigma(t) - \sigma(\infty)]/[\sigma(0^+) - \sigma(\infty)] \quad (3)$$

and σ is the measured quantity (e.g., the instantaneous stress following a step change in deformation.) τ in Eq. (2) is a characteristic relaxation time, whose temperature dependence is often non-Arrhenius (fragile behavior.) Other functional forms, such as power-law relaxation (Richert and Blumen, 1994), have also been used to fit nonsimple exponential behavior. More important than the exact functional form (especially since those used so far are not theoretically based but empirical fits) is the considerable slowing-down of long-time relaxation embodied in KWW-type behavior. This contrasts sharply with the behavior of liquids above the melting point, which is usually well characterized by simple exponential (Debye) relaxation ($\beta \to 1$). The molecular basis of nonsimple exponential relaxation is not fully understood, but the available evidence suggests that this sluggishness is the consequence of the growth of distinct individually relaxing domains (spatial heterogeneity) (Hyde et al., 1990; Richert, 1994; Cicerone and Ediger, 1995, 1996; Mel'cuk et al., 1995; Hurley and Harrowell, 1996; Perera and Harrowell, 1996a,b; Donati et al., 1999a; Bennemann et al., 1999; Wang and Ediger, 1999; Ediger, 2000). Whether or not the individual relaxation in each of these domains is exponential is an important and interesting open question (Vidal Russell and Israeloff, 2000).

B. OPEN QUESTIONS

The entropy of a liquid at its melting temperature is higher than that of the corresponding crystal.[2] However, the heat capacity of a liquid is generally higher than that of the crystal. Thus, the entropy difference between a liquid and its stable crystal decreases upon supercooling. Figure 5 (Kauzmann, 1948) shows the entropy difference between several supercooled liquids and their stable crystals as a function of temperature, at atmospheric pressure. For lactic acid the entropy difference decreases so fast that a modest extrapolation of experimental data predicts its vanishing. In practice, the glass transition intervenes, and the crossing does not occur. This is shown by the dotted horizontal lines in Fig. 5. If the glass transition did not intervene, the liquid entropy would equal the crystal's at a temperature T_K (the Kauzmann temperature.) Below T_K the entropy of the crystal approaches zero as T tends to zero, and hence the entropy of the liquid would become negative

---

[2] See Greer (2000) and Rastogi et al. (1999) for an apparent exception.

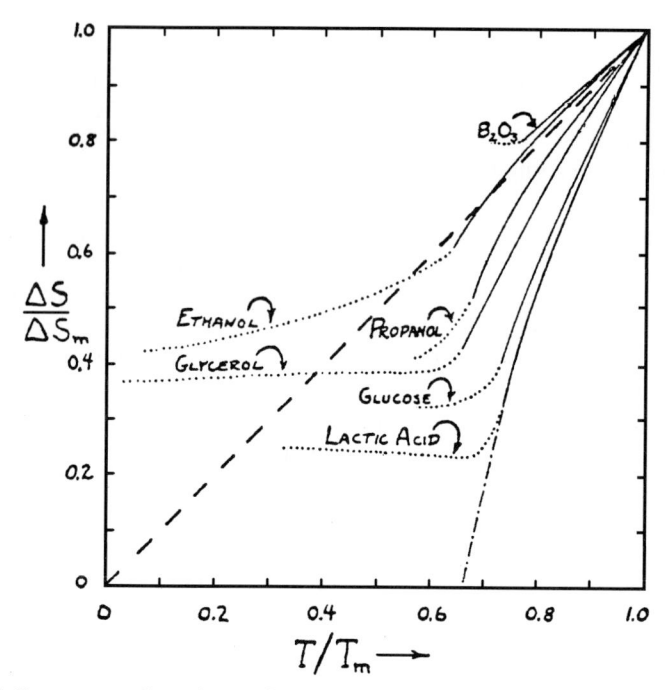

FIG. 5. Temperature dependence of the entropy difference between various supercooled liquids and their stable crystals, ΔS. ΔS_m is the entropy change upon melting, and T_m is the melting temperature. (Reprinted with permission from W. Kauzmann. The nature of the glassy state and the behavior of liquids at low temperatures. *Chem. Rev.* (1948); 43:219. Copyright © 1948, American Chemical Society.)

upon further cooling. This violation of the Third Law of Thermodynamics[3] is known as *Kauzmann's paradox,* although it was first noted by Simon (1931) (Wolynes, 1988).

The entropy crisis described in the preceding paragraph is the result of an extrapolation. With the exception of ^{3}He and ^{4}He (Wilks, 1967),[4] there is no known substance for which a Kauzmann temperature is actually reached. Nevertheless, the extrapolation needed to provoke a conflict with the Third Law is indeed modest for many substances (Angell, 1997), and what intervenes to thwart the imminent crisis is a kinetic phenomenon, the laboratory glass transition. This suggests a connection between the kinetics and

---

[3] Negative entropies are inconsistent with Boltzmann's formula, $S = k \ln \Omega$, where Ω denotes the number of quantum states corresponding to a given energy, volume, and mass.

[4] ^{4}He is a liquid at 0 K and 1 bar (liquid HeII); its equilibrium freezing pressure at 0 K is 26 bar. At this point, the entropies of the liquid and the crystal are equal, and this is therefore a Kauzmann point. The melting curves of both ^{3}He and ^{4}He exhibit pressure minima: these occur at ca. 0.32 K (^{3}He) and 0.8 K (^{4}He). These are also equal-entropy (Kauzmann) points.

the thermodynamics of glasses (Wolynes, 1988), a striking manifestation of which is the fact that fragile glass-formers behave like lactic acid in Fig. 5, their entropy of fusion being rapidly consumed upon supercooling (Ito et al., 1999). Equally intriguing is the fact that for many fragile glass-formers, T_K, a thermodynamic quantity obtained from calorimetric measurements, is close to T_0, a dynamic quantity obtained from transport property measurements (Angell, 1997) [T_0 is the singular temperature where the VTF equation, Eq. (1), predicts complete structural arrest to occur]. Although the validity of extrapolations leading to entropy crises has been questioned (Stillinger, 1988), the situation depicted in Fig. 5 for fragile liquids such as lactic acid prompts inquiry into whether the laboratory glass transition is a kinetically controlled manifestation of an underlying thermodynamic transition, the ideal glass transition (Gibbs and DiMarzio, 1958). The proper role of thermodynamics, and its connection with kinetics, are major open questions in the physics of glasses.

The translational and rotational motion of a Brownian particle immersed in a fluid continuum is well described by the Stokes–Einstein and Debye equations, respectively,

$$D = k_B T / 6\pi \eta a \qquad (4)$$

$$D_r = k_B T / 8\pi \eta a^3, \qquad (5)$$

where D is the particle's translational diffusion coefficient, k_B is Boltzmann's constant, η is the fluid's viscosity, a is the radius of the Brownian particle, and D_r is its rotational diffusion coefficient.[5] Surprisingly, these equations hold down to the molecular level, and they have, accordingly, found widespread application in the interpretation and correlation of data on both tracer and self-diffusion in liquids. If account is taken of boundary conditions and molecular shape effects, the Stokes–Einstein–Debye relations are often accurate to within a factor of 2 (Cicerone and Ediger, 1996) provided $T \geq T_m$. In contrast, it is found experimentally that in supercooled liquids the Stokes–Einstein relationship breaks down around $1.2T_g$ (Fujara et al., 1992; Cicerone and Ediger, 1995, 1996). Below this temperature, translational motion (both probe and self-diffusivity) is faster than predicted by the Stokes–Einstein equation by factors that become at least as high as 100 near T_g (Stillinger and Hodgdon, 1996). Note that this breakdown is in the direction opposite to that which would be predicted by invoking a growth in the effective size of molecules due to increasingly cooperative rearrangements upon supercooling. The inverse relationship between rotational diffusion and viscosity, however, continues to be accurately obeyed. This means that, upon cooling

---

[5] The numerical coefficients 6 and 8 in Eqs. (4) and (5) correspond to no slip at the fluid–particle interface. Other boundary conditions result in different numerical constants [e.g., 4 in Eq. (4) for fluid–particle slip].

below ca. $1.2T_g$, molecules translate increasingly faster than expected based on the known viscosity, and they also translate more for every rotation they execute. Although a plausible interpretation of the data has been offered invoking spatially heterogeneous dynamics (Cicerone and Ediger, 1995, 1996; Stillinger and Hodgdon, 1996), this remains a very active area of research because a definitive explanation of experimental observations does not exist (Wang and Ediger, 1999; Cicerone and Ediger, 1995, 1996; Hinze et al., 1999; Liu and Oppenheim, 1996; Tarjus and Kivelson, 1995).

Sophisticated theoretical tools and experimental protocols exist for the characterization of crystalline structure (Kittel, 1966). The situation is quite different with disordered materials. The quantitative description of glassy structure is an important open problem. Ideally, this description should be based on *structural order parameters* that vary continuously between 0 (complete randomness) and 1 (perfect order). These order parameters should track specific types of order, such as translational order (the tendency of molecules to occupy preferred positions in space) and orientational order (the tendency of anisotropic molecules to adopt a preferred orientation). The development of an analytical framework for quantifying disorder is still in its infancy (Ziman, 1979; Zallen, 1983; Torquato et al., 2000; Truskett et al., 2000). Progress in this area could lead to advances in the early detection of tumors (Hama et al., 1999), the design of transdermal drug delivery systems (Brinon et al., 1999), the prediction and characterization of flow through porous media and packed beds (Bryant and Blunt, 1992; Torquato, 1994; Sahimi, 1995), the efficient handling and processing of powders and granular materials (Shahinpoor, 1980), and the characterization and processing of foods (Blanshard and Lillford, 1993).

Other important unanswered questions were mentioned in Section I.A (origin of energy barriers that give rise to Arrhenius and super-Arrhenius behavior close to T_g, origin of stretched exponential dynamics). Clearly, there are large gaps in our knowledge of disordered solids and the liquids from which they are commonly formed. In this article we review recent theoretical progress toward an improved understanding of glasses and supercooled liquids resulting from the application of the energy landscape formalism and statistical geometry.

C. STRUCTURE OF THIS ARTICLE

In Section II we define energy landscapes, and we present the formalism that relates potential energy minima to the thermal properties of supercooled liquids and glasses. Section III discusses the characterization of voids in dense particle packings, and how this approach, combined with energy minimization techniques, can yield powerful new insights into the mechanical properties of glasses. Also included in this section is a discussion of recent work on the characterization of disorder and its application to model hard-sphere glasses. The intriguing connection between the dynamics and the thermodynamics of the glassy state is discussed in Section IV. It is shown there that understanding the manner in which a supercooled liquid samples its potential energy surface provides powerful insight into stretched exponential behavior, low-temperature breakdown of the Stokes–Einstein behavior, and the connection between entropy and viscous slowdown. Section V introduces the statistical description of an energy landscape and derives its connection to a macroscopic system's thermodynamic properties. It also suggests experimental and computational routes to the investigation of landscape statistics. Section VI summarizes the significant progress that has occurred in recent years through the application of energy landscape and statistical geometry concepts to the understanding of disordered solids and the liquids from which they are formed, and lists major open questions.

II. The Energy Landscape

The interactions operating among the atoms, ions, or molecules of any material system play a dominant role in determining static and dynamic properties of that system. Dilute gases are easy to analyze, at least conceptually, in that interactions occur primarily in isolated small clusters (pairs, triplets, etc.). But the situation is qualitatively different and far more challenging for fluid and solid condensed phases: virtually every particle remains in constant contact with many neighbors, and the system presents a volume-spanning macroscopic cluster. In particular, this is true for the supercooled liquids and glasses that form the subject of this article.

Under these condensed-phase conditions it is natural to consider the full N-body potential energy function $\Phi(\mathbf{r}_1 \ldots \mathbf{r}_N)$ for the material system of interest and to seek to describe the way its details generate the wide variety of collective thermodynamic and kinetic phenomena that have been experimentally observed in condensed matter. For the remainder of this section we suppose that all N particles are the same chemical species and that vectors $\mathbf{r}_i$ comprise all relevant position, orientation, conformation, and vibration coordinates. For the moment, volume V will be constant.

Except for configurations with coincidence of nuclei, Φ is bounded and arbitrarily differentiable in all its coordinates. Therefore it is useful to examine the geometry of the smooth hypersurface generated in the multidimensional space of variables,

$$\mathbf{r}_1, \mathbf{r}_2, \ldots, \mathbf{r}_N, \Phi. \tag{6}$$

This hypersurface constitutes the system's energy landscape. If ν is the number of internal degrees of freedom per particle (orientation, conformation, etc.), then the dimension of the space (6) will be

$$(3 + \nu)N + 1. \qquad (7)$$

Theoretical study of the N-body system now focuses on the Φ hypersurface topography, or, more colloquially, its "rugged landscape." Specific landscape characteristics of interest are the number of minima and their distribution and the nature of saddle points (transition states) throughout the landscape. A schematic illustration of an energy landscape is shown in Fig. 6 (Stillinger, 1995).

The early work of Goldstein (1965, 1969) is a prescient precursor of the topographic viewpoint of condensed phases (Stillinger, 1995) illustrated schematically in Fig. 6. Landscape-based ideas have since found fruitful application to a wide variety of problems, such as protein folding (Frauenfelder et al., 1991; Wolynes, 1992; Abkevich et al., 1994; Chan and Dill, 1994; Frauenfelder and Wolynes, 1994; Saven et al., 1994; Wolynes et al., 1995; Wang et al., 1996; Plotkin et al., 1996, 1997; Becker and Karplus, 1997; Dill and Chan, 1997; Wolynes, 1997), melting and freezing phenomena (LaViolette and Stillinger, 1985a; Patashinski and Ratner, 1997), the mechanical properties of glasses (Lacks, 1998; Malandro and Lacks, 1997, 1999; Utz et al., 2001), shear-enhanced diffusion in liquids and colloidal suspensions (Malandro and

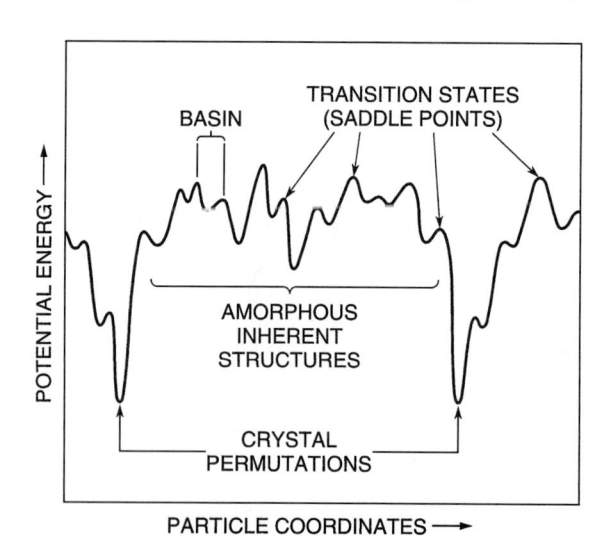

FIG. 6. Schematic illustration of an energy landscape for a many-particle system. The horizontal direction represents all configurational coordinates. (Reprinted with permission from F. H. Stillinger. A topographic view of supercooled liquids and glass formation. *Science* 1995; 267:1935. Copyright © 1995, American Association for the Advancement of Science.)

Lacks, 1998), the dynamics of supercooled liquids (Schulz, 1998; Sastry et al., 1998a; Keyes, 1999), and economic optimization with complex cost functions. A useful collection of papers on this topic can be found in the volume edited by Frauenfelder et al. (1997). The landscape formalism presented in the rest of this section focuses on supercooled liquids and glasses.

Symmetries intrinsic to the N-body system lead to a partitioning of the configuration space into equivalent symmetry sectors. If γ is the symmetry number for each particle, the number of sectors is

$$\Gamma = N! \gamma^N, \tag{8}$$

where the first factor $N!$ accounts for the exchange permutations that are possible with the N identical particles. The potential energy landscape thus consists of many replicas of a primitive topographic parcel. In particular, each distinguishable Φ minimum and saddle point appears Γ times in the configuration space. The Φ local minima offer a conceptually simple way to describe the fine details of the energy landscape. The system configuration $\mathbf{r}_1 \ldots \mathbf{r}_N$ at any one of these minima by definition is one with overall mechanical stability: forces and torques on every particle vanish simultaneously. For this reason these discrete special configurations are often called "inherent structures" (Stillinger and Weber, 1982). It has been demonstrated (Stillinger, 1999) that in the large-system limit, the number of inherent structures in each symmetry sector rises exponentially with N, assuming that the number density N/V (>0) is held fixed.

Identification of inherent structures leads to a natural division of the multidimensional configuration space into nonoverlapping regions, one for each inherent structure, that cover the entire space. The most direct way to accomplish this is to use steepest descent mapping, defined by solutions to the equation set

$$d\mathbf{r}_i/ds = -\partial \Phi / \partial \mathbf{r}_i \quad (1 \leq i \leq N; \quad 0 \leq s), \tag{9}$$

where s is a "progress variable." The locus of all points that connect to a given inherent structure by solutions to Eq. (9) define the "basin of attraction" for that inherent structure (see Fig. 6). These basins contain all configurations that can be viewed as vibrationally distorted versions of the respective inherent structures. Small intrabasin displacements from the inherent structure minimum will be accurately described as harmonic motions, while those of higher amplitude that carry the system configuration close to a shared boundary between two basins will tend to be strongly anharmonic.

The deepest basin in each symmetry sector corresponds to the most nearly perfect crystalline arrangement of the N particles in the available volume V for that crystal structure that is stable at 0 K. If V is conformal with that crystal structure, and N is one of the corresponding "magic number" integers, then the inherent structure at the bottom of these Γ deepest basins will be that of a defect-free crystal. Small displacement then are conventionally described as composed of independent phonons (Ashcroft and Mermin, 1976). The simplest structural excitations from this defect-free state to neighboring higher-lying basins involve localized rearrangents that produce point defects in the crystal and are expected to occur in $O(N)$ different real-space locations or, equivalently, along the same number of distinct directions in the multidimensional configuration space. At the other extreme each symmetry sector will contain one (or more) highest-lying basins. These surround inherent structures that are energetically the least favorable ways to arrange the N particles in space to achieve mechanical stability. In most cases of interest this requirement is expected to produce an amorphous particle deposit.

The potential energy range between the lowest- and the highest-lying basin bottoms is an extensive property of the N-body system. In view of the fact that exponentially many (in N) distinct inherent structures are crowded in between these limits, it is sensible to consider the statistical distribution of basins in a symmetry sector by depth. Let

$$\phi = \Phi/N \tag{10}$$

be an intensive order parameter for classifying basins by depth (Speedy, 1999). We can write the following expression, asymptotically valid in the large system limit, for that distribution (Stillinger, 1999)

$$\frac{d\Omega}{d\phi} = C \exp[N\sigma(\phi)], \tag{11}$$

where $d\Omega$ denotes the number of inherent structures with a depth (on a per particle basis) between ϕ and $\phi \pm d\phi/2$. Here C and $\sigma(\phi)$ are independent of N, with the former a scale factor with dimension inverse energy. If ϕ_c and ϕ_w, respectively, stand for the ϕ values of the lowest-lying (crystalline) and the highest-lying (worst) inherent structures, then $\sigma(\phi)$, the basin enumeration function, will be nonnegative and continuous between these limits, presumably with

$$\begin{aligned}\sigma(\phi_c) &= 0 \\ \sigma(\phi_w) &= 0\end{aligned} \tag{12}$$

and passing through at least one maximum between these limits.

A vibrational partition function can be assigned to each basin α (Debenedetti et al., 1999),

$$Q_{v,\alpha} = \Lambda^{-N}(\beta) \int_{B(\alpha)} d\mathbf{r}_1 \ldots d\mathbf{r}_N \exp\{\beta[\Phi_\alpha - \Phi(\mathbf{r}_1 \ldots \mathbf{r}_N)]\}, \tag{13}$$

where Λ comes from integration over conjugate momenta, $\beta = 1/k_B T$, $B(\alpha)$ is the region occupied by basin α in the multidimensional configuration space, and Φ_α is the potential energy of the embedded inherent structure. While these $Q_{v,\alpha}$ will surely vary from basin to basin, it is their average dependence on order parameter ϕ that is most significant. For that reason, select a narrow range $\phi \pm \varepsilon$ (ε eventually to go to zero after the large-system limit), and compute the vibrational free energy per particle $a^v(\phi, \beta)$ as the following arithmetic mean over this narrow basin fraction:

$$\exp[-N\beta a^v(\beta, \phi)] = \langle Q_{v,\alpha}(\beta) \rangle_{\phi \pm \varepsilon}. \tag{14}$$

With the benefit of the σ and a^v definitions, it becomes possible to express the canonical partition function for the N-particle system as a simple ϕ integral (Stillinger and Weber, 1982),

$$Q(N, V, \beta) = \Lambda^{-N} \int d\mathbf{r}_1 \ldots d\mathbf{r}_N \exp[-\beta \Phi(\mathbf{r}_1 \ldots \mathbf{r}_N)]$$

$$= C \int_{\phi_c}^{\phi_w} d\phi \exp\{N[\sigma(\phi) - \beta\phi - \beta a^v(\beta, \phi)]\} \tag{15}$$

$$\equiv \exp(-\beta A),$$

giving the Helmholtz free energy A as usual.[6] The large-N limit causes the integrand, and thus the integral itself, to be dominated by the neighborhood of the maximum of the bracketed combination appearing in the exponent. Let $\bar{\phi}(\beta)$ denote the value of the order parameter which produces the integrand maximum at the temperature under consideration. The Helmholtz free energy per particle then possesses the simple form

$$-\beta A(\beta)/N = \sigma[\bar{\phi}(\beta)] - \beta\bar{\phi}(\beta) - \beta a^v[\beta, \bar{\phi}(\beta)]. \tag{16}$$

Temperature variations cause $\bar{\phi}$ to shift, thereby accessing different positions of the basin-depth distribution function (11). In any event $\bar{\phi}(\beta)$ identifies the set of basins preferentially inhabited at the chosen temperature. As will be shown in Section V, $\sigma(\bar{\phi})$ for low molecular weight substances in the liquid state tends to fall in the range 1 to 10 (see also Stillinger, 1998; Speedy, 1999).

If N is equal to Avogadro's number, the single-sector distribution (11) will contain an overwhelmingly dominating factor approximately equal to

$$10^{10^{23}}. \tag{17}$$

With any reasonable estimate for the kinetic rate of interbasin transitions, it would take far longer than the age of the Universe for the 1-mol system

---

[6] Both σ and a^v depend on density. The notation in (15) and (16) corresponds to isochoric exploration of the landscape.

to visit all of the basins that are relevant for the given temperature (even if no returns to previously visited basins occurred). Given this situation, it is important to understand that thermal equilibrium on the laboratory time scale involves sampling only a tiny, but *representative*, subset of the basins designated $\sigma(\bar{\phi})$.

Equation (16) for the Helmholtz free energy shows that equilibrium thermodynamic data intrinsically entangles basin numeration (σ) and vibration (a^v) aspects of the N-body system (Stillinger *et al.*, 1998; Stillinger and Debenedetti, 1999). Partial disentanglement becomes possible upon leaving the equilibrium domain. Specifically, very rapid temperature quenches from an equilibrium state at T can trap the system in a basin or small group of neighboring basins contained in $\sigma[\bar{\phi}(\beta)]$. Indeed, the ideal limit of infinitely rapid temperature quench to 0 K would be equivalent to the steepest descent mapping that defines the inherent structures and their basins. In this way the energies and structures of individual inherent structures could in principle be determined.

Crystal nucleation from a pure melt in the neighborhood of its thermodynamic freezing point is kinetically sluggish for most substances, thus permitting at least some amount of liquid supercooling. This circumstance violates the "representative sampling" mentioned above, by excluding those basins whose inherent structures contain some significant amount of crystalline order. Those basins not so excluded correspond to amorphous inherent structures, whose depth distribution could be described by $\sigma_a(\phi)$, where

$$\sigma_a(\phi) \leq \sigma(\phi). \tag{18}$$

A free energy expression for the metastable supercooled liquid, exactly analogous to that shown in Eq. (16), now becomes applicable:

$$-\beta A_a(\beta)/N = \sigma_a[\bar{\phi}_a(\beta)] - \beta\bar{\phi}_a(\beta) - \beta a_a^v[\beta, \bar{\phi}_a(\beta)]. \tag{19}$$

As indicated, the vibrational free energy requires an average over the restricted basin set, and $\bar{\phi}_a(\beta)$ is the statistically preferred depth in that set obtained by maximizing this modified expression. Of course, even this extension (19) breaks down at and below a glass transition temperature.

Although the various considerations covered in this section could be used to determine some basic characteristics of the multidimensional energy landscape, much would remain undetermined. In particular, details about interbasin transition states and about the overall arrangement of the basins of various depths have not yet been illuminated. The time-dependent phenomena considered in Section IV can supply at least some of the desired information.

The development thus far has implicitly assumed constant-volume (isochoric) conditions. Constant-pressure (isobaric) conditions are also experimentally important and can simply be handled by appending volume V, now variable, to the list of configurational coordinates:

$$\mathbf{r}_1, \mathbf{r}_2, \ldots, \mathbf{r}_N, V. \tag{20}$$

At the same time, the pressure–volume product should be appended to the potential energy function

$$\Psi(\mathbf{r}_1 \ldots \mathbf{r}_N, V) = \Phi(\mathbf{r}_1 \ldots \mathbf{r}_N) + pV \tag{21}$$

to yield a "potential enthalpy." The dimension of the augmented configuration space thus increases by unity, and the "rugged landscape" is that of Ψ. All of the preceding considerations adapt to this alternative circumstance, with basins distributed in depth by the intensive order parameter,

$$\psi = \Psi/N, \tag{22}$$

the analogue of Eq. (10).

III. Statistical Geometry and Structure

Unlike crystalline solids, supercooled liquids and glasses are characterized by a topological complexity at the molecular level that eliminates long-range periodic order. This should not be surprising given that experimental protocols for glass formation, by design, frustrate the natural tendency of substances to crystallize at low temperatures. Moreover, since supercooled liquids "fall out of equilibrium" upon vitrification, the details of the resulting glassy structures depend sensitively on the mode of preparation, i.e., on the thermal processing history (Angell, 1995; Debenedetti, 1996).

Although long-range order is noticeably absent in liquids and glasses, species-dependent intermolecular interactions (e.g., "hard-core" repulsion or strong molecular association) inevitably promote the buildup of substantial short-range order in condensed phases, as evidenced by diffraction experiments (Zallen, 1983) and molecular simulation (Angell et al., 1981). Although the molecular-based study of liquids and their mixtures is an important component of modern chemical engineering research (e.g., Theodorou, 1994; Prausnitz, 1996; Gubbins and Quirke, 1996, Debenedetti, 1996; Davis, 1996; Deem, 1998), comparatively less effort has been devoted to the problem of characterizing and quantifying structure in disordered phases.[7]

In this section, we examine recent developments in the description and interpretation of the underlying geometric structure of the liquid and glassy

---

[7] The use of fractal and percolation concepts to characterize pore-space topology (Sahimi, 1993) is a notable exception.

state. In particular, we focus on the application of powerful new theoretical and algorithmic tools to interrogate the nature of molecular ordering and the geometry of the intervening void space in model systems.

A. VOID GEOMETRY AND CONNECTIONS TO THE ENERGY LANDSCAPE

The complex geometry of the pore structure, or alternatively the *void space*, in a material (including its volume, surface area, and connectivity) can play a central role in determining many of its physical properties. Examples include transport processes in porous media (Torquato, 1991; Reiss, 1992); permeability, flow, and diffusion of fluids in packed beds (Thompson and Fogler 1997); mass transfer in polymeric glasses (Greenfield and Theodorou, 1993); adsorption in zeolite crystals (Dodd and Theodorou, 1991); stability and function of proteins and nucleic acids (Liang *et al.*, 1998); bubble nucleation (Shen and Debenedetti, 1999); the thermodynamics of fluids (Speedy, 1980; Sastry *et al.*, 1998b); and the nature of phase transitions in condensed-phase systems (Reiss and Hammerich, 1986; Bowles and Corti, 2000).

Given the diversity of relevant applications, it is not surprising that the characterization of voids in disordered systems has an appreciable history, which can be traced back to primitive "hole" theories of the liquid state (Frenkel, 1955; Ono and Kondo, 1960). While the early theories offer an admittedly rudimentary "lattice" description of voids, recent computational advances permit an exact (and highly efficient) characterization of the continuum void geometry present in particle packings in two (Rintoul and Torquato, 1995) and three dimensions (Sastry *et al.*, 1997a).

An important recent development in the rigorous characterization of disordered materials is the geometric algorithm of Sastry *et al.* (1997a). It allows the identification of disconnected cavities in monodisperse and polydisperse sphere packings, and the exact determination of the volume and surface area of these packings (Fig. 7).[8] This algorithm has proven to be a powerful tool for probing the void geometry of metastable liquids and glasses (Sastry *et al.*, 1997b, 1998b; Shen and Debenedetti, 1999; Vishnyakov *et al.*, 2000; Utz *et al.*, 2001). It should be mentioned in passing that some of the void quantities mentioned above can, at least in principle, be determined using standard Monte Carlo sampling methods (Shah *et al.*, 1989). However,

---

[8] In any sphere packing, it is possible to partition the given volume into occupied and available space. The former is the union of all the exclusion spheres, and the latter is its complement, namely, the volume available for the placement of the center of an additional sphere. The exclusion region of a sphere of diameter σ is a concentric sphere of radius σ. Exclusion spheres can overlap. At a high enough density, the available space is in general composed of disconnected cavities.

THEORY OF SUPERCOOLED LIQUIDS AND GLASSES

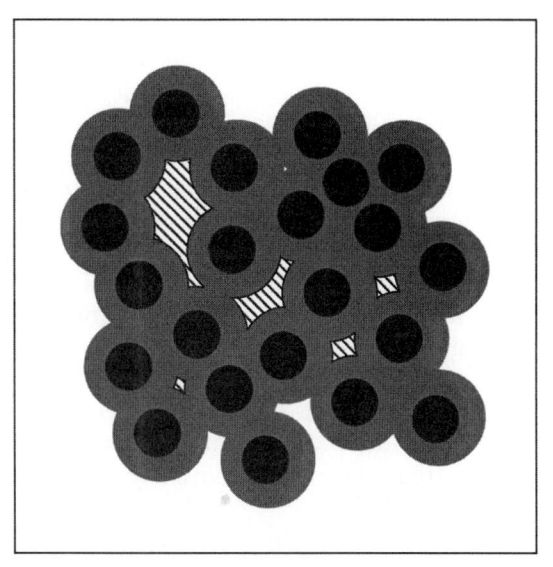

FIG. 7. A random configuration of atoms (black) surrounded by exclusion spheres (gray). The disconnected pockets of space that lie outside of the generally overlapping exclusion spheres are termed "cavities" (cross-hatched). A natural choice for the effective exclusion radius for the Lennard–Jones fluid is $r_{ex} = \sigma$, the Lennard–Jones diameter.

stochastic schemes become unsatisfactory at high density, when the volume fraction of the void space is small (Rintoul and Torquato, 1995).

The geometric algorithm of Sastry *et al.* (1997a) is based on a Voronoi–Delaunay tessellation[9] (Tanemura *et al.*, 1983). It consists of three basic steps.

(a) *Identification of cavities.* This is accomplished by obtaining the percolation clusters of Voronoi edges in the void.
(b) *Identification of polyhedra enclosing the cavities.* The union of Delaunay tetrahedra dual to the Voronoi vertices in a cavity provides an upper bound on the cavity volume.
(c) *Determination of cavity volume and surface area.* This is done by a systematic decomposition of each Delaunay tetrahedron into 24 subunits.

For details on each of these steps, interested readers should consult the original paper (Sastry *et al.*, 1997a).

---

[9] The Voronoi tessellation divides space into polyhedral regions that are closer to the center of a given particle than to any other. Joining pairs of particle centers whose Voronoi polyhedra share a face yields a dual tessellation of space into Delaunay simplices.

The application of this algorithm to explore the morphology of Lennard–Jones inherent structures has yielded new insights into the fundamental nature of the glass transition and suggested a possible and previously unsuspected connection with bubble nucleation in liquids under tension. Sastry et al. (1997b) investigated the inherent structures of a Lennard–Jones system with smoothly truncated interactions. The results from this study, together with previous work on single-component, nonassociating liquids (Weber and Stillinger, 1984; LaViolette and Stillinger, 1985b; Stillinger and Weber, 1985; LaViolette, 1989; Stillinger and Stillinger, 1997), reveal that the packing geometry and energy of the inherent structures are virtually independent of the equilibration temperature. In contrast, it is known that the energy landscape exhibits a rich and highly nontrivial density dependence (LaViolette, 1989; Sastry et al., 1997b; Malandro and Lacks, 1997). It is precisely the exploration of this density dependence that has yielded important new information on the glass transition, metastability, and nucleation (Sastry et al., 1997b; Debenedetti et al., 1999).

In the case of the shifted-force Lennard–Jones system, Sastry et al. (1997b), confirming earlier similar observations by LaViolette (1989) showed that inherent structure morphologies can divided into three distinct intervals in density:

$$\begin{aligned} &\text{A:} \quad 0.99 < \rho\sigma^3 \\ &\text{B:} \quad 0.89 < \rho\sigma^3 < 0.99 \\ &\text{C:} \quad \rho\sigma^3 < 0.89. \end{aligned} \tag{23}$$

Here, σ and ε are the familiar Lennard–Jones parameters, ρ is the number density ($\rho = N/V$), N is the number of molecules, and V is the total volume. As a point of reference, the reduced triple-point density and the critical-point density for the liquid in this model occur at $\rho_{TP}\sigma^3 = 0.815$ and, $\rho_c\sigma^3 = 0.323$, respectively (the corresponding critical pressure and temperature are $P_c\sigma^3/\varepsilon = 0.0805$ and $kT_c/\varepsilon = 0.935$) (Errington, 2000).

Figure 8 illustrates the nonmonotonic density dependence of the average pressure in the inherent structures. In the first of these intervals (A), the inherent structures exist at a positive pressure (although the pressure is smaller in magnitude than for the equilibrated fluid configurations at the same density). Interval B contains inherent structures in tension (i.e., at negative pressure). Note that the magnitude of the tension in these structures increases with decreasing density, approaching a state of maximum isotropic tensile strength at a reduced density of

$$\rho\sigma^3 = 0.89. \tag{24}$$

Interval C also shows inherent structures with negative pressures; however,

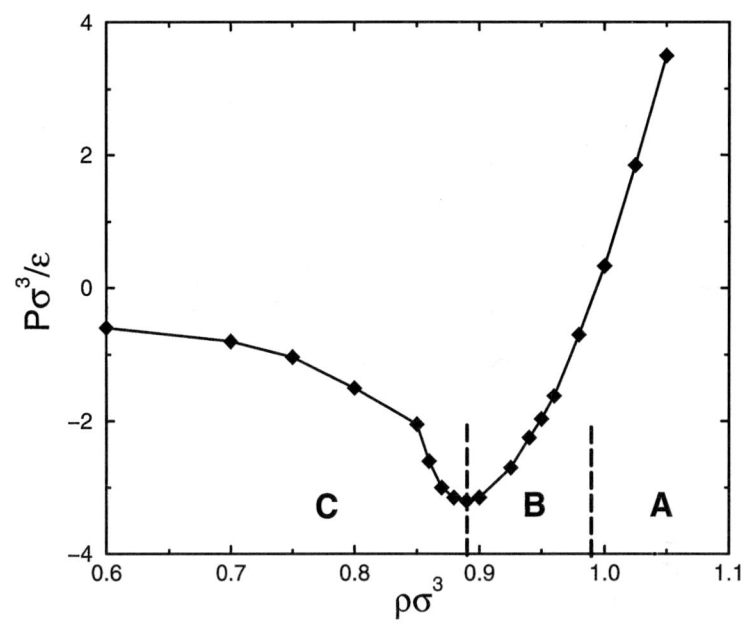

FIG. 8. Density variation of the inherent structure pressure for a fluid with a smoothly truncated Lennard–Jones potential (Sastry et al., 1997b). Regions A, B, and C identify distinguishing density intervals for the inherent structures discussed in the text.

in this density range, the sustained tension is clearly reduced with decreasing density.

A detailed geometric analysis (Sastry et al., 1997b) of the void space in the inherent structures reveals that the density of maximum tension ρ_s has special significance. It represents the density below which the attractive interactions in the system are unable to sustain a mechanically stable packing that is amorphous, isotropic, and statistically homogeneous. This density, known as the Sastry density, is an important material characteristic: since a glass is a liquid arrested at (or close to) a mechanically stable configuration, or inherent structure, we reach the important conclusion that it is not possible to form a homogeneous glass below a material's Sastry density.

Expanding the system to lower densities literally shreds the inherent structures into several densely packed regions threaded by large system-spanning voids. Given that equilibrium liquid configurations in interval C "fracture" as they are continuously deformed into their respective local potential energy minima, it is not surprising that the tension sustained in the resulting heterogeneous inherent structures declines with decreasing density. Although extremely low-density inherent structures ($\rho\sigma^3 < 0.60$) have not been systematically investigated, the rational expectation is that the tension will monotonically diminish as the density approaches zero. The geometry of the resulting mechanically stable structures should be extremely tenuous, perhaps resembling aerogels (Fricke, 1986; Kieffer and Angell, 1988; Stillinger, 2001). Finally, we point out that the structures at $\rho\sigma^3 \approx 0.99$ experience zero pressure and may indeed exhibit features similar to amorphous deposits prepared from low-pressure vapor deposition of single-component substances.

The shape of the pressure versus density curve in Fig. 8 is reminiscent of the metastable pressure isotherms which cross through the vapor–liquid coexistence region, as predicted by the van der Waals equation of state and other mean-field theories (Hirschfelder et al., 1954). In fact, Sastry et al. (1997b) argued that the curve shown in Fig. 8 represents the true zero-temperature limit of such metastable isotherms in the smoothly truncated Lennard–Jones fluid, and hence the extremum of the $p(\rho)$ curve at the Sastry density is the low-temperature limit of the spinodal curve along which the superheated liquid becomes mechanically unstable with respect to the vapor. This identification is supported by mean-field calculations on a number of model systems (Debenedetti et al., 1999), as discussed in more detail in Section V. The important point here is the intriguing connection and possible convergence of the limit of stability upon superheating and the ultimate vitrification limit upon supercooling.

The results of the simple Lennard–Jones fluid invite the following basic question about materials: What aspects of molecular architecture (e.g., shape, symmetry, flexibility, etc.) are crucial in determining the details of the energy landscape and, ultimately, the structural and mechanical features of the liquid and the glass? To shed some light on this important issue, the density dependencies of various properties of the energy landscape have been recently explored for molecular models of ethane, pentane, and cyclopentane (Utz et al., 2001). Interestingly, the molecular liquids were found to be very similar to the Lennard–Jones fluid in one respect: the "equation of state" of the energy landscape, i.e., the relationship between inherent structure pressure and bulk density, appears to be virtually independent of temperature. The extent to which molecular factors such as chain length and branching will alter this picture is a fascinating open question. Details of the void geometry reveal another interesting feature, namely, that the surfaces of the inherent structure cavities that form below ρ_s (the density of maximum tensile strength) are enriched in end groups (Utz et al., 2001). This finding may have important implications for understanding the mechanisms for cavitation in chain-like fluids.

An analogous but much richer behavior is expected to be exhibited by good glass-forming materials. As discussed in more detail in Section IV, the mean inherent structure energy in these systems can show a significant dependence on equilibration temperature (Sastry *et al.*, 1998a). Furthermore, the average inherent structure pressure and the underlying void geometry are expected to be notably temperature dependent, replacing the simple result shown for the Lennard–Jones fluid in Fig. 8 with a family of curves. The prediction, although (to the best of our knowledge) not yet tested experimentally, is that lower equilibration temperatures will result in inherent structures which are more capable of resisting dilatational fracture.

Finally, we note that the specific, directional hydrogen bonds present in liquid water and aqueous solutions confer complexity to their underlying energy landscapes. The wide diversity of inherent structures in water is evidenced experimentally by the existence of multiple crystalline polymorphs (Eisenberg and Kauzmann, 1969), clathrate networks (Davidson, 1973), and high- and low-density amorphous solids (Mishima and Stanley, 1998). Recent simulations of a model for liquid water (Roberts *et al.*, 1999) indicate that its energy landscape is indeed rugged and diverse. Not only are the properties of the inherent structures temperature dependent, but also Roberts *et al.* (1999) demonstrate that in some density ranges the amorphous inherent structures can attain lower potential energies than the ground states of the pure crystalline forms. Although the structural features underlying the energy landscape have not yet been elucidated, this observed behavior suggests the possibility of a microscopic interpretation of many of water's anomalous features, including the observed polyamorphic transition between its glassy phases (Mishima and Stanley, 1998) and the fragile-to-strong transition (Ito *et al.*, 1999) as water is supercooled to its vitreous form.

The systematic investigation of structure, void distribution, and morphology in mechanically stable packings (inherent structures) is still in its infancy. Yet what has been learned so far about the mechanical properties of molecular glasses (Utz *et al.*, 2001), absolute limits to vitrification (Debenedetti *et al.*, 1999), the phase behavior of metastable water (Roberts *et al.*, 1999), and bubble nucleation (Sastry *et al.*, 1997b; Utz *et al.*, 2001) suggests that exploring the connection between statistical geometry and energy landscapes is a powerful route to understanding and eventually predicting the thermal and volumetric properties of supercooled liquids and glasses.

B. Quantifying Molecular Disorder in Equilibrium and Glassy Systems

Although significant theoretical and computational progress has yielded valuable insights into the morphology of void space in supercooled liquids and glasses, much remains to be clarified about the nature of molecular-level disorder present in amorphous systems. Glassy systems are of special interest because they exhibit rich and complex microstructures with pronounced molecular correlations, yet they lack the long-range order characteristic of crystalline solids. In this subsection we focus on recent advances in the description of simple models of disordered condensed phases; however, it should be emphasized that the general problem of describing quantitatively the types of disorder that nature exhibits over a wide variety of length scales (e.g., porous rock, plant structure, tissues) is of enormous scientific and technical importance (Cusack, 1987; Laughlin *et al.,* 2000; Truskett *et al.,* 2000).

At one extreme, a truly random system, by definition, can exhibit no molecular correlations, be they positional, orientational, or conformational—its structure is that of an ideal gas. On the other hand, a perfect periodic crystalline array is a manifestation of perfect order. Our experience with molecular systems in Nature indicates that these extremes exist only as limiting concepts. Between the ideal gas and the perfect crystal lie imperfect gases, liquids (both stable and metastable), liquid crystals, defective crystals, incommensurate structures, quasicrystals, and a variety of structures that organize by nonequilibrium processes, such as glasses and materials formed by irreversible adsorption onto a substrate. Depending on the point of view, all such systems exhibit a certain degree of order (or disorder), and the differences between them can be remarkably subtle. For instance, the problem of distinguishing between the structure of dense glasses and that of polycrystalline materials remains a significant challenge for materials scientists and engineers (Zallen, 1983; Cusack, 1987).

To describe quantitatively the disorder present in a material, it is often convenient to introduce a *structural order parameter*. This term refers to a metric that can detect the development of order in a many-body system, perhaps by employing the tools of pattern recognition (Brostow *et al.*, 1998). In many cases, such a measure is constructed to serve as a "reaction coordinate" for a thermodynamic phase transition (van Duijneveldt and Frenkel, 1992). However, since the form of the order parameter clearly depends on the phenomenon of interest, the development of such measures can be a difficult and subtle matter.

Given that the supercooling of a liquid can lead to structurally distinct possibilities (the stable crystal or a glass), structural order parameters are especially valuable in understanding low-temperature metastability. In particular, it has been demonstrated (van Duijneveldt and Frenkel, 1992) that the *bond-orientational* order parameters introduced by Steinhardt *et al.* (1983) are well suited for detecting crystalline order in computer simulations of simple supercooled liquids. The bond-orientational order parameters are so named because they focus on the spatial orientation of imaginary "bonds" that connect molecules to their nearest neighbors defined as above with the Voronoi–Delaunay tesselation. If these bond orientations persist over a macroscopic distance in the sample (e.g., as they do in a perfect crystal), then the system is said to be *bond-orientationally ordered*.

Of particular interest is the specific bond-orientational order parameter given by

$$Q_6 = \sqrt{\frac{4\pi}{13} \sum_{m=-6}^{6} |\bar{Y}_{6m}|^2}, \quad (25)$$

where $\bar{Y}_{6m}$ represents the spherical harmonic associated with the orientation (θ, ϕ) of a bond in the laboratory reference frame, and the overbar indicates an average over all bonds in the sample. The parameter Q_6 has the desirable property that it vanishes for a completely random system in the infinite-volume limit (Rintoul and Torquato, 1996), whereas it is significantly larger when crystallites are present, attaining its maximum value for space-filling structures ($Q_6^{\text{FCC}} = 0.5745$ in the perfect face-centered cubic (FCC) lattice). Q_6 is also large for a number of alternative crystalline structures, including the body-centered cubic, the simple cubic, and the hexagonal lattices (Steinhardt et al., 1983). This property renders Q_6 an extremely valuable tool for investigating metastability and crystal growth in computer simulations of simple atomic fluids and colloidal suspensions (Rein ten Wolde *et al.*, 1996; Rintoul and Torquato, 1996; Lacks and Wienhoff, 1999; Huitema *et al.*, 1999; Richard *et al.*, 1999). In what follows, we normalize the bond-orientational order parameter Q_6 by its value in the perfect FCC lattice, $\Theta = Q_6/Q_6^{\text{FCC}}$.

In contrast to the bond-orientational order parameters mentioned above, scalar measures for translational order [that is, of the tendency of particles (atoms, molecules) to adopt preferential pair distances in space] have not been well studied. However, a number of simple metrics have been introduced recently (Truskett *et al.*, 2000; Torquato *et al.*, 2000, Errington and Debenedetti, 2001) to capture the degree of spatial ordering in a many-body system. In particular, the structural order parameter τ,

$$\tau = \left| \frac{\sum_{i=1}^{N_c} \left(n_i - n_i^{\text{ideal}} \right)}{\sum_{i=1}^{N_c} \left(n_i^{\text{crystal}} - n_i^{\text{ideal}} \right)} \right|, \quad (26)$$

was constructed to measure the degree of translational order in a system relative to some relevant crystal lattice[10] (Truskett *et al.*, 2000; Torquato *et al.*, 2000). To understand the structural order parameter τ, we consider three systems with the same number density ρ: the system of interest, a completely

---

[10] The structural order parameter τ should not be confused with the relaxation time τ defined in Eq. (2).

random system (i.e., an ideal gas), and the reference crystal lattice. Here, n_i^{crystal} indicates the number of neighbors that are located in a molecule's ith neighbor shell (a distance r_i from the molecule) in the reference crystal lattice. Similarly, n_i^{ideal} refers to the average number of neighbors that are located at a distance r from a molecule [where $r_i - (a\delta/2) < r < r_i + (a\delta/2)$] in the ideal gas. Finally, n_i measures the number of neighbors that lie in that same spherical shell surrounding a molecule in the system of interest. As can be seen from (26), τ quantifies the degree of spatial ordering in the system of interest by "metering" its position between the two natural limits, the ideal gas ($\tau = 0$) and the perfect crystalline lattice ($\tau = 1$). In the above outline, $a = r_1$ (the first nearest-neighbor distance for the reference crystal), δ is a shell width parameter, and N_C is the total number of neighbor shells considered in the reference crystal.

The structural order parameter τ was originally introduced to investigate ordering in the hard-sphere system, for which the appropriate reference crystal structure is the FCC lattice (Torquato et al., 2000). In that investigation, the first seven neighbor shells were considered ($N_c = 7$), allowing for a shell-width parameter of size $\delta = 0.196$. From a practical perspective, it was noted that consideration of more neighbor shells did not result in qualitatively different behavior for τ (Truskett et al., 2000). We mention in passing that the approach outlined above is useful only if the relevant ordered state (i.e., the reference crystal structure) is known a priori. The development of robust structural order parameters that do not require such information, so-called *crystal-independent* measures, is an active area of research (Truskett et al., 2000; Errington and Debenedetti, 2001).

We can gain some insight into the molecular ordering that occurs in condensed phases by studying a map of the structural order parameters (τ, Θ) for equilibrium and nonequilibrium packings in a simple system. Figure 9 represents such an *ordering phase diagram* constructed from molecular dynamics simulations of a collection of 500 identical hard spheres (Truskett et al., 2000; Torquato et al., 2000). Shown on the diagram are the equilibrium fluid, the equilibrium FCC crystal, and a set of glassy structures generated from the Lubachevsky–Stillinger compression protocol (Lubachevsky and Stillinger, 1990; Lubachevsky et al., 1991). In this protocol, glassy hard-sphere packings are produced from low-density sphere configurations by allowing the hard-sphere diameter $\sigma(t)$ to grow linearly in time (at a given rate) during the course of a constant-volume molecular dynamics simulation. The compression protocol terminates when a high-density "jammed" sphere packing is achieved (Lubachevsky and Stillinger, 1990). As a result, the degree of disorder (as measured by τ and Θ) and the final packing fraction ϕ in the glasses depend on the "processing history" through the compression rate, i.e., the diameter growth rate.

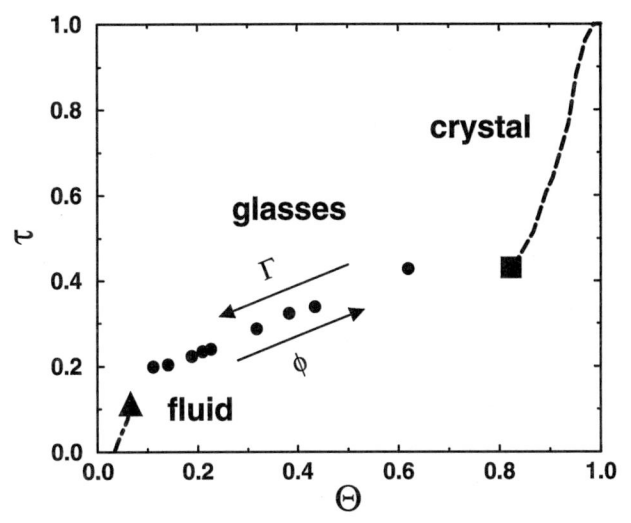

FIG. 9. Two-parameter ordering phase diagram for a system of 500 identical hard spheres (Truskett et al., 2000; Torquato et al., 2000). Shown are the coordinates in structural order parameter space (τ, Θ) for the equilibrium fluid (dot–dashed), the equilibrium FCC crystal (dashed), and a set of glasses (circles) produced with varying compression rates. Here, τ is the translational order parameter from (26) and Θ is the bond-orientational order parameter Q_6 from (25) normalized by its value in the perfect FCC crystal ($\Theta = Q_6/Q_6^{FCC}$). Each circle represents an average of 27 glasses produced by compressions at a given rate Γ. Unlike the equilibrium state points, the degree of ordering (τ, Θ) and the packing fraction ϕ in the glasses are determined by the processing history (in this case, the compression rate Γ). The freezing and melting transitions are indicated by the triangle and the square, respectively.

One striking feature of the ordering phase diagram shown in Fig. 9 is the strong positive correlation that exists between τ and Θ in the equilibrium hard-sphere fluid and crystalline phases. This indicates that entropy (i.e., packing efficiency) promotes an appreciable coupling between translational and bond-orientational order in the hard-sphere system. In addition, we note that there is a discontinuous jump in the structural order parameters across the first-order freezing transition, creating a large gap in order parameter space ($0.15 < \tau < 0.40$ and $0.1 < \Theta < 0.8$) that serves as a "no man's land" for the pure equilibrium phases. Interestingly, we find that sphere packings that exhibit an intermediate order, corresponding to coordinate pairs (τ, Θ) in no man's land, can be generated if we resort to a nonequilibrium (history-dependent) method for preparation, such as the Lubachevsky–Stillinger protocol. The observation that the jammed structures populate a different region of the ordering phase diagram than the equilibrium system indicates that certain nonequilibrium packings can be distinguished statistically from the equilibrium configurations based on structural information alone. In particular, we note that the hard-sphere glasses are not simply solids with the "frozen-in" structure of the liquid.

Examination of the glassy sphere packings produced by the Lubachevsky–Stillinger protocol reveals that the amount of order in the structures can be statistically controlled by the compression rate. In fact, we note that "randomness" is a matter of degree in the jammed hard-sphere structures; i.e., jammed structures with slightly higher packing fractions ϕ can be realized at the expense of small increases in order (τ, Θ). This observation has recently led to a reassessment of the traditional notion that the random close-packed state is the densest possible amorphous sphere packing (Torquato *et al.*, 2000).

Although the ordering phase diagram presented in Fig. 9 was constructed from a highly idealized model system, it suggests challenging scientific questions about the morphology of real materials. For instance, it is clear that there exist large regions of the (τ, Θ) plane, i.e., certain types of molecular ordering, that are statistically inaccessible for a system in equilibrium. Is it possible to understand the relationship between these "inaccessible regions" in order parameter space and the relevant interactions in the system (i.e., the energy landscape)? In other words, can materials which possess a particular morphology be engineered by "tuning" their interactions? Moreover, can we use the ordering phase diagram as a viable guide for quantifying the relationship between disorder and history for realistic nonequilibrium protocols? Although we are far from answering these questions for real materials, the notion of an ordering phase diagram provides a useful conceptual framework for investigating the relationship among microscopic interactions, processing conditions, and the morphology of equilibrium and nonequilibrium systems.

The systematic investigation of order phase diagrams, such as that shown in Fig. 9 but for different molecular interactions, the location of inherent structures in the order plane, and the comparative exploration of order (or disorder) in "computer glasses" generated by different quenching protocols are, we believe, fruitful avenues for research into the nature, classification, and quantification of disorder in glasses and other technologically important nonequilibrium materials.

IV. Landscape Dynamics and Relaxation Phenomena

We presume in what follows that the supercooled liquid or glass of interest can be described by classical mechanics. Assuming that $\mathbf{r}_j$ comprises

Cartesian coordinates for all nuclei of particle j, the isochoric dynamical evolution of the N-particle system follows the Newtonian equations of motion:

$$\mathbf{m}_j \cdot d^2 \mathbf{r}_j / dt^2 = -\nabla_j \Phi \quad (1 \leq j \leq N). \tag{27}$$

The diagonal matrix $\mathbf{m}_j$ specifies the masses of the nuclei. Recall that Φ includes both intraparticle and interparticle potentials, as well as interactions of particles with confining walls that define system volume V. The qualitative nature of the trajectory traced out by the system configuration point,

$$\mathbf{R}(t) \equiv [\mathbf{r}_1(t), \mathbf{r}_2(t), \ldots, \mathbf{r}_N(t)], \tag{28}$$

as it evolves by Eqs. (27) depends strongly on the magnitude of the total energy. If that total energy is high, $\mathbf{R}(t)$ is able to move over the landscape with relatively little hindrance, moving quickly from basin to basin without being forced to pass close to the locations of basin-boundary saddle points. This situation typically describes liquids well above their melting temperatures. A variety of computer simulations (LaViolette and Stillinger, 1985a; Buchner et al., 1992; Moore and Keyes, 1994; Wu and Tsay, 1996; La Nave et al., 2000; Keyes, 1997; Keyes et al., 1997) have established that $\mathbf{R}(t)$ in these high-energy circumstances moves across parts of the landscape at which the Hessian matrix (of second Φ derivatives) has a significant fraction of negative eigenvalues. In other words, the instantaneous normal mode (INM) spectrum for hot liquids exhibits a significant fraction of imaginary frequencies. Mode coupling theory (MCT) describes this situation well (Leutheusser, 1984; Bengtzelius et al., 1984; Götze and Sjögren, 1992, 1995). In this high-temperature situation, a landscape-based description of liquid dynamics is not required, although still applicable in principle.

Lowering the total energy reduces the rate of interbasin transitions, obliges $\mathbf{R}(t)$ during those transitions to pass on average closer to the saddle points, and produces a reduction in the fraction of Hessian eigenvalues that are negative. As noted in Section II, the mean depth $\bar{\phi}$ of the basins visited by $\mathbf{R}(t)$ increases in absolute value as the energy and/or temperature decline, so that smaller and smaller subsets of basins can be visited. In the very low-energy/low-temperature range, interbasin transitions become very infrequent and, at least for solid amorphous materials, will be dominated by low-barrier transitions. These few remaining transitions are conventionally identified as two-level systems, and they have important consequences for heat capacity and sound propagation measurements on those amorphous solids (Phillips, 1972; Anderson et al., 1972).

Whereas MCT appears well suited to describe dynamics and relaxation processes in liquids that are above, at, or moderately below their equilibrium melting points (Götze and Sjögren, 1995), it becomes physically inappropriate for strongly supercooled liquids and the glasses that they form below an experimental (or simulational) glass transition temperature T_g. MCT erroneously predicts a singular temperature T_x below which the N-body system loses ergodicity, i.e., loses the ability to display complete relaxation to equilibrium (or restricted equilibrium for noncrystallizing liquids). One has

$$T_g < T_x \qquad (29)$$

and typically T_x is 10 or 20% higher than T_g (Debenedetti, 1996). This shortcoming arises from the failure of the original MCT (Bengtzelius *et al.*, 1984) to incorporate landscape details and their profound influence on low temperature behavior. More recent elaborations of MCT (Sjögren and Götze, 1991; Götze and Sjögren, 1992) address this problem by incorporating a coupling between density and momentum fluctuations.

A few general observations are in order here about basin shapes and about the transitions that carry the system from one basin to a neighboring one. First, the intrabasin vibrational motions are substantially anharmonic for liquids at and slightly below their melting temperature T_m. One measure of this anharmonicity has already been mentioned, the incidence of negative eigenvalues of the Hessian matrix. Another revealing measure is the mean-square displacement of the configuration point, on a per-particle basis, from the inherent structure, plotted versus the temperature. This would be a straight line proportional to T if the basin interior were exactly harmonic, assuming classical statistics apply. However, as the schematic Fig. 10

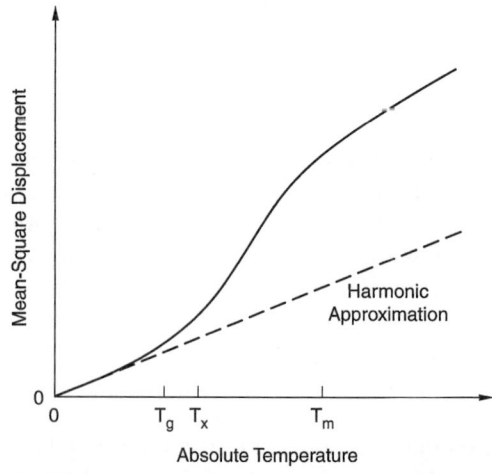

FIG. 10. Schematic illustration of mean-squared atomic displacements versus temperature, measured from the inherent structure, for amorphous-phase basins. The melting, MCT singular, and glass transition temperatures are T_m, T_x, and T_g, respectively.

indicates, the initial harmonic rise for low temperature becomes significantly enhanced at higher T due to strong anharmonicity (Sastry *et al.*, 1998a). This enhancement appears to arise from the contribution of basin "arms" that stretch outward from the inherent structure configuration, while rising only slowly in potential energy with distance (LaViolette and Stillinger, 1986). The anharmonic augmentation of the root-mean-square displacement is substantially larger for amorphous (liquid) basins than for crystalline basins, and this distinction has led to generalization of the venerable Lindemann melting criterion (Lindemann, 1910; Martin and O'Connor, 1977) to include a liquid freezing criterion (LaViolette and Sillinger, 1985a).

Computer simulations for several models (Weber and Stillinger, 1985; Ohmine, 1995) have determined that the elementary transitions between neighboring basins entail shifts of only small local groups of particles. To be precise, the difference between the inherent structures of the two basins involved in a large N-particle system is concentrated on a neighboring set of $O(1)$ particles; the remainder particles experience at most a minor elastic response to the localized repacking (Lacks, 1998). In view of the fact that the number of such localized repacking possibilities is proportional to system size, the number of transition states (saddle points) in the boundary of any basin will be $O(N)$, i.e., an extensive property. So too, then, will be the net kinetic exit rate from any basin at positive temperature.

It is important to bear in mind that the localized nature of the fundamental transitions implies that their respective saddle points lie only an $O(1)$ excitation energy above the bottom of the basin from which the system exits. In contrast, the total excitation energy possessed by the system while it resides in that basin will be an $O(N)$ quantity, even at very low temperatures. The central issue then becomes how long on average the system must wait before the intrabasin dynamics concentrates sufficient energy along the direction of the saddle point to permit the transition to the neighboring basin. This is the N-body analogue of the same issue that arises in chemical kinetics theory of unimolecular decomposition for variable intramolecular excitation energy (Gilbert and Smith, 1990).

At least in the temperature range relevant to liquid supercooling and glass formation, the kinetics of interbasin transitions and resulting relaxation phenomena can be described by a master equation (Stillinger, 1985). Suppose that the conserved system energy is E, and let $P_\alpha(t)$ stand for the probability that the system's configuration point resides in basin α at time t. The master equation takes the form

$$dP_\alpha(t)/dt = \sum_\gamma [K(\gamma \to \alpha|E)P_\gamma(t) - K(\alpha \to \gamma|E)P_\alpha(t)], \qquad (30)$$

where with obvious notation the K's are transition rates between neighboring basins α and γ at system energy E. These $O(1)$ rates are required to satisfy detailed balance conditions to assure that Eq. (30) has time-independent equilibrium solutions. Specifically, for all α,γ pairs, the following equality must hold:

$$M_\alpha(E)K(\alpha \to \gamma|E) = M_\gamma(E)K(\gamma \to \alpha|E). \qquad (31)$$

Here M_α and M_γ are the microcanonical measures of the interiors of basins α and γ at energy E. Equation (31) ensures that total probability is conserved at all times,

$$\sum_\alpha P_\alpha(t) = 1, \qquad (32)$$

even when the system is out of equilibrium.

The general solution to the linear master equation (30) consists of a linear combination of eigenfunctions (equal in number to the number of basins that can be occupied):

$$P_\alpha(t) = \sum_{n\geq 0} A_n \chi_\alpha^{(n)}(E)\exp[-\lambda_n(E)t] \quad (\lambda_n(E) \geq 0). \qquad (33)$$

Initial conditions determine the linear combination coefficients A_n, which express the relative contribution of the nth eigenfunction. Thermal equilibrium at energy E corresponds to a vanishing decay constant, say λ_0, and in connection with earlier remarks, the collection of eigenfunctions $\chi_\alpha^{(0)}(E)$ is strongly concentrated in those basins whose depths are close to the $\bar{\phi}$ for the given E. The terms in expression (33), with $\lambda_n > 0$ $(n > 0)$ describe relaxation toward equilibrium.

A supercooled liquid or amorphous solid can be driven out of (restricted) equilibrium in a wide variety of ways, and the kinetics of relaxation back to (restricted) equilibrium subsequently followed. This variety includes sudden temperature or pressure changes, mechanical working, application of a polarizing electric field, and irradiation with energetic particles. As mentioned in Section I, the linear-response relaxation kinetics of supercooled liquids as monitored by any of several properties is observed to follow a Kohlrausch–Williams–Watts (KWW) stretched exponential form [see Eq. (2)],

$$f(t) = f(0)\exp\{-[t/\tau(T)]^{\beta(T)}\}, \qquad (34)$$

where stretching exponent $\beta(T)$ empirically tends to fall in the range

$$\tfrac{1}{3} \leq \beta(T) \leq 1. \qquad (35)$$

The relaxation time τ in Eq. (35) should not be confused with the translational order parameter defined in Eq. (26). The mean relaxation time defined by $f(T)$ is

$$t_{\text{rel}}(T) = \int_0^\infty t f(t) dt \bigg/ \int_0^\infty f(t) dt = [\Gamma(2/\beta)/\Gamma(1/\beta)]\tau(T) \qquad (36)$$

The tendency is for $\beta(T)$ to decline with declining temperature, and as β passes downward between the limits shown in Eq. (35), the coefficient of τ in Eq. (36) increases from 1 to 60.

By expressing $f(t)$ as a Laplace transform,

$$f(t) = f(0) \int_0^\infty F(\lambda) \exp(-\lambda t) dt, \qquad (37)$$

the stretched exponential formally becomes resolved into individual simple-exponential decays with a range of decay rates. This allows contact with the general solution shown above in Eq. (33), with the understanding that for any property Q the observable relaxation would follow the form

$$Q(t) = \sum_\alpha Q_\alpha P_\alpha(t) = \sum_n \left(\sum_\alpha Q_\alpha A_n \chi_\alpha^{(n)} \right) \exp(-\lambda_n t) \cong \int_0^\infty q(\lambda) \exp(-\lambda t) d\lambda. \qquad (38)$$

Here Q_α is the mean value of property Q averaged over basin α (at energy E), and $q(\lambda)$ is the spectral weight in the continuum limit of the modes with exponential decay constant λ. If $Q(t)$ in fact has the stretched exponential form, then $q(\lambda)$ will be proportional to the Laplace transform $F(\lambda)$, for which both numerical (Lindsey and Patterson, 1980) and analytical (Helfand, 1983) studies are available. In the simple exponential decay limit $\beta = 1$, $F(\lambda)$ reduces to an infinitely narrow Dirac delta function but it broadens as β decreases toward the lower limit $\frac{1}{3}$ to involve a wide range of simple exponential relaxation rates.

An obvious question to ask is what characteristics of the basin distribution and of the kinetic connections between them can produce the wide spectrum of relaxation rates and the associated stretched exponentials. Some hints may come from simple models for supercooling and glass formation, whose kinetics are known to display stretched exponential relaxation functions. Two such models are the Fredrickson–Andersen kinetic Ising model (Fredrickson, 1984) and the tiling model (Weber et al., 1986). Both involve discrete space rather than a continuum, and both view the glass-forming medium as residing in two dimensions rather than three. Nevertheless, these models indicate that as the temperature declines and as deeper and deeper portions of the respective (discrete) potential functions are probed, larger and larger clusters of "particles" must be rearranged for the system to find yet lower potential energy configurations. Furthermore, increasing diversity arises in how these rearrangements can occur, in both size and shape of the affected local regions.

The shear viscosity $\eta(T)$ measures the rate at which a liquid can rearrange in response to an applied shear stress. As explained in Section I, the extent to which the temperature dependence of $\eta(T)$ conforms to, or deviates from, Arrhenius behavior,

$$\eta(T) = \eta_0 \exp(E_\eta/k_B T), \tag{39}$$

forms the basis of the "strong" vs "fragile" classification of glass formers (Angell, 1985). Here η_0 and the activation energy for viscous flow E_η are substance-specific, but temperature-independent, constants at a fixed volume.

Molten silica (SiO_2) is often considered the prototypical strong glass former (Angell, 1988). Its density is relatively constant in the supercooled range (<1700°C), and in that range the Arrhenius form (39) provides a good description for $\eta(T)$. A numerical fit for that viscosity leads to the results (Mackenzie, 1961)

$$\begin{aligned} \eta_0 &\cong 1.6 \times 10^{-13} \text{ P} \\ E_\eta &\cong 180 \text{ kcal/mol}. \end{aligned} \tag{40}$$

The latter stems from a local mechanism that rearranges the silica network structure and, presumably, involves breakage and reformation of Si–O chemical bonds.

The situation is quite different for glass formers at the fragile extreme. o-Terphenyl (OTP; $C_{18}H_{14}$) is a prototypical case and a favorite subject of experimental investigation (Fujara et al., 1992; Cicerone and Ediger, 1993; Wang and Ediger, 1999). Plotting $\ln \eta$ vs $1/T$ produces a graph with a strong upward curvature (see Fig. 4), the slope of which defines a temperature-dependent effective activation energy (Greet and Turnbull 1967; Plazek et al., 1994):

$$E_\eta^{(\text{eff})}(T) = \partial \ln \eta / \partial (1/T). \tag{41}$$

It has been pointed out (Greet and Turnbull, 1967) that this activation energy rises from approximately one-fourth of the heat of vaporization for the liquid above its melting point to approximately five times the heat of vaporization when the temperature is reduced to $T_g \approx -35°C$. This rise by a factor of 20 unambiguously testifies that the OTP energy landscape is strongly heterogeneous. The portions of that landscape accessed at high temperatures permit structural relaxation by surmounting low-energy barriers, presumably each step of which involves only rearranging slightly a small number of molecules. In contrast, the very large $E_\eta^{(\text{eff})}(T \cong T_g)$ can arise only by cooperative rearrangement of a large number of molecules, perhaps in the range 20–100. Furthermore, such rearrangements are unlikely to consist of single elementary transitions between neighboring basins; to pass from one low-lying basin to one of equal or greater depth, it may be required to pass upward in landscape altitude (energy) through a long sequence of elementary transitions before descending far away to a suitably deep basin. The latter concept is supported by the very small effective preexponential factor for OTP near T_g (Plazek *et al.*, 1994),

$$\eta_0^{(\text{eff})}(T \cong T_g) \cong 7 \times 10^{-68} \text{ P}, \tag{42}$$

reflecting a large activation entropy associated with the great diversity of transition pathways possible between a pair of separated deep basins.

Figure 11 attempts to provide a schematic illustration for the landscape topographic differences just discussed for the strong (SiO_2) and fragile (OTP) extremes. Of course, most glass-forming liquids fall between these extremes and should be thought of as interpolating the topographies illustrated. Figure 11 intends to illustrate only one symmetry sector. The strong case involves a single "metabasin" down into which the cooling liquid could configurationally trickle, surmounting barriers but encountering no substantial traps. The contrasting fragile case displays deep and widely separated traps, i.e., a diverse collection of metabasins in the same symmetry sector. The

(a) Strong Glass Formers

(b) Fragile Glass Formers

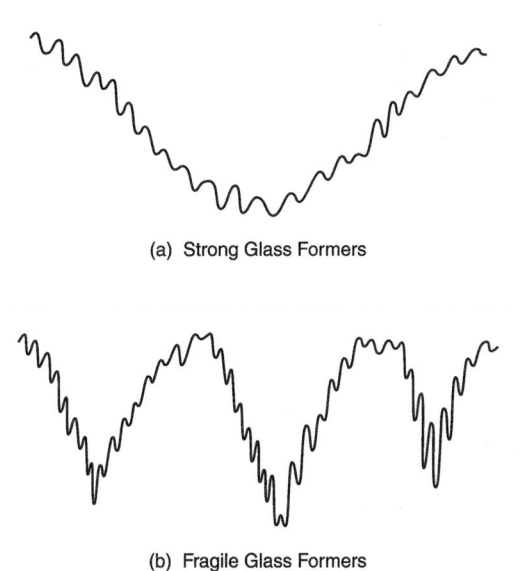

FIG. 11. Schematic illustration of the topographic distinction between energy landscapes for strong and fragile glass formers. Only one symmetry sector is represented. Potential energy increases upward; the horizontal direction represents all configurational coordinates.

former, Fig. 11a, is similar to the configuration space "funnels" that have been postulated for properly folding biological proteins, while the latter, Fig. 11b, is characteristic of biologically useless misfolding proteins (Saven and Wolynes, 1997).

The long-pathway rearrangement processes expected for fragile materials at low temperatures are expected to be rare, to involve a local disruption of the otherwise well-structured amorphous medium, and to be relatively long-lived on the usual molecular time scale. These features all contribute to a substantial lengthening of the mean relaxation time $t_{rel}(T)$, Eq. (36), with declining temperature. Furthermore, the landscape diversity of deep traps and of the configuration space pathways that connect them should produce a broad spectrum of relaxation times, just as required by stretched-exponential relaxation functions, Eq. (34).

This scenario requires strongly supercooled fragile glass formers to be dynamically heterogeneous media, consisting at any instant mostly of nondiffusing well-bonded particles, but with a few local "hot spots" of mobile particles. The phenomenon of low-temperature dynamical heterogeneity has experimental support (Cicerone and Ediger, 1995) and has also been clearly documented by computer simulation (Donati *et al.*, 1999a,b; Bennemann *et al.*, 1999; Perera and Harrowell, 1996a,b). Both techniques confirm that the regions of anomalous mobility grow strongly in mean size with declining temperature.

The Stokes–Einstein relation connects shear viscosity η to the self-diffusion constant D in a liquid [see also Eq. (4)],

$$D(T) = k_B T/[6\pi a \eta(T)]. \tag{43}$$

For liquids composed of relatively compact molecules, this relation describes the temperature dependence of D well above the melting point and even for moderate supercooling, using the measured η and a temperature-independent a that tends to approximate roughly the size of the given molecular structure. However the Stokes–Einstein relation is based on macroscopic hydrodynamics that treats the liquid as a continuum. This hydrodynamic picture clearly contradicts the dynamic heterogeneity that applies to strongly supercooled fragile liquids. Perhaps, then, it should come as no surprise that the Stokes–Einstein relation fails dramatically in those circumstances: $D(T \cong T_g)$ has been observed experimentally to be 10^2 to 10^3 times larger than Eq. (43) predicts, retaining the same a that is applicable at high temperatures (Fujara *et al.*, 1992; Cicerone and Ediger, 1993; Kind *et al.*, 1992; Cicerone and Ediger, 1996). It has been pointed out (Stillinger and Hodgdon, 1994) that this D enhancement can arise from a suitable combination of mobile-region volume fraction, size, and lifetime.

No intrinsic mathematical reason exists to connect the depth distribution of the basins to the kinetic pathways between them. The latter depend on the height of saddle points connecting neighboring basins. However, intermolecular interactions, particularly those for fragile glass formers, constitute a very special class of functions. Considerations similar to those that have suggested the form of Fig. 11b might also suggest that the rugged landscape obeys a statistical scaling relation. This is the basis of the formula suggested long ago by Adam and Gibbs (1965), which connects the kinetic property t_{rel} or, equivalently, η to supercooled liquid thermodynamics:

$$t_{rel}, \eta \cong A \exp(B/T s_{conf}). \tag{44}$$

Here A and B are positive constants, expected to depend on which of t_{rel} or η is involved, and s_{conf} is the molar configurational entropy to be obtained from calorimetric measurements for crystal, liquid, and glass. s_{conf} is a measure of the landscape's "ruggedness" and is related to the basin enumeration function introduced in Section II,[11]

$$s_{conf}/R = \sigma_a[\bar{\phi}_a(T)], \tag{45}$$

where R is the gas constant. For Eq. (44) to hold, one must have

$$E_\eta^{(eff)}(T) \propto 1/\sigma_a[\bar{\phi}_a(T)]; \tag{46}$$

that is, the effective dynamical excitation energy should be inversely proportional to the depth-dependent basin enumeration function σ_α evaluated at the depth populated at the given temperature. Thus, in the Adam–Gibbs picture, the origin of viscous slowdown close to T_g is the dearth of basins (configurations) that the system is able to sample at low temperatures. Furthermore, structural arrest is predicted to occur at the Kauzmann temperature. If the configurational entropy has the form[12] (Debenedetti, 1996; Richert and Angell, 1998),

$$s_{conf} = s_\infty \left(1 - \frac{T_K}{T}\right), \tag{47}$$

where T_K is the Kauzmann temperature (at which s_{conf} presumably vanishes), the Adam–Gibbs relation reads

$$\eta = A \exp\left(\frac{\tilde{B}}{T - T_K}\right). \tag{48}$$

This is the VTF equation [see (1)], with T_0, the temperature of structural arrest, equal to T_K. The remarkable closeness of these two temperatures for several substances [T_0 obtained from relaxation measurements and T_K from

---

[11] The experimental determination of the configurational entropy from calorimetric (heat capacity) measurements is discussed in Section V.

[12] Equation (47) results when the heat capacity difference between the supercooled liquid and the crystal is inversely proportional to the absolute temperature, as is experimentally found to be the case for many substances (Alba et al., 1990).

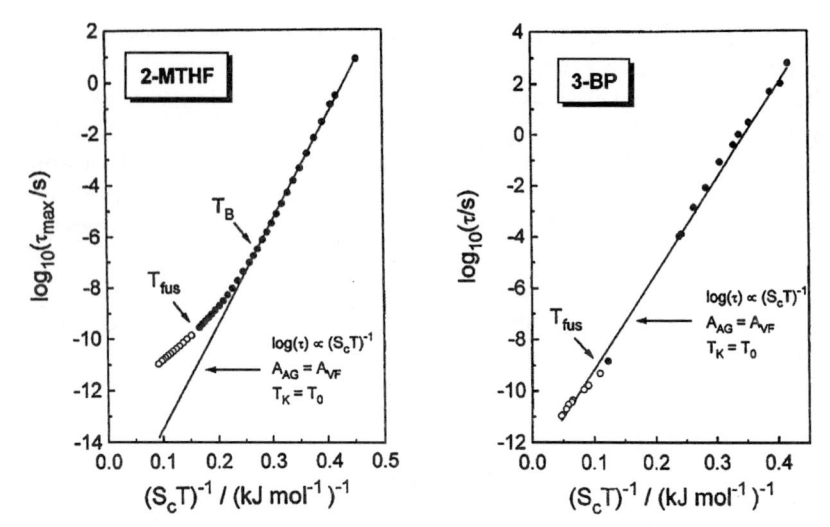

FIG. 12. Adam–Gibbs plots of the dielectric relaxation time of 2-methyltetrahydrofuran (2-MTHF) and 3-bromopentane (3-BP) versus $(Ts_{conf})^{-1}$. The lines are VTF fits, T_{fus} is the fusion temperature, and T_B is the temperature below which the VTF equation applies. A_{AG} and A_{VF} are prefactors in the Adam–Gibbs and VTF equations, respectively. T_K is the calorimetrically determined Kauzmann temperature, and T_0 is the VTF singular temperature, which were set equal in the VTF (line) fits. (Reprinted with permission from R. Richert and C. A. Angell. Dynamics of glass-forming liquids. V. On the link between molecular dynamics and configurational entropy. *J. Chem. Phys.* (1998); 108:9016. Copyright © 1998, American Institute of Physics.)

calorimetric experiments (Angell, 1997)] was mentioned in Section I. This is an example of the apparent connection between dynamics and thermodynamics in glasses (Wolynes, 1988). Figure 12 (Richert and Angell, 1998) shows the dielectric relaxation time for 2-methyltetrahydrofuran (MTHF) and 3-bromopentane (3-BP), plotted in Adam–Gibbs fashion. It can be seen that the theory provides an adequate description of the dielectric relaxation in 3-BP, but less so for MTHF. The Adam–Gibbs theory, in other words, is approximate and not general. Nevertheless, it remains a very useful and popular analytical framework for correlating and extrapolating data. Understanding the connection between kinetic properties and thermodynamics on which the Adam–Gibbs theory is based is arguably the most important problem in the thermophysics of glasses and supercooled liquids.

V. Thermodynamics

The statistical description of an energy landscape needed to make connection with thermodynamic properties is provided by the basin enumeration function, $\sigma(\phi)$, introduced in Section II,

$$\frac{d\Omega}{d\phi} = C \exp[N\sigma(\phi)], \qquad (11)$$

where $d\Omega$ is the number of potential energy minima whose depth, on a per-molecule basis, is $\phi \pm d\phi/2$; N is the number of molecules; and C is constant (Stillinger, 1999). It follows from the above equation that σ, the basin enumeration function, is the configurational entropy per molecule arising from the existence of multiple minima of depth ϕ [see also Eq. (45)],

$$\frac{S_{conf}}{Nk_B} = \sigma. \qquad (49)$$

As explained in Section II, the partition function can be written as a one-dimensional integral over the basin depth ϕ (Stillinger and Weber, 1982). In the thermodynamic limit, the integral in Eq. (15) is dominated overwhelmingly by basin depths in the neighborhood of a particular, temperature-dependent value, which satisfies the extremum condition

$$\frac{\partial}{\partial \phi}(\sigma - \beta\phi - \beta a^v) = 0, \qquad (50)$$

whereupon the Helmholtz energy can be written

$$-\beta A(\beta)/N = \sigma[\bar{\phi}(\beta)] - \beta\bar{\phi}(\beta) - \beta a^v[\beta, \bar{\phi}(\beta)]. \qquad (16)$$

For the limiting case where a^v does not depend on ϕ (i.e., all basins have the same mean curvature at their respective minima), $\bar{\phi}$ follows from the condition

$$\frac{d\sigma}{d\phi} = \frac{1}{k_B T}. \qquad (51)$$

Equation (16) establishes the formal connection between the thermodynamic properties of a system and its energy landscape. In particular, it expresses the Helmholtz energy in terms of an energy contribution, $\bar{\phi}$, arising from the sampling of basins of a particular, temperature-dependent depth; a configurational entropy contribution, σ, due to the existence of an exponential multiplicity of such basins [Eq. (11)]; and a vibrational contribution that arises from the thermal excitation that allows the system to sample basins at a temperature-dependent "height" above the local potential energy minimum.

The basin enumeration function $\sigma(\phi)$ can be obtained from experimental thermodynamic data. Because experiments are commonly done at constant pressure, Eqs. (13), (15), (16), and (50) must be recast in terms of the isobaric, isothermal (N, P, T) ensemble. Following arguments identical to those in the isochoric case, the important result (Stillinger, 1998) is

$$-\beta G(\beta)/N = \sigma[\bar{\psi}(\beta)] - \beta\bar{\psi}(\beta) - \beta g^v[\beta, \bar{\psi}(\beta)], \quad (52)$$

where G is the Gibbs free energy, ψ is a "potential enthalpy" basin depth per particle [see also Eqs. (21) and (22)], and g^v is the intrabasin vibrational Gibbs free energy per particle. To construct $\sigma(\psi)$, we consider the equilibrium between a liquid (l) and its crystal (x) at the melting point (T_m, P_m):

$$-\sigma^{(l)}(\bar{\psi}) + \beta_m\bar{\psi}^{(l)} + \beta_m g^v(\beta, \bar{\psi})^{(l)} = -\sigma^{(x)}(\bar{\psi}) + \beta_m\bar{\psi}^{(x)} + \beta_m g^v(\beta, \bar{\psi})^{(x)}. \quad (53)$$

Because the crystal is confined to a single basin, $\sigma^{(x)} = 0$. If the intrabasin vibrational free energy depends only on temperature, the above equation simplifies to

$$\sigma^{(l)} = \beta_m \left[\psi^{(l)} - \psi^{(x)}\right], \quad (54)$$

$\sigma^{(l)}$ corresponds to the entropy of fusion, and $\psi^{(l)} - \psi^{(x)}$ corresponds to the enthalpy of fusion. For $T < T_m$ and $P = P_m$, we can therefore write

$$\psi^{(l)}(\beta, P_m) - \psi^{(x)}(\beta, P_m) \equiv \Delta\psi = \Delta h_m - \int_T^{T_m} \Delta c_p dT \quad (55)$$

$$\sigma^{(l)}(\beta, P_m) = \frac{\Delta h_m}{k_B T_m} - \frac{1}{k_B} \int_T^{T_m} \frac{\Delta c_p}{T} dT, \quad (56)$$

where $\Delta c_p = c_p^{(l)} - c_p^{(x)}$. Thus, one can calculate the enthalpy excitation profile, $\psi^{(l)} - \psi^{(x)} = f_1(T)$, and the configurational entropy, $\sigma^{(l)} = f_2(T)$, from heat capacity data. Cross-plotting yields the desired basin enumeration function, $\sigma = \sigma(\Delta\psi)$ (Stillinger, 1998; Speedy, 1999). Equations (55) and (56) are valid under the assumption that the vibrational free energies of the crystal and the liquid are equal at the same temperature. Experimentally, this would result in negligible differences between crystalline and glassy heat capacities. Examples of such substances include OTP (Chang and Bestul, 1972) and 3-methylpentane (Takahara et al., 1994). 1-Propanol, in contrast, shows nonnegligible differences between vitreous and crystalline heat capacities (Takahara et al., 1994). These differences have not been taken into account in the calculations shown below.

Figure 13 shows the isobaric basin enumeration functions for 1-propanol and 3-methylpentane, at various pressures in the range 0–2 kbar. Figure 14 shows the corresponding excitation profiles. Figure 15 illustrates schematically the salient characteristics of the isobaric basin enumeration function.

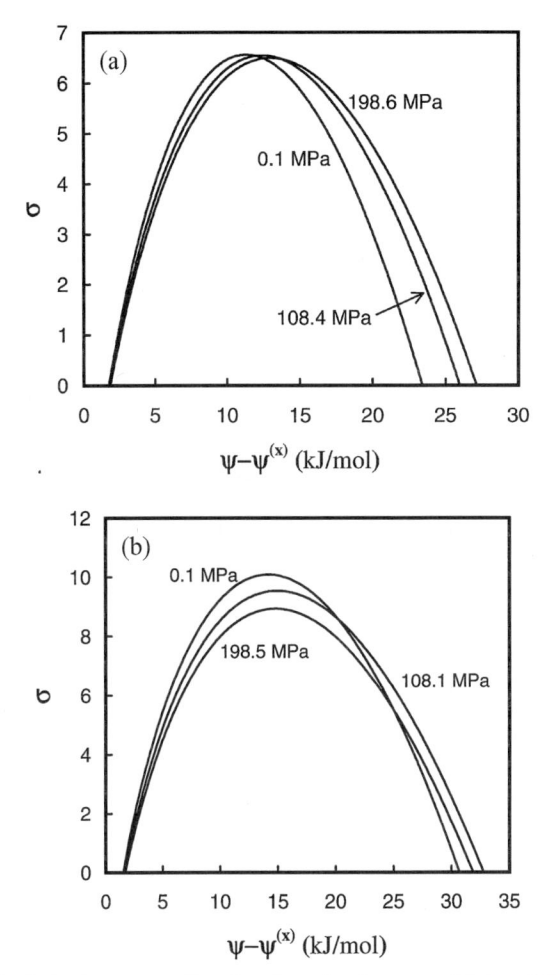

FIG. 13. Isobaric basin enumeration functions for 1-propanol (a) and 3-methylpentane (b) at different pressures. The x axis gives the difference between inherent structure and crystalline "potential enthalpies." Calculations performed according to Eqs. (55) and (56), using the experimental heat capacity data of Takahara *et al.* (1994). (From Lewis, 2000.)

The characteristic temperature-dependent basin depth is obtained from the condition [see also Eq. (51)]

$$\frac{d\sigma(\Delta\psi)}{d\Delta\psi} = \frac{1}{k_B T}. \tag{57}$$

Temperature increases monotonically from the point where $\sigma = 0$ ($d\sigma/d\Delta\psi > 0$ branch) to the top of the curve, which corresponds to the

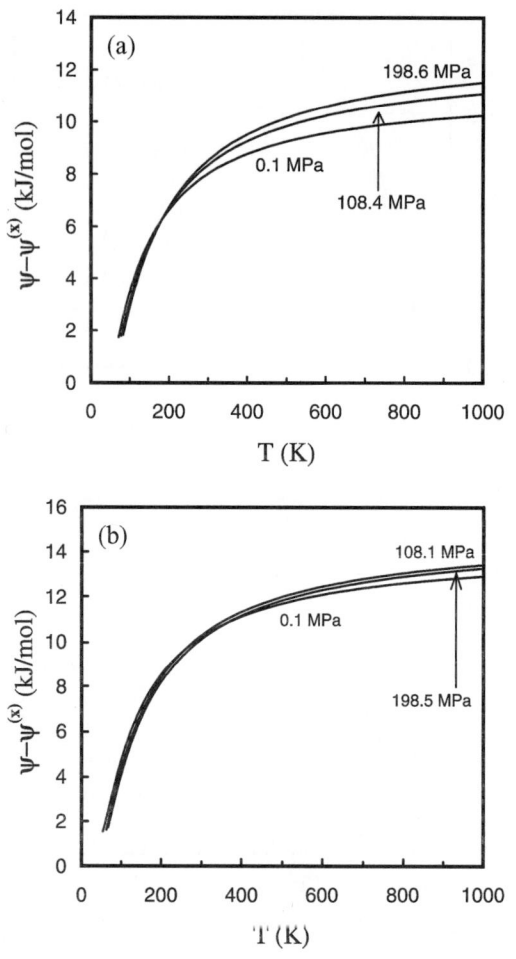

FIG. 14. Isobaric excitation profiles for 1-propanol (a) and 3-methylpentane (b) at different pressures. The y axis gives the "potential enthalpy" difference between the amorphous basin sampled preferentially at a given temperature and the single crystalline basin. Calculations performed according to Eq. (55), using the experimental heat capacity data of Takahara et al. (1994). (From Lewis, 2000.)

limit $T \to \infty$. The condition $\sigma = 0$ implies the existence of a unique basin; it therefore corresponds to the Kauzmann temperature T_K, where the entropy of the supercooled liquid equals that of the stable crystal.[13] When the system is in this unique basin, it has attained the lowest possible energy that a noncrystalline packing can adopt. This condition is called the ideal glass. The

[13] This is contingent on the assumption that $d\sigma/d\psi$ remains finite (see Stillinger, 1988).

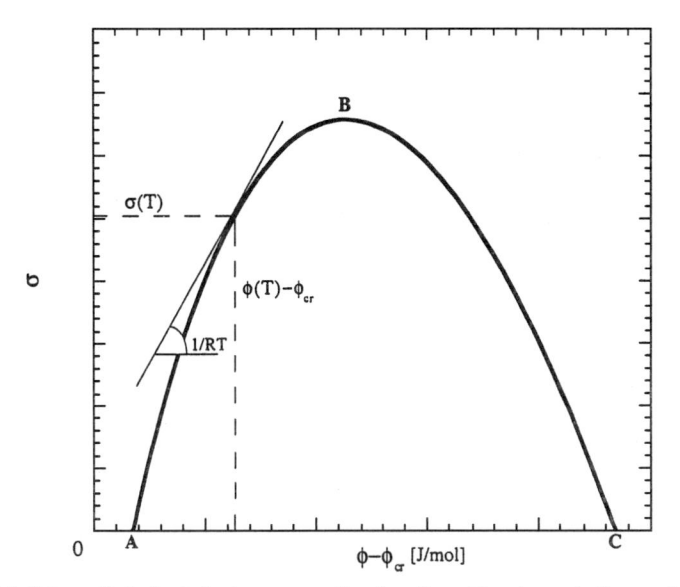

FIG. 15. Schematic isobaric basin enumeration function. Also shown is the graphical construction that yields σ and $\Delta\psi$ at each temperature. See also Eqs. (50) and (57). (From Lewis, 2000.)

negatively sloped portion of the basin enumeration function corresponds to negative temperatures and is therefore not physically relevant for the present purposes.

The similarity between the calculated isobaric excitation profiles shown in Fig. 14 and the isochoric profiles obtained by molecular simulation (Sastry et al., 1998a) is remarkable. The isobaric excitation profiles have a discontinuity (not shown) at T_K: for $T < T_K$ the system remains trapped in the unique (ideal glass) basin, and $\Delta\psi$ is constant. The discontinuity is absent in the simulated isochoric profiles, because the system gets trapped kinetically in a cooling rate-dependent basin and is not able to access the deepest amorphous basin.

The approximately parabolic shape of the curves in Fig. 13 suggests the parametrization

$$\sigma = \sigma_\infty - m(\Delta\psi - \Delta\psi_\infty)^2, \tag{58}$$

where $\Delta\psi_\infty$ is the infinite temperature limit of $\Delta\psi$, and σ_∞ is the corresponding configurational entropy (Fig. 15). Table I (Lewis, 2000) gives the values of, σ_∞, m, and $\Delta\psi_\infty$ for 1-propanol and 3-methylpentane, as a function of pressure. The calculated isobaric basin enumeration functions are asymmetric, and Eq. (58) describes the curves accurately only if different values of m

TABLE I
Gaussian Landscape Parameters for 1-Propanol and 3-Methylpentane Calculated from Heat Capacity Measurements[a]

P (MPa)	σ_∞	$\Delta\psi_\infty$ (kJ/mol)	$100\,m$ (mol/kJ)^2
1-Propanol			
0.1	6.564	11.3	6.8
108.4	6.549	12.4	5.49
198.6	6.499	13.0	4.96
3-Methylpentane			
0.1	10.083	14.2	5.87
108.1	9.532	15.0	5.00
198.5	8.927	14.8	4.90

[a] From Takahara et al. (1994).

are used for positive and negative temperatures. Since only the former have physical significance, it seems prudent at this stage in our understanding of landscape statistics to preserve the simplicity of Eq. (58) rather than to include higher-order terms in an attempt to capture the slightly asymmetric character of the basin enumeration function. Landscapes with parabolic enumeration functions will henceforth be referred to as *Gaussian* (Speedy, 1999; Büchner and Heuer, 1999).

Several useful relations can be written for systems with Gaussian landscapes. The excitation profile satisfies the relation

$$\Delta\psi = \begin{cases} \Delta\psi_\infty - \dfrac{1}{2mk_B T} & (T > T_K) \\ \Delta\psi_\infty - \dfrac{1}{2mk_B T_K} & (T \leq T_K). \end{cases} \quad (59)$$

Combining Eqs. (16), (51), and the isochoric analogue of (58),

$$\sigma = \sigma_\infty - m(\phi - \phi_\infty)^2, \quad (60)$$

leads to the following expression for the Helmholtz free energy:

$$-\beta A/N = \sigma_\infty + \beta\phi_\infty - \beta^2/4m - \beta a^v. \quad (61)$$

The equation of state follows by differentiation,

$$\beta P/\rho = \beta \frac{d\phi_\infty}{d\ln\rho} - \frac{d\sigma_\infty}{d\ln\rho} + \left(\frac{\beta}{2m}\right)^2 \frac{dm}{d\ln\rho} + \left(\frac{\partial\beta a^v}{\partial\ln\rho}\right)_\beta. \quad (62)$$

Thus, all thermodynamic (equilibrium) properties of a macroscopic system with a Gaussian landscape can be calculated from knowledge of $\phi_\infty(\rho)$ (the density dependence of the basin depth per particle in the high-temperature limit), $\sigma_\infty(\rho)$ (the density dependence of the configurational entropy associated with the existence of an exponential multiplicity of basins of a given depth, in the high-temperature limit), $m(\rho)$ (the density dependence of the effective width of the basin distribution), and $a^v(\beta, \rho)$ (the temperature and density dependence of the vibrational free energy). $\phi_\infty(\rho)$ and $m(\rho)$ can be obtained from molecular simulation studies that combine energy minimization and standard particle moves (e.g., Sastry et al., 1998a; Jonsson and Andersen, 1988). Fitting the resulting excitation profiles to the isochoric analogue of Eq. (59) ($T > T_K$) yields ϕ_∞ and m. The vibrational free energy can be obtained from the equation

$$A^v = \sum_{i=1}^{vN} \left[\frac{\hbar \omega_i}{2} + k_B T \ln \left(1 - e^{-\beta \hbar \omega_i} \right) \right], \quad (63)$$

where $\hbar$ is Planck's constant divided by 2π, ω_i are the density-dependent normal mode angular frequencies, and v is the number of degrees of freedom per molecule[14] (Landau and Lifshitz, 1980). The normal mode frequencies are obtained from the Hessian matrix of second derivatives of the energy. For a Gaussian landscape,

$$\sigma_\infty = m \left(\phi_K - \phi_\infty \right)^2, \quad (64)$$

where ϕ_K corresponds to the deepest amorphous basin, that is, to the ideal glass. Thus, σ_∞ can, in principle, be obtained numerically by performing a constrained optimization in which stable packings whose measures of order exceed a specified cutoff are excluded from consideration.

From the preceding discussion it can be concluded that significant progress in understanding the thermodynamics of complex condensed systems under conditions such that each constituent unit (for example, a molecule in the case of a molecular glass former or an amino acid residue in the case of a protein) experiences simultaneous strong interactions with many neighbors can result from a numerical investigation of basic landscape features (ϕ_∞, σ_∞, m) as a function of the density and molecular architecture. Recent studies by Sciortino et al. (1999) and by Büchner and Heuer (1999) demonstrate the powerful insights that can be gained by this approach.

The energy landscape perspective provides fresh insight into, and suggests unexpected connections between, stretched liquids and the glassy state. As shown in Fig. 8, the $p(\rho)$ relationship for mechanically stable packings of simple, atomic systems shown van der Waals-type behavior. We refer to this type of curve as the equation of state of an energy landscape (Debenedetti et al., 1999). The landscapes of ethane, n-pentane, and cyclopentane exhibit similar behavior (Utz et al., 2001). Water's landscape, in contrast, consists

---

[14] Equation (63) involves a proper quantum treatment of harmonic vibrational degrees of freedom.

of a family of $p(\rho)$ curves which depend on the temperature of the equilibrated liquid from which the energy minimization is performed (Roberts *et al.*, 1999). While our present understanding of those features of molecular architecture and interactions that give rise to a temperature-dependent landscape equation of state is far from complete, it is clear that orientation-dependent interactions are sufficient to cause this type of complexity. In what follows we restrict our attention to "simple," that is, temperature-independent, landscape equations of state.

The minimum in the $p(\rho)$ curve corresponds to the maximum tensile strength of which the amorphous form of a given material is capable. As mentioned in Section III, the density at which this occurs is called the Sastry density, an important material characteristic. Below the Sastry density, mechanically stable packings are fissured due to the appearance of large cavities (Sastry *et al.*, 1997b). Thus, this density is a limiting condition, below which amorphous packings can no longer be mechanically stable (i.e., minimum energy) and simultaneously spatially homogeneous. Since a glass is a liquid trapped in a mechanically stable configuration (potential energy minimum), it is not possible to form a homogeneous glass below a material's Sastry density.

The nonmonotonic shape of the $p(\rho)$ curve is reminiscent of the behavior of approximate equations of state, such as the van der Waals equation. It must be understood, however, that whereas the unstable portion of such equations is unphysical, the landscape equation of state is exact. Neighboring states on this curve, however, are not connected by a continuum of thermally equilibrated states. Rather, thermal motion having been removed by construction, the path from one point to another along the landscape equation of state is through the high-temperature equilibrated fluid. This caveat notwithstanding, there appears to be a deep and illuminating relationship between the spinodal curve for the superheated liquid and the Sastry point. The theoretical maximum for the tensile strength of noncrystalline forms of a material is given by the $T = 0$ limit of the spinodal curve along the superheated liquid branch (Debenedetti, 1996). Table II lists the predictions for the Sastry density ρ_s and maximum tensile strength P_s (i.e., the Sastry point), according to three cubic equations of state, and compares these theoretical predictions with our simulation results for the Lennard–Jones fluid (Fig. 8). Of course, none of these equations are accurate representations of the Lennard–Jones fluid. Nevertheless, the clear implication from the simulations is that the Sastry point corresponds to the $T = 0$ limit of the superheated liquid spinodal.

Since the Sastry density is an absolute (thermodynamic) limit to vitrification as $T \to 0$, it follows that the point $(\rho_s, T = 0)$ is the low-temperature termination of the Kauzmann curve. Since materials are vitrified by cooling and/or compression, the Kauzmann curve is expected to have a positive slope

TABLE II

COMPARISON OF THEORETICAL PREDICTIONS OF THE SASTRY POINT
(ρ_s, P_s) ACCORDING TO CUBIC EQUATIONS OF STATE AND SIMULATIONS[a]

	ρ_s/ρ_c	P_s/P_c
van der Waals	3	−27
Soave–Redlich–Kwong[b,c]	3.85	−62.5
Peng–Robinson[c,d]	3.95	−71.5
Simulation[e]	2.76	−39.8

[a] Subscript c denotes the value of a property at the critical point.
[b] From Soave (1972).
[c] An acentric factor $\omega = 0.001$ was used, corresponding to Ar (Reid et al., 1987).
[d] From Peng and Robinson (1976).
[e] Smoothly truncated Lennard–Jones fluid (Sastry et al., 1997b). See Fig. 8.

in the (T, ρ) plane. Our results suggest, therefore, that the low-temperature termination of the Kauzmann curve coincides with the $T \to 0$ limit of the superheated liquid spinodal.[15] The predicted convergence of two such apparently unrelated limits is surprising. It is not obvious why the limit of stability for the superheated liquid should tend to the locus along which the entropies of a deeply supercooled liquid and of its stable crystalline form coincide. Nevertheless, this conclusion from our simulations is supported by mean-field theoretical calculations (Debenedetti et al., 1999). Figure 16 shows the calculated phase diagram of a soft-sphere system whose constituent atoms interact via a pairwise-additive spherically symmetric potential that decays as the inverse ninth power of the interatomic distance, plus a mean-field, van der Waals-type attraction. In addition to the phase coexistence loci, the diagram shows the Kauzmann curve, as well as the spinodal curves corresponding to the liquid–vapor transition. The remarkable feature of the calculation is the low-temperature convergence of the superheated liquid spinodal and the Kauzmann curve. Similar results have also been obtained for the hard-sphere and hard-dumbbell models with mean-field attraction.

In spite of the remarkable agreement between theoretical predictions, on the one hand, and the logical consequence of the simulations, on the other, we believe that the generality of behavior such as shown in Fig. 16 cannot be accepted uncritically but should instead be regarded as a fundamental open question on the properties of disordered materials. This caution is warranted by the approximate, mean-field nature of the calculations, as well as by microscopic critiques of the very notion of a Kauzmann temperature (Stillinger, 1988). Perhaps more significantly, recent experiments on

[15] For a contrasting viewpoint, see Sastry (2000).

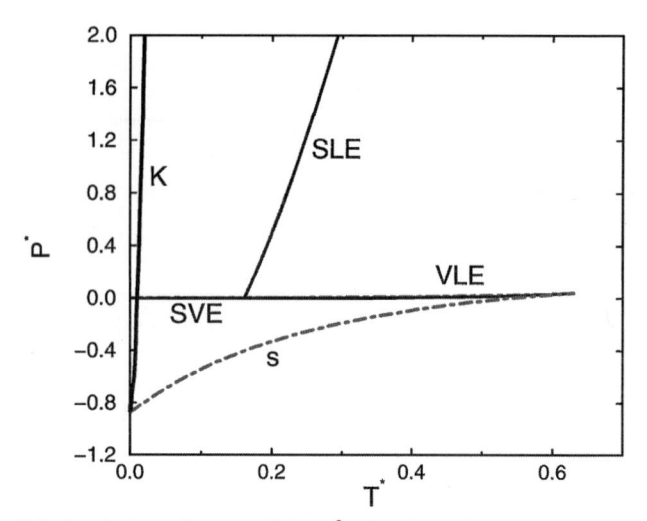

FIG. 16. Calculated phase diagram of the r^{-9} soft-sphere plus mean-field model, showing the vapor–liquid (VLE), solid–liquid (SLE), and solid–vapor (SVE) coexistence loci, the superheated liquid spinodal (s), and the Kauzmann locus (K) in the pressure–temperature plane ($P^* = P\sigma^3/\varepsilon; T^* = k_B T/\varepsilon$). The Kauzmann locus gives the pressure-dependent temperature at which the entropies of the supercooled liquid and the stable crystal are equal. Note the convergence of the Kauzmann and spinodal loci at $T = 0$. See Debenedetti et al. (1999) for details of this calculation.

poly(4-methylphentene-1), which show inverse melting at low temperatures and high pressures, that is, freezing upon heating, call into question the notion of an ideal glass and of the Kauzmann temperature as an absolute limit to supercooling (Rastogi et al., 1999; Greer, 2000).

VI. Conclusion

In glasses and the supercooled liquids from which they are commonly formed, molecules are subject to the simultaneous action of many neighbors. Under these conditions the multidimensional N-body potential energy function $\Phi(\mathbf{r}_1 \ldots \mathbf{r}_N)$, the energy landscape, provides a convenient framework within which to describe the thermophysical properties of this important class of condensed-phase systems. Melting and freezing phenomena (LaViolette and Stillinger, 1985a), complex dynamics in supercooled liquids (Sastry et al., 1998a), the mechanical strength of glasses (Utz et al., 2001; Malandro and Lacks, 1997, 1999), the limits of stability of the liquid state of matter (Debenedetti et al., 1999; Sastry, 2000), and aging phenomena in glasses (Utz et al., 2000; Kob et al., 2000) are some of the important phenomena on which the landscape perspective has yielded useful new insights.

Major open questions where landscape-based ideas should prove helpful include the possible thermodynamic basis for the glass transition (Debenedetti *et al.*, 1999), the relationship between kinetics and thermodynamics of deeply supercooled liquids and glasses (Adam and Gibbs, 1965; Wolynes, 1988), and translation–rotation decoupling and the breakdown of the Stokes–Einstein relationship in supercooled liquids (Fujara *et al.*, 1992). In addition, the reformulation of the thermodynamics of liquids embodied in Eqs. (16), (52), (55), (56), (61), and (62) suggests that understanding basic topological features of a landscape's density-dependent statistics could lead to improved theories of simple and complex liquids. As explained in Section V, landscape statistics can be obtained from experiments, theory, and simulations.

The quantitative description of disorder in liquids and glasses is the second theme of this article. Recent work on the simple hard-sphere system (Torquato *et al.*, 2000; Truskett *et al.*, 2000) shows that it is possible to distinguish equilibrium and nonequilibrium states based on the accessible types of molecular disorder. It is also possible to relate the type of molecular disorder to a glass' processing history. Extending these developments to models with more realistic interactions, and mapping their order phase diagrams, including their inherent structures, are some of the important questions, the answers to which could shed new light on the relationship among molecular interactions, process conditions, and the morphology of glasses. More generally, the problem of quantifying disorder in condensed phases has implications for early detection of tumors, transdermal drug delivery, flow through porous media, and powder engineering.

Glasses are central to a wide variety of commercial processes and technical applications, such as the preservation of labile biochemicals, food processing, the manufacture of optical wave guides, corrosion resistance, photovoltaic cells, and the extrusion and molding of engineering plastics. In spite of their widespread natural occurrence and technical utilization, there are major gaps in the present understanding of the vitreous state of matter. In this article we have presented a geometric viewpoint of glass transition phenomena and supercooling. These ideas have contributed significantly to contemporary knowledge of this important topic. We believe that they are likely to prove equally fruitful in the future.

ACKNOWLEDGMENTS

P.G.D. gratefully acknowledges the support of the U.S. Department of Energy, Division of Chemical Sciences, Geosciences, and Biosciences, Office of Basic Energy Science (Grant DE-FG02-87ER13714). T.M.T. acknowledges the support of the National Science Foundation.

REFERENCES

Abkevich, V. I., Gutin, A. M., and Shakhnovich, E. I., Free energy landscape for protein folding kinetics: Intermediates, traps, and multiple pathways in theory and lattice model simulations. *J. Chem. Phys.* **101,** 6052 (1994).
Adam, G., and Gibbs, J. H., On the temperature dependence of cooperative relaxation properties in glass-forming liquids. *J. Chem. Phys.* **43,** 139 (1965).
Alba, C., Busse, L. E., List, D. J., and Angell, C. A., Thermodynamic aspects of the vitrification of toluene and xylene isomers, and the fragility of light hydrocarbons. *J. Chem. Phys.* **92,** 617 (1990).
Anderson, P. W., Halperin, B. J., and Varma, C. M., Anomalous low-temperature properties of glasses and spin glasses. *Philos. Mag.* **25,** 1 (1972).
Angell, C. A., Strong and fragile liquids. *In* "Relaxations in Complex Systems" (K. Ngai and G. B. Wright, Eds.), Nat. Tech. Info. Ser., U.S. Dept. Commerce, Springfield, VA, 1985, p. 1.
Angell, C. A., Structural instability and relaxation in liquid and glassy phases near the fragile liquid limit. *J. Non-Cryst. Sol.* **102,** 205 (1988).
Angell, C. A., Formation of glasses from liquids and biopolymers. *Science* **267,** 1924 (1995).
Angell, C. A., Landscapes with megabasins: Polyamorphism in liquids and biopolymers and the role of nucleation in folding and folding diseases. *Physica D* **107,** 122 (1997).
Angell, C. A., and Kanno, H., Density maxima in high-pressure supercooled water and liquid silicon dioxide. *Science* **193,** 1121 (1976).
Angell, C. A., Clarke, J. H. R., and Woodcock, L. V., Interaction potentials and glass formation: A survey of computer experiments. *Adv. Chem. Phys.* **48,** 397 (1981).
Angell, C. A., Ngai, K. L., McKenna, G. B., McMillan, P. F., and Martin, S. W., Relaxation in glassforming liquids and amorphous solids. *J. Appl. Phys.* **88,** 3113 (2000).
Ashcroft, N. W., and Mermin, N. D., "Solid State Physics." Saunders College, Philadelphia, 1976.
Becker, O. M., and Karplus, M., The topology of multidimensional potential energy surfaces: Theory and application to peptide structure and kinetics. *J. Chem. Phys.* **106,** 1495 (1997).
Bengtzelius, U., Götze, W., and Sjölander, A., Dynamics of supercooled liquids and the glass transition. *J. Phys. C Sol. State Phys.* **17,** 5915 (1984).
Bennemann, C., Donati, C., Baschnagel, J., and Glotzer, S. C., Growing range of correlated motion in a polymer melt on cooling towards the glass transition. *Nature* **399,** 246 (1999).
Blanshard, J. M. V., and Lillford, P. J. (Eds.), "The Glassy State in Foods." Nottingham University Press, 1993.
Bowles, R. K., and Corti, D. S., Statistical geometry of hard sphere systems: Exact relations for first-order phase transitions in multicomponent systems. *Mol. Phys.* **98,** 429 (2000).
Brinon, L., Geiger, S., Alard, V., Doucet, J., Tranchant, J.-F., and Courraze, G., Percutaneous absorption of sunscreens from liquid crystalline phases. *J. Contr. Rel.* **60,** 67 (1999).
Brostow, W., Chybicki, M., Laskowski, R., and Rybicki, J., Voronoi polyhedra and Delaunay simplexes in the structural analysis of molecular-dynamics-simulated materials. *Phys. Rev. B* **57,** 13448 (1998).
Brüning, R., and Samwer, K., Glass transition on long time scales. *Phys. Rev. B* **46,** 318 (1992).
Bryant, S., and Blunt, M., Prediction of relative permeability in simple porous-media. *Phs. Rev. A* **46,** 2004 (1992).
Büchner, S., and Heuer, A., Potential energy landscape of a model glass former: Thermodynamics, anharmonicities, and finite size effects. *Phys. Rev. E* **60,** 6507 (1999).

Buchner, M., Ladanyi, B. M., and Stratt, R. M., The short-time dynamics of molecular liquids. Instantaneous-normal-mode theory. *J. Chem. Phys.* **97,** 8522 (1992).

Chan, H. S., and Dill, K. A., Transition-states and folding dynamics of proteins and heteropolymers. *J. Chem. Phys.* **100,** 9238 (1994).

Chang, S. S., and Bestul, A. B., Heat capacity and thermodynamic properties of o-terphenyl crystal, glass, and liquid. *J. Chem. Phys.* **56,** 503 (1972).

Chaudhari, P., and Turnbull, D., Structure and properties of metallic glasses. *Science* **199,** 11 (1978).

Cicerone, M. T., and Ediger, M. D., Photobleaching technique for measuring ultraslow reorientation near and below the glass transition—Tetracene in o-terphenyl. *J. Phys. Chem.* **97,** 10489 (1993).

Cicerone, M. T., and Ediger, M. D., Relaxation of spatially heterogeneous dynamic domains in supercooled ortho-terphenyl. *J. Chem. Phys.* **103,** 5684 (1995).

Cicerone, M. T., and Ediger, M. D., Enhanced translation of probe molecules in supercooled o-terphenyl: Signature of spatially heterogeneous dynamics? *J. Chem. Phys.* **104,** 7210 (1996).

Cusack, N. E., "The Physics of Structurally Disordered Matter: An Introduction." Adam Hilger, Bristol, 1987.

Davidson, D. W., *In* "Water, a Comprehensive Treatise" (Franks, F., Ed.), Vol. 2, Chap. 3. Plenum Press, New York, 1973.

Davis, H. T., "Statistical Mechanics of Phases, Interfaces, and Thin Films." VCH, New York, 1996.

Debenedetti, P. G., "Metastable Liquids," Princeton University Press. Princeton, NJ, 1996.

Debenedetti, P. G., Stillinger, F. H., Truskett, T. M., and Roberts, C. J., The equation of state of an energy landscape. *J. Phys. Chem. B* **103,** 7390 (1999).

Deem, M. W., Recent contributions of statistical mechanics in chemical engineering. *AIChE J.* **44,** 2569 (1998).

Dill, K. A., and Chan, H. S., From Levinthal to pathways and funnels. *Nature Struct. Biol.* **4,** 10 (1997).

Dodd, L. R., and Theodorou, D. N., Analytical treatment of the volume and surface area of molecules formed by an arbitrary collection of unequal spheres intersected by planes. *Macromolecules* **72,** 1313 (1991).

Donati, C., Glotzer, S. C., and Poole, P. H., Growing spatial correlations of particle displacements in a simulated liquid on cooling toward the glass transition. *Phys. Rev. Lett.* **82,** 5064 (1999a).

Donati, C., Glotzer, S. C., Poole, P. H., Kob, W., and Plimpton, S. J., Spatial correlations of mobility and immobility in a glass-forming Lennard–Jones liquid. *Phys. Rev. E* **60,** 3107 (1999b).

Ediger, M. D., Spatially heterogeneous dynamics in supercooled liquids. *Annu. Rev. Phys. Chem.* **51,** 99 (2000).

Ediger, M. D., Angell, C. A., and Nagel, S. R., Supercooled liquids and glasses. *J. Phys. Chem.* **100,** 13200 (1996).

Eisenberg, D., and Kauzmann, W., "The Structure and Properties of Water." Oxford University Press, New York, 1969, Chap. 3.

Errington, J. R., Unpublished results (2000).

Errington, J. R., and Debenedetti, P. G., Relationship between structural order and the anomalies of liquid water. *Nature,* **409,** 318 (2001).

Franks, F., Long-term stabilization of biologicals. *Biotechnology* **12,** 253 (1994).

Frauenfelder, H., and Wolynes, P. G., Biomolecules: Where the physics of complexity and simplicity meet. *Phys. Today* **Feb.,** 58 (1994).

Frauenfelder, H., Sligar, S. G., and Wolynes, P. G., The energy landscapes and motions of proteins. *Science* **254,** 1598 (1991).
Frauenfelder, H., Bishop, A. R., Garcia, A., Perelson, A., Schuster, P., Sherrington, D., and Swart, P. J. (eds.), Landscape paradigms in physics and biology. Concepts, structures and dynamics. *Physica D* **107,** Nos. 2–4 (1997).
Fredrickson, G. H., Linear and nonlinear experiments for a spin model with cooperative dynamics. *Ann. N.Y. Acad. Sci.* **484,** 185 (1986).
Frenkel, J., "Kinetic Theory of Liquids." Dover, New York, 1955, p. 174.
Fricke, J. (Ed.), "Aerogels." Spinger, Heidelberg, 1986.
Fujara, F., Geil, B., Sillescu, H., and Fleischer, G., Translational and rotational diffusion in supercooled orthoterphenyl close to the glass transition. *Z. Phys. B Cond. Matter* **88,** 195 (1992).
Fulcher, G. S., Analysis of recent measurements of the viscosity of glasses. *J. Am. Ceram. Soc.* **8,** 339 (1925).
Gibbs, J. H., and DiMarzio, E. A., Nature of the glass transition and the glassy state. *J. Chem. Phys.* **28,** 373 (1958).
Gilbert, R. J., and Smith, S. C., "Theory of Unimolecular and Recombination Reactions." Blackwell Scientific, Oxford, 1990.
Goldstein, M., On the temperature dependence of cooperative relaxation properties in glass-forming liquids—Comment on a paper by Adam and Gibbs. *J. Chem. Phys.* **43,** 1852 (1965).
Goldstein, M., Viscous liquids and the glass transition: A potential energy barrier picture. *J. Chem. Phys.* **51,** 3728 (1969).
Götze, W., and Sjögren, L., Relaxation processes in supercooled liquids. *Rep. Prog. Phys.* **55,** 241 (1992).
Götze, W., and Sjögren, L., The mode coupling theory of structural relaxations. *Transp. Theory Stat. Phys.* **24,** 801 (1995).
Greenfield, M. L., and Theodorou, D. N., Geometric analysis of diffusion pathways in glassy and melt atactic polypropylene. *Macromolecules* **26,** 4561 (1993).
Greer, A. L., Metallic glasses. *Science* **267,** 1947 (1995).
Greer, A. L., Too hot to melt. *Nature* **404,** 134 (2000).
Greet, R. J., and Turnbull, D., Glass transition in o-terphenyl. *J. Chem. Phys.* **46,** 1243 (1967).
Gubbins, K. E., and Quirke, N. (Eds.), "Molecular Simulation and Industrial Applications: Methods, Examples, and Prospects." Gordon and Breach, Amsterdam, 1996.
Guggenheim, E. A., "Thermodynamics," 2nd ed., Interscience, New York, 1950, p. 131.
Hama, Y., Suzuki, K., Shingu, K., Fujimori, M., Kobayashi, S., Usuda, N., and Amano, J., *Thyroid* **9,** 927 (1999).
Helfand, E., On inversion of the Williams-Watts function for large relaxation times. *J. Chem. Phys.* **78,** 1931 (1983).
Hinze, G., Francis, R. S., and Fayer, M. D., Translational-rotational coupling in supercooled liquids: Heterodyne detected density induced molecular alignment. *J. Chem. Phys.* **111,** 2710 (1999).
Hirschfelder, J. O., Curtiss, C. F., and Bird, R. B., "Molecular Theory of Gases and Liquids," John Wiley and Sons, New York, 1954, p. 250.
Huitema, H. E. A., Vlot, M. J., and van der Eerden, J. P., Simulations of crystal growth from Lennard–Jones melt: Detailed measurements of the interface structure. *J. Chem. Phys.* **111,** 4714 (1999).
Hurley, M. M., and Harrowell, P., Non-gaussian behavior and the dynamical complexity of particle motion in a dense two-dimensional liquid. *J. Chem. Phys.* **105,** 10521 (1996).
Hyde, P. D., Evert, T. E., and Ediger, M. D., Nanosecond and microsecond study of probe reorientation in orthoterphenyl. *J. Chem. Phys.* **93,** 2274 (1990).

Ito, K., Moynihan, C. T., and Angell, C. A., Thermodynamic determination of fragility in liquids and a fragile-to-strong liquid transition in water. *Nature* **398,** 492 (1999).
Jonsson, H., and Andersen, H. C., Icosahedral ordering in the Lennard-Jones crystal and glass. *Phys. Rev. Lett.* **60,** 2295 (1988).
Kauzmann, W., The nature of the glassy state and the behavior of liquids at low temperatures. *Chem. Rev.* **43,** 219 (1948).
Keyes, T., Instantaneous normal mode approach to liquid state dynamics. *J. Phys. Chem. A* **101,** 2921 (1997).
Keyes, T., Dependence of supercooled liquid dynamics on elevation in the energy landscape. *Phys. Rev. E* **59,** 3207 (1999).
Keyes, T., Vijayadamodar, G. V., and Zurcher, U., An instantaneous normal mode description of relaxation in supercooled liquids. *J. Chem. Phys.* **106,** 4651 (1997).
Kieffer, J., and Angell, C. A., Generation of fractal structures by negative pressure rupturing of SiO_2 glass. *J. Non-Cryst. Sol.* **106,** 336 (1988).
Kind, R., Liechti, O., Korner, N., Hulliger, J., Dolinsek, J., and Blinc, R., Deuteron-magnetic-resonance study of the cluster formation in the liquid and supercooled-liquid state of 2-cyclooctylamino-5-nitropyridine. *Phys. Rev. B* **45,** 7697 (1992).
Kittel, C., "Introduction to Solid State Physics," 7th ed., Wiley, New York, 1966.
Kob, W., Sciortino, F., and Tartaglia, P., Aging as dynamics in configuration space. *Europhys. Lett.* **49,** 590 (2000).
Kohlrausch, R., Theorie des elektrischen rückstandes in der leidener flasche. *Ann. Phys. Chem. (Leipzig)* **91,** 179 (1874).
Lacks, D. J., Localized mechanical instabilities and structural transformations in silica glass under high pressure. *Phys. Rev. Lett.* **80,** 5385 (1998).
Lacks, D. J., and Wienhoff, J. R., Disappearences of energy minima and loss of order in polydisperse colloidal systems. *J. Chem. Phys.* **111,** 398 (1999).
LaNave, E., Scala, A., Starr, F. W., Sciortino, F., and Stanley, H. E., Instantaneous normal mode analysis of supercooled water. *Phys. Rev. Lett.* **84,** 4605 (2000).
Landau, L. D., and Lifshitz, E. M., "Statistical Physics. Part 1," Vol. 5 of Course on Theoretical Physics. 3rd ed., Chap. 6, Pergamon Press, Oxford, 1980.
Laughlin, R. B., Pines, D., Schmalian, J., Stojkovic, B. P., and Wolynes, P., The middle way. *Proc. Natl. Acad. Sci. USA* **97,** 32 (2000).
Laughlin, W. T., and Uhlmann, D. R., Viscous flow in simple organic liquids. *J. Phys. Chem.* **76,** 2317 (1972).
LaViolette, R. A., Amorphous deposits with energies below the crystal energy. *Phys. Rev. B* **40,** 9952 (1989).
LaViolette, R. A., and Stillinger, F. H., Multidimensional geometric aspects of the solid-liquid transition in simple substances. *J. Chem. Phys.* **83,** 4079 (1985a).
LaViolette, R. A., and Stillinger, F. H., Consequences of the balance between the repulsive and attractive forces in dense, non-associated liquids. *J. Chem. Phys.* **82,** 3335 (1985b).
LaViolette, R. A., and Stillinger, F. H., Thermal disruption of the inherent structure of simple liquids. *J. Chem. Phys.* **85,** 6027 (1986).
Leutheusser, E., Dynamical model of the liquid-glass transition. *Phys. Rev. A* **29,** 2765 (1984).
Lewis, C. P., "Towards a Thermodynamic Understanding of the Glass Transition," Senior thesis, Dept. Chem. Eng., Princeton University, Princeton, NJ, 2000.
Liang, J., Edelsbrunner, H., Fu, P., Sudhakar, P. V., and Subramaniam, S., Analytical shape computation of macromolecules. I. Molecular area and volume through alpha shape. *Proteins* **33,** 1 (1998).
Lindemann, F. A., Über die berechnung molekularer eigenfrequenzen. *Phys. Z.* **11,** 609 (1910).
Lindsey, C. P., and Patterson, G. D., Detailed comparison of the William-Watts and Cole-Davidson functions. *J. Chem. Phys.* **73,** 3348 (1980).

Liu, C. Z.-W., and Oppenheim, I., Enhanced diffusion upon approaching the kinetic glass transition. *Phys. Rev. E* **53**, 799 (1996).
Lubachevsky, B. D., and Stillinger, F. H., Geometric properties of random disk packings. *J. Stat. Phys.* **60**, 561 (1990).
Lubachevsky, B. D., Stillinger, F. H., and Pinson, E. N., Disks vs. spheres: Contrasting properties of random packings. *J. Stat. Phys.* **64**, 501 (1991).
Mackenzie, J. D., Viscosity-temperature relationship for network liquids. *J. Am. Ceram. Soc.* **44**, 598 (1961).
Malandro, D. L., and Lacks, D. J., Volume dependence of potential energy landscapes in glasses. *J. Chem. Phys.* **107**, 5804 (1997).
Malandro, D. L., and Lacks, D. J., Molecular-level instabilities and enhanced self-diffusion in flowing liquids. *Phys. Rev. Lett.* **81**, 5576 (1998).
Malandro, D. L., and Lacks, D. J., Relationships of shear-induced changes in the potential energy landscape to the mechanical properties of ductile glasses. *J. Chem. Phys.* **110**, 4593 (1999).
Martin, C. J., and O'Connor, D. A., An experimental test of Lindemann's melting law. *J. Phys. C Sol. State Phys.* **10**, 3521 (1977).
Mel'cuk, A. I., Ramos, R. A., Gould, H., Klein, W., and Mountain, R. D., Long-lived structures in fragile glass-forming liquids. *Phys. Rev. Lett.* **75**, 2522 (1995).
Mishima, O., and Stanley, H. E., The relationship between liquid, supercooled and glassy water. *Nature* **396**, 329 (1998).
Moore, P., and Keyes, T., Normal mode analysis of liquid CS_2: Velocity correlation functions and self-diffusion constants. *J. Chem. Phys.* **100**, 6709 (1994).
Moynihan, C. T., Macedo, P. B., Montrose, C. J., Gupta, P. K., DeBolt, M. A., Dill, J. F., Dom, B. E., Drake, P. W., Easteal, A. J., Elterman, P. B., Moeller, R. P., Sasabe, H., and Wilder, J. A., in "The Glass Transition and the Nature of the Glassy State " (M. Goldstein and R. Simha, Eds.), *Ann. N.Y. Acad. Sci.* **279**, 15 (1976).
Ohmine, I., Liquid water dynamics: collective motions, fluctuation, and relaxation. *J. Phys. Chem.* **99**, 6767 (1995).
Ono, S., and Kondo, S., "Handbuch der Physik." Springer, Berlin, 1960, Vol. 10, p. 135.
Patashinski, A. Z., and Ratner, M. A., Inherent amorphous structures and statistical mechanics of melting. *J. Chem. Phys.* **106**, 7249 (1997).
Peng, D.-Y., and Robinson, D. B., A new two-constant equation of state. *Ind. Eng. Chem. Fundam.* **15**, 59 (1976).
Perera, D. N., and Harrowell, P., Measuring diffusion in supercooled liquids: The effect of kinetic inhomogeneities. *J. Chem. Phys.* **104**, 2369 (1996a).
Perera, D. N., and Harrowell, P., Consequence of kinetic inhomogeneities in glasses. *Phys. Rev. E* **54**, 1652 (1996b).
Phillips, W. A., Tunneling states in amorphous solids. *J. Low Temp. Phys.* **7**, 351 (1972).
Plazek, D. J., Bero, C. A., and Chay, I.-C., The recoverable compliance of amorphous materials. *J. Non-Cryst. Sol.* **172–174**, 181 (1994).
Plotkin, S. S., Wang, J., and Wolynes, P. G., Correlated energy landscape model for finite, random heteropolymers. *Phys. Rev. E* **53**, 6271 (1996).
Plotkin, S. S., Wang, J., and Wolynes, P. G., Statistical mechanics of a correlated energy landscape model for protein folding funnels. *J. Chem. Phys.* **106**, 2932 (1997).
Prausnitz, J. M., Molecular thermodynamics: Opportunities and responsibilities. *Fluid Phase Equil.* **116**, 12 (1996).
Rastogi, S., Höhne, G. W. H., and Keller, A., Unusual pressure-induced phase behavior in crystalline poly (4-methylpentene-1): Calorimetric and spectroscopic results and further implications. *Macromolecules* **32**, 8897 (1999).

Reid, R. C., Prausnitz, J. M., and Poling, B., "The Properties of Gases and Liquids," 4th ed., Appendix A, McGraw–Hill, New York, 1987.
Rein ten Wolde, P., Ruiz-Montero, M. J., and Frenkel, D., Numerical calculation of the rate of crystal nucleation in a Lennard-Jones system at moderate undercooling. *J. Chem. Phys.* **104,** 9932 (1996).
Reiss, H., Statistical geometry in the study of fluids and porous media. *J. Phys. Chem.* **96,** 4736 (1992).
Reiss, H., and Hammerich, A. D., Hard spheres—Scaled particle theory and exact relations on the existence and structure of the fluid-solid phase transition. *J. Phys. Chem.* **90,** 6252 (1986).
Richard, P., Oger, L., Troadec, J.-P., and Gervois, A., Geometrical characterization of hard-sphere systems. *Phys.-Rev. E* **60,** 4551 (1999).
Richert, R., Geometrical confinement and cooperativity in supercooled liquids studied by solvation dynamics. *Phys. Rev. B* **54,** 762 (1996).
Richert, R., and Angell, C. A., Dynamics of glass-forming liquids. V. On the link between molecular dynamics and configurational entropy. *J. Chem. Phys.* **108,** 9016 (1998).
Richert, R., and Blumen, A., Disordered systems and relaxation, *in* "Disorder Effects on Relaxational Processes" (R. Richert and A. Blumen, Eds.), p. 1. Springer, Berlin, 1994.
Rintoul, M. D., and Toquato, S., Algorithm to compute void statistics for random arrays of disks. *Phys. Rev. E* **52,** 2635 (1995).
Rintoul, M. D., and Torquato, S., Computer simulations of dense hard-sphere systems. *J. Chem. Phys.* **105,** 9258 (1996).
Roberts, C. J., Debenedetti, P. G., and Stillinger, F. H., Equation of state of the energy landscape of SPC/E water. *J. Phys. Chem. B* **103,** 10258 (1999).
Sahimi, M., Flow phenomena in rocks—From continuum models to fractals, percolation, cellular-automata, and simulated annealing. *Rev. Mod. Phys.* **65,** 1393 (1993).
Sahimi, M., "Flow and Transport in Media and Fractured Rock." VCH, Weinheim, Germany, 1995.
Sastry, S., Liquid limits: Glass transition and liquid-gas spinodal boundaries of metastable liquids. *Phys. Rev. Lett.* **85,** 590 (2000).
Sastry, S., Corti, D. S., Debenedetti, P. G., and Stillinger, F. H., Statistical geometry of particle packings. I. Algorithm for exact determination of connectivity, volume, and surface areas of void space in monodisperse and polydisperse sphere packings. *Phys. Rev. E* **56,** 5524 (1997a).
Sastry, S., Debenedetti, P. G., and Stillinger, F. H., Statistical geometry of particle packings. II. Weak spots in liquids. *Phys. Rev. E* **56,** 5533 (1997b).
Sastry, S., Debenedetti, P. G., and Stillinger, F. H., Signatures of distinct dynamical regimes in the energy landscape of a glass-forming liquid. *Nature* **393,** 554 (1998a).
Sastry, S., Truskett, T. M., Debenedetti, P. G., Torquato, S., and Stillinger, F. H., Free volume in the hard sphere liquid. *Mol. Phys.* **95,** 289 (1998b).
Saven, J. G., and Wolynes, P. G., Local signals in the entropy landscape of collapsed helical proteins. *Physica D* **107,** 330 (1997).
Saven, J. G., Wang, J., and Wolynes, P. G., Kinetics of protein folding: The dynamics of globally connected rough energy landscapes with biases. *J. Chem. Phys.* **101,** 11037 (1994).
Schulz, M., Energy landscape, minimum points, and non-Arrhenius behavior of supercooled liquids. *Phys. Rev. B* **57,** 11319 (1998).
Sciortino, F., Kob, W., and Tartaglia, P., Inherent structure entropy of supercooled liquids. *Phys. Rev. Lett.* **83,** 3214 (1999).
Shah, V. M., Stern, S. A., and Ludovice, P. J., Estimation of the free volume in polymers by means of a Monte Carlo technique. *Macromolecules* **22,** 4660 (1989).

Shahinpoor, M., Statistical mechanical considerations on the random packing of granular materials. *Powder Technol.* **25,** 163 (1980).
Shen, V. K., and Debenedetti, P. G., A computational study of homogeneous liquid-vapor nucleation in the Lennard–Jones fluid. *J. Chem. Phys.* **111,** 3581 (1999).
Simon, F., Über den zustand der unterkühlten flüssigkeiten und glässer. *Z. Anorg. Allg. Chem.* **203,** 219 (1931).
Sjögren, L., and Götze, W., α-Relaxation near the glass transition. *J. Non-Cryst. Sol.* **131–133,** 153 (1991).
Soave, G. S., Equilibrium constants from a modified Redlich–Kwong equation of state. *Chem. Eng. Sci.* **27,** 1197 (1972).
Speedy, R. J., Statistical geometry of hard-sphere systems. *J. Chem. Soc. Faraday Trans. II* **76,** 693 (1980).
Speedy, R. J., Relations between a liquid and its glasses. *J. Phys. Chem.* B **103,** 4060 (1999).
Steinhardt, P. J., Nelson, D. R., and Ronchetti, M., Bond-orientational order in liquids and glasses. *Phys. Rev.* B **28,** 784 (1983).
Stillinger, F. H., Role of potential energy scaling in the low-temperature relaxation behavior of amorphous materials. *Phys. Rev.* B **32,** 3134 (1985).
Stillinger, F. H., Supercooled liquids, glass transitions, and the Kauzmann paradox. *J. Chem. Phys.* **88,** 7818 (1988).
Stillinger, F. H., A topographic view of supercooled liquids and glass formation. *Science* **267,** 1935 (1995).
Stillinger, F. H., Enumeration of isobaric inherent structures for the fragile glass former o-terphenyl. *J. Phys. Chem.* B **102,** 2807 (1998).
Stillinger, F. H., Exponential multiplicity of inherent structures. *Phys. Rev.* E **59,** 48 (1999).
Stillinger, F. H., Inherent structures enumeration for low-density materials. *Phys. Rev. E,* **63,** 011110-1 (2001).
Stillinger, F. H., and Debenedetti, P. G., Distinguishing vibrational and structural equilibration contributions to thermal expansion. *J. Phys. Chem.* B **103,** 4052 (1999).
Stillinger, F. H., and Hodgdon, J. A., Translation-rotation paradox for diffusion in fragile glass-forming liquids. *Phys. Rev.* E **50,** 2064 (1994).
Stillinger, F. H., and Hodgdon, J. A., Reply to "Comment on 'Translation-rotation paradox for diffusion in fragile glass-forming liquids.'" *Phys. Rev.* E **53,** 2995 (1996).
Stillinger, F. H., and Stillinger, D. K., Negative thermal expansion in the Gaussian core model. *Physica A* **244,** 358 (1997).
Stillinger, F. H., and Weber, T. A., Hidden structure in liquids. *Phys. Rev.* A **25,** 978 (1982).
Stillinger, F. H., and Weber, T. A., Inherent structure-theory of liquids in the hard-sphere limit. *J. Chem. Phys.* **83,** 4767 (1985).
Stillinger, F. H., Debenedetti, P. G., and Sastry, S., Resolving vibrational and structural contributions to isothermal compressibility. *J. Chem. Phys.* **109,** 3983 (1998).
Takahara, S., Yamamuro, O., and Suga, H., Heat capacities and glass transitions of 1-propanol and 3-methylpentane under pressure. New evidence for the entropy theory. *J. Non-Cryst. Sol.* **171,** 259 (1994).
Tammann, G., and Hesse, W., Die abhängigkeit der viskosität von der temperatur bei unterkühlten flüssigkeiten. *Z. Anorg. Allg. Chem.* **156,** 245 (1926).
Tanemura, M., Ogawa, T., and Ogita, N., A new algorithm for 3-dimensional Voronoi tessellation. *J. Comput. Phys.* **51,** 191 (1983).
Tarjus, G., and Kivelson, D., Breakdown of the Stokes–Einstein relation in supercooled liquids. *J. Chem. Phys.* **103,** 3071 (1995).
Theodorou, D. N., Symposium in print—Molecular modeling. *Chem. Eng. Sci.* **49,** 2715 (1994).

Thompson, K. E., and Fogler, H. S., Modeling flow in disordered packed beds from pore-scale fluid mechanics. *AIChE J.* **43,** 1377 (1997).
Torquato, S., Random heterogeneous media: Microstructure and improved bounds on effective properties. *Appl. Mech. Rev.* **44,** 37 (1991).
Torquato, S., Unified methodology to quantify the morphology and properties of inhomogeneous media. *Physica A* **207,** 79 (1994).
Torquato, S., Truskett, T. M., and Debenedetti, P. G., Is random close packing of spheres well defined? *Phys. Rev. Lett.* **84,** 2064 (2000).
Truskett, T. M., Torquato, S., and Debenedetti, P. G., Towards a quantification of disorder in materials: Distinguishing equilibrium and glassy sphere packings. *Phys. Rev. E* **62,** 993 (2000).
Uhlmann, D. R., A kinetic treatment of glass formation. *J. Non-Cryst. Sol.* **7,** 337 (1972).
Utz, M., Debenedetti, P. G., and Stillinger, F. H., Atomistic simulation of aging and rejuvenation in glasses. *Phys. Rev. Lett.* **84,** 1471 (2000).
Utz, M., Debenedetti, P. G., and Stillinger, F. H., Isotropic tensile strength of molecular glasses *J. Chem. Phys.* **114,** 10049 (2001).
van Duijneveldt, J. S., and Frenkel, D., Computer simulation study of free energy barriers in crystal nucleation. *J. Chem. Phys.* **96,** 4655 (1992).
Vidal Russell, E., and Israeloff, N. E., Direct observation of molecular cooperativity near the glass transition. *Nature* **408,** 695 (2000).
Vishnyakov, A., Debenedetti, P. G., and Neimark, A. V., Statistical geometry of cavities in a metastable confined fluid. *Phys. Rev. E* **62,** 538 (2000).
Vogel, H., Das temperatur-abhängigkeitsgesetz der viskosität von flüssigkeiten. *Phys. Zeit.* **22,** 645 (1921).
Wang, C.-Y., and Ediger, M. D., How long do regions of different dynamics persist in supercooled o-terphenyl? *J. Phys. Chem. B* **103,** 4177 (1999).
Wang, J., Onuchic, J., and Wolynes, P., Statistics of kinetic pathways on biased rough energy landscapes with applications to protein folding. *Phys. Rev. Lett.* **76,** 4861 (1996).
Weber, T. A., and Stillinger, F. H., Inherent structures and distribution-functions for liquids that freeze into bcc crystals. *J. Chem. Phys.* **81,** 5089 (1984).
Weber, T. A., and Stillinger, F. H., Interactions, local order, and atomic rearrangement kinetics in amorphous nickel-phosphorus alloys. *Phys. Rev. B* **32,** 5402 (1985).
Weber, T. A., Fredrickson, G. H., and Stillinger, F. H., Relaxation behavior in a tiling model for glasses. *Phys. Rev. B* **34,** 7641 (1986).
Wilks, J., "The Properties of Liquid and Solid Helium." Clarendon Press, Oxford, 1967.
Williams, G., and Watts, D. C., Non-symmetrical dielectric relaxation behavior arising from a simple empirical decay function. *Trans. Faraday Soc.* **66,** 80 (1970).
Wolynes, P. G., Aperiodic crystals: Biology, chemistry and physics in a fugue with stretto, *in* "Proceedings of the International Symposium on Frontiers in Science" (S. S. Chan and P. G. Debrunner, Eds.), *AIP Conf. Proc.* **180,** 39 (1988).
Wolynes, P. G., Randomness and complexity in chemical physics. *Acc. Chem. Res.* **25,** 513 (1992).
Wolynes, P. G., Entropy crises in glasses and random heteropolymers. *J. Res. Natl. Inst. Stand. Tech.* **102,** 187 (1997).
Wolynes, P. G., Onuchic, J. N., and Thirumalai, D., Navigating the folding routes. *Science* **267,** 1619 (1995).
Wu, T., and Tsay, S., Instantaneous normal mode analysis of liquid Na. *J. Chem. Phys.* **105,** 9281 (1996).
Zallen, R., "The Physics of Amorphous Solids." Wiley, New York, 1983.
Zarzycki, J., "Glasses and the Vitreous State." Cambridge University Press, Cambridge, 1991.
Ziman, J. M., "Models of Disorder. The Theoretical Physics of Homogeneously Disordered Systems," Cambridge University Press, Cambridge, 1979.

A STATISTICAL MECHANICAL APPROACH TO COMBINATORIAL CHEMISTRY

Michael W. Deem

Chemical Engineering Department, University of California, Los Angeles, California 90095-1592

I. Introduction   81
II. Materials Discovery   83
   A. The Space of Variables   84
   B. Library Design and Redesign   85
   C. Searching the Variable Space by Monte Carlo   87
   D. The Simplex of Allowed Compositions   89
   E. Significance of Sampling   91
   F. The Random Phase Volume Model   92
   G. Several Monte Carlo Protocols   94
   H. Effectiveness of the Monte Carlo Strategies   95
   I. Aspects of Further Development   96
III. Protein Molecular Evolution   97
   A. What Is Protein Molecular Evolution?   98
   B. Background on Experimental Molecular Evolution   100
   C. The Generalized NK Model   102
   D. Experimental Conditions and Constraints   104
   E. Several Hierarchical Evolution Protocols   105
   F. Possible Experimental Implementations   109
   G. Life Has Evolved to Evolve   111
   H. Natural Analogs of These Protocols   113
   I. Concluding Remarks on Molecular Evolution   115
IV. Summary   117
   References   118

I. Introduction

The goal of combinatorial chemistry is to find compositions of matter that maximize a specific material property. When combinatorial chemistry is applied to materials discovery, the desired property may be superconductivity, magnetoresistance, luminescence, ligand specificity, sensor response, or catalytic activity. When combinatorial chemistry is applied to proteins, the desired property may be enzymatic activity, fluorescence, antibiotic resistance, or substrate binding specificity. In either case, the property to be optimized, the figure of merit, is generally an unknown function of the variables and can be measured only experimentally.

Combinatorial chemistry is, then, a search over a multidimensional space of composition and noncomposition variables for regions with a high figure of merit. A traditional synthetic chemist would carry out this search by using chemical intuition to synthesize a few initial molecules. Of these molecules, those that have a favorable figure of merit would be identified. A homologous series of compounds similar to those best starting points would then be synthesized. Finally, of the compounds in these homologous series, that with the best figure of merit would be identified as the optimal material.

If the space of composition and noncomposition variables is sufficiently large, novel, or unfamiliar, the traditional synthetic approach may lead to the identification of materials that are not truly the best. It is in this case that combinatorial chemistry becomes useful. In combinatorial chemistry, trial libraries of molecules are synthesized instead of trial molecules. By synthesizing and screening for figure of merit an entire library of 10^2–10^5 molecules instead of a single molecule, the variable space can be searched much more thoroughly. In this sense, combinatorial chemistry is a natural extension of traditional chemical synthesis. Intuitive determination of the individual molecules to synthesize is replaced by methods for design of the molecular libraries. Likewise, synthesis of homologous compounds is replaced by redesign of the libraries for multiple rounds of parallel screening experiments.

While the combinatorial approach attempts to search composition space broadly, an exhaustive search is usually not possible. It would take, for example, a library of 9×10^6 compounds to search a five-component system at a mole fraction resolution of 1%. Similarly, it would take a library of $20^{100} \approx 10^{130}$ proteins to search exhaustively the space of all 100-amino acid protein domains. Clearly, a significant aspect to the design of a combinatorial chemistry experiment is the design of the library. The library members should be chosen so as to search the space of variables as effectively as possible, given the experimental constraints on the library size.

The task of searching composition space in combinatorial chemistry for regions with a high figure of merit is very similar to the task of searching configuration space by Monte Carlo computer simulation for regions with a low free energy. The space searched by Monte Carlo computer simulation is often extremely large, with 10^4 or more continuous dimensions. Yet, with recent advances in the design of Monte Carlo algorithms, one is able to locate reliably the regions with low free energy even for fairly complicated molecular systems.

This chapter pursues the analogy between combinatorial chemistry and Monte Carlo computer simulation. Examples of how to design libraries for both materials discovery and protein molecular evolution will be given. For materials discovery, the concept of library redesign, or the use of previous experiments to guide the design of new experiments, will be introduced. For molecular evolution, examples of how to use "biased" Monte Carlo to search the protein sequence space will be given. Chemical information, whether intuition, theoretical calculations, or database statistics, can be naturally incorporated as an a priori bias in the Monte Carlo approach to library design in combinatorial chemistry. In this sense, combinatorial chemistry can be viewed as an extension of traditional chemical synthesis, one ideally suited for chemical engineering contributions.

II. Materials Discovery

A variety of materials have been optimized or developed to date by combinatorial methods. Perhaps the first experiment to gather great attention was the demonstration that inorganic oxide high-T_c superconductors could be identified by combinatorial methods (Xiang et al., 1995). By searching several 128-member libraries of different inorganic oxide systems, the known compositions of superconducting BiSrCaCuO and YBaCuO were identified. Since then, many demonstrations of finding known materials and discoveries of new materials have appeared. Known compositions of giant magnetoresistant materials have been identified in libraries of various cobalt oxides (Briceño et al., 1995). Blue and red phosphors have been identified from large libraries of 25,000 different inorganic oxides (Danielson et al., 1988, 1997; Wang et al., 1998). Polymer-based sensors for various organic vapors have been identified by combinatorial methods (Dickinson and Walt, 1997). Catalysts for the oxidation of CO to CO_2 have been identified by searching ternary compounds of Pd, Pt, and Rh or Rh, Pd, and Cu (Weinberg et al., 1998; Cong et al., 1999). Phase diagrams of zeolitic materials have been mapped out by a combinatorial "multiautoclave" (Akporiaye et al., 1998). Novel enantioselective catalysts have been found by searching libraries of transition metal–peptide complexes (Cole et al., 1996). Novel phosphatase catalysts were found by searching libraries of carboxylic acid-functionalized polyallylamine polymers (Menger et al., 1995). New catalysts and conditions for C—H insertion have been found by screening of ligand–transition metal systems (Burgess et al., 1996). A new catalyst for the conversion of methanol in a direct methanol fuel cell was identified by searching the quaternary compositon space of Pt, Ir, Os, and Ru (Reddington et al., 1998). Finally, a novel thin-film high-dielectric compound that may be used in future generation of DRAM chips was identified by searching through over 30 multicomponent, ternary oxide systems (van Dover et al., 1998).

The task of identifying the optimal compound in a materials discovery experiment can be reformulated as one of searching a multidimensional space, with the material composition, impurity levels, and synthesis conditions as variables. Present approaches to combinatorial library design and screening invariably perform a grid search in composition space, followed by a " steepest-ascent" maximization of the figure of merit. This procedure becomes inefficient in high-dimensional spaces or when the figure of merit is not a smooth function of the variables. *Indeed, the use of a grid search is what has limited essentially all current combinatorial chemistry experiments to quaternary compounds,* i.e., *to searching a space with three variables.* What is needed is an automated, yet more efficient, procedure for searching composition space.

An analogy with the computer simulation technique of Monte Carlo allows us to design just such an efficient protocol for searching the variable space (Falcioni and Deem, 2000). In materials discovery, a search is made through the composition and noncomposition variables to find good figure-of-merit values. In Monte Carlo, a search is made through configuration space to find regions of low free energy. By using insight gained from the design of Monte Carlo methods, the search in materials discovery can be improved.

A. THE SPACE OF VARIABLES

Several variables can be manipulated to seek the material with the optimal figure of merit. Material composition is certainly a variable. But also, film thickness (van Dover *et al.,* 1998) and deposition method (Novet *et al.,* 1995) are variables for materials made in thin-film form. The processing history, such as temperature, pressure, pH, and atmospheric composition, is a variable. The guest compositon or impurity level can greatly affect the figure of merit (Cong *et al.,* 1999). In addition, the "crystallinity" of the material can affect the observed figure of merit (van Dover *et al.,* 1998). Finally, the method of nucleation or synthesis may affect the phase or morphology of the material and so affect the figure of merit (Helmkamp and Davis, 1995; Zones *et al.,* 1998).

There are important points to note about these variables. First, a small impurity composition can cause a big change in the figure of merit, as seen by the rapid variation of catalytic activity in the Cu/Rh oxidation catalyst (Cong *et al.,* 1999). Second, the phases in thin film are not necessarily the same as those in bulk, as seen in the case of the thin-film dielectric, where the optimal material was found outside the region where the bulk phase forms (van Dover *et al.,* 1998). Finally, the "crystallinity" of the material can affect the observed figure of merit, again as seen in the thin-film dielectric example (van Dover *et al.*, 1998).

B. LIBRARY DESIGN AND REDESIGN

The experimental challenges in combinatorial chemistry appear to lie mainly in the screening methods and in the technology for the creation of the libraries. The theoretical challenges, on the other hand, appear to lie mainly in the library design and redesign strategies. It is this second question that is addressed by the analogy with Monte Carlo computer simulation.

Combinatorial chemistry differs from usual Monte Carlo simulations in that several simultaneous searches of the variable space are carried out. That is, in a typical combinatorial chemistry experiment, several samples, e.g., 10,000, are synthesized and screened for figure of merit at one time. With the results of this first round, a new set of samples can be synthesized and screened. This procedure can be repeated for several rounds, although current materials discovery experiments have not systematically exploited this feature.

Pursuing the analogy with Monte Carlo, each round of combinatorial chemistry corresponds to a move in a Monte Carlo simulation. Instead of tracking one system with many configurational degrees of freedom, however, many samples are tracked, each with several composition and noncomposition degrees of freedom. Modern experimental technology is what allows for the cost-effective synthesis and screening of multiple sample compositions.

The technology for materials discovery is still in the developmental stage, and future progress can still be influenced by theoretical considerations. In this spirit, I assume that the composition and noncomposition variables of each sample can be changed independently, as in spatially addressable libraries (Akporiaye *et al.*, 1998; Pirrung, 1997). This is significant, because it allows great flexibility in how the space can be searched with a limited number of experimental samples.

Current experiments uniformly tend to perform a grid search on the composition and noncomposition variables. It is preferable, however, to choose the variables statistically from the allowed values. It is also possible to consider choosing the variables in a fashion that attempts to maximize the amount of information gained from the limited number of samples screened, via a quasi-random, low-discrepancy sequence (Niederreiter, 1992; Bratley *et al.*, 1994). Such sequences attempt to eliminate the redundancy that naturally occurs when a space is searched statistically, and they have several favorable theoretical properties. An illustration of these three approaches to materials discovery library design is shown in Fig. 1.

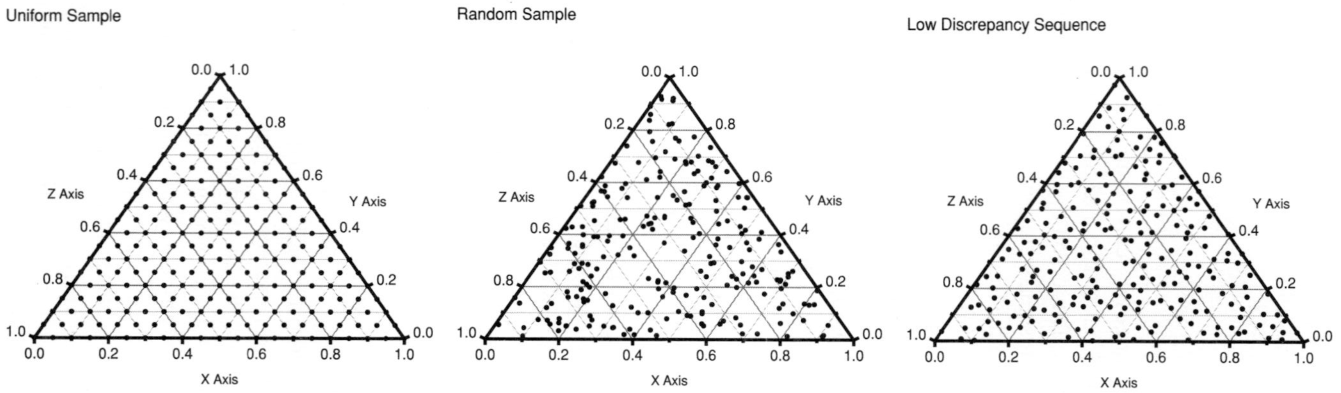

FIG. 1. The grid, random, and low-discrepancy sequence approaches to designing the first library in a materials discovery experiment with three compositional variables. The random approach breaks the regular pattern of the grid search, and the low-discrepancy sequence approach avoids overlapping points that may arise in the random approach.

Information about the figure-of-merit landscape in the composition and noncomposition variables can be incorporated by multiple rounds of screening. One convenient way to incorporate this feedback as the experiment proceeds is by treating the combinatorial chemistry experiment as a Monte Carlo in the laboratory. This approach leads to sampling the experimental figure of merit, E, proportional to exp (βE). If β is large, then the Monte Carlo procedure will seek out values of the composition and noncomposition variables that maximize the figure of merit. If β is too large, however, the Monte Carlo procedure will get stuck in relatively low-lying local maxima. The first round is initiated by choosing the composition and noncomposition variables statistically from the allowed values. The variables are changed in succeeding rounds as dictated by the Monte Carlo procedure.

Several general features of the method for changing the variables can be enumerated. The statistical method of changing the variables can be biased by concerns such as material cost, theoretical or experimental a priori insight into how the figure of merit is likely to change, and patentability. Both the composition and the noncomposition variables will be changed in each round. Likely, it would be desirable to have a range of move sizes for both types of variables. The characteristic move size would likely best be determined by fixing the acceptance ratio of the moves, as is customary in Monte Carlo simulations (Frenkel and Smit, 1996). In addition, there would likely be a smallest variable change that would be significant, due to experimental resolution limitations in the screening step. Finally, a steepest-ascent optimization to find the best local optima of the figure of merit would likely be beneficial at the end of a materials discovery experiment driven by such a Monte Carlo strategy.

C. Searching the Variable Space by Monte Carlo

Two ways of changing the variables are considered: a small random change of the variables of a randomly chosen sample and a swap of a subset of the variables between two randomly chosen samples. Swapping is useful when there is a hierarchical structure to the variables. The swapping event allows for the combination of beneficial subsets of variables between different samples. For example, a good set of composition variables might be combined with a particularly good impurity composition. Or a good set of composition variables might be combined with a good set of processing variables. These moves are repeated until all the samples in a round have been modified. The values of the figure of merit for the proposed new samples are then measured. Whether to accept the newly proposed samples or to keep the current samples for the next round is decided according to the detailed balance acceptance criterion. For a random change of one sample, the Metropolis acceptance probability is applied

$$p_{\text{acc}}(c \to p) = \min\{1, \exp[\beta(E_{\text{proposed}} - E_{\text{current}})]\}. \tag{1}$$

Proposed samples that increase the figure of merit are always accepted; proposed samples that decrease the figure of merit are accepted with the Metropolis probability. Allowing the figure of merit occasionally to decrease is what allows samples to escape from local maxima. Moves that lead to invalid values of the composition or noncomposition variables are rejected.

For the swapping move applied to samples i and j, the modified acceptance probability is applied

$$p_{\text{acc}}(c \to p) = \min\left\{1, \exp\left[\beta\left(E^i_{\text{proposed}} + E^j_{\text{proposed}} - E^i_{\text{current}} - E^j_{\text{current}}\right)\right]\right\}. \tag{2}$$

Figure 2a shows one round of a Monte Carlo procedure. The parameter β is not related to the thermodynamic temperature of the experiment and should be optimized for best efficiency. The characteristic sizes of the random changes in the composition and noncomposition variables are also parameters that should be optimized.

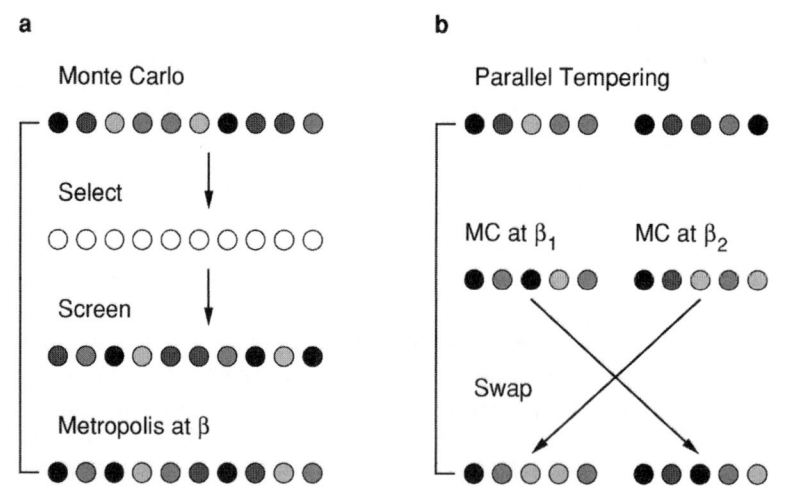

FIG. 2. Schematic of the Monte Carlo library design and redesign strategy (from Falcioni and Deem, 2000). (a) One Monte Carlo round with 10 samples: an initial set of samples, modification of the samples, measurement of the new figures of merit, and the Metropolis criterion for acceptance or rejection of the new samples. (b) One parallel tempering round with five samples at β_1 and five samples at β_2. In parallel tempering, several Monte Carlo simulations are performed at different temperatures, with the additional possibility of sample exchange between the simulations at different temperatures.

If the number of composition and noncomposition variables is too great, or if the figure of merit changes with the variables in a too-rough fashion, normal Monte Carlo will not achieve effective sampling. Parallel tempering is a natural extension of Monte Carlo that is used to study statistical (Geyer, 1991), spin glass (Marinari *et al.*, 1998), and molecular (Falcioni and Deem, 1999) systems with rugged energy landscapes. Our most powerful protocol incorporates the method of parallel tempering for changing the system variables. In parallel tempering, a fraction of the samples are updated by Monte Carlo with parameter β_1, a fraction by Monte Carlo with parameter β_2, and so on. At the end of each round, samples are randomly exchanged between the groups with different β's, as shown in Fig. 2b. The acceptance probability for exchanging two samples is

$$p_{\text{acc}}(c \to p) = \min\{1, \exp[-\Delta\beta \Delta E]\}, \quad (3)$$

where $\Delta\beta$ is the difference in the values of β between the two groups, and ΔE is the difference in the figures of merit between the two samples. It is important to notice that this exchange step does not involve any extra screening compared to Monte Carlo and is, therefore, "free" in terms of experimental costs. This step is, however, dramatically effective at facilitating the protocol to escape from local maxima. The number of different systems and the temperatures of each system are parameters that must be optimized.

To summarize, the first round of combinatorial chemistry consists of the following steps: constructing the initial library of samples, measuring the initial figures of merit, changing the variables of each sample a small random amount or swapping subsets of the variables between pairs of samples, constructing the proposed new library of samples, measuring the figures of merit of the proposed new samples, accepting or rejecting each of the proposed new samples, and performing parallel tempering exchanges. Subsequent rounds of combinatorial chemistry repeat these steps, starting with making changes to the current values of the composition and noncomposition variables. These steps are repeated for as many rounds as desired, or until maximal figures of merit are found.

D. The Simplex of Allowed Compositions

The points to be sampled in materials discovery are the allowed values of the composition and noncomposition variables. Typically, the composition variables are specified by the mole fractions. Since the mole fractions sum to one, sampling on these variables requires special care.

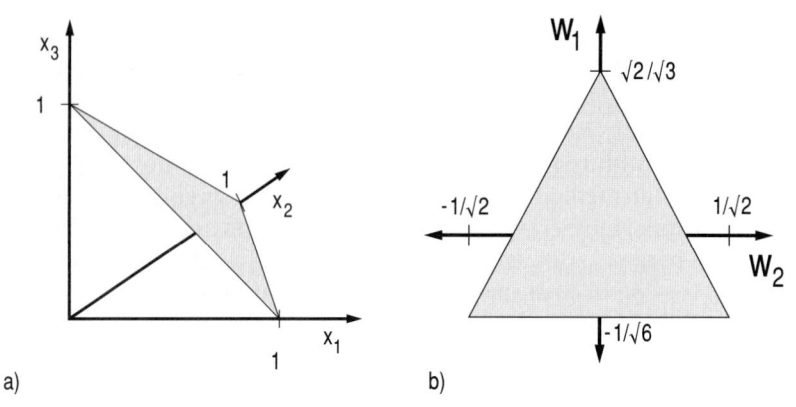

FIG. 3. The allowed composition range of a three-component system is shown in (a) the original composition variables, x_i, and (b) the Gram–Schmidt variables, w_i.

In particular, the specification or modification of the d mole fraction variables, x_i, is done in the $(d-1)$-dimensional hyperplane orthogonal to the d-dimensional vector $(1, 1, \ldots, 1)$. This procedure ensures that the constraint $\sum_{i=1}^{d} x_i = 1$ is maintained. This subspace is identified by a Gram–Schmidt procedure, which identifies a new set of basis vectors, $\{\mathbf{u}_i\}$, that span this hyperplane. Figure 3 illustrates the geometry for the case of three composition variables.

The new basis set is identified as follows. First, $\mathbf{u}_d$ is defined to be the unit vector orthogonal to the allowed hyperplane

$$\mathbf{u}_d = \left(\frac{1}{\sqrt{d}}, \frac{1}{\sqrt{d}}, \ldots, \frac{1}{\sqrt{d}} \right). \tag{4}$$

The remaining $\mathbf{u}_i$, $1 \leq i < d$, are chosen to be orthogonal to $\mathbf{u}_d$, so that they lie in the allowed hyperplane. Indeed, the $\mathbf{u}_i$ form an orthonormal basis for the composition space. This orthonormal basis is identified by the Gram–Schmidt procedure. First, the original composition basis vectors are defined

$$\begin{aligned} \mathbf{e}_1 &= (1, 0, \ldots, 0, 0) \\ \mathbf{e}_2 &= (0, 1, \ldots, 0, 0) \\ &\vdots \\ \mathbf{e}_{d-1} &= (0, 0, \ldots, 1, 0). \end{aligned} \tag{5}$$

Each $\mathbf{u}_i$ is identified by projecting these basis vectors onto the space orthogonal to $\mathbf{u}_d$ and the $\mathbf{u}_j$, $j < i$, already identified

$$\mathbf{u}_1 = \frac{\mathbf{e}_1 - (\mathbf{e}_1 \cdot \mathbf{u}_d)\mathbf{u}_d}{|\mathbf{e}_1 - (\mathbf{e}_1 \cdot \mathbf{u}_d)\mathbf{u}_d|}$$

$$\mathbf{u}_2 = \frac{\mathbf{e}_2 - (\mathbf{e}_2 \cdot \mathbf{u}_d)\mathbf{u}_d - (\mathbf{e}_2 \cdot \mathbf{u}_1)\mathbf{u}_1}{|\mathbf{e}_2 - (\mathbf{e}_2 \cdot \mathbf{u}_d)\mathbf{u}_d - (\mathbf{e}_2 \cdot \mathbf{u}_1)\mathbf{u}_1|} \qquad (6)$$

$$\vdots$$

$$\mathbf{u}_i = \frac{\mathbf{e}_i - (\mathbf{e}_i \cdot \mathbf{u}_d)\mathbf{u}_d - \sum_{j=1}^{i-1}(\mathbf{e}_i \cdot \mathbf{u}_j)\mathbf{u}_j}{|\mathbf{e}_i - (\mathbf{e}_i \cdot \mathbf{u}_d)\mathbf{u}_d - \sum_{j=1}^{i-1}(\mathbf{e}_i \cdot \mathbf{u}_j)\mathbf{u}_j|}.$$

A point in the allowed composition range is specified by the vector $\mathbf{x} = \sum_{i=1}^{d} w_i \mathbf{u}_i$, with $w_d = 1/\sqrt{d}$. Note that the values w_i are related to the composition values x_i by a rotation matrix, since the Gram–Schmidt procedure simply identifies a rotated basis for the composition space

$$\mathbf{x} = R\mathbf{w}, \qquad (7)$$

where R_{ij} is given by the ith component of $\mathbf{u}_j$. Each of the numbers w_i, $1 \leq i < d$, is to be varied in the materials discovery experiment. Not all values of w_i are feasible, however, since the constraint $x_i \geq 0$ must be satisfied. Feasible values are identified by transforming the w_i to the x_i by Eq. (7), and then checking that the compositon variables are nonnegative. The constraint that the composition variables sum to unity is automatically ensured by the choice $w_d = 1/\sqrt{d}$.

E. Significance of Sampling

Sampling the figure of merit by Monte Carlo, rather than global optimization by some other method, is favorable for several reasons. First, Monte Carlo is an effective stochastic optimization method. Second, simple global optimization may be misleading since concerns such as patentability, cost of materials, and ease of synthesis are not usually included in the experimental figure of merit. Moreover, the screen that is most easily performed in the laboratory, the "primary screen," is usually only roughly correlated with the true figure of merit. Indeed, after finding materials that look promising based upon the primary screen, experimental secondary and tertiary screens are usually performed to identify that material which is truly optimal. Third, it might be advantageous to screen for several figures of merit at once. For example, it might be profitable to search for reactants and conditions that lead to the synthesis of several zeolites with a particularly favorable property, such as the presence of a large pore. As another example, it might be useful to search for several electrocatalysts that all possess a useful property, such as being able to serve as the anode or cathode material in a particular fuel cell.

For all of these reasons, sampling by Monte Carlo to produce several candidate materials is preferred over global optimization.

F. THE RANDOM PHASE VOLUME MODEL

The ultimate test of new, theoretically motivated protocols for materials discovery is, of course, experimental. To motivate such experimentation, the effectiveness of these protocols is demonstrated by combinatorial chemistry experiments where the experimental screening step is replaced by figures of merit returned by the random-phase volume model. The random phase volume model is not fundamental to the protocols; it is introduced as a simple way to test, parameterize, and validate the various searching methods.

The random phase volume model relates the figure of merit to the composition and noncomposition variables in a statistical way. The model is fast enough to allow for validation of the proposed searching methods on an enormous number of samples yet possesses the correct statistics for the figure-of-merit landscape.

The composition mole fractions are nonnegative and sum to unity, and so the allowed compositions are constrained to lie within a simplex in $d - 1$ dimensions. For the familiar ternary system, this simplex is an equilateral triangle, as shown in Fig. 3b. Typically, several phases will exist for different compositions of the material. The figures of merit will be dramatically different between each of these distinct phases. To mimic this expected behavior, the composition variables are grouped in the random phase volume model into phases centered around N_x points $\mathbf{x}_\alpha$ randomly placed within the allowed composition range. The phases form a Voronoi diagram (Sedgewick, 1988), as shown in Fig. 4.

The random phase volume model is defined for any number of composition variables, and the number of phase points is defined by requiring the average spacing between phase points to be $\xi = 0.25$. To avoid edge effects, additional points are added in a belt of width 2ξ around the simplex of allowed compositions. The number of phase points for different grid spacing is shown in Table I.

The figure of merit should change dramatically between composition phases. Moreover, within each phase α, the figure of merit should also vary with $\mathbf{y} = \mathbf{x} - \mathbf{x}_\alpha$ due to crystallinity effects such as crystallite size, intergrowths, defects, and faulting (van Dover et al., 1998). In addition, the noncomposition variables should also affect the measured figure of merit. The noncomposition variables are denoted by the b-dimensional vector $\mathbf{z}$, with

TABLE I
NUMBER OF PHASE POINTS AS A FUNCTION OF
DIMENSION AND SPACING

ξ	d	Number of points
0.1	3	193
0.1	4	1,607
0.1	5	12,178
0.1	6	81,636
0.2	3	86
0.2	4	562
0.2	5	3,572
0.2	6	20,984
0.25	3	70
0.25	4	430
0.25	5	2,693
0.25	6	15,345
0.3	3	59
0.3	4	353
0.3	5	2,163
0.3	6	12,068
0.35	3	53
0.35	4	306
0.35	5	1,850
0.35	6	10,234

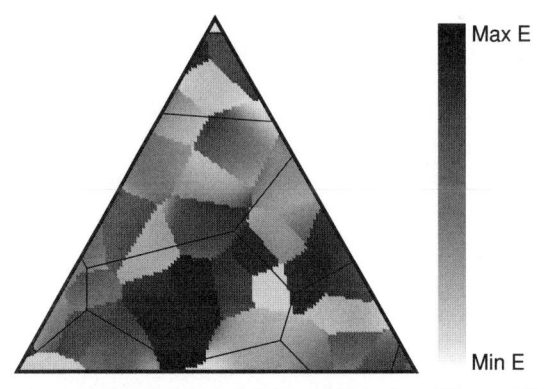

FIG. 4. The random phase volume model (from Falcioni and Deem, 2000). The model is shown for the case of three composition variables and one noncomposition variable. The boundaries of the **x** phases are evident by the sharp discontinuities in the figure of merit. To generate this figure, the **z** variable was held constant. The boundaries of the **z** phases are shown as thin dark lines.

each component constrained to fall within the range $[-1, 1]$ without loss of generality. There can be any number of noncomposition variables. The figure of merit depends on the composition and noncomposition variables in a correlated fashion. In particular, how the figure of merit changes with the noncomposition variables should depend on the values of the compositon variables. To mimic this behavior within the random phase volume model, the noncomposition variables also fall within N_z noncomposition phases defined in the space of composition variables. There are a factor of 10 fewer noncomposition phases than composition phases.

The functional form of the model when $\mathbf{x}$ is in composition phase α and noncomposition phase γ is

$$E(\mathbf{x}, \mathbf{z}) = U_\alpha + \sigma_x \sum_{k=1}^{q} \sum_{i_1 \geq \ldots \geq i_k = 1}^{d} f_{i_1 \ldots i_k} \xi_x^{-k} A_{i_1 \ldots i_k}^{(\alpha k)} y_{i_1} y_{i_2} \ldots y_{i_k} \\ + \frac{1}{2} \left(W_\gamma + \sigma_z \sum_{k=1}^{q} \sum_{i_1 \geq \ldots \geq i_k = 1}^{b} f_{i_1 \ldots i_k} \xi_z^{-k} B_{i_1 \ldots i_k}^{(\gamma k)} z_{i_1} z_{i_2} \ldots z_{i_k} \right), \tag{8}$$

where $f_{i_1 \ldots i_k}$ is a constant symmetry factor, ξ_x and ξ_z are constant scale factors, and U_α, W_γ, $A_{i_1 \ldots i_k}^{(\alpha k)}$, and $B_{i_1 \ldots i_k}^{(\gamma k)}$ are random Gaussian variables with unit variance. In more detail, the symmetry factor is given by

$$f_{i_1 \ldots i_k} = \frac{k!}{\prod_{i=1}^{l} o_i!}, \tag{9}$$

where l is the number of distinct integer values in the set $\{i_1, \ldots, i_k\}$, and o_i is the number of times that distinct value i is repeated in the set. Note that $1 \leq l \leq k$ and $\sum_{i=1}^{l} o_i = k$. The scale factors are chosen so that each term in the multinomial contributes roughly the same amount: $\xi_x = \xi/2$ and $\xi_z = (\langle z^6 \rangle / \langle z^2 \rangle)^{1/4} = (3/7)^{1/4}$. The σ_x and σ_z are chosen so that the multinomial, crystallinity terms contribute 40% as much as the constant, phase terms on average. For both multinomials $q = 6$. As Fig. 4 shows, the random phase volume model describes a rugged figure-of-merit landscape, with subtle variations, local maxima, and discontinuous boundaries.

G. Several Monte Carlo Protocols

Six ways of searching the variable space are tested with increasing numbers of composition and noncomposition variables. The total number of samples whose figure of merit will be measured is fixed at $M = 100,000$, so that all protocols have the same experimental cost. The single-pass protocols grid, random, and low-discrepancy sequence (LDS) are considered.

For the grid method, the number of samples in the composition space is $M_x = M^{(d-1)/(d-1+b)}$ and the number of samples in the noncomposition space is $M_z = M^{b/(d-1+b)}$. The grid spacing of the composition variables is $\zeta_x = (V_d/M_x)^{1/(d-1)}$, where

$$V_d = \frac{\sqrt{d}}{(d-1)!} \tag{10}$$

is the volume of the allowed composition simplex. Note that the distance from the centroid of the simplex to the closest point on the boundary of the simplex is

$$R_d = \frac{1}{[d(d-1)]^{1/2}}. \tag{11}$$

The spacing for each component of the noncomposition variables is $\zeta_z = 2/M_z^{1/b}$. For the LDS method, different quasi-random sequences are used for the composition and noncomposition variables. The feedback protocols Monte Carlo, Monte Carlo with swap, and parallel tempering are considered. The Monte Carlo parameters were optimized on test cases. It was optimal to perform 100 rounds of 1000 samples with $\beta = 2$ for $d = 3$ and $\beta = 1$ for $d = 4$ or 5, and $\Delta x = 0.1 R_d$ and $\Delta z = 0.12$ for the maximum random displacement in each component. The swapping move consisted of an attempt to swap all of the noncomposition values between the two chosen samples, and it was optimal to use $P_{\text{swap}} \simeq 0.1$ for the probability of a swap versus a regular random displacement. For parallel tempering it was optimal to perform 100 rounds with 1000 samples, divided into three subsets: 50 samples at $\beta_1 = 50$, 500 samples at $\beta_2 = 10$, and 450 samples at $\beta_3 = 1$. The 50 samples at large β essentially perform a "steepest-ascent" optimization and have smaller $\Delta x = 0.01 R_d$ and $\Delta z = 0.012$.

H. Effectiveness of the Monte Carlo Strategies

The figures of merit found by the protocols are shown in Fig. 5. The single-round protocols, random and low-discrepancy sequence, find better solutions than does grid in one round of experiment. Interestingly, the low-discrepancy sequence approach fares no better than does random, despite the desirable theoretical properties of low-discrepancy sequences.

The multiple-round, Monte Carlo protocols appear to be especially effective on the more difficult systems with larger numbers of composition and noncomposition variables. That is, the Monte Carlo methods have a tremendous advantage over one-pass methods, especially as the number of variables increases, with parallel tempering the best method. The Monte Carlo

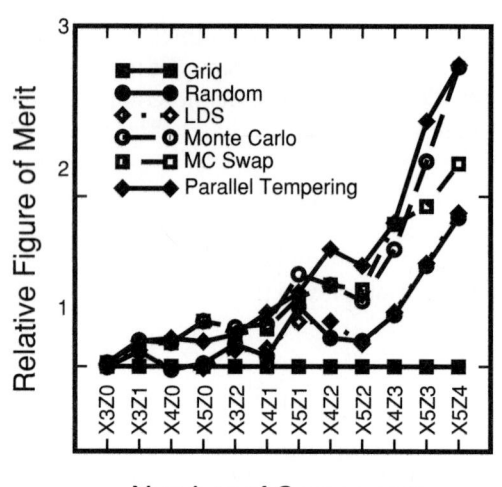

FIG. 5. The maximum figure of merit found with different protocols on systems with different number of composition (**x**) and noncomposition (**z**) variables (from Falcioni and Deem, 2000). The results are scaled to the maximum found by the grid searching method. Each value is averaged over scaled results on 10 instances of the random phase volume model with different random phases. The Monte Carlo methods are especially effective on systems with a larger number of variables, where the maximal figures of merit are more difficult to locate.

methods, in essence, gather more information about how best to search the variable space with each succeeding round. This feedback mechanism proves to be effective even for the relatively small total sample size of 100,000 considered here. It is expected that the advantage of the Monte Carlo methods will become even greater for larger sample sizes. Note that in cases such as catalytic activity, sensor response, or ligand specificity, the experimental figure of merit would likely be exponential in the values shown in Fig. 5, so that the success of the Monte Carlo methods would be even more dramatic. A better calibration of the parameters in Eq. (8) may be possible as more data become available in the literature.

I. Aspects of Further Development

The space of composition and noncomposition variables to search in materials discovery experiments can be forbiddingly large. Yet, by using Monte Carlo methods, one can achieve an effective search with a limited number of experimental samples.

Efficient implementations of the Monte Carlo search strategies are feasible with existing library creation technology. Moreover "closing the loop" between library design and redesign is achievable with the same database technology currently used to track and record the data from combinatorial chemistry experiments. These multiple-round protocols, when combined with appropriate robotic automation, should allow the practical application of combinatorial chemistry to more complex and interesting systems.

Many details need to be worked out to flesh out the proposed protocols for materials discovery. For example,

1. How rough are real figures of merit, and can the random phase volume model be calibrated better?
2. Can more of the hierarchical structure of the variables be identified? and
3. What are the best methods of manipulating the variables in the Monte Carlo?

Additional questions, some of which this chapter has begun to answer, include how does the proximity to the global optimum scale with the number of samples and with the algorithm by which they are selected? What is the best set of samples to choose for an optimal result, chosen "all at once" or in stages or sequentially? What is the minimum number of samples required to make a Monte Carlo-based algorithm attractive as the driver?

III. Protein Molecular Evolution

The space to be searched in protein combinatorial chemistry experiments is extremely large. Consider, for example, that a relatively short 100-amino acid protein domain were to be evolved. The number of possible amino acid sequences of this length is $20^{100} \approx 10^{130}$, since there are 20 naturally occurring amino acid residues. Clearly, all of these sequences cannot be synthesized and then screened for figure of merit in the laboratory. Some means must be found for searching this space with the 10^4 or so proteins that can be screened per day experimentally.

A hierarchical decomposition of the protein space can provide an effective searching procedure. It is known from protein structural biology that proteins are encoded by DNA sequences, DNA sequences code for amino acids, amino acids arrange into secondary structures, secondary structures arrange into domains, domains group to form protein monomers, and protein monomers aggregate to form multiprotein complexes. By sampling on each level of this hierarchy, one is able to search the sequence space much more effectively. In this chapter, search strategies making use of the DNA, amino acid, and secondary structure hierarchy will be described. With this approach, functional protein space has a large, yet manageable, number of dimensions. That is, in a 100-amino acid protein domain there are approximate 10 secondary structures of 5 types (helices, loops, strands, turns, and others), roughly yielding the potential for $\approx 10^7$ basic protein folds. Organization into secondary structural classes represents a dramatic reduction in the complexity of sequence space, since there are $\approx 10^{170}$ different DNA sequences and $\approx 10^{130}$ different amino acid sequences in this space.

Sampling on the different levels of protein structure is analogous to combination of different move types in a Monte Carlo simulation. A variety of moves, from small, local moves to large, global moves, are often incorporated in the most successful Monte Carlo simulations. While protein molecular evolution is carried out in the laboratory, and Monte Carlo simulations are carried out *in silico,* the parallels are striking. One of the most powerful new concepts in Monte Carlo is the idea that moves should be "biased" (Frenkel and Smit, 1996). That is, small moves, such as the Metropolis method, sample configuration space rather slowly. Larger moves are preferred, since they sample the space more rapidly. Large moves are usually rejected, however, since they often lead the system into a region of high energy. So that the large moves will be more successful, a bias toward regions that look promising is included. Such biased Monte Carlo simulations have been a factor of 10^5 to 10^{10} times more efficient than previous methods, and they have allowed the examination of systems previously uncharacterizable by molecular simulation techniques (Frenkel *et al.,* 1992; Frenkel and Smit, 1992; de Pablo *et al.,* 1992; Smit and Maesen, 1995).

In this section, the possibility of evolving protein molecules by strategies similar to biased Monte Carlo will be explored. The large moves of Monte Carlo are implemented by changing an evolving protein at the secondary structure level. These evolutionary events will be biased, in that the amino acid sequences inserted will be chosen so that they code for viable secondary structures. The concept of bias also applies at the amino acid level, where different DNA sequences coding for the same amino acids can lead to different propensities for future evolution.

A. What Is Protein Molecular Evolution?

Protein molecular evolution can be viewed as combinatorial chemistry of proteins. Since protein sequence space is so large, most experiments to date have sought to search only small regions. A typical experiment seeks to optimize the figure of merit of an existing protein. For example, an improvement in the selectivity or activity of an enzyme might be sought. Alternatively, an expansion in the operating range of an enzyme might be sought to higher temperatures or pressures. This improvement would be achieved by changing, or evolving, the amino acid sequence of the enzyme.

A more ambitious goal would be the *ab initio* evolution of a protein with a specific function. That is, nothing would be known about the desired molecule, except that it should be a protein and that an experimental screen for the desired figure of merit is available. One might want, for example, an enzyme that catalyzes an unusual reaction. Or one might want a protein that binds a specific substrate. Or one might want a protein with an unusual fluorescence spectrum. The *ab initio* evolution of a protein has never been accomplished before. Such a feat would be remarkable. Natural biological diversity has evolved despite the essentially infinite complexity of protein sequence. Replication of this feat in the laboratory would represent substantial progress, and mimicking this feat of Nature is a current goal in the molecular evolution field. Indeed, the protocols described in this section are crafted with just this task in mind.

A still more ambitious goal would be the evolution of a multiprotein complex. This is a rather challenging task, due to the increased complexity of the space to be searched. The task can be made manageable by asking a rather general evolutionary question. One can seek to evolve, for example, a multiprotein complex that can serve as the coat protein complex for a virus. Since there are many proteins that may accomplish this task, this evolutionary task may not be as specific and difficult as it might seem at first.

The most ambitious goal for laboratory evolution that has been imagined is the evolution of new life forms. Evolution on this scale requires changes not only at the secondary structure scale, but also at the domain, protein, and protein pathway scale. Due to their simplicity, viruses or phage would be the most likely targets of such large-scale evolution attempts. It is unclear how new life forms would be distributed in terms of pathogenicity, and so such experiments should be approached with caution.

The hierarchical decomposition of sequence space will allow effective molecular evolution if there are many proteins with a high value of any particular figure of merit. That is, if only 1 of 10^{130} small protein domains exhibits a high score on a particular figure of merit, this protein is unlikely to be identified. On the other hand, if many proteins score highly on the figure of merit, only a subset of these molecules need be sampled. This same issue arises in conventional Monte Carlo simulations. Sampling all of configuration space is never possible in a simulation, yet ensemble averages and experimental behavior can be reproduced by sampling representative configurations. That life on our planet has evolved suggests that there is a great redundancy in protein space (Kauffman, 1993), and so one may hope to search this space experimentally with sufficiently powerful moves.

B. Background on Experimental Molecular Evolution

There are some constraints on molecular evolution as it is carried out in the laboratory. There are constraints arising from limitations of molecular biology, i.e., only certain types of moves are possible on the DNA that codes for the protein. There are also constraints arising from technical limitations, i.e., only a certain number of proteins can be screened for figure of merit in a day.

Existing approaches to the evolution of general proteins are essentially limited to changes at the single-base level. Somewhat more sophisticated methods are available for evolution of antibodies, but this is a special case that is not considered here. The first type of evolutionary change that is possible in the laboratory is a base substitution. Base substitutions are naturally made as DNA is copied or amplified by PCR. The rate at which base substitutions, or mistakes in the copying of the template DNA, are made can be adjusted by varying experimental conditions, such as the manganese and magnesium ion concentrations. It is important to note that these base substitutions are made without knowledge of the DNA sequences of the evolving proteins. Equally important is that these base substitutions are made without the use of a chemical synthesizer. These changes are made naturally within the context of efficient molecular biology methods. Another means of modifying an existing protein is to use random or directed mutagenesis to change specific DNA bases so that they code for random or specified amino acids. The approach requires both that the DNA sequence be known and that it subsequently be synthesized by chemical means. Such a laborious approach is not practical in high-throughput evolution experiments, where typically 10^4 proteins are simultaneously modified and evolved per day.

Since the average length of a human gene is roughly 1800 bases, base-by-base point mutation will achieve significant evolution only very slowly. More significantly, the figure-of-merit landscape for protein function is typically quite rugged. Base mutation, therefore, invariably ceases evolution at a local optima of the figure of merit. Base mutation can be viewed as an experimental method for local optimization of protein figures of merit.

Much of the current enthusiasm for protein molecular evolution is due to the discovery of DNA shuffling by Pim Stemmer in 1994. DNA shuffling is a method for evolving an existing protein to achieve a higher figure of merit. The great genius of Stemmer was to develop a method for combining beneficial base mutations that is naturally accomplished with the tools of molecular biology and that does not require DNA sequencing or chemical synthesis. The method is successful because combination of base mutations that were individually beneficial is likely to lead to an even higher figure of merit than is achieved by either mutation alone. Of course, this will not

TABLE II
GENES AND OPERONS EVOLVED BY DNA SHUFFLING (FROM PATTEN ET AL., 1997)[a]

System	Improvement	Size	Mutations
TEM-1 β-lactamase	Enzyme activity 32,000-fold	333 aa	6 aa
β-Galactosidase	Fucosidase activity 66-fold	1333 aa	6 aa
Green fluorescence protein	Protein folding 45-fold	266 aa	3 aa
Antibody	Avidity >400-fold	233 aa	34 aa[b]
Antibody	Expression level 100-fold	233 aa	5 aa
Arsenate operon	Arsenate resistance 40-fold	766 aa	3 aa
Alkyl transferase	DNA repair 10-fold	166 aa	7 aa
Benzyl esterase	Antibiotic deprotection 150-fold	500 aa	8 aa
tRNA synthetase	Charging of engineered tRNA 180-fold	666 aa	ND

[a] aa, amino acids; ND, not determined.
[b] This was a case of family shuffling (Crameri et al., 1998), so most of these changes were between homologous amino acids.

always be true, but the extent to which it is true is the extent to which DNA shuffling will be an effective technique. DNA shuffling, combined with base mutation, is the current state of the art experimental technique for protein molecular evolution.

Table II lists a few of the protein systems that have been evolved by the Stemmer group. Evidently, DNA shuffling is highly effective at improving the function of an existing protein, much more effective than is simple base mutation. The specificity of an enzyme can even be altered, as in the conversion of a β-galactosidase into a fucosidase. A rough median of the improvement factors is about 100. Most important, however, is that all of these improvements were achieved with a relatively small number of amino acid changes. On average, only 6 amino acids were altered, of roughly 400 total residues in the protein. As with base mutation, then, DNA shuffling is able to search sequence space only locally. After a small number of amino acid changes, DNA shuffling produces a protein with a locally rather than globally optimal figure of merit.

The current state-of-the-art experimental techniques for protein molecular evolution can be viewed as local optimization procedures in protein sequence space. Alternatively, they can be viewed as experimental implementations of a simple, or Metropolis, Monte Carlo procedure. By using our intuition regarding the design of powerful, biased Monte Carlo algorithms, we can develop more powerful experimental protocols for molecular evolution.

Interestingly, theoretical treatments of evolution, whether in Nature or in the laboratory, tend to consider only the effects of point mutation (Kauffman, 1993; Volkenstein, 1994). Indeed, interesting theories regarding the evolutionary potential of point mutations have been developed. As shown experimentally, however, point mutation is incapable of significantly evolving proteins at substantial rate. Even the more powerful technique of DNA shuffling searches protein space merely locally. Only with the inclusion of more dramatic moves, such as changes at the level of secondary structures, can protein space be searched more thoroughly.

C. The Generalized NK Model

To validate the molecular evolution protocols to be presented, a model that relates amino acid sequence to protein function is needed. Of course, the real test of these protocols should be experimental, and I hope that these experiments will be forthcoming. To stimulate interest in the proposed protocols, their effectiveness will be simulated on a model of protein function. Such a model would seem to be difficult to construct. It is extremely difficult to determine the three-dimensional structure of a protein given the amino acid sequence. Moreover, it is extremely difficult to calculate any of the typical figures of merit given the three-dimensional structure of a protein.

It is fortunate that a model that relates figure of merit to amino acid sequence for a specific protein is not needed. The requirement is simply a model that produces figure-of-merit landscapes in sequence space that are analogous to those that would be measured in the laboratory on an ensemble of proteins. This type of model is easier to construct, and a random energy model can be used to accomplish the task.

The generalized NK model is just such a random energy model. The NK model was first introduced to model combinatorial chemistry experiments on peptides (Kauffman and Levin, 1987; Kauffman, 1993; Kauffman and MacReady, 1995). It was subsequently generalized to account for secondary structure in real proteins (Perelson and Macken, 1995). The model was further generalized to account for interactions between the secondary structures and for the presence of a binding pocket (Bogarad and Deem, 1999).

This generalized NK model assigns a unique figure of merit to each evolving protein sequence. This model, while a simplified description of real proteins, captures much of the thermodynamics of protein folding and ligand binding. The model takes into account the formation of secondary structures via the interactions of amino acid side chains as well as the interactions between secondary structures within proteins. In addition, for specificity, the figure of merit is assumed to be a binding constant, and so the model includes a contribution representing binding to a substrate. The combined ability to fold and bind substrate is what will be optimized or evolved. That is, the direction of the protein evolution will be based upon the figure of merit returned by this generalized NK model. This generalized NK model contains several parameters, and a reasonable determination of these parameters is what allows the model to compare successfully with experiment.

The specific energy function used as the selection criterion in the molecular simulations is

$$U = \sum_{\alpha=1}^{M} U_\alpha^{sd} + \sum_{\alpha>\gamma=1}^{M} U_{\alpha\gamma}^{sd-sd} + \sum_{i=1}^{P} U_i^c. \qquad (12)$$

This energy function is composed of three parts: secondary structural subdomain energies (U^{sd}), subdomain–subdomain interaction energies (U^{sd-sd}), and chemical binding energies (U^c). Each of these three energy terms is weighted equally, and each has a magnitude near unity for a random sequence of amino acids. In this NK-based simulation, each different type of amino acid behaves as a completely different chemical entity; therefore, only $Q = 5$ chemically distinct amino classes are considered (e.g., negative, positive, polar, hydrophobic, and other). Interestingly, restricted alphabets of amino acids not only are capable of producing functional proteins (Kamtekar et al., 1993; Riddle et al., 1997) but also may have been used in the primitive genetic code (Miller and Orgel, 1974; Schuster and Stadler, 1998). The evolving protein will be a relatively short, 100-amino acid protein domain. Within this domain will be roughly $M = 10$ secondary structural subdomains, each $N = 10$ amino acids in length. The subdomains belong to one of $L = 5$ different types (e.g., helices, strands, loops, turns, and others). This gives L different (U^{sd}) energy functions of the NK form (Kauffman and Levin, 1987; Kauffman, 1993; Kauffman and MacReady, 1995; Perelson and Macken, 1995),

$$U_\alpha^{sd} = \frac{1}{[M(N-K)]^{1/2}} \sum_{j=1}^{N-K+1} \sigma_\alpha(a_j, a_{j+1}, \ldots, a_{j+K-1}). \qquad (13)$$

The degree of complexity in the interactions between the amino acids is parameterized by the value of K. Low values of K lead to figure-of-merit landscapes upon which evolution is easy, and high values of K lead to extremely rugged landscapes upon which evolution is difficult. Combinatorial chemistry experiments on peptides have suggested the value of $K = 4$ as a reasonable one (Kauffman and MacReady, 1995). Note that the definition of K here is one greater than the convention in (Kauffman and Levin, 1987; Kauffman, 1993; Kauffman and MacReady, 1995). The quenched, unit-normal random number σ_α in Eq. (13) is different for each value of its argument for each of the L classes. This random form mimics the complicated amino acid side-chain interactions within a given secondary structure. The energy of interaction between secondary structures is given by

$$U_{\alpha\gamma}^{\text{sd-sd}} = \left[\frac{2}{DM(M-1)}\right]^{1/2} \times \sum_{i=1}^{D} \sigma_{\alpha\gamma}^{(i)}\left(a_{j_1}^\alpha, \ldots, a_{j_{K/2}}^\alpha; a_{j_{K/2+1}}^\gamma, \ldots, a_{j_K}^\gamma\right). \quad (14)$$

The number of interactions between secondary structures is set at $D = 6$. Here the unit-normal weight, $\sigma_{\alpha\gamma}^{(i)}$, and the interacting amino acids, $\{j_1, \ldots, j_K\}$, are selected at random for each interaction (i, α, γ). The chemical binding energy of each amino acid is given by

$$U_i^c = \frac{1}{\sqrt{P}}\sigma_i(a_i). \quad (15)$$

The contributing amino acid, i, and the unit-normal weight of the binding, σ_i, are chosen at random. A typical binding pocket is composed of five amino acids, and so the choice of $P = 5$ is made.

D. Experimental Conditions and Constraints

A typical protein evolution experiment starts with an initial protein sequence. This sequence is then copied to a large number of identical sequences. All of these sequences are evolved, or mutated, in parallel. After one round of mutation events, the proteins are screened for figure of merit. This screening step is typically the rate-limiting step, and so the efficiency of this step determines how many proteins can be evolved in parallel. For typical figures of merit, 10,000 proteins can be screened in a day. If selection, that is, use of a screen based upon whether an organism lives or dies, were performed instead, 10^9–10^{15} proteins could be screened in a day. Selection is a special case, however, and so the more conservative case of screening 10,000 proteins per day is considered.

After the screen, the proteins are ranked according to their measured value of the figure of merit. Typically, the top x percent of the sequences is kept for the next round of mutation. The parameter x is to be adjusted experimentally. In the simulated evolutions, the value of $x = 10\%$ was always found to be optimal. Other methods for selecting the proteins to keep for the next round have been considered. For example, keeping proteins proportional to exp $(-\beta U)$ has been considered. This strategy seems to work less well than the top x percent method. The main reason seems to be that in the top x percent method, the criterion for selecting which sequences to keep adjusts naturally with the range of figures of merit found in the evolving sequences. After the top x percent sequences are selected, they are copied back up to a total of 10,000 sequences. These sequences are the input for the next round of mutation and selection.

In the simulated molecular evolutions, the experiment is continued for 100 rounds. This is a relatively large number of rounds to carry out experimentally. With the most powerful protocols, however, it is possible to evolve proteins *ab initio*. This feat has not been achieved to date in the laboratory. To mimic this feat of Nature, one should be willing to do some number of rounds.

E. Several Hierarchical Evolution Protocols

1. Amino Acid Substitution

To obtain a baseline for searching fold space, molecular evolution is first simulated via simple mutagenesis (see Fig. 6a). Simulated evolutions by amino acid substitution lead to significantly improved protein energies, as shown in Table III. These evolutions always terminated at local energy minima, however. This trapping is due to the difficulty of combining the large number of correlated substitutions necessary to generate new protein folds. Increasing the screening stringency in later rounds did not improve the binding constants of simulated proteins, most likely due to the lack of additional selection criteria such as growth rates. Although only nonconservative mutations were directly simulated, conservative and synonymous neutral mutations are not excluded and can be taken into account in a more detailed treatment. Indeed, the optimized average mutation rate of 1 amino acid substitution/sequence/round is equivalent to roughly 1–6 random base substitutions/round.

2. DNA Shuffling

DNA shuffling improves the search of local fold space via a random yet correlated combination of homologous coding fragments that contain limited numbers of beneficial amino acid substitutions. As in experimental evolutions (Stemmer, 1994; Crameri *et al.*, 1998; Zhang *et al.*, 1997; Moore *et al.*, 1997), the simulated shuffling improved protein function significantly better than did point mutation alone (see Table III and Fig. 6b). However, local

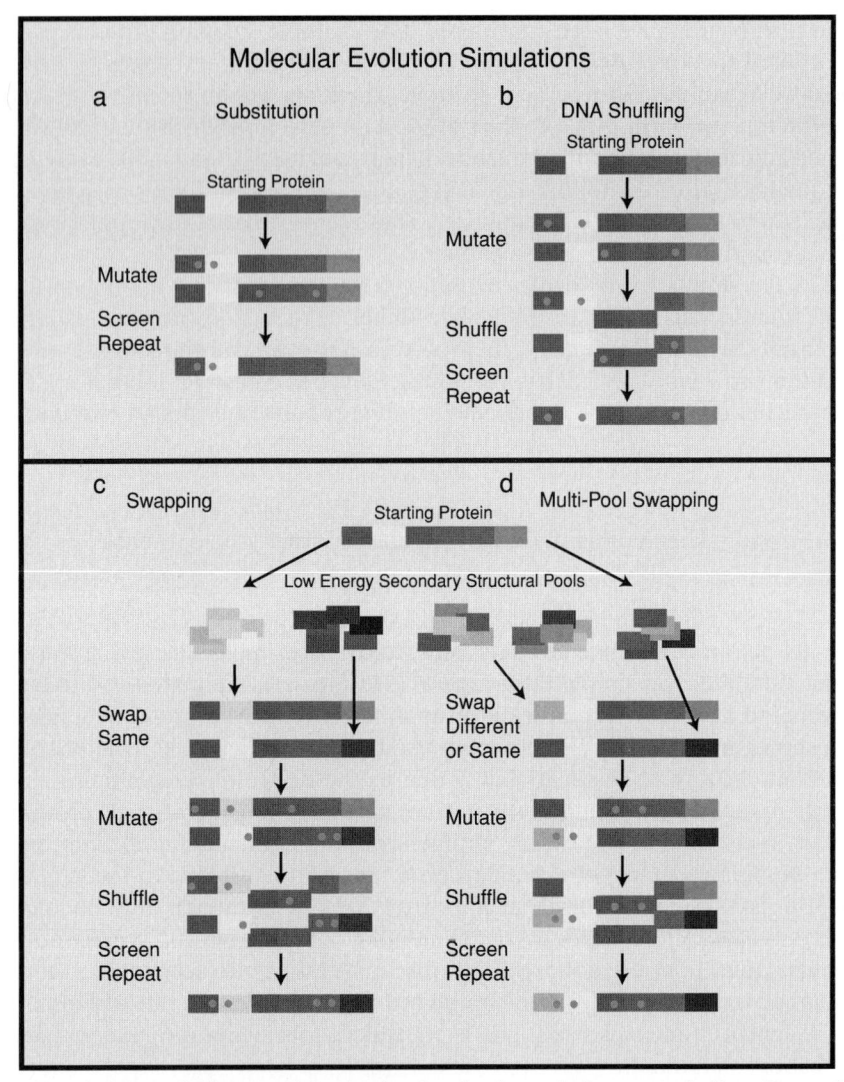

FIG. 6. Schematic diagram of the simulated molecular evolution protocols (from Bogarad and Deem, 1999). (a) Simulation of molecular evolution via base substitution (substitutions are represented by orange dots). (b) Simulated DNA shuffling showing the optimal fragmentation length of two subdomains. (c) The hierarchical optimization of local space searching: The 250 different sequences in each of the five pools (e.g., helices, strands, turns, loops, and others) are schematically represented by different shades of the same color. (d) The multipool swapping model for searching vast regions of tertiary fold space is essentially the same as in c except that now sequences from all five structural pools can be swapped into any subdomain. Multipool swapping allows for the formation of new tertiary structures by changing the type of secondary structure at any position along the protein.

TABLE III
RESULTS OF MONTE CARLO SIMULATION OF THE EVOLUTION PROTOCOLS
(FROM BOGARAD AND DEEM, 1999)[a]

Evolution method	U_{start}	$U_{evolved}$	$k_{binding}$
Amino acid substitution	−17.00	−23.18	1
DNA shuffling	−17.00	−23.83	100
Swapping	0	−24.52	1.47×10^4
Mixing	0	−24.88	1.81×10^5
Multipool swapping[b]	0	−25.40	8.80×10^6

[a]The starting polypeptide energy of −17.00 comes from a protein-like sequence (minimized U^{sd}), and 0 comes from a random initial sequence of amino acids. The evolved energies and binding constants are median values. The binding constants are calculated from $k_{binding} = ae^{-bU}$, where a and b are constants determined by normalizing the binding constants achieved by point mutation and shuffling to 1 and 100, respectively.
[b]Note that the energies and binding constants achieved via multipool swapping represent typical best-evolved protein folds.

barriers in the energy function also limit molecular evolution via DNA shuffling. For example, when the screen size was increased from 10,000 to 20,000 proteins per round, no further improvement in the final evolved energies was seen. Interestingly, the optimal simulated DNA shuffling length of 20 amino acids (60 bases) is nearly identical to fragment lengths used in experimental protocols (Crameri *et al.*, 1998).

3. Single-Pool Swapping

In Nature, local protein space can be rapidly searched by the directed recombination of encoded domains from multigene pools. A prominent example is the creation of the primary antibody repertoire in an adaptive immune system. These events are generalized by simulating the swapping of amino acid fragments from five structural pools representing helices, strands, loops, turns, and others (see Fig. 6c). During the swapping step, subdomains were randomly replaced with members of the same secondary structural pools with an optimal probability of 0.01/subdomain/round. The simulated evolution of the primary fold is limited by maintaining the linear order of swapped secondary structure types. The addition of the swapping move was so powerful that it was possible to achieve binding constants 2 orders of magnitude higher than in shuffling simulations (see Table III). Significantly, these improved binding constants were achieved starting with 10–20 times less minimized structural subdomain material.

4. Mixing

Parallel tempering is a powerful statistical method that often allows a system to escape local energy minima (Geyer, 1991). This method simultaneously simulates several systems at different temperatures, allowing systems at adjacent temperatures to swap configurations. The swapping between high- and low-temperature systems allows for an effective searching of configuration space. This method achieves rigorously correct canonical sampling, and it significantly reduces the equilibration time in a simulation. Instead of a single system, a larger ensemble with n systems is considered in parallel tempering, and each system is equilibrated at a distinct temperature T_i, $i = 1, \ldots, n$. The procedure in parallel tempering is illustrated in Fig. 7.

The system with the lowest temperature is the one of our interest; the higher-temperature systems are added to aid in the equilibration of the system of interest. In addition to the normal Monte Carlo moves performed in each system, swapping moves are proposed that exchange the configurations between two systems i and $j = i + 1$, $1 \leq i < n$. The higher-temperature systems are included solely to help the lowest-temperature system to escape from local energy minima via the swapping moves. To achieve efficient sampling, the highest temperature should be such that no significant free energy barriers are observed. So that the swapping moves are accepted with a reasonable probability, the energy histograms of systems adjacent in the temperature ladder should overlap.

In Nature, as well, it is known that genes, gene fragments, and gene operons are transferred between species of different evolutionary complexity (i.e., at different "temperatures"). By analogy, limited population mixing is performed among several parallel swapping experiments by randomly

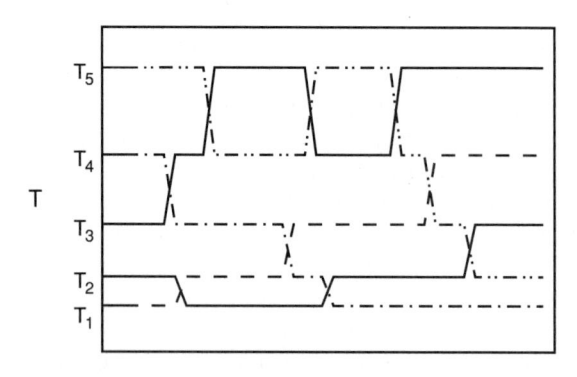

FIG. 7. A schematic drawing of the swapping taking place during a parallel tempering simulation (from Falcioni and Deem, 1999).

exchanging evolving proteins at an optimal probability of 0.001/protein/ round. These mixing simulations optimized local space searching and achieved binding constants $\approx 10^5$ higher than did base substitution alone (see Table III). Improved function is due, in part, to the increased number of events in parallel experiments. Indeed, mixing may occur in Nature when the evolutionary target function changes with time. That is, in a dynamic environment with multiple selective pressures, mixing would be especially effective when the rate of evolution of an isolated population is slower than the rate of environmental change. It has also been argued that spatial heterogeneity in drug concentration, a form of "spatial parallel tempering," facilitates the evolution of drug resistance (Kepler and Perelson, 1998).

5. Multipool Swapping

The effective navigation of protein space requires the discovery and selection of tertiary structures. To model the large-scale search of this space, a random polypeptide sequences was used as a starting point, and the swapping protocol was repeated. Now, however, secondary structures from all five pools were permitted to swap in at every subdomain (see Fig. 6d). This multipool swapping approach evolved proteins with binding constants $\approx 10^7$ better than did amino acid substitution of a protein-like starting sequence (see Table III). This evolution was accomplished by the random yet correlated juxtaposition of different types of low-energy secondary structures. This approach dramatically improved specific ligand binding while efficiently discovering new tertiary structures (see Fig. 8). Optimization of the rate of these hierachical molecular evolutionary moves, including relaxation of the selection criteria, enabled the protein to evolve despite the high rate of failure for these dramatic swapping moves. Interestingly, of all the molecular evolutionary processes modeled, only multipool swapping demonstrated chaotic behavior in repetitive simulations. This chaotic behavior was likely due to the discovery of different model folds that varied in their inherent ability to serve as scaffolds for ligand specific binding.

F. POSSIBLE EXPERIMENTAL IMPLEMENTATIONS

An important motivation of this work was that the proposed protocols must be experimentally feasible. Indeed, the ultimate test of the effectiveness of these protocols will be experimental. It is hoped that the search of large regions of protein space apparently possible with these methods will identify new protein folds and functions of great value to basic, industrial, and medical research.

FIG. 8. Schematic diagram representing a portion of the high-dimensional protein composition space (from Bogarad and Deem, 1999). The three-dimensional energy landscape of protein fold 1 is shown in cutaway. The arcs with arrowheads represent the ability of a given molecular evolution process to change the composition and so to traverse the increasingly large barriers in the energy function. The smallest arc represents the ability to evolve improved fold function via point mutation; then, in increasing order, DNA shuffling, swapping, and mixing. Finally, the multipool swapping protocol allows an evolving system to move to a different energy landscape representing a new tertiary fold (bottom).

The main technical challenge posed by the swapping protocols is the nonhomologous recombination necessary to swap the DNA that codes for different secondary structures into the evolving proteins. One approach would be to generate multiple libraries of synthetic oligonucleotide pools (Mandecki, 1990; Stemmer et al., 1995) encoding the different secondary subdomain structures. Asymmetric, complementary encoded linkers with embedded restriction sites would make the assembly, shuffling, and swapping steps possible.

Alternatively, the techniques of ITCHY (Ostermeier et al., 1999a) and SCRATCHY (Ostermeier et al., 1999b) may be used to accomplish the nonhomologous swapping of secondary structures required within our protocol.

Finally, exon shuffling may be used to perform the nonhomologous swapping events. In this case, the pools of secondary structures would be encoded with exons of a living organism, such as *E. coli*. There is precedent for such use of exon shuffling, at both the DNA (Fisch *et al.*, 1996) and the RNA (Moran *et al.*, 1999) level.

G. Life Has Evolved to Evolve

Although the focus has been on the higher levels of the evolution hierarchy, because that is where the biggest theoretical and experimental gap lies, all levels are important. In particular, the details of how point mutation assists protein evolution are important.

DNA base mutation leads only indirectly to changes in protein expression. How mutations occur in the bases and how these mutations lead to codon changes, and so amino acid changes, is not purely random. The inherent properties of the genetic code and biases in the mechanisms of DNA base substitution are perfectly suited for the "neutral" search of local space. Previously, the genetic code has been presented as a nodal or hypercubic structure to illustrate these relationships (Maeshiro and Kimura, 1998; Jiménez-Montaño *et al.*, 1996). It seems preferable to view the standard genetic code quantitatively as a 64×64 two-dimensional matrix. Seeds of this approach can be found in Kepler's work regarding evolvability in immunoglobulin genes (Cowell *et al.*, 1999; Kepler and Bartl, 1998; Kepler, 1997). The values in this matrix are the probabilities of a specific codon mutating to another by a single base change under error-prone conditions, e.g., mutator strains of bacteria, error-prone PCR, or somatic hypermutation. Assuming each base mutates independently in the codon, this matrix can be calculated from a simpler 4×4 matrix of base mutation probabilities. The base mutation matrix can be extracted from available experimental data (Smith *et al.*, 1996). A synonymous transition probability can be defined for each codon, which is the probability per replication of a base change that leads to a codon that codes for the same amino acid. A conservative transition probability can further be defined, which is the probability per replication that a base change leads to a conservative mutation. Finally, a nonconservative transition probability can be defined, which is the probability per replication that a base change leads to a nonconservative mutation. The conservative and nonconservative mutation probabilities can be viewed as defining the evolutionary potential of each codon: codons with high conservative and nonconservative mutation rates can be said to exhibit a high evolutionary potential. These mutation tendencies are shown in Fig. 9. In general, amino acids that exhibit a dramatic functional property, such as the charged residues, the ringed residues, cysteine, and tryptophan, tend to mutate at higher nonconservative rates

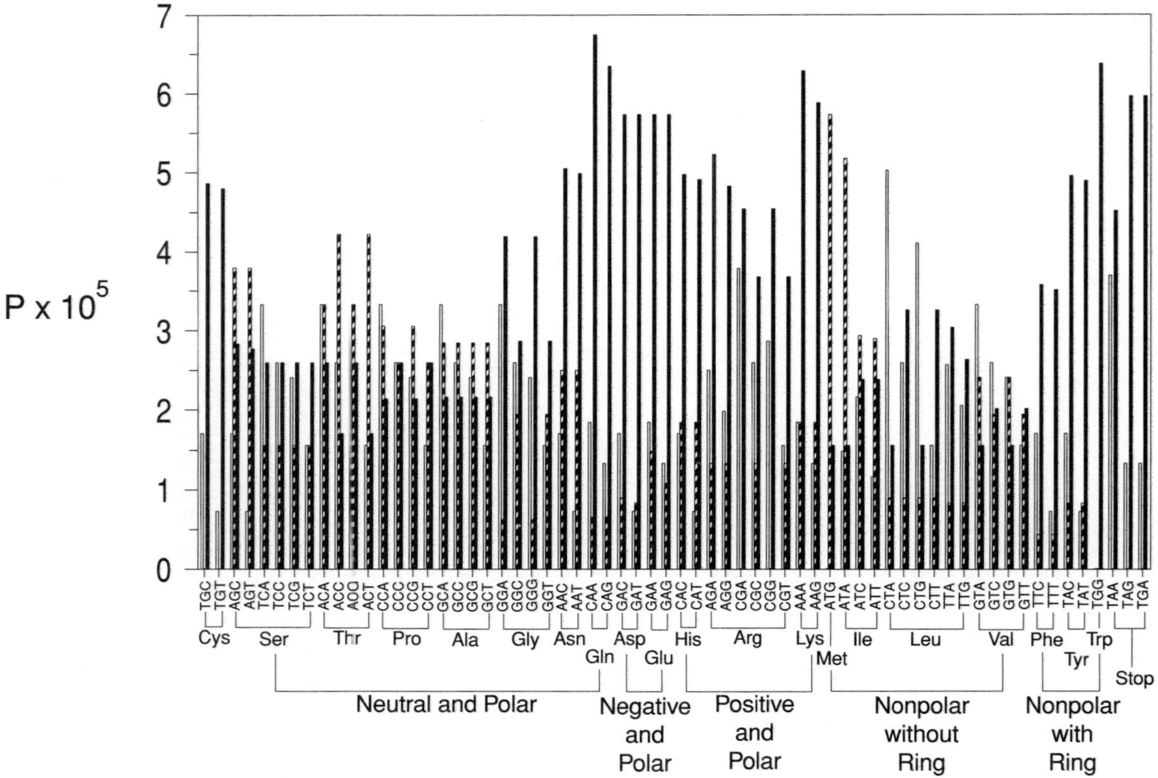

FIG. 9. Shown are the probabilities of a given codon mutating in a synonymous (open), conservative (hatched), and non-conservative (filled) way in one round of replication under error-prone conditions. The codons are grouped by the amino acid encoded, and the amino acids are grouped by category.

that allow for the possible deletion of the property. Amino acids that are more generic, such as the polar neutral residues and the nonpolar nonringed residues, tend to mutate at higher conservative rates that allow sequence space to be searched for similar favorable contacts.

There is a connection between condon usage and DNA shuffling. In the most successful DNA shuffling experiments by the Stemmer group, codon assignments of the initial coding sequences are optimized for expression. This assignment typically increases the nonsynonymous mutation rates described above. In particular, this assignment tends to increase the conservative mutation rate for the generic amino acids and the nonconservative mutation rate for the dramatically functional amino acids. Codon usage, then, has already implicitly been used to manipulate mutation rates. Explicit consideration of the importance of codon assignment would be an interesting amino acid level refinement of existing molecular evolution protocols. Optimized protocol parameters can be identified, taking into account the detailed codon usage information. Similarly, the codon potential matrix can be used in the design of the pools of secondary structures in the swapping-type molecular evolution protocols. That is, DNA can be chosen that codes for the secondary structures that (i) tends not to mutate, (ii) tends to mutate to synonymous sequences, (iii) tends to mutate to conservative sequences, or (iv) tends to mutate to non-conservative sequences.

In Nature there are numerous examples of exploiting codon potentials in ongoing evolutionary processes (Kepler, 1997). In the V regions of encoded antibodies, high-potential serine codons such as AGC are found predominantly in the encoded CDR loops, while the encoded frameworks contain low-potential serine codons such as TCT. Unfortunately, antibodies and drugs are often no match for the hydrophilic, high-potential codons of "error-prone" pathogens. The dramatic mutability of the HIV gp120 coat protein is one such example. One can envision a scheme for using codon potentials to target disease epitopes that mutate rarely (i.e., low-potential) and unproductively (i.e., become stop, low-potential, or structure-breaking codons). Such a therapeutic scheme is quite simple, and so could be quite generally useful against diseases that otherwise tend to become drug resistant.

H. Natural Analogs of These Protocols

During the course of any evolutionary process, proteins become trapped in local energy minima. Dramatic moves, such as swaps and juxtaposition, are needed to break out of these regions. Dramatic moves are usually deleterious, however. The evolutionary success of these events depends on population size, generation time, mutation rate, population mixing, selective pressure or freedom, such as successful genome duplications or the establishment of set-aside cells (Davidson et al., 1995), and the mechanisms that transfer low-energy, encoded structural domains.

By using the analogy with Monte Carlo to design "biased" moves for molecular evolution, a swapping-type move has been derived that is similar to several mechanisms of natural evolution. Viruses and transposons, for example, have evolved large-scale integration mechanisms (Pennisi, 1998). Exon shuffling is also a generator of diversity, and a possible scenario is that exon shuffling generated the primordial fold diversity (Gilbert, 1978; Gilbert et al., 1997; Netzer and Hartl, 1997). Alternatively, random swapping by horizontal transfer (Lawrence, 1997), rearrangement, recombination, deletion, and insertion can lead to high in-frame success rates in genomes with high densities of coding domains and reading frames, as in certain prokaryotes and mitochondria.

While inter- and intraspecies exchange of DNA is often thought to occur primarily on the scale of genes and operons, shorter exchange often occurs. Indeed, the most prevalent exchange length within *E. coli* is of the order of several hundred to a thousand base pairs (Syvanen, 1997). Similarly, analysis of the evolution of vertebrate cytochrome *c* suggests that transfer of segments significantly smaller than a single gene must have occurred (Syvanen, 1997).

Indeed, swapping mutations leading to significant diversity are not rare in Nature. *Neisseria meningitidis* is a frequent cause of meningitis in sub-Saharan Africa (Hobbs et al., 1997). The Opa proteins are a family of proteins that make up part of the outer coat of this bacteria. These proteins undergo some of the same class switching and hypervariable mutations as antibody domains. A significant source of diversity also appears to have come from interspecies transfer with *Neisseria gonorrhoeae* (Hobbs et al., 1997). Both the surface coat proteins and the pilon proteins of *N. gonorrhoeae* undergo significant homologous recombination to produce additional diversity. It appears generally true that the intra- and interspecies transfer of short segments of genes is common in *E. coli, Streptococci,* and *Neisseria*. Rapid evolution of diversity such as this obviously poses a significant challenge for therapeutic protocols.

Three dramatic examples of use of swapping by Nature are particularly notable. The first is the development of antibiotic resistance. It was originally thought that no bacteria would become resistant to penicillin due to the many point mutations required for resistance. Resistance occurred, however, within several years. It is now known that this resistance occurred through the swapping of pieces of DNA between evolving bacteria (Shapiro, 1992, 1997). One mechanism of antibiotic resistance was incorporation of genes coding for β-lactamases. These genes, which directly degrade β-lactam antibiotics, appear to be relatively ancient, and their incorporation is a relatively simple example of a swapping-type event. These β-lactamases have continued to evolve, however, in the presence of antibiotic pressure, both by point mutation and by shuffling of protein domains via exon shuffling (Maiden, 1998; Medeiros, 1997). The bacterial targets of penicillin, the penicillin-binding proteins, are modular proteins that have undergone significant structural evolution since the introduction of penicillin (Massova and Mobashery, 1999; Goffin and Ghuysen, 1998). This evolution was of a domain-shuffling form, and it is a more sophisticated example of a Natural swapping-type move. Multidrug resistance is, of course, now a major, current health care problem. The creation of the primary antibody repertoire in vertebrates is another example of DNA swapping (of genes, gene segments, or pseudo-genes). Indeed, the entire immune system mostly likely evolved from a single transposon insertion some 450 million years ago (Plasterk, 1998; Agrawal et al., 1998). This insertion, combined with duplication and subsequent mutation of a single membrane spanning protein, mostly likely lead to the class switching apparatus of the primary repertoire. Finally, the evolution of *E. coli* from *Salmonella* occurred exclusively by DNA swapping (Lawrence, 1997). None of the phenotypic differences between these two species is due to point mutation. Moreover, even the observed rate of evolution due to DNA swapping, 31,000 bases/million years, is higher than that due to point mutation, 22,000 bases/million years. Even though a DNA swapping event is less likely to be tolerated than is a point mutation, the more dramatic nature of the swapping event leads to a higher overall rate of evolution. This is exactly the behavior observed in the simulated molecular evolutions.

I. Concluding Remarks on Molecular Evolution

DNA base substitution, in the context of the genetic code, is ideally suited for the generation, diversification, and optimization of local protein space (Miller and Orgel, 1974; Maeshiro and Kimura, 1998). However, the difficulty of making the transition from one productive tertiary fold to another limits evolution via base substitution and homologous recombination alone. Nonhomologous DNA recombination, rearrangement, and insertion allow for the combinatorial creation of productive tertiary folds via the novel combination of suitable structures. Indeed, efficient search of the high-dimensional fold space requires a hierarchical range of mutation events.

This section has addressed from a theoretical point of view the question of how protein space can be searched efficiently and thoroughly, either in the laboratory or in Nature. It was shown that point mutation alone is incapable of evolving systems with substantially new protein folds. It was further demonstrated that even the DNA shuffling approach is incapable of evolving substantially new protein folds. The Monte Carlo simulations demonstrated that nonhomologous DNA "swapping" of low-energy structures is a key step in searching protein space.

More generally, the simulations demonstrated that the efficient search of large regions of protein space requires a hierarchy of genetic events, each encoding higher order structural substitutions. It was shown how the complex protein function landscape can be navigated with these moves. It was concluded that analogous moves have driven the evolution of protein diversity found in Nature. The proposed moves, which appear to be experimentally feasible, would make an interesting addition to the techniques of molecular biology. An especially important application of the theoretical approach to molecular evolution is modeling the molecular evolution of disease.

There are many experimental applications of the technology for molecular evolution (Patten *et al.*, 1997). Perhaps some of the most significant are in the field of human therapeutics. Molecular evolution can be used directly to improve the performance of protein pharmaceuticals. Molecular evolution can be used indirectly to evolve small molecule pharmaceuticals by evolving the pathways that code for small molecule synthesis in *E. coli.* Molecular evolution can be used for gene therapy and DNA vaccines. Molecular evolution can be used to produce recombinant protein vaccines or viral vaccines. Finally, molecular evolution can be used to create modified enzymatic assays in drug screening efforts. The ability to develop new assays that do not infringe on competitors' techniques is an important ability for large pharmaceutical companies, given the current complex state of patent claims. There is a similar range of applications of molecular evolution in the field of biotechnology. As shown in Table II, many of the tools of molecular biology can be improved or modified through the use of molecular evolution.

A wide variety of pest organisms and parasites, including fungi, weeds, insects, protozoans, macroparasites, and bacteria, have used evolutionary processes to evade chemical control. The range of evolutionary events exhibited by these organisms is similar in spirit to the hierarchy of moves present in the molecular evolution protocol (see Fig. 6). Bacteria provide one of the most pressing examples of the problems posed by an evolving disease (a "moving target"). Although there undoubtedly have been many selective pressures upon bacteria, the novel pressure with the largest impact in the last half-century has been the worldwide use of antibiotics. This background presence of antibiotics has led to the development of antibiotic resistance in many species of bacteria. Indeed, multidrug resistance is now a major health care issue, with some strains resistant to all but one, or even all, known antibiotics.

Interestingly, there is another strong pressure on evolving bacteria, that of the vertebrate immune system. This pressure is thought to be responsible for mosaic, or modular as a result of swapping-type events, genes found in species of bacteria not naturally genetically competent, such as *E. coli* and *S. pyogenes* (Dowson *et al.*, 1997). In these cases, the long-standing, strong selective pressure due to the interaction with the immune system likely led to genetic exchange.

Kepler and Perelson (1998) have noted that a spatial heterogeneity in the concentration of a drug can facilitate evolved resistance in a disease organism. This occurs because regions of low drug concentration provide a "safe harbor" for the disease, where replication and mutation can occur. The regions of high disease concentration provide the selective pressure for the evolution. Explicit examples of this mode of evolution include the role of spatial heterogeneity in the spread of insecticide resistance, noncompliance to antibiotic regimes in the rise of resistance in the tuberculosis bacterium, and heterogeneity within the body of the protease inhibitor indinavir in the rise of resistant HIV-1 strains (Kepler and Perelson, 1998). As noted above, this type of evolution is a spatial example of parallel tempering, a technique that has proven to be very powerful at sampling difficult molecular systems with many and large energy barriers. This analogy with parallel tempering suggests that heterogeneities must be of great and intrinsic importance in natural evolution.

Qualitative changes in protein space such as those modeled here allow viruses, parasites, bacteria, and cancers to evade the immune system, vaccines, antibiotics, and therapeutics. All of these pathogens evolve, to a greater or lesser degree, by large, swapping-type mutations. The successful design of vaccines and drugs must anticipate the evolutionary potential of both local and large space searching by pathogens in response to therapeutic and immune selection. The addition of disease specific constraints to simulations such as these should be a promising approach for predicting pathogen plasticity. Indeed, infectious agents will continue to evolve unless we can force them down the road to extinction.

IV. Summary

Significant opportunities exist for the application of ideas from statistical mechanics to the burgeoning area of combinatorial chemistry. While combinatorial chemistry was not invented by researchers in the field of statistical mechanics, it is fair to say that perhaps it should have been! The design of effective experimental methods for searching composition space is similar in concept to the design of effective Monte Carlo methods for searching configuration space. Optimization of the parameters in combinatorial chemistry protocols is analogous to the integration of various types of moves in Monte Carlo simulation. It is notable that one of the strongest present chemical-engineering proponents of combinatorial chemistry in the solid state, Henry Weinberg at SYMYX Technologies, has taught graduate statistical mechanics for the last 25 years! Applications of combinatorial chemistry abound in the fields of catalysis, sensors, coatings, microelectronics, biotechnology, and human therapeutics. Hopefully, statistical mechanics will have a significant role to play in shaping these new methods of materials design.

ACKNOWLEDGMENTS

It is a pleasure to acknowledge the contributions of my collaborators Leonard D. Bogarad, Marco Falcioni, and Taison Tan. This work was supported by the National Science Foundation.

REFERENCES

Agrawal, A., Eastman, Q. M., and Schatz, D. G., Transposition mediated by RAG1 and RAG2 and its implications for the evolution of the immune system. *Nature* **394,** 744–751 (1998).
Akporiaye, D. E., Dahl, I. M., Karlsson, A., and Wendelbo, R., Combinatorial approach to the hydrothermal synthesis of zeolites. *Angew. Chem. Int. Ed.* **37,** 609–611 (1998).
Bogarad, L. D., and Deem, M. W., A hierarchical approach to protein molecular evolution. *Proc. Natl. Acad. Sci. USA* **96,** 2591–2595 (1999).
Bratley, P., Fox, B. L., and Niederreiter, H., Algorithm-738-programs to generate Niederreiter's low-discrepancy sequences. *ACM Trans. Math. Software* **20,** 494–495 (1994).
Briceño, G., Chang, H., Sun, X., Schultz, P. G., and Xiang, X.-D., A class of cobalt oxide magnetoresistance materials discovered with combinatorial synthesis. *Science* **270,** 273–275 (1995).
Burgess, K., Lim, H.-J., Porte, A. M., and Sulikowski, G. A., New catalysts and conditions for a C-H insertion reaction identified by high throughput catalyst screening. *Angew. Chem. Int. Ed.* **35,** 220–222 (1996).
Cole, B. M., Shimizu, K. D., Krueger, C. A., Harrity, J. P. A., Snapper, M. L., and Hoveyda, A. H., Discovery of chiral catalysts through ligand diversity: Ti-catalyzed enantioselective addition of TMSCN to *meso* epoxides. *Angew. Chem. Int. Ed.* **35,** 1668–1671 (1996).
Cong, P., Doolen, R. D., Fan, Q., Giaquinta, D. M., Guan, S., McFarland, E. W., Poojary, D. M., Self, K., Turber, H. W., and Weinberg, W. H., High-throughput synthesis and screening of combinatorial heterogeneous catalyst libraries. *Angew. Chem. Int. Ed.* **38,** 484–488 (1999).
Cowell, L. G., Kim, H. J., Humaljoki, T., Berek, C., and Kepler, T. B., Enhanced evolvability in immunoglobulin V genes under somatic hypermutation. *J. Mol. Evol.* **49,** 23–26 (1999).
Crameri, A., Raillard, S. A., Bermudez, E., and Stemmer, W. P. C., DNA shuffling of a family of genes from diverse species accelerates directed evolution. *Nature* **391,** 288–291 (1998).

Danielson, E., Devenney, M., Giaquinta, D. M., Golden, J. H., Haushalter, R. C., McFarland, E. W., Poojary, D. M., Reaves, C. M., Weinberg, W. H., and Wu, X. D., A rare-earth phosphor containing one-dimensional chains identified through combinatorial methods. *Science* **279,** 837–839 (1988).

Danielson, E., Golden, J. H., McFarland, E. W., Reaves, C. M., Weinberg, W. H., and Wu, X. D., A combinatorial approach to the discovery and optimization of luminescent materials. *Nature* **389,** 944–948 (1997).

Davidson, E. H., Peterson, K. J., and Cameron, R. A., Origin of bilaterian body plans— Evolution of developmental regulatory mechanisms. *Science* **270,** 1319–1325 (1995).

de Pablo, J. J., Laso, M., and Suter, U. W., Estimation of the chemical potential of chain molecules by simulation. *J. Chem. Phys.* **96,** 6157 (1992).

Dickinson, T. A., and Walt, D. R., Generating sensor diversity through combinatorial polymer synthesis. *Anal. Chem.* **69,** 3413–3418 (1997).

Dowson, C. G., Barcus, V., King, S., Pickerill, P., Whatmore, A., and Yeo, M., Horizontal gene transfer and the evolution of resistance and virulence determinants in *Streptococcus*. *J. Appl. Micro. Biol. Symps. Supl.* **83,** 42S–51S (1997).

Falcioni, M., and Deem, M. W., A biased Monte Carlo scheme for zeolite structure solution. *J. Chem. Phys.* **110,** 1754–1766 (1999).

Falcioni, M., and Deem, M. W., Library design in combinatorial chemistry by Monte Carlo methods. *Phys. Rev. E* **61,** 5948–5952 (2000).

Fisch, I., Kontermann, R. E., Finnern, R., Hartley, O., Solergonzalez, A. S., Griffiths, A. D., and Winter, G., A strategy of exon shuffling for making large peptide repertoires displayed on filamentous bacteriophage. *Proc. Natl. Acad. Sci. USA* **93,** 7761–7766 (1996).

Frenkel, D., and Smit, B., Unexpected length dependence of the solubility of chain molecules. *Mol. Phys.* **75,** 983 (1992).

Frenkel, D., and Smit, B., *"Understanding Molecular Simulation: From Algorithms to Applications."* Academic Press, San Diego, 1996.

Frenkel, D., Mooij, C. G. A. M., and Smit, B., Novel scheme to study structural and thermal properties of continuously deformable molecules. *J. Phys. Condens. Matter* **4,** 3053 (1992).

Geyer, C. J., Markov chain Monte Carlo maximum likelihood. In "Computing Science and Statistics: Proceedings of the 23rd Symposium on the Interface." American Statistical Association, New York, 1991, pp. 156–163.

Gilbert, W., Why genes in pieces? *Nature* **271,** 501 (1978).

Gilbert, W., DeSouza, S. J., and Long, M., Origin of genes. *Proc. Natl. Acad. Sci. USA* **94,** 7698–7703 (1997).

Goffin, C., and Ghuysen, J.-M., Multimodular penicillin-binding proteins: An enigmatic family of orthologs and paralogs. *Microbiol. Mol. Biol.* **62,** 1079–1093 (1998).

Helmkamp, M. M., and Davis, M. E., Synthesis of porous silicates. *Annu. Rev. Mater. Sci.* **25,** 161–192 (1995).

Hobbs, M. M., Seiler, A., Achtman, M., and Cannon, J. G., Microevolution within a clonal population of pathogenic bacteria: Recombination, gene duplication and horizontal genetic exchange in the *opa* gene family of *Neisseria meningitidis*. *Mol. Microbiol.* **12,** 171–180 (1997).

Jiménez-Montaño, M. A., de la Mora-Basánez, C. R., and Pöschel, T., The hypercube structure of the genetic code explains conservative and non-conservative amino acid substitutions *in vivo* and *in vitro*. *BioSystems* **39,** 117–125 (1996).

Kamtekar, S., Schiffer, J. M., Xiong, H. Y., Babik, J. M., and Hecht, M. H., Protein design by binary patterning of polar and nonpolar amino acids. *Science* **262,** 1680–1685 (1993).

Kauffman, S. A., "The Origins of Order." Oxford University Press, New York, 1993.

Kauffman, S. A., and MacReady, W. G., Search strategies for applied molecular evolution. *J. Theor. Biol.* **173,** 427–440 (1995).

Kauffman, S., and Levin, S., Towards a general theory of adaptive walks on rugged landscapes. *J. Theor. Biol.* **128,** 11–45 (1987).
Kepler, T. B., Codon bias and plasticity in immunoglobulins. *Mol. Biol. Evol.* **14,** 637–643 (1997).
Kepler, T. B., and Bartl, S., Plasticity under somatic mutation in antigen receptors. *Curr. Topics Micrbiol.* **229,** 149–162 (1998).
Kepler, T. B., and Perelson, A. S., Drug concentration heterogeneity facilitates the evolution of drug resistance. *Proc. Natl. Acad. Sci. USA* **95,** 11514–11519 (1998).
Lawrence, J. G., Selfish operons and speciation by gene transfer. *Trends Microbiol.* **5,** 355–359 (1997).
Maeshiro, T., and Kimura, M., Role of robustness and changeability on the origin and evolution of genetic codes. *Proc. Natl. Acad. Sci. USA* **95,** 5088–5093 (1998).
Maiden, M. C. J., Horizontal genetic exchange, evolution, and spread of antibiotic resistance in bacertia. *Clin. Infect. Dis.* **27,** S12–S20 (1998).
Mandecki, W., A method for construction of long randomized open reading frames and polypeptides. *Protein Eng.* **3,** 221–226 (1990).
Marinari, E., Parisi, G., and Ruiz-Lorenzo, J., Numerical simulations of spin glass systems. In *Spin Glasses and Random Fields,* "Directions in Condensed Matter Physics" (A. Young, ed.), World Scientific, Singapore, Vol. 12, 1998, pp. 59–98
Massova, I., and Mobashery, S., Structural and mechanistic aspects of evolution of β-lactamases and penicillin-binding proteins. *Curr. Pharm. Design* **5,** 929–937 (1999).
Medeiros, A. A., Evolution and dissemination of β-lactamases accelerated by generation of β-lactam antiobiotics. *Clin. Infect. Dis.* **24,** S19–S45 (1997).
Menger, F. M., Eliseev, A. V., and Migulin, V. A., Phosphatase catalysis develped *via* combinatorial organic chemistry. *J. Org. Chem.* **60,** 6666–6667 (1995).
Miller, S., and Orgel, L., "The Origin of Life on Earth." Prentice Hall, London, 1974.
Moore, J. C., Jin, H.-M., Kuchner, O., and Arnold, F. H., Strategies for the *in-vitro* evolution of protein function—Enzyme evolution by random recombination of improved sequences. *J. Mol. Biol.* **272,** 336–347 (1997).
Moran, J. V., DeBerardinis, R. J., and Kazazian, H. H., Exon shuffling by L1 retrotransposition. *Science* **283,** 1530–1534 (1999).
Netzer, W. J., and Hartl, F. U., Recombination of protein domains facilitated by co-translational folding in eukaryotes. *Nature* **388,** 343–349 (1997).
Niederreiter, H., "Random Number Generation and Quasi-Monte Carlo Methods," Society for Industrial and Applied Mathematics, Philadelphia, 1992.
Novet, T., Johnson, D. C., and Fister, L., Interfaces, interfacial reactions and superlattice reactants. *Adv. Chem. Ser.* **245,** 425–469 (1995).
Ostermeier, M., Nixon, A. E., and Benkovic, S. J., Incremental truncation as a strategy in the engineering of novel biocatalysts. *Bioorg. Med. Chem.* **7,** 2139–2144 (1999a).
Ostermeier, M., Shim, J. H., and Benkovic, S. J., A combinatorial approach to hybrid enzymes independent of DNA homology. *Nature Biotech.* **17,** 1205–1209 (1999b).
Patten, P. A., Howard, R. J., and Stemmer, W. P. C., Applications of DNA shuffling to pharmaceuticals and vaccines. *Curr. Opin. Biotech.* **8,** 724–733 (1997).
Pennisi, E., How the genome readies itself for evolution. *Science* **281,** 1131–1134 (1998).
Perelson, A. S., and Macken, C. A., Protein evolution on partially correlated landscapes. *Proc. Natl. Acad. Sci. USA* **92,** 9657–9661 (1995).
Pirrung, M. C., Spatially addressable combinatorial libraries. *Chem. Rev.* **97,** 473–488 (1997).
Plasterk, R., V(D)J recombination: Ragtime jumping. *Nature* **394,** 718–719 (1998).
Reddington, E., Sapienza, A., Gurau, B., Viswanathan, R., Sarangapani, S., Smotkin, E. S., and Mallouk, T. E., Combinatorial electrochemistry: A highly parallel, optical screening method for discovery of better electrocatalysts. *Science* **280,** 1735–1737 (1998).

Riddle, D. S., Santiago, J. V., Brayhall, S. T., Doshi, N., Grantcharova, V. P., Yi, Q., and Baker, D., Functional rapidly folding proteins from simplified amino acid sequences. *Nature Struct. Biol.* **4,** 805–809 (1997).

Schuster, P., and Stadler, P. F., Sequence redundancy in biopolymers: A study on RNA and protein structures. In *Viral Regulatory Structure and Their Degeneracy* (G. Myers, ed.), Addison–Wesley, New York, 1998, pp. 163–186.

Sedgewick, R., "Algorithms" 2nd ed., Addison–Wesley, New York, 1998.

Shapiro, J. A., Natural genetic engineering in evolution. *Genetica* **86,** 99–111 (1992).

Shapiro, J. A., Genome organization, natural genetic engineering and adaptive mutation. *Trends Genet.* **13,** 98–104 (1997).

Smith, B., and Maesen, T. L. M., Commensurate 'freezing' of alkanes in the channels of a zeolite. *Nature* **374,** 42 (1995).

Smith, D. S., Creadon, G., Jena, P. K., Portanova, J. P., Kotzin, B. L., and Wysocki, L. J., Di- and trinucleotide target preferences of somatic mutagenesis in normal and autoreactive B cells. *J. Immunol.* **156,** 2642–2652 (1996).

Stemmer, W. P. C., Rapid evolution of a protein *in-vitro* by DNA shuffling. *Nature* **370,** 389–391 (1994).

Stemmer, W. P. C., Crameri, A., Ha, K. D., Brennan, T. M., and Heyneker, H. L., Single-step assembly of a gene and entire plasmid from large numbers of oligodeoxyribonucleotides. *Gene* **164,** 49–53 (1995).

Syvanen, M., Horizontal gene transfer: Evidence and possible consequences. *Annu. Rev. Genet.* **28,** 237–261 (1997).

van Dover, R. B., Schneemeyer, L. F., and Fleming, R. M., Discovery of a useful thin-film dielectric using a composition-spread approach. *Nature* **392,** 162–164 (1998).

Volkenstein, M. V., "Physical Approaches to Biological Evolution." Springer-Verlag, New York, 1994.

Wang, J., Yoo, Y., Gao, C., Takeuchi, I., Sun, X., Chang, H., Xiang, X.-D., and Schultz, P. G., Identification of a blue photoluminescent composite material from a combinatorial library. *Science* **279,** 1712–1714 (1998).

Weinberg, W. H., Jandeleit, B., Self, K., and Turner, H., Combinatorial methods in homogeneous and heterogeneous catalysis. *Curr. Opin. Chem. Biol.* **3,** 104–110 (1998).

Xiang, X.-D., Sun, X., Briceño, G., Lou, Y., Wang, K.-A., Chang, H., Wallace-Freedman, W. G., Chang, S.-W., and Schultz, P. G., A combinatorial approach to materials discovery. *Science* **268,** 1738–1740 (1995).

Zhang, J.-H., Dawes, G., and Stemmer, W. P. C., Directed evolution of a fucosidase from a galactosidase by DNA shuffling and screening. *Proc. Natl. Acad. Sci. USA* **94,** 4504–4509 (1997).

Zones, S. I., Nakagawa, Y., Lee, G. S., Chen, C. Y., and Yuen, L. T., Searching for new high silica zeolites through a synergy of organic templates and novel inorganic conditions. *Micropor. Mesopor. Mat.* **21,** 199–211 (1998).

FLUCTUATION EFFECTS IN MICROEMULSION REACTION MEDIA

Venkat Ganesan and Glenn H. Fredrickson

Department of Chemical Engineering, University of California–Santa Barbara, Santa Barbara, California 93106

I. Introduction	123
II. Reactions in the Bicontinuous Phase	127
A. Diffusion Equations	127
B. Objectives	128
C. Mean-Field Analysis	129
D. Renormalization Group Theory	132
E. Discussion	134
F. Summary	135
III. Reactions in the Droplet Phase	136
A. Outline	136
B. Fluctuations of the Droplet Phase	137
C. Diffusion Equation and Perturbation Expansion	139
D. Consideration of Temporal Regimes	141
E. Intermediate Times	143
F. Short Time Regime	143
G. Effect of the Péclet Number	144
H. Discussion	145
I. Other Effects	146
J. Summary	146
References	147

I. Introduction

The field of chemical engineering reactor design and kinetics, a traditional chemical engineering forté, has undergone a rapid transformation in the past two decades on both theoretical and practical fronts. On the theoretical front, the validity of classical theoretical ideas embodying *mean field* chemical rate expressions have been challenged, necessitating for resolution the injection of radically new ideas and concepts drawn from statistical mechanics and field theory [1–6]. These studies have led to a recognition of the importance of fluctuations and the role of spatial dimensionality in influencing chemical reaction kinetics. On the practical front, the focus is slowly moving away from classical reactors and catalysts and toward utilizing novel microstructured materials as catalysts and as reaction media for effecting reactions. For instance, nanoporous materials and self-assembled systems are presently being advanced as effective catalyst materials for their ability to provide compartmentalized spaces with a high surface-to-volume ratio [7]. It is therfore becoming imperative to gain a fundamental understanding of the reaction kinetics and mechanisms in such materials. However, most of the above-mentioned theoretical studies have so far dealt with the purported existence of a homogeneous continuum (which can be envisioned as the interior of a CSTR or a batch reactor), lacking the fine microscale structure characterizing these "novel" self-assembled materials. It is therefore of practical interest for chemical engineers to utilize the fundamental theoretical concepts developed so far to glean insights into the chemical kinetics of reactions effected in microstructured media. This article summarizes some of our own research focusing on this goal. While the example we have chosen is quite specific, the ideas embodied herein and the statistical mechanical tools and the issues thereby confronted possess a generic basis transcending the specificity of the scenario.

As previously mentioned, there has been a recent surge of interest in applications exploiting self-assembled systems such as micellar solutions, microemulsions, and lipid bilayers as reaction media [8, 9]. Among these systems, microemulsions have emerged as a popular choice for a number of diverse applications (cf., Refs. 10 and 11 and the references cited therein). Microemulsions represent thermodynamically stable dispersions of water in oil (or vice versa), owing their stability to the presence of an amphiphile at the interface between the constituents [12]. The phase behavior of these ternary mixtures has been extensively studied and can be considered well understood [12–15]. Thermodynamic studies have shown that, depending on the concentration of the different constituents, the structure of microemulsions can range from a dispersion of spherical droplets (at dilute concentrations of either oil or water) to complex interweaving bicontinuous networks of oil and water (at higher concentrations) [16]. The observed phase behavior of these microemulsions has presented interesting opportunities for the creation of novel materials by appropriate reactions within these media. A number of efforts are presently under way to "fix" these *nanoscale* structures of the microemulsion by reactions within the system. For instance, polymerization reactions have been carried out in microemulsions to produce the nanoscale analog of the latex particles conventionally produced by emulsion polymerization techniques [17]. Bicontinuous networks have also been fixed to create reticulated frameworks envisaging biomineralization applications [18]. Numerous other applications utilizing reactions in these microemulsion phases have also been suggested and implemented in contexts relating to catalysis, drug delivery, etc.

Considering the practical utility of such systems, the study of chemical reactivity in these systems is bound to possess important ramifications of relevance to the chemical engineering design of reactors envisioning the use of such media. In this context, it is pertinent to observe that most of the applications utilizing microemulsions as reaction media fall broadly into two generic classes: (i) those involving reactions within either or both (and possibly also at the interface) phases and (ii) those involving only reactions at the interface. It is to be noted here that there have been a few investigations on problems within the former category, focusing specifically on the effect of confining the reactants to "restricted" spaces [20–22]. In this work, we direct our considerations to the problems encompassed in the latter category. Within such a framework, we examine the effects (if any) of an oft-neglected issue in the consideration of reactions in these complex fluid systems, viz., the impact of thermal fluctuations upon the kinetics of interfacial reactions. Based on our (brief) description (in the preceding paragraph) of the different possible structures in these systems, it would be erroneous to conclude that the different phase structures observed in these systems are rigid and static. On the contrary, the surfactant interfaces that define the structure of the microemulsion possess bending moduli of the order of a few k_bT (where k_b represents the Boltzmann constant and T denotes the temperature), thereby making them highly susceptible to the influence of thermal fluctuations [23, 24]. Indeed, microemulsion phases owe their existence to the destruction of more ordered mesophases by thermal fluctuations. One may speculate that these fluctuations which dynamically modulate the interfaces might also influence the kinetics of reactions occurring on these "fluctuating" surfaces. It is to the examination of such effects that our analysis is directed in this two-part article. Our analysis is purely at the level of "transport processes"—"chemistry" issues relating to the possible change in mechanism of the reaction, etc., are completely ignored. Within a broader framework, our analysis identifies some important features in systems involving thermal fluctuations and diffusion limited reactions—a class of problems which, to our knowledge, has not been addressed heretofore in the literature. We also hope that this article will serve to highlight for the chemical engineering community some of the *nonclassical* issues encountered in reaction kinetics and the field-theoretic tools to resolve them.

In the following analysis, our primary interest centers on reactions of the type $A + B \to \emptyset$, wherein one of the reactants, say A, resides predominantly in the water phase, and the other reactant, B, resides in the oil phase. The product ∅ is assumed to be inert, chemically (i.e., does not react with A or B) and physicochemically (i.e., does not impede the transport of A or B or destabilize the microemulsion). In such a scenario, the reaction between A and B occurs predominantly at the oil–water interface. In this work, we also suppose that the reaction kinetics of such a reaction is of first order with respect to both the reactants, i.e., the rate of the reaction r is given by $-r = \lambda c_A c_B$, where λ denotes the second-order rate constant, and c_A and c_B denote the concentrations of A and B, respectively. Such an assumption is motivated purely by considerations of analytical tractability (in fact, we are forced to simplify this nonlinear kinetics even further in the second part of this work). Further, we focus specifically upon the bicontinuous and the droplet phases of the microemulsions (Fig. 1). As expounded in the Introduction both these structures have been utilized widely in applications, thereby rendering our analysis practically relevant. Moreover, the method of analysis employed for spherical microemulsions possesses a number of features generic to the comparable analysis of diffusion and reaction within ordered lyotropic phases. Thus, the qualitative features unearthed in the second part of this article (Section III) are expected to be equally valid for considerations pertaining to reactions carried out in lamellar phases, etc. Further, in both parts of the article (Sections II and III) we ignore the convective transport of the reactants (over and above that arising from pure molecular diffusion) arising as a consequence of the convective flow field generated due to fluctuations of the interface. While such effects are potentially interesting, they considerably complicate the analysis. The simplification arising from the neglect of convective transport can be construed as a low Péclet number limit of the physical situation. With such a scenario in mind, we henceforth focus on the purely diffusive transport of the solute species and their subsequent mutual anhiliation at the fluctuating interface.

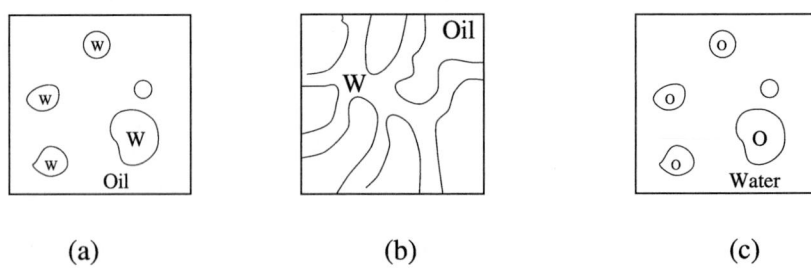

FIG. 1. The different microstructures of the microemulsion phase: (a) the water-in-oil droplet phase, (b) the bicontinuous phase, (c) the oil-in-water droplet phase.

In the following text we summarize the pertinent details of the analysis and the accompanying results. A reader who is interested in an elaboration of these details can refer to the detailed version of our previous article [25].

II. Reactions in the Bicontinuous Phase

A. DIFFUSION EQUATIONS

As expounded in Section I, we consider reactions of the type

$$A + B \to \emptyset, \qquad -r = \lambda c_A c_B, \tag{1}$$

where the reactants A and B possess contrasting affinities toward oil and water. Such an affinity is conveniently modeled within the framework of a diffusion equation by the inclusion of a biasing potential dependent upon the concentration of oil (or water). In the symmetric case involving equal oil and water concentrations (as we assume throughout this part of the article), the difference in the local concentrations of water and oil can instead be associated with an order parameter $\phi(\mathbf{x})$ [26], with the affinity of A and B then determined by the sign of $\phi(\mathbf{x})$. Without loss of generality, if we assume that reactant A has an affinity to water ($\phi > 0$) and B to oil ($\phi < 0$), we can formulate the transport and reaction of A and B in terms of a set of reaction–diffusion equations governing the concentrations c_A and c_B,

$$\frac{\partial c_A}{\partial t} = D_A \nabla^2 c_A - \lambda c_A c_B - \mu_A \nabla \cdot (c_A \nabla \phi), \tag{2}$$

$$\frac{\partial c_B}{\partial t} = D_B \nabla^2 c_A - \lambda c_A c_B + \mu_B \nabla \cdot (c_B \nabla \phi). \tag{3}$$

In the above equations D_A and D_B denote the diffusivities of A and B, respectively. The strengths of the affinities of the reactants A and B toward the solvents are quantified by μ_A and μ_B (each ≥ 0), respectively. While the above equation represents the most general structure for the diffusion equations appropriate for such a scenario, in an attempt to focus on the essential physics in the following analysis, we consider the symmetric case corresponding to $D_A = D_B = D$ and $\mu_A = \mu_B = \mu$.

To complete the description of the diffusion and reaction of solutes A and B in the fluctuating potential field $\phi(\mathbf{x}, t)$, we still need to specify the statistics of the fluctuations of $\phi(\mathbf{x}, t)$. To this purpose we conveniently adopt the "disordered fluid" model of the bicontinuous phase of the microemulsion [27] in conjunction with a conserved model for the dynamics of the fluctuations of potential field [28]. Within such a description, the *Gaussian* statistics of the potential field $\phi(\mathbf{x}, t)$ can be computed as[1]

$$\langle\phi(\mathbf{k}, \omega)\rangle = 0; \quad \langle\phi(\mathbf{k}, \omega)\phi(\mathbf{k}', \omega')\rangle = \frac{\Delta^2 k^2 (2\pi)^d \delta^d(\mathbf{k}+\mathbf{k}')(2\pi)\delta(\omega+\omega')}{[\omega^2 + (uk^2 + vk^4 + wk^6)^2]}. \quad (4)$$

In the above equation Δ denotes a phenomenological constant specifying the strength of the fluctuations. Further, u, v, and w represent phenomenological parameters characterizing the equilibrium structure of the bicontinuous microemulsion. These parameters can be discerned experimentally by employing light-scattering techniques. In Eq. (4) we have displayed the results in terms of $\phi(\mathbf{k}, \omega)$, denoting the Fourier transform of $\phi(\mathbf{x}, t)$ and defined as

$$\phi(\mathbf{x}, t) = \int \frac{d^d \mathbf{k}}{(2\pi)^d} \int_{-\infty}^{\infty} \frac{d\omega}{2\pi} e^{-i\mathbf{k}\cdot\mathbf{x} - i\omega t} \phi(\mathbf{k}, \omega).$$

B. Objectives

The above set of equations, (1)–(4) (with $D_A = D_B = D$ and $\mu_A = \mu_B = \mu$), represents the field-theoretic formulation of the transport and reaction of solutes A and B in a bicontinuous phase of microemulsion. In view of the fact that the fluctuations corresponding to the order parameter ϕ possess a vanishing mean value, i.e., $\langle\phi\rangle = 0$, it is reasonable to query if one could, at sufficiently long length and time scales (to allow for the self-averaging of the fluctuating potential), infer that the effect of the fluctuations of the microemulsion proves irrelevant in impacting the transport and the reaction of solutes A and B. On the other hand, since the concentrations of the reactants are coupled to the fluctuations of the ϕ field, it also appears intuitive that the dynamic distribution of solutes A and B, and therefore the fluctuations of the microemulsion, should exert a discernible influence on the instantaneous reaction rate. In the former case, it should therefore be possible to quantify the decay rates of reactants A and B without any reference whatsoever to the fluctuating "potential" $\phi(\mathbf{x}, t)$. In the event that such a quantification is indeed possible, one could infer that the microscale fluctuations of the bicontinuous phase have no discernible effect on the overall transport and reaction processes. On the other hand, uncovering a nonclassical decay rate for the reactants would instead indicate a nontrivial behavior arising from the presence of fluctuations of the bicontinuous phase.

Mean-field scaling analysis provides a "quick and easy" way to discern the answers to the questions raised above. However, the results of such an

---

[1] Note that a Gaussian field is uniquely specified by its first two moments.

analysis are not always accurate and would typically need to be corroborated by more rigorous analysis employing renormalization group theory (RG). Nevertheless, mean-field analysis enables one to glean physical insights into the results and also understand the regimes of their validity. In the following text, we elaborate the details of a self-consistent mean-field analysis which allows one to discern the asymptotic decay kinetics. Subsequently, we indicate the results of the RG calculations which confirm the results predicted by mean-field theory.

C. Mean-Field Analysis

At the outset we consider the case wherein $\phi(\mathbf{x}, t) = 0$ [29, 30] to elucidate the role of fluctuations and the (mean-field) manner in which they can be discerned. This also serves to illustrate the underlying basis for the *anomalous mean-field* (AMF) kinetics encountered in reaction-diffusion systems.[2] Subsequently, we extend the arguments to the case wherein the potential $\phi(\mathbf{x}, t)$ influences the dynamics.

1. $\phi(\mathbf{x}, t) = 0$ (AMF Kinetics)

At $t = 0$, the mean densities n_{0A} and n_{0B} of reactants A and B are identical (denoted n_0) and spatially homogeneous. However, the randomness embodied in the distribution of A and B allows for fluctuations in the number of particles within a volume V to an order $(n_0 V)^{1/2}$. After a time t, the particles have had a chance to diffuse and mutually anhilate other particles within a volume $(Dt)^{d/2}$ around themselves. Therefore, after the elapse of the time t, the volume $(Dt)^{d/2}$ retains only the initial imbalance resulting from fluctuations. This corresponds to the presence of $[n_0(Dt)^{d/2}]^{1/2}$ particles in the volume $(Dt)^{d/2}$, and therefore a density $[n_0/(Dt)^{d/2}]^{1/2}$, yielding $n(t) \approx n_0^{1/2}(Dt)^{-d/4}$.

2. In the Presence of the Potential $\phi(\mathbf{x}, t)$

It is evident from the above argument that the temporal variation of the mean-squared displacement of the reactants determines the asymptotic decay rates. For instance, if $\langle\langle r^2 \rangle\rangle \sim t^\delta$ (where the notation $\langle\langle \cdots \rangle\rangle$ is used to denote ensemble averages over the different particles of a reactant species), then the concentrations decay as $n_0^{1/2} t^{-d\delta/4}$. Therefore it behooves us to determine the exponent δ for diffusion in the fluctuating potential field.

---

[2] This is in contrast to the classical kinetics, which, for a second-order reaction of the type considered in this article, would predict $c_A(t), c_B(t) \sim t^{-1}$ when $c_A(t=0) = c_B(t=0)$.

In the following we outline a self-consistent Flory-type argument to enable the computation of δ. This type of argument has also been used in contexts relating to the wandering exponent of polymers [31] and diffusion in turbulent velocity fields [32] and has provided results reasonably in accord with more rigorous calculations.

The starting point for this analysis is the Langevin equation for the motion of a particle in the potential field ϕ (in the absence of inertial terms)

$$\frac{d\mathbf{r}}{dt} = -\nabla\phi + \boldsymbol{\xi}(t). \tag{5}$$

$\boldsymbol{\xi}$ represents a random thermal noise with statistics,

$$\langle \boldsymbol{\xi}(t) \rangle = 0; \qquad \langle \boldsymbol{\xi}(t)\boldsymbol{\xi}(t') \rangle = 2D\mathbf{I}\delta(t-t'), \tag{6}$$

where D represents the diffusivity of the particle and $\mathbf{I}$ denotes the idemfactor (or unit tensor). The above Langevin equation in conjunction with the noise statistics (6) is formally equivalent to the diffusion equation (2) [or (3)] in the absence of reaction terms ($\lambda = 0$).

We commence our analysis by considering the motion of the particle in the fluctuating potential ϕ and in the absence of the thermal noise $\boldsymbol{\xi}$. In such a case, the mean-squared displacement of the particle, denoted $\sigma^2(t)$, can be expressed in terms of a heuristic generalization of the Kubo formula[3]

$$\sigma^2(t) \sim t \int_0^t d\tau \langle \langle \nabla\phi(\mathbf{r}(\tau), \tau) \cdot \nabla\phi(\mathbf{r}(0), 0) \rangle_\phi \rangle_\mathbf{r}. \tag{7}$$

(In the following our interest centers on scaling laws, and therefore we omit the numerical prefactors that accompany the expressions.) In the above equation, we use the notation $\langle \cdots \rangle_\phi$ to denote an average over the statistics of the fluctuating field ϕ and $\langle \cdots \rangle_\mathbf{r}$ to denote a comparable average over an ensemble of particle displacements.

The results utilized so far to describe the dynamics of the particle are accurate and involve no assumptions [except insofar as the validity of Eq. (7), which posits that sufficient time has elapsed to assure stationarity properties]. However, to evaluate the averages embodied in Eq. (7) we require the statistics of the paths of the particle—which is unknown. Consequently, we invoke the assumption that the statistics of the particle paths are Gaussian with a variance $\sigma^2(t)$ to be determined *self-consistently* from Eq. (7). This assumption is equivalent to the claim that the statistics of the particle trajectories can be determined from the knowledge of the first two moments

---

[3]Since $\langle \phi \rangle = 0$, there is no net drift of the particle.

of the trajectory. Within such an assumption, the probability distribution $P[\mathbf{r}(t)]$ for the displacement of the particle $\mathbf{r}(t)$ can be written

$$P[\mathbf{r}(t)] = \frac{1}{[2\pi\sigma^2(t)]^{d/2}} \exp\left[-\frac{\mathbf{r}^2(t)}{2\sigma^2(t)}\right]. \tag{8}$$

Utilizing the above assumption and Eq. (4) we obtain

$$\langle\langle\nabla\phi(\mathbf{r}(\tau),\tau)\cdot\nabla\phi(\mathbf{r}(0),0)\rangle_\phi\rangle_\mathbf{r}$$
$$\sim \int d^d\mathbf{r}\, P[\mathbf{r}(t)] \int \frac{d^d\mathbf{k}}{(2\pi)^d} \frac{k^2 e^{-i\mathbf{k}\cdot\mathbf{r}} e^{-k^2(u+vk^2+wk^4)t}}{u+vk^2+wk^4}, \tag{9}$$

The quadratures involving $\mathbf{r}$ in the above equation can be explicitly evaluated to yield

$$\langle\langle\nabla\phi(\mathbf{r}(\tau),\tau)\cdot\nabla\phi(\mathbf{r}(0),0)\rangle_\phi\rangle_\mathbf{r} \sim \int k^{d+1} dk \frac{e^{-k^2(u+vk^2+wk^4)t - k^2\sigma^2/2}}{u+vk^2+wk^4}. \tag{10}$$

The above expression can be evaluated asymptotically for $t \to \infty$ by Laplace's method to yield

$$\langle\langle\nabla\phi(\mathbf{r}(\tau),\tau)\cdot\nabla\phi(\mathbf{r}(0),0)\rangle_\phi\rangle_\mathbf{r} \sim \left(\frac{\sigma^2}{2} + ut\right)^{-(d+2)/2}. \tag{11}$$

Using Eq. (7) in conjunction with the above result we obtain,

$$\frac{d}{dt}\left(\frac{\sigma^2}{t}\right) \sim \left(\frac{\sigma^2}{2} + ut\right)^{-(d+2)/2}. \tag{12}$$

If $\sigma^2/t \gg u$ as $t \to \infty$ (to be verified *a posteriorily*), then

$$\frac{d}{dt}\left(\frac{\sigma^2}{t}\right) \sim t^{-(d+2)/2}\left(\frac{\sigma^2}{t}\right)^{-(d+2)/2}, \tag{13}$$

i.e.,

$$\sigma^2 \sim t^{4/(d+4)}. \tag{14}$$

However, this result is not consistent with the assumption employed in reducing the differential equation (12) to Eq. (13). Therefore we set $\sigma^2/t \ll u$ as $t \to \infty$ to obtain

$$\frac{d}{dt}\left(\frac{\sigma^2}{t}\right) \sim t^{-(d+2)/2}, \tag{15}$$

i.e.,

$$\sigma^2 \sim t^{(2-d)/2}. \tag{16}$$

The above self-consistent analysis suggests that the asymptotic dynamics of a particle in the fluctuating potential field satisfies $\sigma^2(t) \sim t^{(2-d)/2}$. Note that the above analysis has neglected the thermal noise $\xi(t)$. If we append the thermal noise to the above dynamics, we obtain $\sigma^2(t) \sim t^{(2-d)/2} + t \sim t$ as $t \to \infty$ ($\forall d > 0$). Therefore, we obtain the surprising result that the thermal noise dominates the diffusion dynamics of the particle in the potential field. Thus, the mean-squared displacement of the particle is expected to be identical to that in the absence of the potential ϕ, and therefore the reaction kinetics is expected to be unmodified from its value in the absence of the potential, viz., $c(t) \sim t^{-d/4}$.

Note, however, that the conclusions drawn are valid only in the asymptotic long-time limit ($t \to \infty$). At short time scales, the particle can be envisioned to execute a random walk in a potential field wherein the term corresponding to $w\nabla^6\phi$ dominates the dynamics of the fluctuations of the potential field. In such a case, one can repeat the above analysis by utilizing the following functional form for the correlations of ϕ

$$\langle \phi(\mathbf{k}, \omega)\phi(\mathbf{k}', \omega') \rangle = \frac{\Delta^2 k^2 (2\pi)^d \delta^d(\mathbf{k} + \mathbf{k}')(2\pi)\delta(\omega + \omega')}{\omega^2 + w^2 k^{12}}. \quad (17)$$

Such an exercise yields $\sigma^2(t) \sim t^{(4-d)/2}$ and therefore $c(t) \sim t^{-d(4-d)/8}$. However, this result is valid only for times t such that $\sigma^2(t) > Dt$ (to dominate thermal diffusion) and $\sigma^2(t) < 1/q_m^2$, where q_m denotes the wavelength at the which the dynamics of the fluctuations of ϕ crosses over to the hydrodynamic form. In situations encountered in practice, this crossover length is of the order of few hundred angstroms,[4] indicating therefore that this regime might be practically unobservable.[5] Thus, mean-field arguments suggest that $c_A, c_B \sim n_0^{1/2} t^{-d/4}$, for $t \to \infty$.

D. RENORMALIZATION GROUP THEORY

Renormalization group theory (RG) provides a natural framework for identifying the behavior of a system at the longest length and time scales [33]. RG enables one to relate the corresponding changes in the magnitudes of the parameters of a model to a change in the length scale of observation. The results of a RG analysis are typically expressed in the form of a flow equation yielding $\psi(l)$, where ψ represents the parameter and l the (variable) length scale of the observation. Synchronous with a rescaling of the length

---

[4] Scattering experiments on bicontinuous microemulsions indicate a peak at a wavelength $q = q^* \sim 10^{-2}$ Å^{-1} [15, 27]. The regime in which the above functional form for the potential is valid corresponds to $q \gg q^*$.

[5] Other factors might also preclude the existence of this regime, for instance, the validity of the Kubo formula for such short time scales.

FLUCTUATION EFFECTS IN MICROEMULSION REACTION MEDIA   133

by a factor b, the time also gets rescaled as $t \to tb^z$, where z is termed the dynamical exponent. The dynamical exponent z embodies the interplay among diffusion, the fluctuations of the potential, and possibly also the reaction. It is evident from straightforward dimensional analysis that the exponent δ in the preceding section is related to z by

$$\delta = \frac{2}{z}. \tag{18}$$

The accompanying discussion in the preceding section then suggests that the asymptotic decay kinetics is of the form $c_A, c_B(t) \sim t^{-d/2z}$.[6] Thus, determining the dynamical exponent z also determines the functional form of the asymptotic decay kinetics. This is the philosophy underlying the analysis we effected. However, in the following we eschew the details of the RG calculations, confining ourselves to a discussion of the resulting flow equations and the manner in which they tie into the mean-field analysis presented. A reader who is interested in the details of the RG calculations is advised to refer to our article [25].

To implement a RG analysis, we need to reformulate the diffusion equations (2) and (3) as a path integral. Such an objective is realized using the generating functional formalism of Martin, Siggia, and Rose (MSR) [34]. In this framework, the field theory corresponding to the above diffusion equations is reexpressed in terms of a generating functional $\mathcal{Z}$. Subsequently, a diagrammatic expansion of the generating functional is employed to discern the functional forms of the renormalized parameters. We employ the (by now) standard momentum space renormalization procedure of Wilson [33, 39]. Subsequently, the length and time scales within the generating functional are rescaled so as to regenerate an action identical in form to the original, with, however, the renormalized parameters replacing the bare (or unrenormalized) parameters. Carrying out this procedure for an infinitesimal rescaling of length enables one to generate (differential) flow equations relating the renormalization of parameters to the renormalization of length.

As indicated at the outset of the RG analysis, our primary interest centers upon the dynamical exponent z. The relevance or irrelevance of the fluctuations of the potential $\phi(\mathbf{x}, t)$ is circumscribed by the explicit value of the dynamical exponent z. If the fluctuations of the potential have no effect on the transport and reaction kinetics of the solute, then the parameter μ in the above equations will flow successively toward smaller values on rescaling, eventually vanishing at the longest length scales ($l \to \infty$). In such a case, we can set $z = 2$, thereby leaving D and u invariant upon rescaling. This would then preserve the kinetics as obtained in the absence of the potential $\phi(\mathbf{x}, t) = 0$. In the following (refer to Ref. 25 for details) we discuss the flow

---

[6] In the case of diffusion in the absence of the potential and reactions, we have $z = 2$, thereby yielding the results in Refs. 29 and 30.

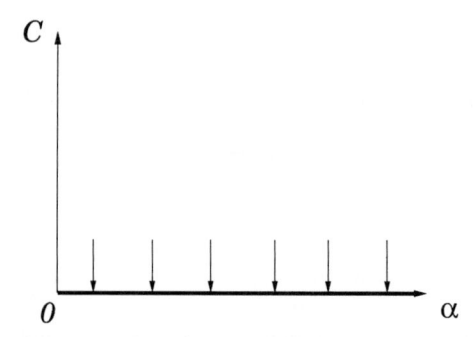

FIG. 2. The phase-space flow diagram corresponding to Eqs. (20) and (21).

equations for n_0, $\alpha = u/D$, and $C = \mu^2 \Delta^2 / D^3 d$, where d denotes the spatial dimensionality of the reaction media

$$\frac{\partial \ln n_0}{\partial l} = d, \tag{19}$$

$$\frac{\partial \ln C}{\partial l} = -d - C \frac{7 - 3\alpha}{2\alpha(1+\alpha)^2}, \tag{20}$$

$$\frac{\partial \ln \alpha}{\partial l} = -C \frac{3}{4(1+\alpha)^2}. \tag{21}$$

The above set of flow equations possesses a line of stable fixed points at $C = 0$ (cf. Fig. 2), implying that at the longest length scales the renormalized system flows toward one wherein C vanishes. This result indicates that the coupling strength β vanishes upon renormalizing, thereby suggesting that the fluctuations of the microemulsion are indeed *irrelevant* in impacting the long-time kinetics of the reaction.

E. DISCUSSION

In this section, we summarize our results and discuss their physical implications on the reaction kinetics in the bicontinuous phases. *The RG analysis confirmed our mean-field arguments and thereby predicts that the dynamic fluctuations of the microemulsion will be irrelevant in impacting the hydrodynamic behavior (long length and time scale) of the system. Thus, the kinetics of the reaction $A + B \to 0$ is expected to follow the anomalous mean-field regime (AMF), one wherein concentrations decay as $c_A, c_B \sim (n_0)^{1/2}(Dt)^{-d/4}$ at long times.* This constitutes the central result of this part of the article.

While this result conforms to our scaling analysis, it nevertheless runs counterintuitive to our physical expectations. For instance, in the extreme limit wherein reactants A and B exhibit a "strong coupling" to the solvents, one might expect the solutes to be confined to the respective phases. In such a scenario, the reaction would occur exclusively at the fluctuating interface. This leads one to expect that confinement of the reaction to a fractal manifold, and the accompanying fluctuations of the interfacial area, would nontrivially impact on the kinetics of the reaction. However, our results in the previous section seem to nullify such an expectation. Some features of our analysis can be identified *a posteriorily* as possibly responsible for this (surprising) result.

(i) The model employed for the reaction and diffusion utilizes a potential field to model the background fluctuations. Our self-consistent mean-field arguments showed that diffusion of the solutes in this "short-ranged" fluctuating potential field is irrelevant compared to random thermal Brownian motion. Indeed, for cases involving long-ranged potentials whose fluctuation statistics exhibit correlations which diverge with the system length, the diffusion of the reactant species exhibits anomalous dynamics, thereby leading to nontrivial reaction kinetics. For instance, reactions effected in a critical fluid can therefore be expected to possess nontrivial decay dynamics.

(ii) Our RG analysis was restricted to the long-time limit. It might, however, be that the fluctuations of the microemulsion do impact on the kinetics at shorter times. Indeed, as demonstrated in Section II.C, such short time regimes do exist but are not discernible by the asymptotic analysis carried out in the present study.

F. SUMMARY

In this part of the article we have considered the effect of the thermal fluctuations of the bicontinuous phase on the reaction kinetics in such phases. We employed a field theoretical ("disordered fluid") model to quantify the structure and the fluctuations of the microemulsion. Subsequently, we analyzed the diffusion and anhilation of a pair of species, A and B, that are preferentially attracted to oil and water. The analysis indicated that, unexpectedly, the fluctuations of the potential are irrelevant in impacting the kinetics in the hydrodynamic limit. This result is counterintuitive to the expectations derived both on physical grounds and based on analogies to other, comparable studies. Subsequently, we also identified some physical reasons possibly responsible for this surprising result. These arguments suggest the imperativeness of treating the "strong-coupling" case, at least within some limits, to discern more carefully the effect of the fluctuating interfaces.

Furthermore, an analysis encompassing the intermediate time scales in addition to the long-time asymptotics would be of value in identifying effects of relevance to practical applications. In Section III, we consider another limiting case of reactions in microemulsions, namely, that involving the droplet phase of the microemulsion. In this case, because of the geometrical simplifications, we are able to account for the entire range possible for the time scales and, also, analyze the strong-coupling limit of the system.

III. Reactions in the Droplet Phase

A. Outline

In this section we provide a brief outline of this part of the article, simultaneously highlighting the assumptions invoked in the subsequent analysis.

As stated earlier, our interest centers on bimolecular reactions of the type $A + B \to \emptyset$, wherein the solute A resides in water and B in oil—necessitating that the anhiliation occurs exclusively at the fluctuating surfaces of the droplets. However, in contrast to the analysis detailed in Section II, it does not appear possible simultaneously to render an exact analytical treatment of this problem and to realize the objectives stated above. Therefore, to focus on the elusive issues pertaining to the effect of fluctuations, we restrict our considerations to a simplified scenario wherein the reaction proceeds at an infinitely fast rate, thereby enabling the droplets to serve as a *perfect* sink (albeit one which is fluctuating). In such a case, it suffices to focus on the reaction $A \to \emptyset$ occurring on the surface of the fluctuating droplets. In addition, a boundary condition requiring the concentration of A to vanish at the surface of the fluctuating droplets imposes the "perfect-sink" nature of the fluctuating interfaces. Within this framework, we analyze the diffusion-limited reactive transport processes of solute A.

We initiate our analysis in Section III.B by providing a brief overview of the dynamical aspects of fluctuations of the droplets of the microemulsion. These features have been analyzed in seminal studies by Safran and Milner [23, 24], allowing us thereby to extract the requisite results. Section III.C commences our analysis by considering the case wherein the concentration of droplets can be termed dilute, thereby enabling us to restrict our considerations to the transport and reaction on a *single* fluctuating sink. In relation to the scenario encountered in practice, such an assumption can be deemed reasonable since the droplet phase of the microemulsion occurs only at very low concentrations of oil (or water) [12]. The ensuing "single-sink" analysis can be construed as a generalization of the Smoluchowski's problem of reaction and diffusion on a sink [36] to one wherein the sinks are allowed to fluctuate. However, even within this simplified case, an exact analytical solution proves elusive, and so we are forced to restrict our analysis to the case wherein the fluctuation amplitudes are small in relation to the mean radius of the droplet. In such a scenario, we can employ (boundary) perturbation techniques [37] to effect the solution of the corresponding equations. The explicit results of such an analysis constitute the main feature of this part of the article. Based on these results, we identify different regimes corresponding to intermediate and long time scales and thereby answer some of the questions left unattended in Section II. In closing, we also briefly highlight the manner in which some of the assumption invoked in this article can be relaxed.

Consistent with the philosophy adopted in II, we eschew elaborating the manifold algebraic details. A reader who is interested in the details of the solution may want to refer to our previous article [25].

B. Fluctuations of the Droplet Phase

In this section, we briefly outline the model adopted for quantifying the fluctuations of the microemulsion droplets. The statistics of these dynamical fluctuations have been studied by Milner and Safran [24] (cf. Safran [23] for an exposition of the equilibrium aspects) in the dilute limit. They addressed the linearized hydrodynamics of the fluctuations of the droplets in the creeping flow limit assuming that the fluids internal and external to the droplets possess identical viscosities. The fluctuations of the microemulsions droplets arise from the presence of an excess area over and above that of a sphere of equivalent volume. In the case wherein one considers an isolated droplet in solution (i.e., at dilute concentrations and for time scales during which the collisions and exchange of material between the droplets can be ignored), this excess area ΔA remains a specified constant during the fluctuations of the droplet. However, imposing this constraint on the excess areas within a dynamical formulation proves cumbersome, and *so a Lagrange multiplier γ is invoked to account for this constraint.* Further, in the case of microemulsion droplets the statistics of the excess areas possessed by a droplet abstracted from the equilibrium solution (of droplets) can be discerned based on equilibrium energetic considerations [24]. Thus, one can identify the (constant) excess area possessed by our fluctuating droplet with mean excess area $\langle \Delta A \rangle$ possessed by a droplet in the microemulsion solution. This enables one to determine the explicit value of this Lagrange multiplier γ based upon the equilibrium statistics of the fluctuations. Such an exercise has been effected in Ref. 24, to which we refer the reader interested in the pertinent details.

In the following, we quote the results derived in Ref. 24 for the statistics of the fluctuations of the droplets.

In this and the following analysis, we assume that the microemulsion droplets possess a mean radius R. However, this mean radius can in general be different from the spontaneous radius of these droplets R_s. The instantaneous shape of the droplet can be parametrized as $r = R'(\theta, \phi, t)$, with R' expressed in terms of an expansion in spherical harmonics as

$$R'(\theta, \phi, t) = R\left[1 + \sum_{l>1,m} \hat{a}_{lm}(t) Y_{lm}(\theta, \phi)\right], \quad (22)$$

with $\hat{a}_{lm}(t)$ representing the amplitudes of the different modes of fluctuation. The fluctuation mode corresponding to $l = 0$ is restrained by the conservation requirement imposed upon the volume of the droplets and, so, is not included in the above expression [24]. Further, the mode corresponding to $l = 1$ signifies spatial translations of the center of the droplet and is, therefore, irrelevant for considerations pertaining to the fluctuations of a single sphere. For concisencess, we henceforth resort to the notation $\sum_{lm}$ to denote $\sum_{l>1,m}$. In the above expression, Y_{lm} denotes the spherical harmonic function [38] of order l,m, which is assumed in this work to be appropriately normalized such that

$$\int_0^{2\pi} d\phi \int_0^{\pi} \sin\theta d\theta\, Y_{lm}(\theta, \phi) Y_{pq}(\theta, \phi) = \delta_{lp}\delta_{mq},$$

to obviate bookkeeping cumbersome algebraic factors.

Within the framework of the above parameterization, the statistics of the fluctuation amplitudes $\hat{a}_{lm}(t)$ was determined in Ref. 24 as

$$\langle \hat{a}_{lm}(t) \rangle = 0; \quad \langle \hat{a}_{lm}(t) \hat{a}_{l'm'}(0) \rangle = \delta_{ll'}\delta_{mm'} \exp(-\omega_{lm} t) M_{lm}, \quad (23)$$

where the static mean-squared fluctuation amplitude, denoted M_{lm}, is of the functional form,

$$M_{lm} = \frac{k_B T}{\kappa} = \{(l+2)(l-1)[l(l+1) - 4u + 2u^2 - \gamma]\}^{-1} \quad (24)$$

with κ representing the bending moduli of the surfactant interfaces and u denoting the ratio R/R_s. The term γ in the above expression denotes the Lagrange multiplier discussed in the preceding paragraph. The corresponding frequencies of fluctuations of the different modes ω_{lm} have been determined as

$$\omega_{lm} = \frac{\kappa}{\eta R^3} \frac{[l(l+1) - \gamma - 4u + 2u^2][l(l+1)(l+2)(l-1)]}{(2l+1)(2l^2 + 2l - 1)} \quad (25)$$

where η denotes the viscosity of the surrounding medium.

C. Diffusion Equation and Perturbation Expansion

In this section we commence our analysis of the diffusion-limited reaction of solute A. As expounded in Section III.A, we initially restrict our considerations to the scenario wherein the concentration of the droplets can be construed as dilute. In such a case, it suffices to focus on the diffusion and reaction of A in the presence of a single fluctuating sink. In view of the spherical symmetry exhibited by the problem, we formulate the transport and reaction of A in terms of a diffusion equation expressed in spherical coordinates (r, θ, ϕ)

$$\frac{\partial c}{\partial t} = D\left[\frac{1}{r^2}\frac{\partial}{\partial r}\left(r^2\frac{\partial c}{\partial r}\right) + \frac{1}{r^2 \sin\theta}\frac{\partial}{\partial \theta}\left(\sin\theta \frac{\partial c}{\partial \theta}\right) + \frac{1}{r^2 \sin^2\theta}\frac{\partial^2 c}{\partial \phi^2}\right], \quad (26)$$

In the above equation we use the symbols c and D to denote, respectively, the concentration and the diffusivity of A. The above conservation equation is to be supplemented by the boundary condition embodying the "perfect-sink" nature of the fluctuating surfaces [cf. Eq. (22)]:

$$c = 0 \quad \text{on} \quad r = R[1 + \sum_{lm} \hat{a}_{lm}(t) Y_{lm}(\theta, \phi)], \quad (27)$$

Further, for convenience, we consider the scenario wherein we maintain a fixed source of the solute A at $r = \infty$, i.e.,

$$c(r, \theta, \phi) = \bar{c} \quad \text{as} \quad r \to \infty \quad (28)$$

and

$$c(r, \theta, \phi, t = 0) = \bar{c}. \quad (29)$$

The conservation equation (26) supplemented by the boundary conditions (27) and (28) represents the mathematical formulation of the diffusion and reaction of solute A in the presence of fluctuating microemulsion droplets in the limit of dilute concentration of these droplets.

An exact analytical solution of the above equations does not appear possible due to the complicated boundary condition (27). However, in the limit wherein the fluctuation amplitudes are low in relation to the radii of the droplets, i.e.,

$$\sqrt{\frac{k_b T}{\kappa}} \equiv \epsilon \ll 1, \quad (30)$$

we can obtain an explicit solution for the solute concentration c quantified in the form of a perturbation series with fluctuation amplitudes ϵ serving as the perturbation parameter. Effecting a transformation of the variables

$\hat{a}_{lm} = a_{lm}\epsilon$ to render the magnitudes of the fluctuation amplitudes explicit enables us to look for a solution of the form

$$c(r, \theta, \phi, t) - \bar{c} = \sum_{i=0}^{\infty} \epsilon^i c_i(r, \theta, \phi, t), \tag{31}$$

The diffusion equations and the corresponding boundary conditions satisfied at each order of perturbation by the different c_i's can be obtained by substituting the above expansion (31) into Eqs. (26)–(28). Subsequently, the solution at each order of the series can be determined sequentially based on the solutions at the preceding orders.

In lieu of the detailed spatial and temporal information embodied within the concentration field $c(r, \theta, \phi, t)$, it is of interest to compute a global temporal measure of the reaction kinetics so as to discern the explicit effect of the fluctuations. Consequently, in addition to the concentration field c, we also compute the averaged time-dependent consumption rate of A, denoted $J(t)$ and defined as

$$J(t) = \frac{d}{dt}\left\langle \int_V d^3\mathbf{r}\, c(r, \theta, \phi, t) \right\rangle, \tag{32}$$

In the above equation, V denotes the exterior volume enclosed between the surface of the fluctuating droplet and $r = \infty$. In contrast to the explicit spatial dependence embodied in the concentration field c, $J(t)$ embodies a purely temporal characteristic that manifests the spatial features of the concentration field in an averaged manner. Thereby, a comparison of the values of $J(t)$ provides a convenient measure for quantifying the explicit dynamical effects of the fluctuations of the sinks. One can transform the above expression by expanding the above integral and the concentration field c as a power series in ϵ, thereby enabling us to utilize the solutions of the diffusion equation at each order.

In the following we do not elaborate the details of the solution procedure but, instead, quote our final results [up to $\mathcal{O}(\epsilon^2)$] for the quantity $J(t)$:

$$J(t) = J_0(t) + \epsilon J_1(t) + \epsilon^2 J_2(t) + \mathcal{O}(\epsilon^3), \tag{33}$$

where

$$J_0(t) = 4\pi D R\bar{c}\left(1 + \frac{R}{\sqrt{\pi Dt}}\right). \tag{34}$$

$$J_1(t) = 0, \tag{35}$$

and

$$J_2(s) = \bar{c}RDg(R, s) - \frac{3\chi\bar{c}R^3}{2} + \bar{c}R\sqrt{\frac{s}{D}}\left[Rg(R, s)D - \frac{\chi R^3}{2}\right], \tag{36}$$

where

$$g(R, s) = \frac{1}{2}\left\{\left(\frac{1}{s} + \frac{R}{\sqrt{sD}}\right)\left(\chi + \sum_{lm} M_{lm} p(s)\frac{K_{l+3/2}[p(s)] + K_{l-1/2}[p(s)]}{K_{l+1/2}[p(s)]}\right) - \chi\left(\frac{2}{s} + \frac{2R}{\sqrt{sD}} + \frac{R^2}{D}\right)\right\} \qquad (37)$$

and $J_2(s)$ represents the Laplace transform of $J_2(t)$.

Two important features can be discerned from the above expressions.

1. In the limit $t \to \infty$, the expression for $J_0(t)$ corresponds to the classical Smoluchowski expression for the (diffusion-limited) reaction rate expression in the presence of a dilute concentration of rigid spherical sinks [36]. It is therefore evident that the steady-state limit of the higher-order terms in the expansion to (33) provides the corrections to the reaction rate arising from the dynamic fluctuations of the sink.

2. The main feature of the above expression (37) is the absence of secular (or diverging) temporal corrections to the zeroth-order field. The fastest-diverging term in the above expression for $s \to 0$ evolves as $1/s$, indicating that the presence of the dynamic fluctuations influences the zeroth-order field only in a perturbative manner at the longest time scales. This result is consistent with the prediction outlined in Section II, wherein we claimed that in the hydrodynamic regime (long times) the dynamics (quantified in terms of the decay exponent of the reactants) is unmodified by the presence of the dynamic fluctuations. However, the fact that we have an exact expression (albeit in a perturbative sense) for the concentration field, supposedly valid at all times, enables us to discern the explicit effect of the fluctuations at intermediate time scales. Such effects might prove of relevance to applications wherein the "long-time" limit might in fact prove unachievable in practice. In the next section, we explicitly analyze the behavior of the consumption field $J(t)$ in different temporal regimes to discern the specific effect of the dynamic fluctuations of the microemulsion droplet.

D. Consideration of Temporal Regimes

As argued earlier in the text, $J_2(t)$ serves as an appropriate spatially independent measure of the impact of the dynamical fluctuations on the kinetics of the reaction. The absence of secular terms in $J_2(t)$ [cf. Eq. (36)] enables us to treat it as a purely perturbative correction to $J_0(t)$. In such a case, we can discern the impact of the dynamical fluctuations by analyzing the temporal features of $J_2(t)$, with the implicit specification that ϵ is small enough to maintain the perturbative nature of the corrections.[7] The following discussion is devoted to effecting such an analysis, deferring a discussion of the implications to the next section.

To focus on the main dynamical features, in the following analysis we concentrate on the mode $l = 2$ and ignore considerations arising from the presence of other modes $l \neq 2$. This assumption does not modify the qualitative features of the results subsequently unearthed. Further, our assumption is physically justified based on the results of Safran [23], who showed that the fluctuations of the mode $l = 2$ possesses the maximal amplitude (represented by M_{lm}). In this limit we can set $\sum_m M_{2m} = \chi$ and further use the unadorned notation ω to denote ω_{2m}. Furthermore, within the same approximation, $\epsilon^2 \chi = \langle \Delta A \rangle / R^2$, where $\langle \Delta A \rangle$ represents the mean excess area (in an ensemble of microemulsion droplets) over and above a perfect sphere of radius R.

1. Steady State

The steady-state limit of expression (36) can be obtained by adopting the dual limit $s \ll \omega$ and $s \ll D/R^2$. Then

$$p(s) = R\sqrt{\frac{\omega}{D}} \tag{38}$$

whose explicit magnitude is $\ll 1$ in view of our assumption of small Péclet numbers (refer to Section I). Simplifying the resulting expression for $J_2(t)$, we obtain

$$J_2(s) \longrightarrow \frac{2\bar{c}RD\chi}{s}, \tag{39}$$

As indicated in Section III.C, the steady-state value of $J(t)$ embodies the (Smoluchowski's) diffusion-limited reaction rate. On appending the above contribution to (34), we obtain the leading order corrections to the diffusion-limited reaction rate arising from the dynamical fluctuations of the spherical sinks

$$\begin{aligned} J(t \to \infty) &= 4\pi D R \bar{c} \left(1 + \epsilon^2 \frac{\chi}{2\pi}\right) \\ &= 4\pi D R \bar{c} \left(1 + \frac{\langle \Delta A \rangle}{2\pi R^2}\right), \end{aligned} \tag{40}$$

---

[7] If $J_2(t)$ had, in contrast, possessed secular terms, then as $t \to \infty$ it would be impossible to choose an ϵ small enough to maintain the perturbative nature of the correction—necessitating a renormalization procedure.

FLUCTUATION EFFECTS IN MICROEMULSION REACTION MEDIA 143

We conclude from the above expression that the dynamical fluctuations always *enhances* the reaction rate by a magnitude proportional to the excess area contained in the droplets. While this result may seem logical in view of the enhanced flux resulting from the excess area, it is to be noted here that in deriving the result we employed the assumption $R\sqrt{\omega/D} \ll 1$. In fact, in the next section, we consider the opposite limit corresponding to $R\sqrt{\omega/D} \gg 1$, wherein the steady-state correction embodies nontrivial contributions dependent on $R\sqrt{\omega/D}$.

E. Intermediate Times

In view of the assumption we have invoked regarding the magnitude of the Péclet number (requiring $\omega R^2/D \ll 1$), the intermediate time regime corresponds to times much longer than the diffusive time scale of the solute species but still short compared to the time scales corresponding to the fluctuations of the droplets, i.e., $\omega \ll s \ll D/R^2$. In such a case,

$$p(s) = R\sqrt{\frac{s}{D}} \ll 1. \tag{41}$$

Simplifying the resulting expression for $J_2(t)$, we obtain

$$J_2(s) \longrightarrow \frac{2\bar{c}RD\chi}{s}. \tag{42}$$

The above expression is identical to that obtained in the steady-state limit (39). This indicates that the diffusion-limited reaction processes reaches a steady state in the time scale comparable to the diffusion of the solute, despite the fact that the fluctuations of the sink have not relaxed within such times.

F. Short Time Regime

For times s such that $s \gg D/R^2$ and $s \gg \omega$, neither the diffusion of the solute nor the fluctuations of the sink have relaxed. In such a case, the leading order contribution to J_2 can be obtained by imposing this dual limit on Eq. (36). This yields

$$J_2(s) \longrightarrow -\chi R^4 \bar{c}\sqrt{\frac{s}{D}}. \tag{43}$$

Thus we have,

$$J(s) \sim 4\pi D R \bar{c} \frac{R}{\sqrt{sD}} - \langle \Delta A \rangle R^2 \bar{c}\sqrt{\frac{s}{D}}. \tag{44}$$

The above result suggests that the dynamical fluctuations possess a nontrivial effect on the kinetics of consumption of solute A in the initial time regime. Explicitly, the perturbative correction to the consumption rate J evolves as $1/t^{3/2}$, in contrast to the $1/t^{1/2}$ behavior embodied in $J_0(t)$ [cf. Eq. (34)]. However, it is to be noted that to maintain the perturbative nature of the correction, ϵ should be small enough. In other words, the above result is valid only for times $t > t^*$, where

$$t^* \sim \frac{\langle \Delta A \rangle}{D}. \tag{45}$$

G. Effect of the Péclet Number

So far, the entirety of our analysis has invoked the assumption corresponding to small Péclet numbers wherein it is justified to ignore the effect of the convective transport altogether. Quantitatively this assumption required that the time scales of fluctuation and diffusion satisfy the following criterion: $\omega R^2/D \ll 1$. However, practically encountered situations might not strictly satisfy this assumption.[8] It is therefore of interest to discern the explicit effect of the fluctuations in the opposite limit, wherein $\omega R^2/D \gg 1$. In this case, if instead of solving the convective-diffusion equation (as necessitated by the magnitude of the Péclet number and the fact that the problem becomes one involving singular perturbation in the Péclet number), we assume that the above expressions (36) are still applicable, then we can repeat the analysis outlined in the preceding section under limits consistent with the limit $\omega R^2/D \gg 1$. However, such an analysis is not rigorously valid and hence the results so obtained must be interpreted with caution and at a qualitative level only. The results of such an analysis indicate the following.

(i) The steady-state dynamics embodied in (39) is unchanged qualitatively. However, the explicit correction to the steady-state reaction rate now scales as $\langle \Delta A \rangle \sqrt{\omega/R^2 D}$. Therefore, it can be inferred that while the correction is proportional to the excess area (as is intuitive), it also embodies nontrivial contributions arising from the ratio of the time scales for diffusion and fluctuation.

(ii) The intermediate time regime in the limit $s \ll \omega$ and $s \gg D/R^2$ (times much before the equilibration of the diffusive transport) contrasts with the dynamics in the above-outlined intermediate time regime. The explicit functional form of the kinetics is a bit more

---

[8] An order-of-magnitude estimate suggests that the Péclet number is $\mathcal{O}(1)$ for realistic situations.

complicated and depends on the magnitudes of the ratios ω/s and $\omega R^2/D$. In the case where the former ratio is of a greater magnitude (shorter times), then the consumption rate exhibits anomalous kinetics similar to that unearthed in the short time regime of the preceding section. In the opposite limit, wherein $\omega R^2/D \gg \omega/s$, the perturbative correction matches the functional form of $J_0(t)$ and exhibits a $1/t^{1/2}$ behavior.
(iii) The short-time behavior ($s \gg \omega, D/R^2$) matches qualitatively and quantitatively the expressions obtained in the previous section [cf. Eq. (44)].

H. Discussion

The preceding sections have outlined the analysis of the transport and reaction of the solute in the presence of fluctuating spherical sinks (representing the droplets of microemulsion). In the following, we summarize the main results and discuss their implications in the context of Section II. The analysis of the leading order corrections [$\mathcal{O}(\epsilon^2)$] to the concentration field and the consumption rates leads to the following conclusions.

1. *The long-time dynamical evolution of the consumption rate is unaffected by the presence of the dynamical fluctuations of the sink. However, the explicit magnitude of the reaction rate is enhanced by the fluctuations.* This result is consistent with our prediction in Section II that the dynamics (or decay exponent) of the reaction is unaffected by the fluctuations of the interfaces. However, the correction to the reaction rate was shown to embody nontrivial corrections arising from the fluctuations of the droplets. The occurrence of a steady state and the dynamical features embodied in the intermediate and short time scales (see below) suggest that the exponentially decaying (in time) correlation structure possessed by the fluctuations of the microemulsion media is possibly responsible for this result. This leads us to speculate that if the correlation structure of the fluctuation amplitudes possesses no intrinsic time scale (i.e., exhibits a power-law correlation behavior), then the long-time kinetics of the microemulsion will possess a nontrivial secular structure manifesting the fluctuation dynamics.

2. *The steady state was reached in a time scale set by the diffusion of the solute species.* This result was a manifestation of the low Péclet number limit assumed to be applicable in our physical situation. As indicated in the previous section, the opposite limit ($\omega R^2/D \gg 1$) involves a different dynamical evolution in the intermediate time regime.

3. *The correction to the consumption rate in the short time regime (corresponding to $t \ll R^2/D$) embodies an anomalous kinetic behavior.* This result is again consistent with our original hypothesis that the short-time dynamics might manifest a profound influence arising from the fluctuations of the interface. In fact, this result might prove of significance to practical applications wherein the steady-state or the long-time limits prove "too long" to realize.

I. Other Effects

The above discussion summarizes the main results obtained in the previous sections, emphasizing their relationship to the results of the preceding article which motivated this analysis. However, as motivated in the text, study of reactions in the droplet and other ordered lyotropic phases possesses enough practical applications to justify an independent study, and therefore we indicate briefly some effects which need to be accounted for in future work to impact on such applications.

1. The main assumption involved in the preceding analysis involves envisioning the droplet phase as dilute and, therefore, eschews considerations such as collisions and the exchange of material arising between droplets.
2. Competition effects between the sinks might manifest even in the absence of a translational mobility.
3. At higher concentrations of sinks, one might also expect shape transitions to ordered surfactant phases of microemulsion.
4. Finally, our analysis was restricted to the case wherein the amplitudes of fluctuation were low.

In all of the preceding contexts, it might prove expedient to employ a different solution technique to the analysis of the situation. The path integral technique of Kardar and co-workers appear promising in this regard [39].

J. Summary

In Section III, we have analyzed a simple model for discerning the impact of the fluctuations of the droplet phase of the microemulsion on the reaction kinetics in such media. The results of this analysis are consistent with those outlined in Section II. We have delineated the different temporal regimes to identify the explicit effect of the dynamical fluctuations of the sink. Finally, we have also elucidated some possible directions for future study and our own preliminary results in that context.

Our analyses in Sections II and III highlight some of the novel issues arising from the interplay between thermal fluctuations and diffusion-limited reactions in complex fluid media—an issue unaddressed heretofore in the literature. Most of the results unearthed herein lend themselves to experimental verification. For future work, it will be of interest to study the effects arising from relaxing the assumptions invoked in this study (see the preceding section).

This article was meant to highlight for the chemical engineering community the "nonclassical" issues pertaining to reaction kinetics and some of the statistical mechanical tools employed to resolve them. To maintain brevity we eschewed detailing the intermediate steps of the analysis. A reader who is interested in the accompanying details is advised to refer to Ref. 25.

ACKNOWLEDGMENTS

We are grateful to Drs. R. Golestanian, S. Ramanathan, and F. Drolet for useful input and encouragement. This work was supported by the National Science Foundation under Award DMR-9870785.

REFERENCES

1. M. Doi, *J. Phys. A* **9,** 1465 (1976).
2. L. Peliti, *J. Phys. A* **19,** L365 (1986).
3. B. P. Lee, *J. Phys. A* **27,** 2633 (1994).
4. J.-M. Park and M. W. Deem, *Phys. Rev. E* **57,** 3618 (1998).
5. J.-M. Park and M. W. Deem, *Eur. Phys. J. B* **10,** 35 (1999).
6. W. J. Chung and M. W. Deem, *Physica* **A265,** 486 (1999).
7. D. Zhao, J. Feng, Q. Huo, N. Melosh, G. H. Fredrickson, B. F. Chmelka, and G. D. Stucky, *Science* **279,** 548 (1998).
8. P. L. Luisi, "Kinetics and Catalysis in Microheterogeneous Systems." Marcel Dekker, New York, 1991.
9. M. P. Pileni (ed.), "Structure and Reactivity in Reversed Micelles." Elsevier, Amsterdam, 1989.
10. R. Johannsson, M. Almgren, and R. Schomacker, *Langmuir* **9,** 1269 (1994).
11. M. A. Lopezquintela and J. Rivas, *J. Coll. Int. Sci.* **158,** 446 (1993).
12. W. M. Gelbart, A. Ben-Shaul, and D. Roux (eds.), "Micelles, Membranes, Microemulsions and Monolayers." Springer-Verlag, New York, 1994.

13. J. Israelachvili, "Intermolecular and Surface Forces." Academic Press, San Diego, 1991.
14. S. A. Safran, "Statistical Thermodynamics of Surfaces, Interfaces, and Membranes." Addison–Wesley, Reading, MA, 1994.
15. G. Gompper and M. Schick, in "Phase Transitions and Critical Phenomena." Vol. 16 (C. Domb amd J. Lebowitz, Eds.), Academic, London, 1994.
16. P. G. deGennes and C. Taupin, *J. Phys. Chem.* **86,** 2294 (1982).
17. K. M. Lusvardi, K. V. Schubert, and E. W. Kaler, *Ber. Bunsen. Phys. Chem.* **100,** 373 (1996).
18. D. Walsh, J. D. Hopwood, and S. Mann, *Science* **264,** 1576 (1994).
19. L. Garcia-Rio, J. R. Leis, and J. C. Mejuto, *J. Phys. Chem.* **100,** 10981 (1996).
20. A. V. Barzykin and M. Tachiya, *Phys. Rev. Lett.* **73,** 3479 (1994).
21. C. Oldfield, *J. Chem. Soc. Faraday Trans.* **87,** 2607 (1991).
22. P. Argyrakis, G. Duprtail, and P. Lianos, *J. Chem. Phys.* **95,** 3808 (1991).
23. S. A. Safran, *J. Chem. Phys.* **78,** 2073 (1983).
24. S. T. Milner and S. A. Safran, *Phys. Rev. A* **36,** 4371 (1987).
25. V. Ganesan and G. H. Fredrickson, *J. Chem. Phys.* **113,** 2901 (2000).
26. G. Gompper and S. Zschocke, *Phys. Rev. A* **46,** 4836 (1992).
27. M. Teubner and R. Strey, *J. Chem. Phys.* **87,** 3195 (1987).
28. P. C. Hohenberg and B. I. Halperin, *Rev. Mod. Phys.* **49,** 435 (1977).
29. D. Toussaint and F. Wilczek, *J. Chem. Phys.* **78,** 2642 (1983).
30. K. Kang and S. Redner, *Phys. Rev. Lett.* **52,** 955 (1984).
31. X.-H. Wang and K.-L. Wang, *Phys. Rev. E* **49,** 5853 (1994).
32. X.-H. Wang and K.-L. Wang, *Mod. Phys. Lett. B* **9,** 801 (1995).
33. D. J. Amit, "Field Theory, the Renormalization Group, and Critical Phenomena." World Scientific, Singapore, 1984.
34. P. Martin, E. D. Siggia, and H. A. Rose, *Phys. Rev. A* **8,** 423 (1973).
35. M. Kardar, Statistical mechanics of fields, unpublished notes (1998).
36. M. V. Smoluchwoski, *Phys. Z.* **17,** 557 (1916); *Phys. Z.* **17,** 585 (1916); *Z. Phys. Chem.* **92,** 129 (1917).
37. H. Brenner, *Chem. Eng. Sci.* **19,** 519 (1964).
38. P. M. Morse and H. Fesbach, "Methods of Theoretical Physics, Part I." McGraw–Hill, New York (1953).
39. R. Golestanian and M. Kardar, *Phys. Rev. A* **58,** 1713 (1998).

MOLECULAR DYNAMICS SIMULATIONS OF ION–SURFACE INTERACTIONS WITH APPLICATIONS TO PLASMA PROCESSING

David B. Graves and Cameron F. Abrams

Department of Chemical Engineering, University of California–Berkeley, Berkeley, California 94720

I. Introduction	149
A. Plasma Processing	149
B. Length Scales in Plasma Processing	152
C. The Nature of Plasma–Surface Interactions	153
D. Ion–Surface Interactions in Plasma Processing	155
II. Use of Molecular Dynamics to Study Ion–Surface Interactions	156
A. Simulation Procedure	156
III. Mechanisms of Ion-Assisted Etching	161
A. Experimental Studies of Ion-Assisted Etching Mechanisms	161
B. Molecular Dynamics Studies of Ion-Assisted Etching Mechanisms	164
C. Ion–Surface Scattering Dynamics	172
D. Ion–Surface Interactions with both Deposition and Etching: CF_3^+/Si	180
IV. Concluding Remarks	198
References	199

I. Introduction

A. PLASMA PROCESSING

The topic of this article is the use of molecular dynamics (MD) simulations of positive ion–surface interactions for insights into the chemical and physical processes that occur at surfaces immersed in glow discharge plasmas. To understand the significance of this topic, it is necessary to have some background in the technology and its major industrial application (Lieberman and Lichtenberg, 1994). The term "plasma" in this context refers to an ionized gas. Irving Langmuir is generally credited with first using the term to refer to ionized gases. The study of the physics of ionized gases has received a great deal of attention since then (Chen, 1984). Plasma physics is indeed a large and well-established field, but it generally deals with the fully ionized, usually very hot and magnetized plasmas that occur in stellar environments or in thermonuclear fusion. Fully ionized, magnetized plasmas are subject to many instabilites, and turbulence is often important. The focus of much of plasma physics research is on creating and stabilizing very hot, fully ionized plasmas. Controlled thermonuclear fusion for power generation has been limited by the difficulties in confining the plasma at sufficiently high temperatures for a sufficient length of time. The types of plasmas used for the semiconductor materials processing applications are quite different from the plasmas commonly associated with the term "plasma physics." It should be noted, however, that when the plasmas used for fusion interact with confining walls, some of the same considerations that can dominate semiconductor processing are present (Post and Behrish, 1986). Compared to plasma physics, the field of plasma materials processing has generally received much less attention, at least in terms of the fundamental science.

In plasma processing of semiconductors, the plasmas are usually highly nonequilibrium, and the gas is only weakly to partially ionized. It is understood, of course, that the term "semiconductor" refers not only to semiconducting materials but also to the conductors and dielectrics used in semiconductor devices. The term "nonequilibrium" here means that the charged species (electrons and ions) generally have a much higher average kinetic energy than the neutral species. A partially ionized gas is a mixture of neutral and charged species. The plasmas are usually created by application of electric and magnetic fields to a relatively low-pressure gas. In most cases, the gas pressure is between a few millitorrs and several torrs. The fields can range from dc to microwave frequencies. Under the right conditions the neutral, mostly insulating gas can be "broken down" and becomes conductive—due mainly to the creation of a sufficient concentration of free electrons in the gas. By passing currents through the gas, the charged species are heated, thereby sustaining the plasma. The peak neutral gas temperature can vary from near room temperature to over 2000 K. The average electron kinetic energy can approach 10 eV/electron (equivalently, over 100,000 K), and positive ions near the boundary of the plasma can reach energies of the same order. Ions accelerated by the relatively large electric fields in the electrical boundary layer of the plasma (i.e., the "sheath") can reach an energy as high as the applied voltage before impacting the surface. Typical conditions in plasma processing are listed in Table I. The reason for the big difference between neutral and charged species energies is that charged species are lost to walls faster than they are able to transfer to neutrals via collisions the energy

TABLE I
Typical Conditions in Nonequilibrium Plasmas

Gas pressure	1–1000 mTorr
Gas temperature	300–2000 K (0.03–0.15 eV/atom)
Degree of ionization ($N_{plasma}/N_{neutral}$)	10^{-6}–10^{-2}
Average electron energy	1–10 eV
Average ion energy in plasma	0.05–1 eV
Average ion energy impacting surface	10–1000 eV
Typical plasma dimension	0.1–1 m

they receive from the applied fields. At higher neutral gas pressures, this is no longer the case, and the neutral gas can become quite hot. A common example is arc welding.

Plasma processing of semiconductors relies on a combination of chemical and physical processes at surfaces exposed to the plasma. Figure 1 is a sketch of a typical plasma chamber used for processing the surface of a silicon wafer. In the example in Fig. 1, the plasma is excited using an external "stovetop" coil, powered using a radiofrequency current (RF1). Since the RF current in the external coil induces a current in the plasma, the power coupling is "inductive." The wafer has a separate, capacitive RF power applied to it (RF2), which allows independent control of ion energy impacting the surface.

Wafers are usually processed one at a time in the chamber. Gas flows into the chamber and is pumped into the exhaust. External power is applied electrically and the plasma is generated and sustained. The plasma modifies the surface of the wafer in two basic ways. First, the inlet gases are dissociated via electron-impact molecular dissociation into reactive free radicals

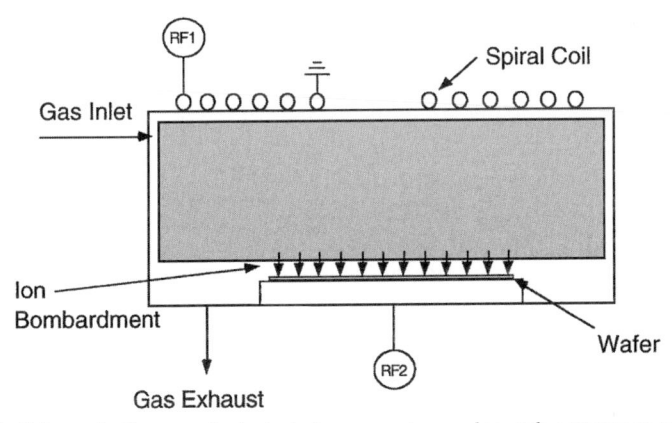

Fig. 1. Schematic diagram of a typical plasma reactor used to etch patterns on wafers.

(e.g., F, CF, CF_2, from CF_4). These radicals generally adsorb at all surfaces within the chamber, including the wafer. In addition, energetic positive ions, accelerated by the strong electric fields at the boundary of the plasma, impact surfaces. The combination of adsorbed radicals and ion impact promotes surface chemical reaction at low temperatures. Relatively low surface temperatures are a major advantage of plasma processes when processing temperature-sensitive microelectronic devices. Material can be removed, in which case the plasma etches the surface, or material can be deposited, in which case the plasma deposits a thin film. In some cases, plasmas are used to simply modify the surface functional groups or remove surface debris. Plasmas are commonly used for all of these applications in semiconductor manufacturing.

B. Length Scales in Plasma Processing

A proper understanding of plasma processing and semiconductor device manufacturing requires a consideration of the range of length scales involved. Semiconductor products such as microprocessors or memory chips are discrete devices and are manufactured on silicon wafers. Each wafer is processed to make a set of nominally identical chips, or "die." The diameter of the wafer for leading edge semiconductor factories, or fabs, has steadily increased, from 50 mm in 1970 to 300 mm in 2001. The "wafer scale" determines the size of the equipment, typically of the order of a meter. However, there are several other natural length scales: the scale of the chip or die (~1 cm), the scale of the critical dimensions within the device (~0.1 μm), and the scale of the dimensions that can affect device properties (~1 Å). The chip scale is important since the transistors must be nominally identical across the entire chip. The other two scales affect the behavior of an individual transistor. As in many other areas of chemical processing, treating multiple length scales can be challenging.

Chips are generally made up of sets of interconnected solid state transistors, or switches. The solid-state transistors, most commonly metal–oxide–semiconductor field effect transistors (MOSFETs), consist of source, gate, and drain. In manufacturing these devices, a complex set of processes is used, involving lithographic pattern transfer, implantation, diffusion, thin-film growth, deposition, polishing, etching, and surface cleaning. Over 300 individual steps are typically used in current leading edge device manufacturing processes (Wolf and Tauber, 2000). The characteristic dimensions of these features are now of the order of 100 nm or 0.1 μm. In etching, one must ensure that features are controlled to within this dimensional constraint. For example, the current nominal width of the gate electrode for a MOSFET is

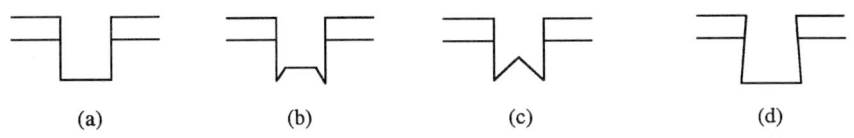

FIG. 2. Examples of several etch profiles: (a) near-straight wall; (b) microtrenches; (c) broad microtrenches; (d) reentrant shape.

about 200 nm, but this width must be controlled to within about ±10 nm. Control of dimensions on the feature length scale is one of the most important tasks faced by equipment and process engineers.

During plasma etching (or deposition in some cases), ion–surface interactions strongly affect the shape evolution of features—indeed, this is why the technology is used. There are examples (circa 2000) of the use of plasma etching to etch features with lateral dimensions of the order of 10 nm. Etch anisotropy (i.e., etching in one preferred direction) and etch selectivity (etching only one material) are important attributes. As noted above, it has proven possible to etch very small features indeed if the mask can be defined—this has been part of the reason that lithography has been and will probably continue to be viewed as the major limiting factor in future developments. However, the plasma etch step can introduce many shape artifacts, the mechanisms for which are imperfectly understood. Other known or suspected effects include feature sidewall charging deflecting ion trajectories, electrical currents from the plasma causing electrical damage to nascent devices, and stress localization in features promoting or retarding etching. Figure 2 shows sketches of some common etch profile shapes. As features shrink and as demands on device performance increase, processing technologies are under increasing pressure to achieve compositional and structural control over atomic dimensions. The consequence is that plasma processes must be understood and controlled sufficiently well that near-atomic dimensional control can be achieved over macroscopic length scales. The ratio of the smallest controlled length scale ($\sim$1 Å) to the wafer size ($\sim$300–400 mm) will then approach 9 to 10 orders of magnitude.

C. THE NATURE OF PLASMA–SURFACE INTERACTIONS

One of the most important aspects of plasma–surface processing is that *the plasma itself often strongly modifies the near-surface region.* An example of this is shown schematically in Figs. 3a and b, in the case of etching. Figure 3a is a depth profile of an originally crystalline silicon surface that has been exposed to a chlorine plasma. (Chlorine plasma is commonly used to etch silicon and other materials.) The plasma has created various silicon

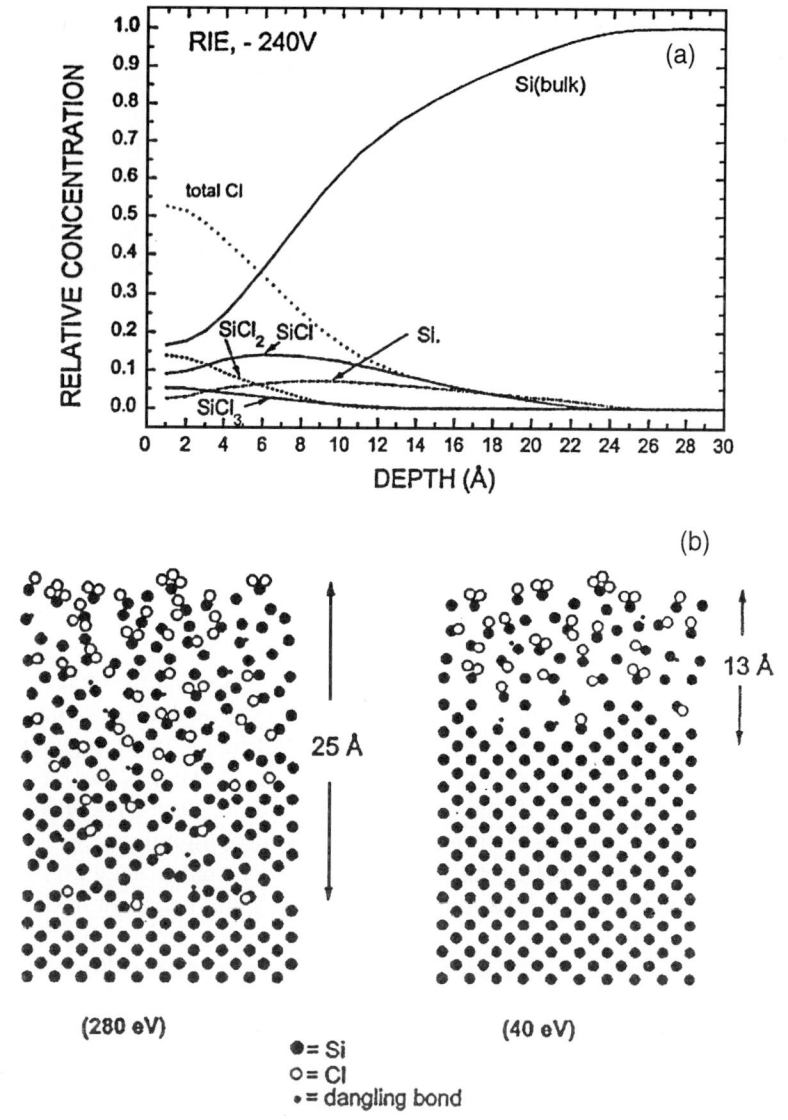

FIG. 3. (a) Steady-state depth composition profile of an originally crystalline silicon surface that has been exposed to a chlorine plasma, obtained from angle-resolved X-ray photoelectron spectroscopy. (b) Corresponding side-view schematics of near-surface atomic coordination: left, 280-eV ions; right, 40-eV ions. (From Layadi et al., 1997.)

chloride products, as well as Si dangling bonds, to a depth of about 30 Å. The depth profile was obtained by angle-resolved X-ray photoelectron spectroscopy. Figure 3b is a sketch at the atomistic level of the near-surface region as envisioned by the authors (Layadi et al., 1997). For the purposes of this discussion, the most important aspect to note is that the plasma has strongly altered the chemical and physical characteristics of the near-surface region (several nanometers). Similar arguments apply to plasma deposition of thin films, as well. Any fundamental understanding of plasma–surface chemistry must come to grips with the profound changes induced at surfaces by the plasma. This complicates studies of plasma–surface chemistry since every time the plasma environment changes, the surface changes. The surface chemistry in general also changes. It is argued that this fundamental fact renders plasma–surface chemistry unusually difficult, and it has a correspondingly complicating effect on simulations of reactive ion–surface interactions.

D. Ion–Surface Interactions in Plasma Processing

The topic of this article is the use of MD to simulate the effects of energetic, often reactive and molecular, positive ions on surfaces exposed to plasmas. Positive ions naturally impact surfaces that form their boundaries since, for the most part, all positive ions created within the plasma are lost to walls rather than to volume recombination with electrons or negative ions (Lieberman and Lichtenberg, 1994). Thin boundary layers called "sheaths" form at walls, retarding electron motion out of the plasma and accelerating positive ions. It is often the case that the final acceleration across the sheath results in few or essentially no ion–neutral collisions, with the result that ions impact surfaces with energies of the order of hundreds to even thousands of volts, depending on the magnitude of externally applied surface voltages.

Chemical bonds are typically of the order of several electron volts, so it is obvious that ions from plasmas can dramatically affect surface chemistry. In its simplest manifestation, unreactive (e.g., Ar^+) ions impact surfaces, resulting in sputtering, or the physical removal of surface atoms by a process analogous to "sandblasting." Sputtering is a very common technique to deposit thin films: material sputtered from one surface will generally condense to form a thin film on another surface. Virtually any material can be sputtered, and sputter deposition has been widely exploited in a variety of applications, including microelectronics manufacturing, optical films, corrosion resistance, and antireflection coatings and for films for decorative purposes. More commonly, ion–surface interactions occur within plasmas that contain chemically reactive species. Plasmas can be used for chemical vapor deposition, for etching or removal of material, or for cleaning or surface modification. In all cases, ion–surface interactions clearly play a role, and often this is a crucially important role.

One motivation for MD studies of ion–surface interactions is to develop a systematic understanding of plasma–surface interactions. Reactive free radicals created in the plasma, as detailed very ably in the article in this volume on SiH_x radical–surface chemistry by Maroudas and co-workers, are one important part of the chemistry induced by plasmas at surfaces being processed. However, the energy deposited by ions, and often the associated directionality of those ions when they interact with surface microfeatures, usually plays the most dramatic role. The ion flux in all applications of interest here is relatively small, and the time and length scales associated with a single ion impact are sufficiently short and small, respectively, that ion impacts do not overlap in time. It happens, fortuitously, that a single ion impact under conditions most commonly relevant to plasma–surface interactions can be treated with a collection of atoms numbering no more than several thousand, on a time scale usually less than 5 ps. In some cases, a few hundred atoms followed for 1 ps or less is sufficient. With semiempirical interatomic potentials, this means that thousands of impacts can be simulated on personal computers (circa 2000) in reasonable times. The combination of increasingly inexpensive computing, the development of reasonably accurate interatomic potentials for some materials of common interest in semiconductor manufacturing, and the fact that plasma processing mainly involves the cumulative effects of individual ion impacts has made technologically relevant studies of reactive ion–surface interaction feasible at a modest cost.

II. Use of Molecular Dynamics to Study Ion–Surface Interactions

A. SIMULATION PROCEDURE

In plasma processing, singly charged positive ions impact surfaces at energies between about 5 and 1000 eV. A variety of chemical and physical events occur that dramatically affect surface chemistry. The basic procedure in simulating this process is summarized in this section. Details can be found elsewhere (Barone, 1995; Helmer, 1998; Abrams, 2000). It should be noted that studies of ion sputtering of surfaces using classical trajectory simulations date back to the late 1960s and the pioneering efforts of Harrison (Harrison et al., 1968, 1978; Harrison, 1988; Smith et al., 1989, 1990). Although some of the early studies did not use periodic boundary conditions and they focused on relatively simple physical sputtering, many of the basic ideas for reactive ion–surface trajectory simulations were developed at this time. Garrison and co-workers extensively studied physical sputtering of silicon and various metals by rare-gas ions (Galijatovic et al., 1996; Garrison et al., 1987). Many other authors have also used MD to study sputtering processes (Urbassek, 1997; Coronell et al., 1998; Kress et al., 1999). More recent efforts to use MD to study energetic species–surface interactions include etching and other plasma–surface processing studies (Feil et al., 1993; Barone and Graves, 1995a,b, 1996; Athavale and Economou, 1995; Hanson et al., 1997, 1998; Helmer and Graves, 1997, 1998; Kubota et al., 1998; Abrams and Graves, 1998, 1999, 2000a–d; Wijesundara et al., 2000).

Ion–surface interactions are simulated by launching an energetic ion at a surface consisting of a few hundred to a few thousand atoms, as illustrated in Fig. 4. A collection of several hundred to several thousand atoms, referred to as the "cell" or "surface layer," represents the semiinfinite surface. A sketch of a copper surface is shown in Fig. 4. The lateral boundaries are periodic to simulate the semiinfinite surface with as small a set of atoms as possible, and the bottom two layers are fixed in space to anchor the cell. The "ion" is initially placed outside the range of interaction with the surface and is then launched with the desired energy, angle of impact, and impact location on the surface and allowed to interact with the surface atoms. The "collision cascade" occurs as the ion impacts the surface and transfers kinetic energy to the atoms in the cell. Some surface atoms may fly off the surface (termed "sputtering"), and others may move on the surface of the cell or to some other position within the cell. The snapshot in Fig. 4 at 1 ps illustrates sputtered atoms leaving the surface. The ion may reflect or imbed in the cell. One is generally interested in simulating the effects of many ions over surfaces

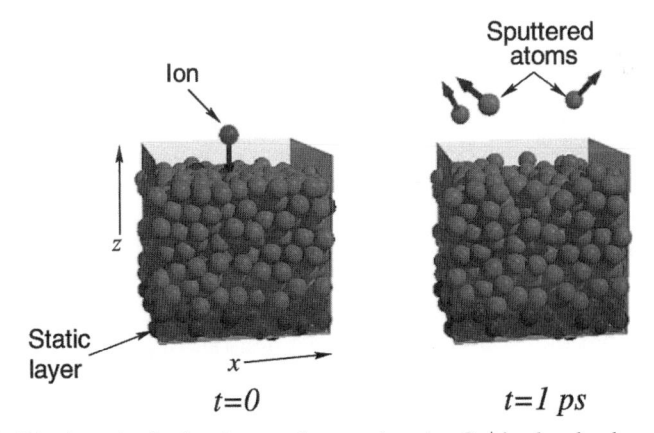

FIG. 4. Side view of a simulated copper layer undergoing Cu^+ ion bombardment and sputtering at time 0 (left) and after 1 ps has elapsed (right). (From Abrams, 2000.)

of macroscopic dimensions, so to obtain statistically significant results, ion impact points are selected at random and tens to hundreds of impacts are simulated. In some cases, the surface is allowed to change after each impact, and in other cases, it is more appropriate to use the same initial conditions for each trajectory. If the proper set of interatomic potentials is used, chemical bond breaking and formation can be simulated.

The system is described using classical mechanics. Each atom is treated separately with interatomic potentials describing the interactions between atoms, including bonding interactions. The interatomic potential is described as $U = U(\{\mathbf{r}_i\})$, in which each atom's position is represented as the vector $\mathbf{r}_i$, and the scalar potential U includes all interatomic interactions. By taking the gradient of the potential with respect to position, $\partial U/\partial \mathbf{r}_i = -m_i \partial^2 \mathbf{r}_i/\partial t^2$, one obtains the set of N coupled equations of motion of the N atoms. The system is simulated by numerically solving the coupled equations of motion of all atoms, as described in MD texts (Allen and Tildsley, 1987; Haile, 1992; Rappaport, 1995). Tully (1980) addresses the adequacy of a classical description in studies of gas–surface interactions and concludes that for atoms heavier than C, quantum effects such as tunneling can be ignored. Rappaport (1995) points out the practical fact that MD simulations have been successful, sometimes surprisingly so, in predicting phenomena that can be measured. The main issue in MD simulations is the accuracy of the interatomic potentials. Some function describing the potential energy as a function of relative atomic distances must be developed. This function must be capable of reproducing all of the important interactions between atoms including bond breaking and forming.

One approach would be to describe electrons via quantum mechanics and the Born–Oppenheimer approximation, assuming that nuclear motion is decoupled from electron dynamics and can be described classically. At each point in time, the total system wavefunction is computed, resulting in the potential energy surface U. From this, the instantaneous forces on all nuclei can be computed from the gradient, and the nuclear positions and velocities can be advanced. Obviously appealing in its rigor, this approach is completely impractical at present for studies of reactive ion–surface interactions. As discussed in Section III, it is necessary to simulate several thousand trajectories for of the order of hundreds to thousands of atoms to obtain results for a single set of conditions (ion type, ion energy, surface materials, etc.). Further, in the MD simulations used for ion–surface simulations, it is assumed that the "ion" is in fact a fast neutral. In part, this assumption is made because we believe that for most systems of interest, the ion recombines via the emission of an Auger electron within a few angstroms of the surface (Helmer, 1998). However, this assumption is necessary because potential energy surfaces are not generally available for charged species.

In addition to assuming that the "ion" is a fast neutral (although the term "ion" will continue to be used to describe the energetic species impacting the surface from the plasma), we neglect any but the ground electronic state. In particular, as Helmer (1998) points out, the recombination process may well result in an excited molecular species that can dissociate before impacting the surface. Reflected species could pick up an electron as it leaves the surface. And image forces in the surface can deflect the motion of charged species near the surface. The procedure described here has neglected all of these potential effects, and we rely on the fact that for kinetic energies above a few electron volts, the ion–surface interaction process is dominated by the kinetic energy exchanged upon impact. There will no doubt be systems for which these approximations will not be adequate, but we must leave their consideration to future work.

Maroudas and co-workers have described a hierarchical scheme for atomistic simulations involving the use of electronic structure calculations to develop and test semiempirical potentials that are in turn used for MD simulations. These results can sometimes be used to develop elementary step transition probabilities for use in dynamic Monte Carlo schemes. With Monte Carlo techniques, the well-known length and time scale limitations of MD can be greatly extended. This hierarchical approach appears to have great promise for the development of simulation strategies that will allow studies of a wide range of practical surface and thin-film chemical and physical processes.

For the purposes of the present article, rather than list the mathematical forms for the interatomic potentials used in the results that follow, we simply cite the original references. The interested reader can consult these papers and the references cited therein for more detailed information. Barone and Graves (1995a,b) and Barone (1995) used three-body Stillinger–Weber (1989) potentials for Si–Si, F–F, and Si–F interactions. Ar was treated with repulsive Moliere pair potentials. Helmer and Graves (1997, 1998) and Helmer (1998) also used the potentials developed by Feil *et al.* (1993) for Si–Cl and Cl–Cl interactions, by adjusting parameter values in the associated Stillinger–Weber potentials. Abrams and Graves (1998, 1999, 2000a–d) and Abrams (2000) used the potentials developed by Tanaka *et al.* (2000) for C–F, C–C, and F–F interactions, based on the ideas of Brenner (1990, 1992) and Tersoff (1988, 1989). Abrams introduced potentials for the Si-C-F system, using the Tanaka *et al.* (2000) formalism and, also, using results from Beardmore and Smith (1996).

The surface layer in Fig. 4 will be heated as a result of the energetic ion impact, since the lateral boundaries are periodic and the bottom layers are fixed. The kinetic energy transferred from the impinging ion to the surface will also result in bonds broken, movement of atoms on and within the layer, sputtering, and so on. Unless some means to remove the energy from the layer is incorporated into the simulation, the layer temperature will rise. In contrast, for the real case of an ion impacting an essentially semiinfinite surface, the local "temperature" near the impact point will rise, then decrease as the energy is dissipated to the surface. Neither the *NVE* (constant number, volume, and energy) nor the *NVT* (constant temperature) cases is entirely appropriate. Various means have been employed to deal with this problem by different investigators (see, e.g., Barone, 1995). It would seem that the most realistic way to deal with this is to use a simulation cell sufficiently large and an ion energy sufficiently low that the simulated system approaches the actual case of a semiinfinite layer that can dissipate the ion impact kinetic energy without heating. This avoids the difficulty of trying to simulate the dynamic process of energy transfer in a finite cell either with boundary layers that remove heat or through a scheme to rescale all atomic velocities throughout the cell based on some arbitrary algorithm.

Obviously, the size of the cell, both laterally and in depth, is a crucially important factor with respect to possible artifacts related to the finite system size. In addition, since the simulation time scales approximately as the cube of the number of atoms in the cell, it is important to establish just how large the system must be to avoid artifacts. The best procedure seems to be the most straightforward—increase the cell size until the results stop changing significantly. Since results must be collected and analyzed statistically, this can be time-consuming, but there is no alternative that avoids arbitrary schemes. Similar comments apply to the choice of time taken per trajectory simulation. In some cases, such as sputtering (or ion scattering), when the sputtered (scattered) species leaves the surface within a few tenths of a picosecond or less (Harrison, 1988), one needs very short integration times. However, if chemistry is important, several picoseconds or more may be needed for the reactions to occur. Of course, there will also be processes that will occur on much longer time scales due to normal thermally induced diffusion and reaction processes, but these must be dealt with using other approaches, in general.

Another important question relates to how to simulate the relatively long times between ion impacts. If one brings an ion into the layer, follows the collision cascade trajectory for several picoseconds, then cools the layer to room temperature and, finally, allows another ion to impact the layer, does this not increase the ion flux by many orders of magnitude above that observed experimentally? In fact, we argue that another interpretation of the simulation scheme is more appropriate. After the layer is cooled to 300 K, by bringing another ion into the layer immediately, we are in effect assuming that nothing happens to the layer between the time the layer has cooled and the time another ion hits. Clearly, this is an approximation, since the layer will relax, and diffusion and some reaction may occur to some extent during the milliseconds to seconds between impacts. However, the main processes in most cases occur during the brief but intense times following ion impact, and by incorporating the cumulative effects of the ion impacts, we expect that most of the important processes have been included. Future work will include adding, for example, Monte Carlo methods to treat more rigorously the relatively infrequent events between ion impacts.

III. Mechanisms of Ion-Assisted Etching

A. Experimental Studies of Ion-Assisted Etching Mechanisms

The mechanisms of plasma or ion-assisted etching have long been a subject of interest and even controversy. It has been known since the classic experiment of Coburn and Winters (1979) that plasma etching is usually the synergistic result of the combination of ion bombardment and neutral chemical reaction. The best-known example is silicon etching by fluorine and argon ions. In a vacuum beam system, the etch rate of a silicon film on a quartz crystal microbalance is measured vs. time. A beam of XeF_2 supplies the F atoms since XeF_2 decomposes readily at the surface. At first, only the XeF_2 beam hits the silicon, and at room temperature a small but measurable etch rate is observed. However, when the Ar^+ beam (e.g., 450 eV) is added, the etch rate jumps dramatically. Finally, when only the ion beam hits the surface, one observes the relatively low physical sputtering rate. This effect is referred to as the "ion-neutral synergism," but the basic idea is simple: the ion beam supplies the energy for F to etch Si to form the thermodynamically favored product, SiF_4. The room-temperature surface is evidently insufficient to scale the activation energy barrier. This simple fact, coupled with the naturally directional nature of ion flux to the surface—recall that ions are accelerated perpendicular to the surface by the natural and applied electric fields at surfaces—results in the characteristically anisotropic etching pattern that is so important in semiconductor manufacturing. Indeed, ion-assisted etching is capable of remarkable degrees of pattern transfer fidelity, as noted above.

Arguments about mechanisms all recognize that etching must consist of several steps occurring in series: adsorption of etchant, then creation of etch product, followed by desorption of the etch product. What are the respective roles of neutral and ions? The most common explanation is that the neutral etchant (the F atoms in the case above) penetrates the top few atomic layers of the silicon, breaking Si–Si bonds in the process. The insertion of F into the near-surface silicon lattice is then thought to weaken this layer so as to make it easier for the bombarding ions to sputter the top surface region. In this scenario, ions play the role of promoting the third step in the etching sequence: product desorption. We term this mechanism "chemically assisted physical sputtering." Another mechanism that had been proposed is that ion bombardment, by damaging the near-surface region, creates additional adsorption sites for F atoms, which results in greater sputtering and etching. In this case, ions play the role of increasing the rate of the first step, as well as the third step. We call this "damage-enhanced etching." A third mechanism is that ion bombardment greatly increases the rate of creation of etch products and that these products desorb thermally, long after the ion impact. This mechanism envisions ion enhancement of the middle step: formation of the etch product. We call this mechanism "chemical etching."

Experimental studies of etch mechanisms tend to be indirect: the ion interacts with the surface over length scales of tens to hundreds of angstroms and time scales of the order of 10^{-12} s. Direct observation is therefore quite difficult. A detailed discussion of the surface science aspects of etching was presented by Winters and Coburn (1992). The experimental evidence relied primarily on the energy distribution of products and on the delay of surface emission of products following ion bombardment. If products were mainly physically sputtered, they should follow the energy distribution for sputtered products that had been established both experimentally and theoretically (Sigmund, 1969). In addition, sputtered products leave the surface on the time scale of the energy deposition: about 10^{-12} s. If products were chemically sputtered, they should leaves the surface thermalized at the surface temperature. The damage-enhanced etching model was easily disproved (at least for Si etched with F) by Winters and Coburn (1992) by simply preexposing a silicon film on a quartz microbalance to Ar^+, then seeing if subsequent F uptake was increased compared to that with no preexposure to the ion beam. No effect was observed so this mechanism was rejected for Si etching by F. However, these authors point out that for W, the effects of ion preexposure can be dramatic in enhancing the halogen uptake.

Evidence has been presented for both chemically assisted physical sputtering and chemical sputtering from the energy distributions of products and the time dependence of their desorption. Figure 5 is a plot of measured etch product flux versus energy (Oostra et al., 1986). SiF_3^+ measured in a mass spectrometer reflects evolved SiF_4 from the surface etched with SF_6 and 3-keV Ar^+. At 50 K, the surface is sufficiently cold that only physically sputtered products emerge, and the characteristic "cascade" energy distribution is seen. The cascade energy distribution for flux vs. energy follows the following relation

$$f_{cc} = \frac{K_{cc} 2 U_0 E}{(U_0 + E)^3},$$

where f_{cc} is the fractional probability density distribution (eV m^2 s)$^{-1}$, K_{cc} is

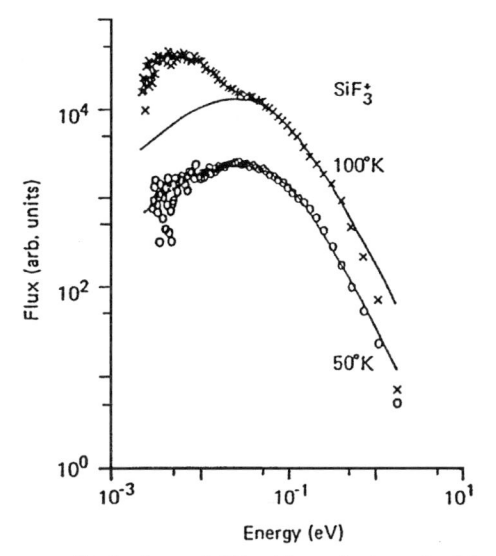

FIG. 5. Kinetic energy distributions of SiF$_4$ etch products evolved from a silicon surface exposed to 3-keV Ar$^+$ ions and 5 × 10^{16} SF$_6$ molecules/cm^2 s, at two surface temperatures, 50 and 100 K. Solid curves represent collision cascade distributions with a surface binding energy (U_0) of 0.05 eV. (From Osstra et al., 1986.)

a normalization constant, E is the energy (eV), and U_0 is a fitted parameter, termed the "binding energy" of the product. The integral of f_{cc} over all energy is the total flux. The solid lines in Fig. 5 are from cascade theory, and clearly the data fit the theory ($U_0 = 0.05$ eV) for a 50 K surface. At 100 K, however, there is a deviation from cascade theory at low energies. This part of the measurement turns out to fit the Maxwell–Boltzmann distribution, expressed as a flux probability density,

$$f_{\mathrm{MB}} = K_{\mathrm{MB}} E \exp\left(\frac{-E}{k_{\mathrm{B}} T}\right),$$

where f_{MB} is the fractional probability density distribution (eV m^2 s)$^{-1}$, K_{MB} is a normalization constant, E is the energy (eV), and T is the species temperature, assumed to be the surface temperature. The obvious interpretation of Fig. 5 is that at 50 K, the surface temperature is too low to allow chemically sputtered products to desorb thermally, so only physically sputtered products are observed. However, at 100 K, some of the species that are left in a weakly bound state after the ion impact have equilibrated to the surface temperature and have simply thermally desorbed.

Another important piece of experimental evidence for chemically sputtered products was presented by Winters and Coburn (1992), reproduced here as Fig. 6. These measurements, also made in a vacuum beam apparatus,

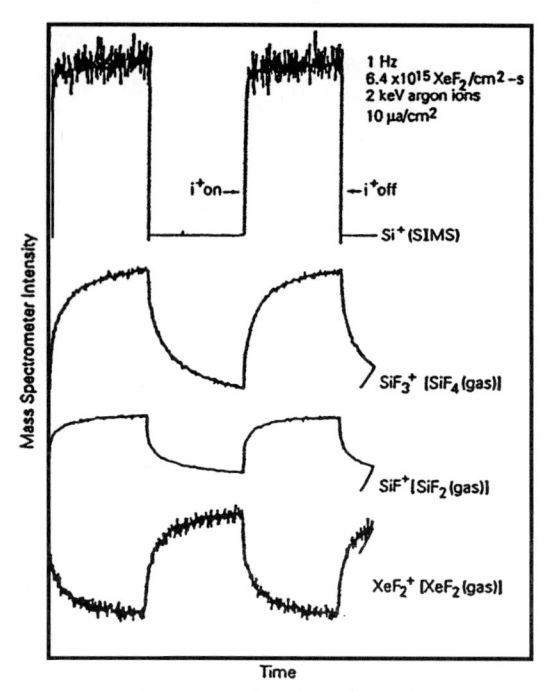

FIG. 6. Mass spectrometric intensity vs time for etch products and reflected XeF$_2$, with a square wave modulated ion beam. Purely physically sputtered products would have the same time dependence as the Si$^+$ SIMS signal. The longer time scales evident in the plots for the etch products (SiF$_3^+$ and SiF$^+$) suggest chemical sputtering. (From Winters and Coburn, 1992.)

are of various species evolving from a surface exposed to a square wave modulated ion beam and a flux of XeF$_2$. The Si$^+$ signal, representing sputtered species, is completely in phase with the ion beam: physically sputtered products leave the surface within 10^{-12} s and therefore no delay in the signal is to be expected. The signals for SiF$_3^+$ (from dissociative ionization of SiF$_4$ in the spectrometer), SiF$^+$ (SiF$_2$), and XeF$_2^+$ (XeF$_2$) all show significant time constants superimposed on the 1-Hz square wave frequency. This is clear indirect evidence for chemically sputtered products that have left the surface long after being created by ion bombardment.

B. MOLECULAR DYNAMICS STUDIES OF ION-ASSISTED ETCHING MECHANISMS

Ion-enhanced etching mechanisms using MD were addressed by Barone and Graves (1995a,b, 1996) in several papers. In the first paper, the Stillinger–Weber potentials were used for Si–Si, Si–F, and F–F interactions. Si–Ar and F–Ar interactions were modeled with a purely repulsive Moliere pair potential. In the second paper, interactions of both F and Cl were modeled with Si. In this paper, the modified SW potentials proposed by Feil et al. (1993). For Cl–Cl and Si–Cl were used.

In the first paper, Barone and Graves (1995a) studied a simplified system: silicon layers with different amounts of prefluorination were exposed to Ar^+ bombardment. The prefluorinated silicon layers were created by adding F atoms at random to the Si lattice during the collision cascade induced by Ar^+ bombardment. Silicon fluoride products that were physically sputtered could easily be detected since they left the layer during the several picosecond trajectory simulation. However, chemically sputtered products would desorb thermally, long after the ion bombardment event occurred. These species were detected by removing SiF_x clusters from the layer one by one and noting the system energy difference before and after their removal. This gives an approximation to the surface binding energy of the clusters. Simple first-order thermal desorption theory was used to determine if the cluster would desorb in the relatively long time between ion impacts. The average time between impact in a layer of area A subjected to an ion flux j_+ is simply $(Aj_+)^{-1}$. The layer areas are of the order of 1000 $Å^2$ or 10^{-13} cm^2, and typical experimental ion fluxes (even in high-flux plasma etch tools) are usually no more than 10^{17} ions/cm^2 s. This yields a minimum time between impacts of about 10^{-4} s. From thermal desorption theory, if the surface binding energy is less than about 0.8 eV, one would expect the product to desorb before the next ion impacted. Figures 7a and b show the results of the bond energies for various surface species with average F/Si contents of 0.39 and 0.91, respectively. No weakly bound clusters were observed for F/Si = 0.39, but significant numbers were found for F/Si = 0.91. The weakly bound species in the Stillinger–Weber Si–F and F–F potentials came from F in SiF_x products forming a weakly attractive bond with another F in the layer. Stillinger and Weber had put this weak long-range attraction in their F–F potentials when they were modeling liquid F_2, and this feature remained when they developed the Si–F potentials (Helmer, 1998). From the limited set of simulation results, it is not possible to say if weakly bound products would have been predicted, at least in similar quantities, if this somewhat fortuitous feature in the F–F potential had not been included. However, it suggests that for studies of chemically assisted etching, more quantitative models will require better approximations for the weakly attractive forces responsible for physisorption and the associated processes of diffusion and desorption.

Figure 8 shows the averaged etch yields from each mechanism as a function of the square root of ion energy. The projected threshold for etching is about 5 eV. Figure 9 is a plot of the predicted flux distribution vs

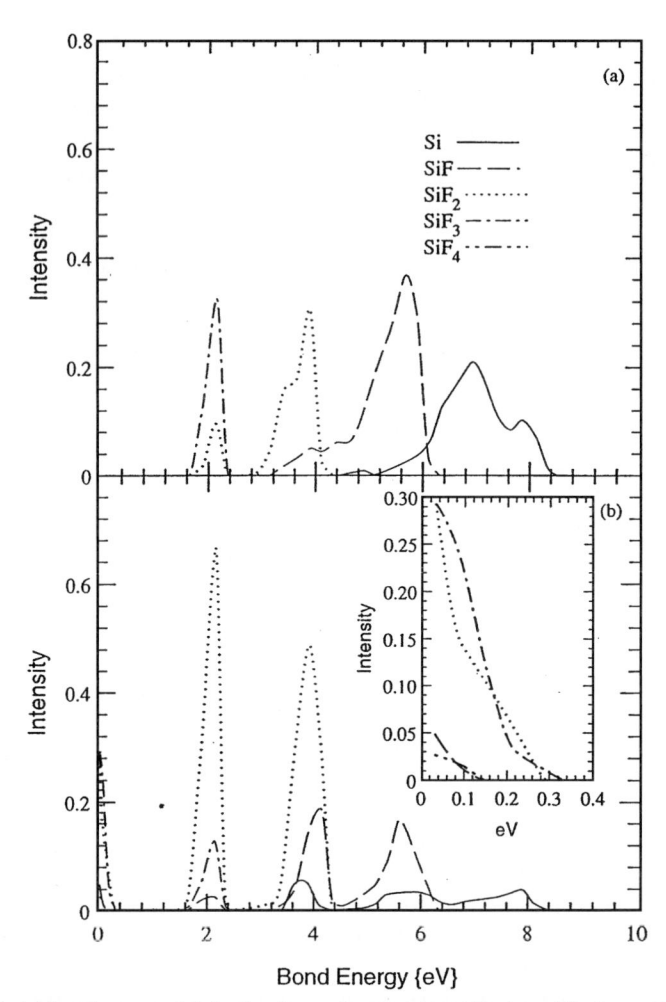

FIG. 7. (a) Bond energy distribution for surface species at the top of the layer with F/Si = 0.39, averaged over 300 impacts, 300 K; (b) Similar distribution for layer with F/Si = 0.91. Note the weakly bound species near 0.1–0.3 eV. (From Barone and Graves, 1995a.)

product energy from the simulation, for 200-eV Ar$^+$ impacting all layers with F/Si > 0.39. The observed sputtered species were SiF, SiF$_2$, and SiF$_3$ (symbols), and their energy distribution was fitted with a linear cascade distribution with $U_0 = 0.85$ eV. The dashed, dotted, and dot–dash lines are from calculations of a Maxwell–Boltzmann distribution of etched species, leaving the surface at 300 K, with a yield scaled to the observed ratio of physically sputtered to chemically sputtered species as shown in Fig. 9. The qualitatitive similarity to Fig. 5 of the measured flux distribution of products

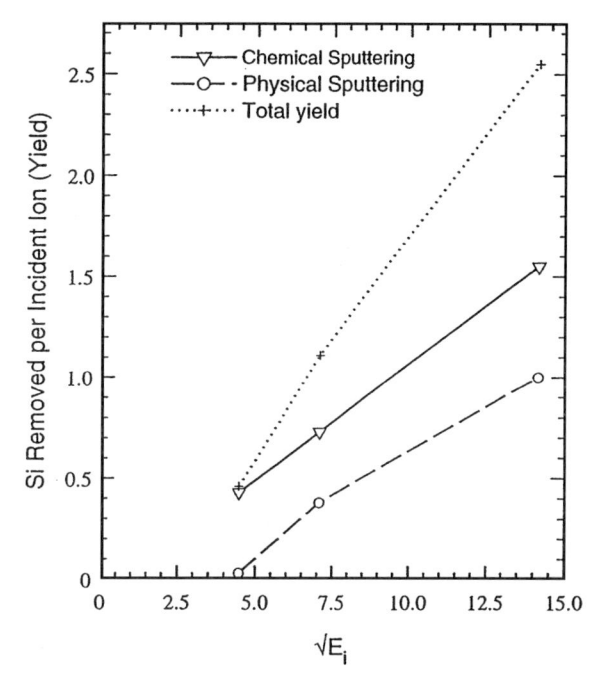

FIG. 8. Etch yield vs ion energy, obtained from averaging MD simulation results, F/Si = 0.91. (From Barone and Graves, 1995a.)

at 100 K is clear: etched products both physically and chemically sputtered would be expected to show a bimodal flux distribution.

These results show that one of the key points in simulating ion-assisted etching is that the Si surface layer must have the proper amount of halogen mixed into it. It does not suffice simply to have the top Si surface coated with a monolayer of halogen. Feil *et al.* (1993) showed that having a monolayer of Cl on Si, then impacting the layer with Ar^+, does not lead to much ion-assisted etching of the underlying Si. The ions mostly sputter the adsorbed halogen, with no increase in Si sputtering—indeed a decrease is often observed in the Si sputtering yield under these conditions. In fact, this point was made by Winters and Coburn (1992), as they concluded that chemically assisted physical sputtering is not generally an important mechanism in ion-assisted etching of Si by halogens. The MD simulations, however, did show that physical sputtering rates can be increased by halogen incorporation, but the halogen needs to be mixed into the Si so as to break Si–Si bonds, thereby weakening the lattice. This raises the question of how to incorporate into the simulation a realistic method to mix the halogen into the layer. Feil *et al.* (1993) proposed that ion bombardment would mix a monolayer of Cl into the

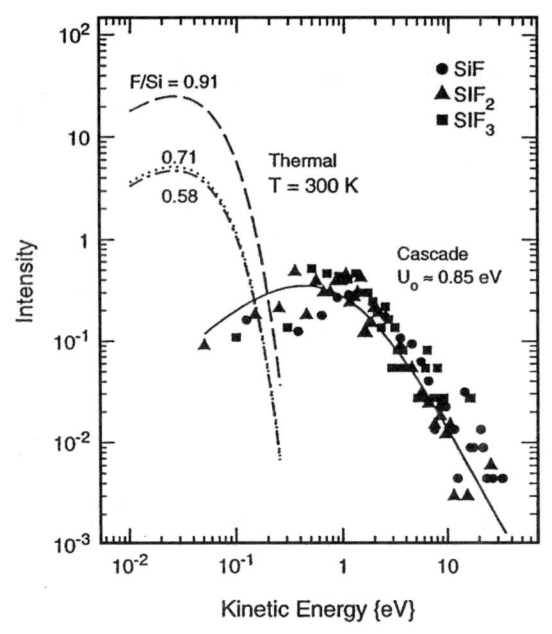

FIG. 9. Kinetic energy distributions for physically sputtered products obtained from the simulation (symbols). Data taken from all simulated layers with F/Si > 0.39 for the chemically sputtering results, represented here as the dashed, dot–dash, and dotted lines, assuming a Maxwell–Boltzmann kinetic energy distribution and 300 K. Solid-line fit to the symbols is collision cascade model with a fitted value of U_0. (From Barone and Graves, 1995a.)

Si, but as noted above, this was not observed. In their second approach to the problem, Barone and Graves (1995b) impacted the Si layer with energetic F or Cl (modeling F^+ and Cl^+). This approach turned out to halogenate the Si near-surface region very rapidly. Indeed, although not emphasized in the original publications, this point is worth stressing: halogen incorporation into Si is most efficiently accomplished through direct halogen ion impact. Direct ion–surface chemistry is important and has probably received less attention than it deserves in the plasma processing community.

In their attempt to study steady state etching of Si by energetic F or Cl, Barone and Graves followed the simulation strategy in Fig. 10. An initially crystalline or amorphised Si layer is impacted with energetic F at normal incidence and some selected energy at a randomly selected lateral location of the top of the layer. Each impact is followed for more than 1 ps, the layer is cooled back to 300 K, weakly bound species are searched for and removed if appropriate, and the statistics are updated. Another random impact location is selected and the process is repeated until it is judged that enough statistics have been collected. It should be noted that there are two stages in the simulation. The first is the approach to a steady-state coverage of halogen.

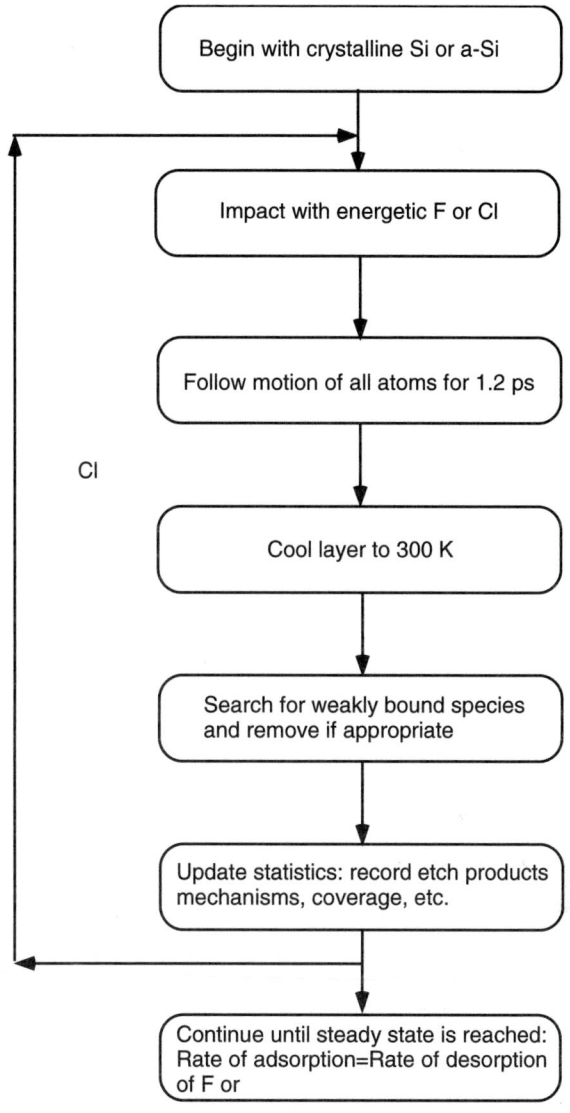

MOLECULAR DYNAMICS SIMULATIONS OF ION–SURFACE   169

```
┌─────────────────────────────────┐
│  Begin with crystalline Si or a-Si │
└─────────────────────────────────┘
                │
                ▼
┌─────────────────────────────────┐
│   Impact with energetic F or Cl  │
└─────────────────────────────────┘
                │
                ▼
┌─────────────────────────────────┐
│ Follow motion of all atoms for 1.2 ps │
└─────────────────────────────────┘
                │
                ▼
┌─────────────────────────────────┐
│        Cool layer to 300 K       │
└─────────────────────────────────┘
                │
                ▼
┌─────────────────────────────────┐
│  Search for weakly bound species │
│     and remove if appropriate    │
└─────────────────────────────────┘
                │
                ▼
┌─────────────────────────────────┐
│ Update statistics: record etch products │
│    mechanisms, coverage, etc.    │
└─────────────────────────────────┘
                │
                ▼
┌─────────────────────────────────┐
│ Continue until steady state is reached: │
│ Rate of adsorption=Rate of desorption │
│            of F or              │
└─────────────────────────────────┘
```

FIG. 10. Simulation strategy for silicon etching with energetic F or Cl. (From Barone and Graves, 1995b.)

Figure 11a shows the average F coverage vs F exposure (in equivalent monolayers), for each of three ion energies. The 10-eV F appears to reach steady-state coverage at about 2 ML after 10- to 15-ML exposure. The 25- and 50-eV ions result in ~3- to 4-ML steady-state coverage. Figure 11b shows the effect of increasing the size of the layer—the fluctuations are reduced with

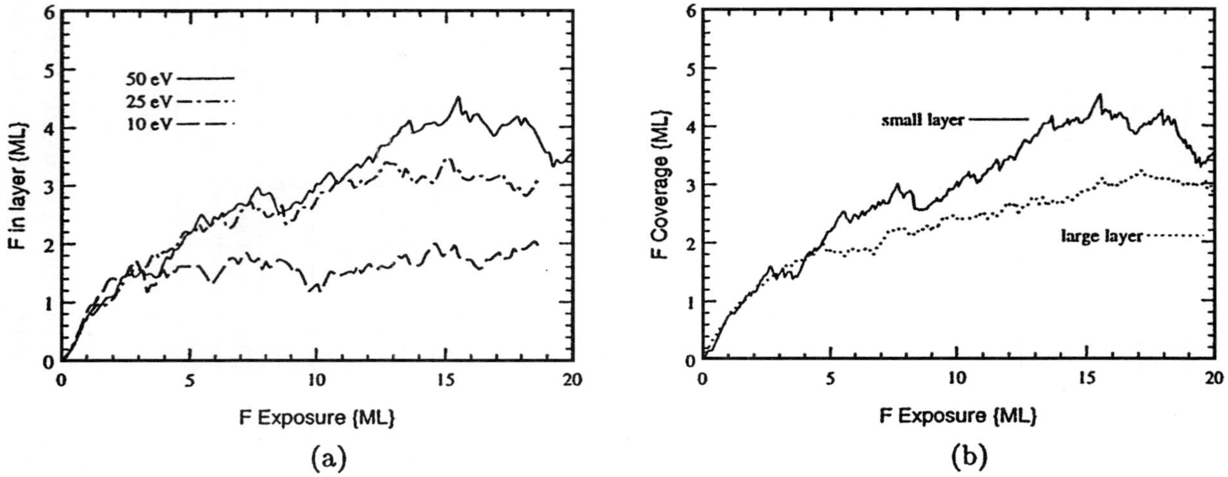

FIG. 11. Fluorine coverage vs F$^+$ fluence: (a) 10-, 25-, and 50-eV F impact onto smaller layer; (b) comparison of uptake vs fluence for 50-eV F impact onto smaller and larger layer. (From Barone and Graves, 1995b.)

the larger layer but the steady-state coverage appears to be about the same. Figures 12a–d show side views of the layers at various stages of fluorination. Note that the F extends rather deeply into the layer and that the halogen appears to clump in crevices and cracks in the Si lattice. Similar results were obtained with the Cl simulations.

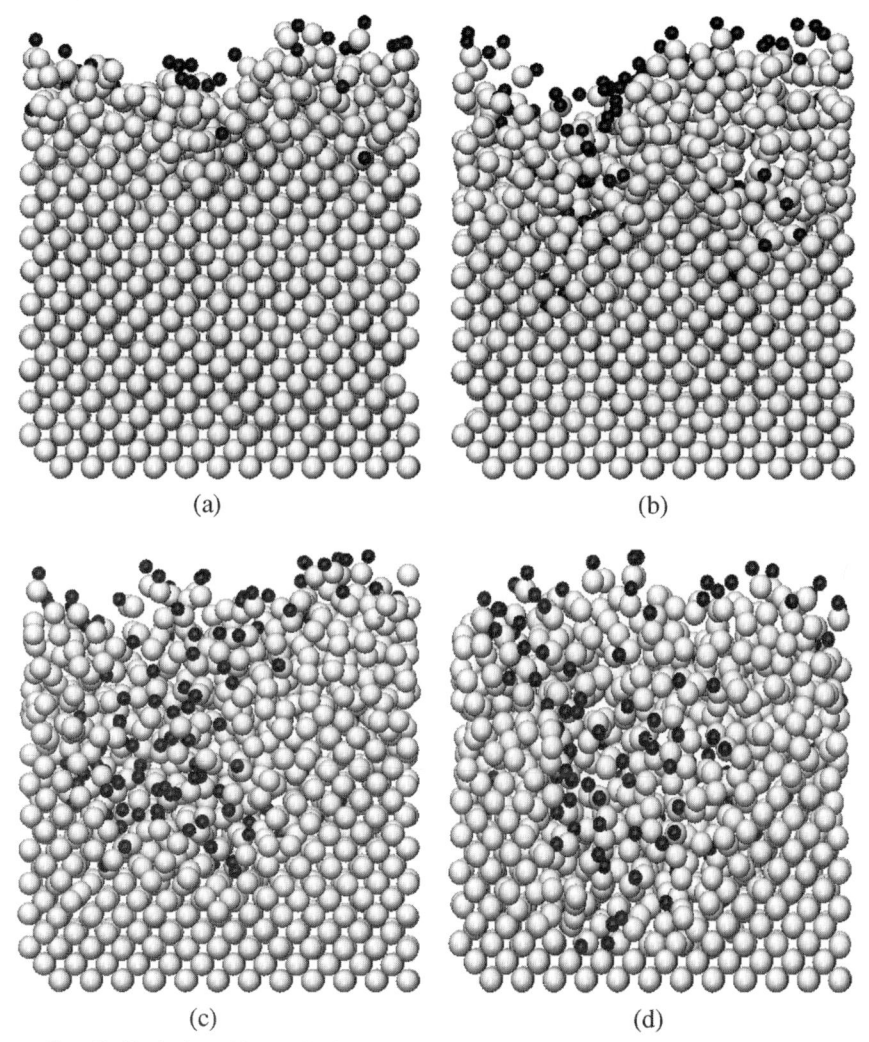

FIG. 12. Evolution of fluorosilyl layer formed from 50-eV F bombardment (a) after 1.39-ML F fluence; (b) after 6.94-ML F fluence; (c) after 13.9-ML F fluence; (d) after 20.8-ML F fluence. (From Barone and Graves, 1995b.)

The second stage of the simulation is to collect enough statistics after the steady-state coverage has been reached so that firm conclusions regarding mechanisms can be established. It turned out that for the cases shown here, chemically enhanced physical sputtering was the dominant form of etching, followed by direct abstraction and chemical sputtering. Direct abstraction is defined to result when the incident reactive species (F or Cl) leaves the surface during the 1-ps collision cascade bound to a Si atom. Sputtering of the adsorbed halogen turned out to be the most important effect limiting the coverage of the halogen at steady state. Other researchers following similar simulation strategies obtained similar results, although quantitative results differed (Athavale and Economou, 1995; Hanson *et al.*, 1997, 1998).

To summarize this section, let us attempt to put the results of these simulations and experiments into some perspective. MD simulations, using approximate semiempirical classical interatomic potentials, have been shown to be very valuable in studying details of events occurring during the rapid collision cascade following ion impact. The major qualitative results of etch mechanisms identified by experimentalists were confirmed by the simulations. However, the simulations provided valuable additional details of the processes. This includes observations such as the much greater rate of halogen incorporation from energetic halogen ions compared to adsorbed halogen impacted by rare-gas ions, the depth and spatial distribution of halogen in the Si near-surface, and the dynamic events leading to steady-state coverage. The side view of the mixed layer of Cl and Si illustrated in Figs. 13a–d can be compared with the view proposed by Layadi *et al.* (1997) (Fig. 3b) in their interpretation of the near-surface silicon region exposed to a chlorine plasma. Furthermore, movies made of trajectories and of layer evolution were found to be very useful for visualizing the relevant processes. It may be that the biggest contribution to the plasma processing research community from MD simulations has been the unprecedented opportunity to develop intuition of dynamic events at surfaces exposed to reactive plasmas using visualization of multiple trajectories.

C. Ion–Surface Scattering Dynamics

As noted above, plasma processing in semiconductor manufacturing generally involves ion and neutral interactions with submicron features. Ions accelerated across the plasma sheaths impact the wafer surface at near-normal incidence but encounter topography at the feature level. This is illustrated in Fig. 14. The ion impacts both the feature bottom and the sidewall. The ion reflects from the sidewall with some energy and angle before striking another part of the feature. If one is interested in predicting the shape evolution occurring during etching, then information about ion reflection is

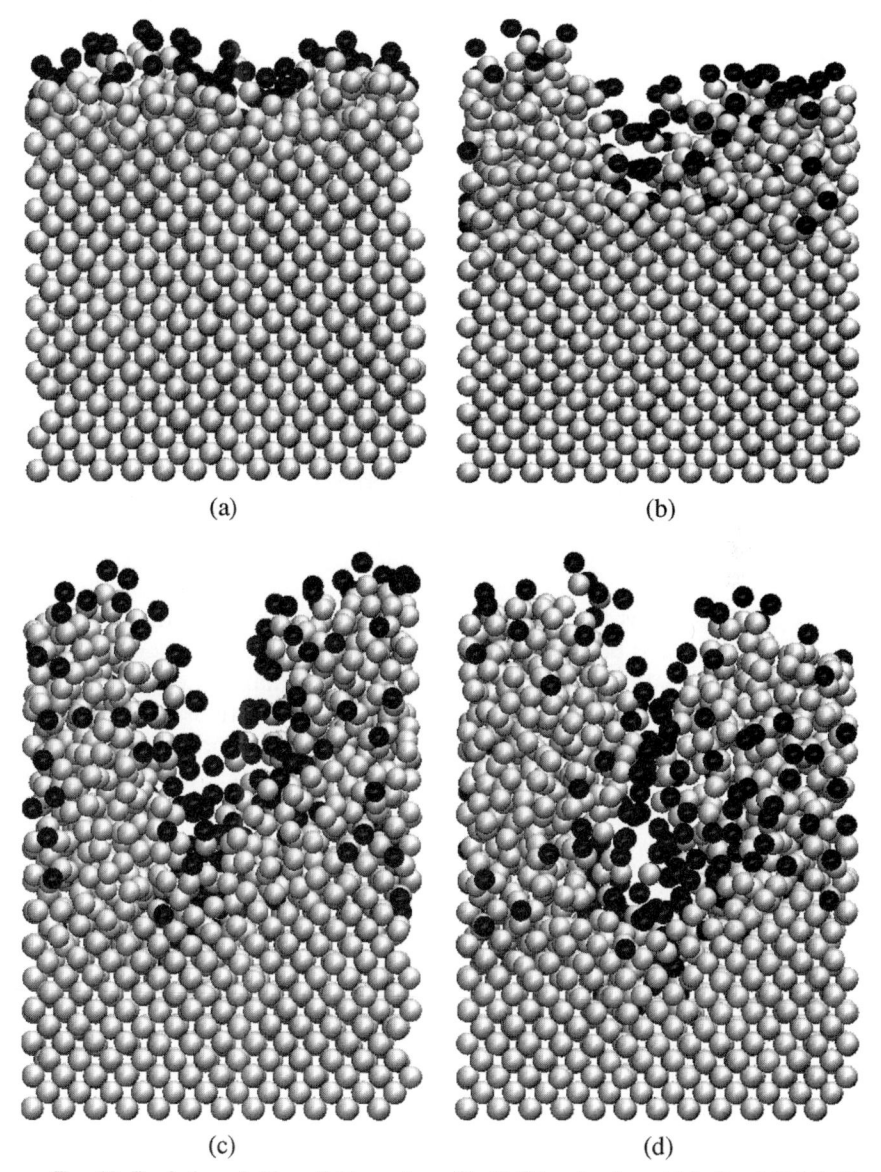

FIG. 13. Evolution of chlorosilyl layer from 50-eV Cl bombardment (a) after 1.39-ML Cl fluence, (b) after 7.9-ML Cl fluence, (c) after 13.9-ML Cl fluence, and (d) after 20.8-ML fluence. (From Barone and Graves, 1995b.)

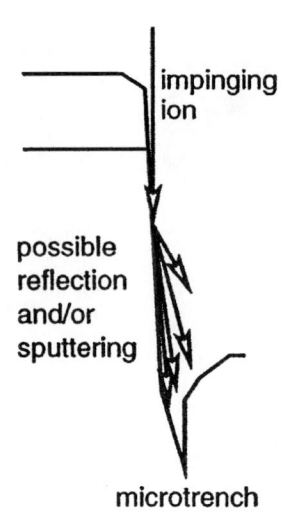

FIG. 14. Schematic of ion impact and scattering on feature sidewall and bottom during plasma etching.

necessary. Indeed, as the feature aspect ratio (depth to width) increases, sidewall interactions become more important. Most profile simulators have simply assumed either that ions reflect at the specular angle, with no loss of kinetic energy, or that reflecting ions are ignored. Both assumptions can introduce significant errors into profile shape evolution models. MD simulations offer an opportunity to study systematically the nature of ion reflection from surfaces. This section focuses on two aspects of ion reflection: the role of atomic-scale roughness and the use of elastic binary collision theory as an approximate model for ion reflection from feature sidewalls.

1. Ion–Surface Scattering: Role of Surface Roughness

The model surfaces chosen for this study are shown in Figs. 15 and 16. A bare silicon surface, a silicon surface with a single monolayer of Cl chemisorbed (Fig. 15), and a silicon surface with about 2.3 monolayers of Cl mixed into the top ~20 Å were chosen (Fig. 16). The potentials were Feil–Stillinger–Weber, and the simulations are described by Helmer and Graves (1998). For the results shown here, Ar^+ and Cl^+ are used as the incident ions.

The coordinate system we use for the ion scattering trajectories is shown in Fig. 17. The incident angle is defined with respect to the surface normal, and is denoted θ_i, with the reflected polar angle θ_r and reflected azimuthal angle ϕ_r. It is necessary to impact the ion many times on a given surface, varying the impact location each time and collecting statistics on the scattering event at a given incident energy, angle, ion type, and surface. For

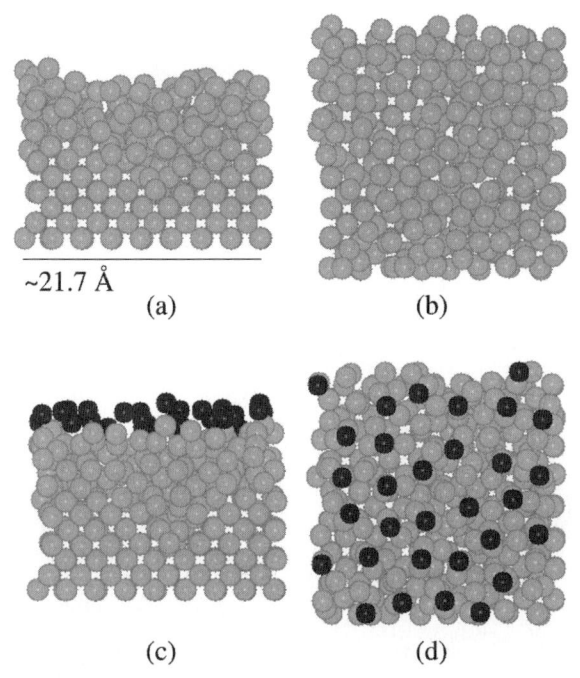

MOLECULAR DYNAMICS SIMULATIONS OF ION–SURFACE 175

FIG. 15. Side and top views of two surface layers: (a) bare Si (side); (b) bare Si (top); (c) 1-ML Si–Cl (side); (d) 1-ML Si–Cl (top). (From Helmer and Graves, 1998.)

each incident ion, the reflected (neutralized) ion comes off at a different (in general) energy and angle. We display the results of these trajectory simulations in several ways. The first is illustrated in Fig. 18, from simulations of an Ar$^+$ ion (85° from normal incident angle) impacting a bare silicon surface at 50 eV. In Fig. 18a, each of the several hundred trajectory simulations is represented as a dot on the polar plot. Each dot on the plot represents a separate trajectory. The scatter plot of several hundred trajectories gives an idea of the distribution of reflected energies and angles. The radial lines represent the fraction of incident energy retained by the incident ion, so that the outermost radial line corresponds to an E_r/E_i (the ratio of reflected-to-incident ion energy) $= 1$. The origin corresponds to an $E_r/E_i = 0$. In this plot, the polar angle θ_r is shown projected on the incident plane defined by the ion vector and the z axis, and therefore no information about azimuthal scattering is shown. A specular reflection would result in a scattering angle equal to the incident angle (85°), with no loss of incident energy, so the dot would be on the outer radial line at 85°. Figure 18a shows that a number of trajectories come close to being specular. Data on the distribution of azimuthal angles ϕ_r are shown in Fig. 18b, projecting the reflected ion exit trajectory vector onto the surface—viewing from "above," so to speak.

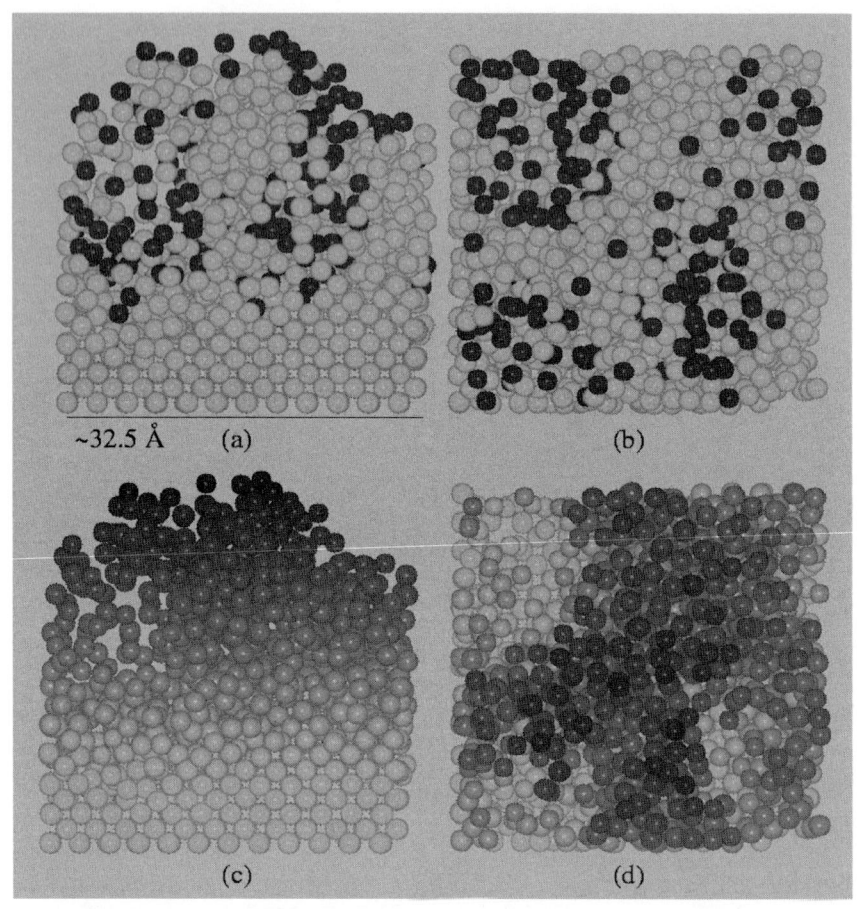

FIG. 16. Views of 2.3-ML Si–Cl surface layers: (a) side view; (b) top view—gray spheres Si, dark spheres Cl atoms; (c) side view with atoms shaded according to vertical position; (d) top view with depth shading. (From Helmer and Graves, 1998.)

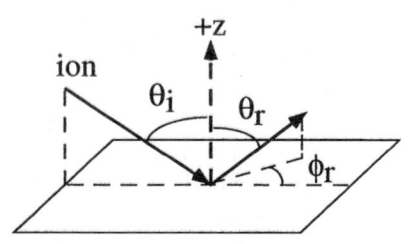

FIG. 17. The coordinate system and angle definition: incident, Θ_i; reflected polar, Θ_r; and reflected azimuthal, ϕ_r. (From Helmer and Graves, 1998.)

(a) E_r/E_i and Polar Angle Θ_r

(b) E_r/E_i and Azimuthal Angle ϕ_r

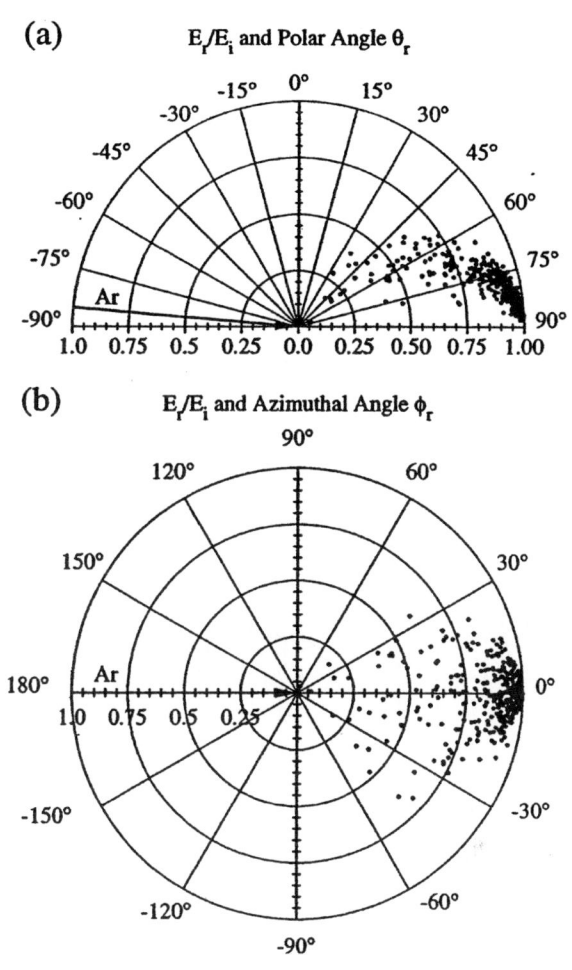

FIG. 18. The angles and reflected energy fractions of Ar reflected from a bare Si surface: (a) E_r/E_i; Θ_r; and (b) E_r/E_i; Θ_r. Each point represents the result of a single impact with $E_i = 50$ eV and $\Theta_i = 85°$. (From Helmer and Graves, 1998.)

It can be seen clearly from these figures that even at an 85° incident angle and on a fairly smooth surface, there is a significant amount of nonspecular scattering. Figures 19a and b illustrate the effects of polar angle scattering from the 1-ML Cl-covered Si surface and the 2.3-ML covered surface, respectively. In these plots, a Cl^+ ion is used at 50-eV incident energy and an 85° angle of incidence. Scattering from the relatively smooth 1-ML surface is fairly specular, but scattering data from the rougher surface show a big deviation from specular. The main point to be drawn from these limited

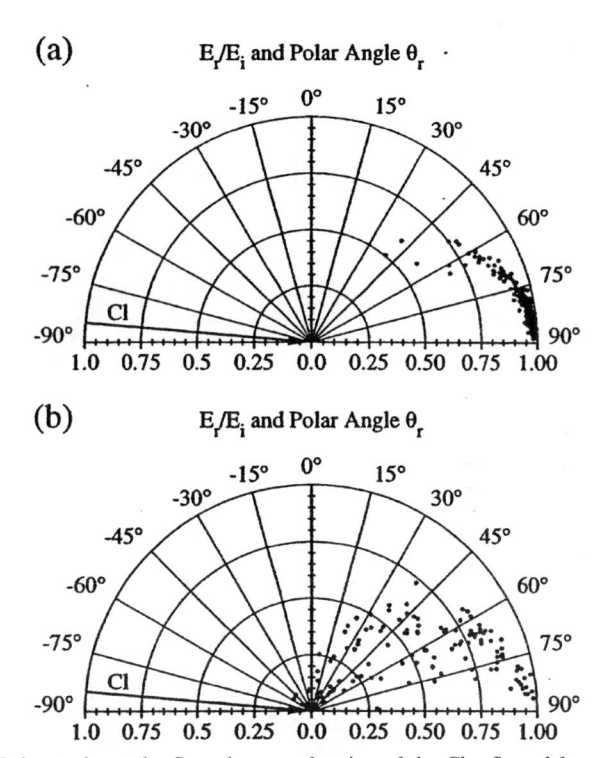

FIG. 19. Polar angles and reflected energy fraction of the Cl reflected from the (a) 1-ML Si–Cl surface and (b) 2.3-ML Si–Cl surfaces. (From Helmer and Graves, 1998.)

examples is that the state of the surface—especially its roughness—will significantly affect scattering, even at near-oblique incidence angles such as 85°. The consequences for profile evolution have been examined for several cases of practical interest, but those results are not presented here (e.g., Vyvoda et al., 1999; Vyvoda, 1999). It is important to recognize that feature shape evolution models must take into account surface roughness and other variables affecting scattering such as the identity of the incident ion, the composition of the underlying material, and any species adsorbed near the surface. This adds to the already very challenging task of constructing self-consistent, quantitative feature profile evolution models. Indeed, the MD simulations demonstrate that whenever any of the relevant variables change (ion type, ion energy, angle of incidence, and surface composition or roughness), the scattering results in general change significantly. While it is not difficult to amass huge amounts of scattering results from trajectory simulations, there are few theories with which we can interpret and digest these data. One approach that has shown some success and may form the basis for future attempts to simplify this problem is described next.

2. Ion–Surface Scattering: Binary Collision Theory

If one assumes that the ion–surface interaction can be described in terms of one (or two) binary, elastic collisions between isolated bodies, some simplifications of the ion scattering problem can be made. Of course, the accuracy of the approximation may be checked by comparing them to MD simulations. Details of this theory and its application to ion–surface scattering can be found elsewhere (Helmer and Graves, 1998; Helmer, 1998), but a brief synopsis is provided here. A relation between the scattered ion energy and the angle can be obtained by equations that represent conservation of linear momentum and kinetic energy, assuming that no energy is lost to electronic excitation (an excellent approximation for the conditions of interest here). The results are presented in terms of the total scatter angle α rather than the polar and azimuthal angles. The total scattering angle is the angle the scattered ion takes from the projected incident angle vector. A total scattering angle of 0° would correspond to the ion continuing on its initial trajectory. Note that for the case of an ion grazing the surface at an 85° incident angle, a total scattering angle of 10° represents specular scattering. One parameter in the model is the ratio of the ion mass to the surface atom mass μ. For Ar^+ on Si, $\mu = 1.43$, and for Cl^+ on Cl-covered Si, $\mu = 1.0$.

Figure 20 shows a plot of the energy fraction versus the total scattering angle for several cases. Note that binary collision (BC) theory simply relates

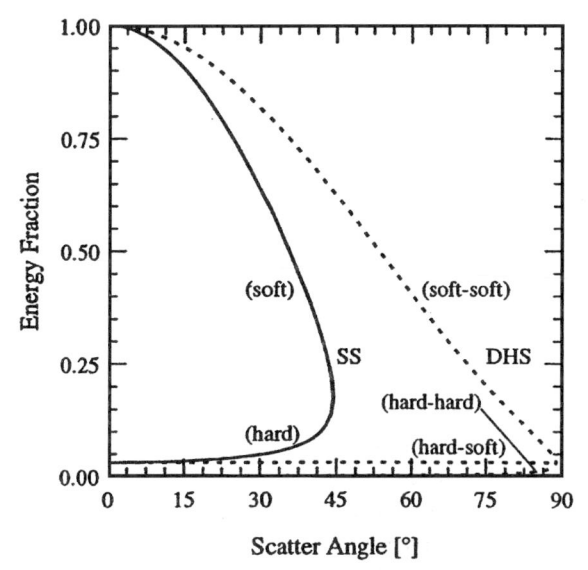

FIG. 20. The reflected energy fraction is shown as a function of the total scatter angle for both the single-scatter and the double half-scatter BC models. Mass ratios $\mu = 1.43$. (From Helmer and Graves, 1998.)

the energy fraction retained in the scattering ion and the angle of scattering. It says nothing about the distribution of scattering angles. The solid line denoted SS represents a single binary collision, with the two branches indicating a "soft" collision and a "hard" collision. Extension of this to two collisions, each of which scatters the ion through half of the total scattering ["double half-scatter" (DHS)], is also shown in Fig. 20, with three branches: two "soft" collisions (soft–soft), a hard and a soft collision (hard–soft), and two hard collisions (hard–hard). For the cases considered here, with ions impacting at energies between 20 and 100 eV, for Ar^+ and Cl^+ onto bare Si and Cl-covered Si, the MD results tended to be between the theoretical results from BC theory assuming a single soft collision and those assuming two soft collisions in double half-scattering. This is shown in Figs. 21a and b. Figure 21a is for the case Ar^+, 50 eV, 85°, and bare Si. The histogram plots the relative frequency, and the symbols are from the MD simulations providing the average E_r/E_i as a function of the scatter angle. The data lie in the region between SS and DHS results. Figure 21b shows the same results for the rough surface (Cl^+, 50 eV, 85°, 2.3-ML Cl). The histogram of scatter angle frequency shows a broader scatter angle distribution, as shown previously for the same data plotted in Fig. 19b. It also demonstrates that the MD data fall between the SS and the DHS results using BC theory. Of course, the relatively good agreement between the MD and the BC theory results does not imply that the collisions are truly binary—detailed analysis of the ion trajectories in the MD simulations indicates that some trajectories were dominated by a single encounter, but many others were not. Nevertheless, the simplifications offered by BC theory make it attractive for future efforts to simplify profile evolution codes while retaining some physics and avoiding complete reliance on nonphysical adjustable parameters. One must, of course, come up with some procedure to select the distribution of scattering angles to use this theory.

D. Ion–Surface Interactions with both Deposition and Etching: CF_3^+/Si

In this section, we return to the question of ion-assisted etching mechanisms, but now with the added complication of having a component of the incident ion (C) that is depositing as well as a component (F) that tends to etch the surface material. This situation is actually quite common in practical etching applications because it is often necessary to have such a depositing species to promote etch selectivity. The best-known and most widely used example is fluorocarbon-containing plasmas used to etch dielectric materials, such as SiO_2 and Si_3N_4, but that must be selective with respect to a layer

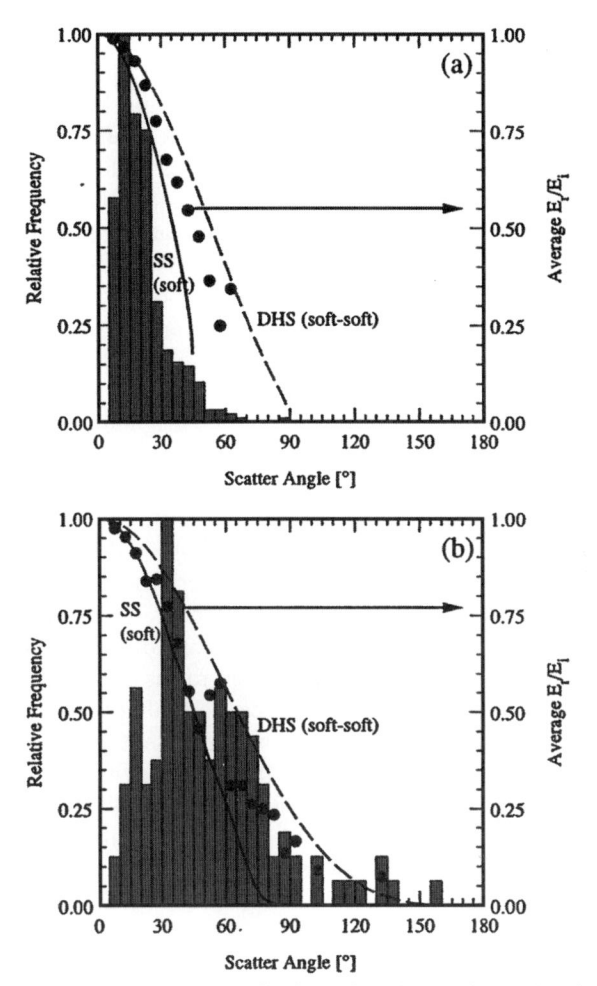

FIG. 21. The histograms show the distributions of total scattering angle a for the reflected atoms. (a) Ar+, 50 eV, 85°, bare Si; and (b) Cl+, 50 eV, 85°, 2.3-ML Si–Cl. Filled circles represent the average reflected energy fraction corresponding to the reflection angle. Single-scatter (SS) soft–soft branch and double half-scatter (DHS) soft–soft branch model predictions for (a) $\mu = 1.43$ and (b) $\mu = 1.0$ also shown for comparison with E_r/E_i data from simulation. (From Helmer and Graves, 1998.)

of silicon below the dielectric film. For SiO_2 etching, fluorocarbon plasmas often work well because the C in the incident ion is volatilized by the O in the film, and the Si by the F in the incident ion. Of course, in plasma reactors, neutral species play an important role as well, but many of the same principles hold for neutral chemistry. To simplify matters, we consider just the case of fluorocarbon ions onto silicon surfaces. In this case, the C in the incident ion tends to be more depositing, and even if the net effect is etching, the etch rate of the underlying Si tends to be slower with C present. A plasma etch chemistry that etches the overlying SiO_2 film readily, but the underlying Si film more slowly, is often desirable in practice. A similar situation can also occur when etch products (e.g., silicon chlorides or fluorides) are ionized in the plasma and return to the surface. The silicon tends to be depositing and the halogen tends to etch. The net effect can be either deposition or etching, depending on the conditions. It was the purpose of the studies summarized here to attempt to learn more about the mechanisms that govern such situations, at least under fairly idealized conditions.

In this case, partly because carbon can form many kinds of bonds with halogen and silicon (sp, sp^2, sp^3), the Stillinger–Weber interatomic potentials were replaced by a set of potentials developed originally by Brenner (1990, 1992) for the carbon–hydrogen system. Tanaka et al. (2000) describe the procedure used to modify the parameters in the C–H system to C–F. Abrams describes the development of this set of interatomic potentials (Abrams and Graves, 1999; Abrams, 2000), including the addition of Si to allow studies of C–F–Si atoms interacting. The simulation procedures followed here are very similar to those described earlier and are treated in much greater detail elsewhere (Abrams and Graves, 1999, 2000a–d; Abrams, 2000). Abrams reports on a comparison between the different forms of potentials for several cases in which some overlap exists. In all cases, qualitative results are robust, but quantitative results generally differ with different forms of the potential energy function or when parameters within any given set are altered. This is, of course, to be expected. In this section, results from MD simulations of CF_3^+ onto silicon surfaces are reported, including the development of a steady-state etching layer, and some information that can be compared to experiment.

1. CF_3^+ Etching of Silicon

In the simulations described here, an initially bare Si layer of 512 atoms, with a lateral area of about 500 $Å^2$ and a depth of about 20 Å, was used. Periodic lateral boundaries and a fixed bottom layer were used as before. The top Si layer was "amorphized" by repeated impacts of 200-eV Ar before exposure to CF_3^+. The goal was to simulate an ion beam experiment in which a beam of monoenergetic CF_3^+ is directed at a silicon surface, with no other species impacting the surface. The initial transient associated with the buildup of a mixed C–F–Si layer at the surface would be followed by a period of steady-state etching in which the composition and thickness of the top mixed layer would not change. Each ion impact was followed for 0.2 ps, and tests were made to be certain that the results did not depend on the length of the impact trajectory time. Clusters of material that were weakly bound to the surface were identified and removed between ion impacts, and species that desorbed during the 0.2-ps trajectory simulation or that were retained within the layer that were strongly bound were recorded. Again, the assumption is made that the only thing that happens between the relatively infrequent ion impacts is the thermal desorption of weakly bound clusters (of any size) and dissipation of the heat added by the ion bombardment. Of course, this is not strictly correct since some species will diffuse and/or react during this time, but since the layer returns to room temperature very quickly after the ion impact, any thermally activated processes are likely to be relatively slow. In any case, this is the simplest set of assumptions that seem consistent with the major identified processes and is therefore a reasonable place to begin the study.

Figures 22 and 23 illustrate the simulated evolution in C, F, and Si content of the layers for 100-eV CF_3^+ impacts at normal incidence, as a function of ion fluence (Abrams and Graves, 1999). Ion fluence is simply the product of ion flux and time, or the number of ions that have hit the simulated surface. Sometimes the unit of monolayer is used, which refers to a number of ions equal to the number of atoms in the topmost layer. These figures also show the statistical repeatability of three runs, including results from a cell with twice the surface area ($\sim$1000 Å^2). It can be clearly seen from Fig. 22 that C and F build up rapidly, with an initial sticking probability of unity, then after a fluence of 500 impacts (equivalent to about 10 monolayers or 10^{16} ions cm^{-2}), the layer composition reaches steady state. Later, evidence is

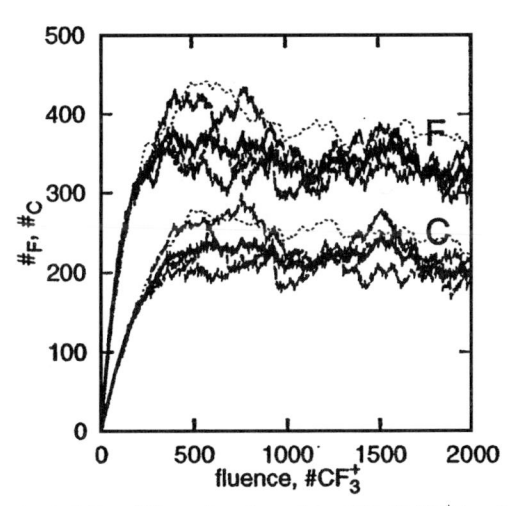

FIG. 22. The content of C and F as a function of the 100-eV CF_3^+ ion fluence. The narrow curves denote each statistically equivalent data set, while the thickest curve is the average. The dotted curve is data from the larger (1000-Å^2) surface, scaled by a factor of 0.5 in both dimensions, and is not included in the average. (From Abrams and Graves, 1999.)

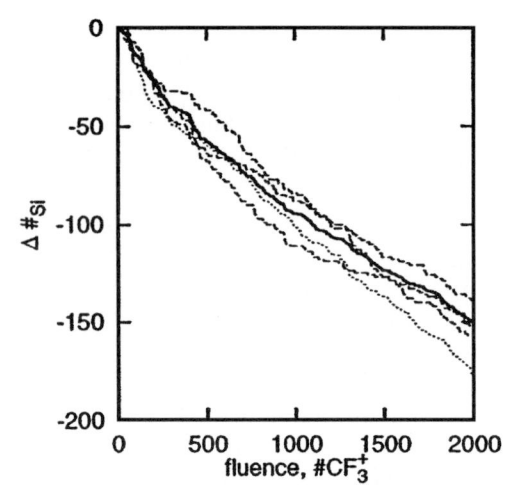

FIG. 23. The change in Si content of the surface as a function of the ion fluence. The thickest curve is the average of the three data sets. At steady state, the slope of this line is the etch yield of Si, 0.06 Si/ion. The dotted curve is data from the larger (1000- Å^2) surface, scaled by a factor of 0.5 in both dimensions, and is not included in the average. (From Abrams and Graves, 1999.)

presented that the F content of the layer governs the C and F saturation characteristic shown in Fig. 22. Figure 23 shows that the Si content of the layer drops at a rate corresponding to an etch yield (Si atoms per incident ion) of 0.06—a relatively low yield. A snapshot of the atomic configuration in the cell after 1000 ion impacts at 100 eV and normal incidence is presented in Fig. 24. The top 20 Å or so consists of a mixed "fluorocarbosilyl" layer over essentially pure Si. The net silicon etching must occur from Si moving up from below, leaving the top of the layer after mixing in the fluorocarbon material that forms there. This result is in qualitative agreement with the experimental measurements by Coburn et al. (1977), Thompson and Helms (1990), and Sikola et al. (1996). It appears that the MD simulation at least qualitatively reproduces the behavior of steady-state Si etching through a fluorocarbon overlayer.

The role of CF_3^+ ion energy in the composition profiles of C, F, and Si is shown in Figs. 25a–d (Abrams and Graves, 2000d). Figure 25a corresponds to an ion energy of 25 eV, and the thickness of the C and F overlayer can be seen to be about 5–10 Å. The Si extends into the mixed F and C region. As energy is increased, up to 200 eV (Fig. 25d), several trends emerge. The F appears to be reduced near the surface, peaking deeper into the layer than the C. In addition, the mixed layer thickness has increased from 5–10 to about 15 Å or more. The Si profile is accordingly more spread out, and the F/C ratio in the mixed layer has decreased from about 2.5 to 1.5.

FIG. 24. Snapshot of the atomic configuration after a fluence of 1000 CF_3^+ at 100 eV and normal incidence. A side view is shown, with the positive z-direction up and the positive x-direction right. The periodic transverse boundaries of the cell are rendered as translucent planes; the boundary facing the viewer is cut away. Si, dark gray; C, gray; F, white. (From Abrams and Graves, 1999.)

Figure 26 plots the etch yield (Si atom/ion) and C layer thickness as a function of ion energy from 25 to 200 eV. The etch yield can be fit to the commonly used phenomenological form with a square root dependence on ion energy and a threshold energy. Results from the larger cell, showing little cell dependence, are also shown in Fig. 26.

These results seem characteristic of many ion-assisted etching systems that contain depositing material (C in this case). Initially, the C and F build up rapidly in the top 10–15 Å of the Si surface, mixing with the underlying Si. As the F content increases, the sticking probability decreases, until as much C and F is removed as is deposited. Under these conditions, some Si from the underlying layer is also lost. As shown later, it is possible (for example, with CF^+) that the fluorocarbon film continues to build and steady-state deposition occurs rather than etching. The key seems to be what near-surface composition corresponds to a steady state, given the ion type, energy, angle of impact, and underlying material. In the case of CF_3^+, there is enough F coming in with the ion that steady state requires Si to be mixed into the near-surface

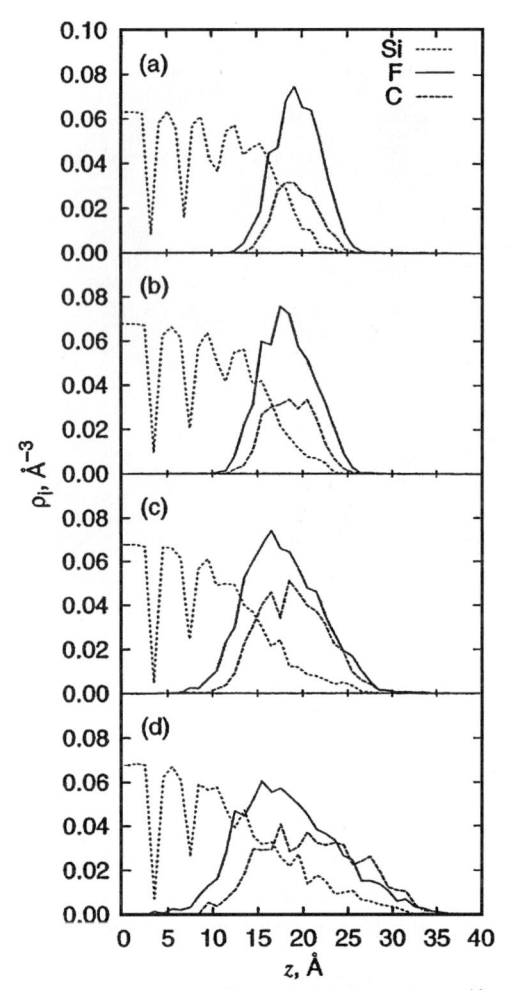

FIG. 25. Depth profiles for Si, F, and C after a fluence of 2×10^{16} ions/cm^2 CF_3^+ (1000 impacts) for E_i = (a) 25 eV, (b) 50 eV, (c) 100 eV, and (d) 200 eV. $z = 0$ corresponds to the simulation cell bottom. (From Abrams and Graves, 2000d.)

composition. This of course corresponds to etching, and to a balance between impacting C and F and C and F being removed. In ion-induced deposition, the impacting ion cannot remove all of the material deposited, and a film builds up.

2. Product Distributions in CF_3^+ Etching of Silicon

A common problem in modeling plasma tools is determining what is coming from walls bounding the plasma. Molecular gases entering the tool

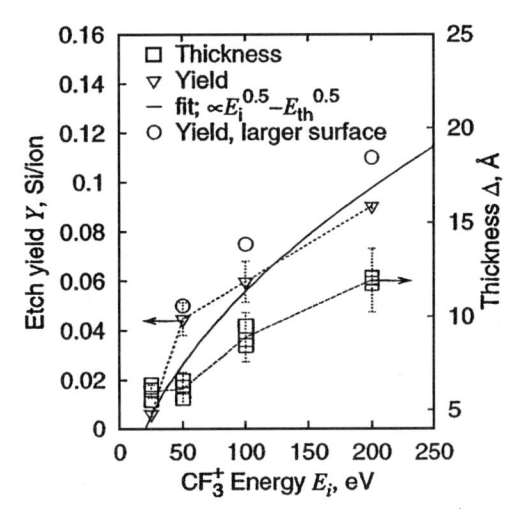

FIG. 26. Etch yield Y and fluorocarbosilyl layer thickness Δ vs CF_3^+ ion energy E_i. The error bars in layer thickness denote the average fluctuation in this quantity during steady state. Etch yield error bars denote the standard deviation among three independent runs. The solid curve is a best fit of $Y = a(E_i^{0.5} - E_{th}^{0.5})$, where $E_{th} = 25$ eV and $a = 0.01$ (eV$^{-0.5}$). Circles are from single runs on a surface with twice the lateral surface area to show the effects of cell size. (From Abrams and Graves, 2000d.)

can be dissociated, and these dissociation products can react in the gas phase and at surfaces. In addition to these sources of chemical species in the plasma, ion impact at surfaces induces a considerable flow of neutral species back into the gas phase. Etch products (i.e., species containing Si that leave the surface) are an obvious example, and it is well known that etch products can play a very important role in plasma tool chemistry. For one thing, these species, or those derived from them by dissociation or ionization, can return to the wafer surface and redeposit. In some cases (depending on applied power, gas pressure, and flow rate among other variables), the flow of material from walls can match or even exceed gas flow into and out of the plasma reactor. Little is known or understood about how ions, especially reactive molecular ions, contribute to this mass flow from walls. The results in this section provide some insight into surface emission product composition and kinetic energy distributions from CF_3^+ impacts onto Si. Of course, in a real tool, many types of ions, in addition to many neutral species, would impact the surface, not just CF_3^+. However, examination of this simpler, but reasonably representative case should prove helpful.

Table II summarizes the majority of species removed from Si surfaces under CF_3^+ impact at energies of 25–200 eV (Abrams and Graves, 2000c). Stastistics were collected after the surfaces reached steady state. The table lists the average yields (number per incident ion) and standard deviation.

TABLE II
BREAKDOWN OF MAJOR SPUTTERED SPECIES AS AVERAGE YIELD PER ION DURING
STEADY-STATE ETCHING FOR ION INCIDENT ENERGIES OF 25, 50, 100, AND 200 eV[a]

Molecule	$E_i = 25$ eV	$E_i = 50$ eV	$E_i = 100$ eV	$E_i = 200$ eV
F	0.051 ± 0.005	0.482 ± 0.012	0.987 ± 0.012	1.240 ± 0.020
F_2	0.001 ± 0.001	0.088 ± 0.002	0.270 ± 0.010	0.282 ± 0.011
C	0.000 ± 0.000	0.000 ± 0.000	0.023 ± 0.004	0.067 ± 0.003
CF	0.001 ± 0.001	0.067 ± 0.008	0.192 ± 0.017	0.217 ± 0.006
CF_2	0.085 ± 0.008	0.484 ± 0.010	0.275 ± 0.006	0.180 ± 0.006
CF_3	0.739 ± 0.005	0.211 ± 0.005	0.036 ± 0.003	0.017 ± 0.001
CF_4	0.091 ± 0.001	0.063 ± 0.005	0.012 ± 0.001	0.006 ± 0.001
C_mF_x	0.022 ± 0.002	0.046 ± 0.003	0.116 ± 0.005	0.118 ± 0.003
m	2.15 ± 0.03	2.34 ± 0.03	2.56 ± 0.06	2.47 ± 0.05
x	5.13 ± 0.05	3.70 ± 0.13	2.45 ± 010	1.99 ± 0.06
SiF	0.000 ± 0.000	0.000 ± 0.000	0.003 ± 0.002	0.009 ± 0.002
SiF_2	0.000 ± 0.000	0.002 ± 0.001	0.007 ± 0.003	0.017 ± 0.003
SiF_3	0.000 ± 0.000	0.008 ± 0.001	0.012 ± 0.003	0.019 ± 0.003
SiF_4	0.000 ± 0.000	0.014 ± 0.003	0.012 ± 0.002	0.013 ± 0.004
SiC_mF_x	0.006 ± 0.001	0.019 ± 0.002	0.019 ± 0.002	0.023 ± 0.002
m	1.62* ± 0.25	1.68 ± 0.17	1.89 ± 0.15	1.88 ± 0.10
x	7.20 ± 0.51	5.91 ± 0.31	4.51 ± 0.15	3.47 ± 0.21
SiC_nF_x	0.000 ± 0.000	0.001 ± 0.001	0.004 ± 0.001	0.005 ± 0.001
n		4.39 ± 2.56	7.10 ± 0.67	6.81 ± 1.07
x		7.11 ± 4.27	7.34 ± 0.64	5.19 ± 0.33

[a]High molecular weight molecules were split into groups according to $1 < m < 5$ and $n > 4$, and average stoichiometries of these groups are given. Errors given are standard deviations among the four runs (Abrams and Graves, 2000c), except that only three runs were done for $E_i = 25$ eV.

Figure 27 shows a series of plots for each of the main C- and F-containing products leaving the surface as a function of the ion energy, as well as the total yield for that species. The main ejection mechanisms are sputtering (S), reflection (R), and abstraction (A). As ion energy increases, the dominant ejection mechanism shifts and is different for different species. At 25 eV, the main product leaving the surface is simply reflected CF_3. As energy is raised to 50 eV, the ion fragments more on impact and CF_2 and F replace CF_3 as the main species. By 100 eV, about half of the incoming C leaves the surface in the form of CF and CF_2. F and F_2 also leave the surface with a yield of 1 and 0.25, respectively. Furthermore, a significant fraction of these species is from sputtering and abstraction, not reflection. In other words, the ejected F, F_2, CF, and CF_2 were part of the surface before the ion impact. There is mixing and exchange with the surface, and material leaving the surface is not simply reflected ions. Of course, under steady-state

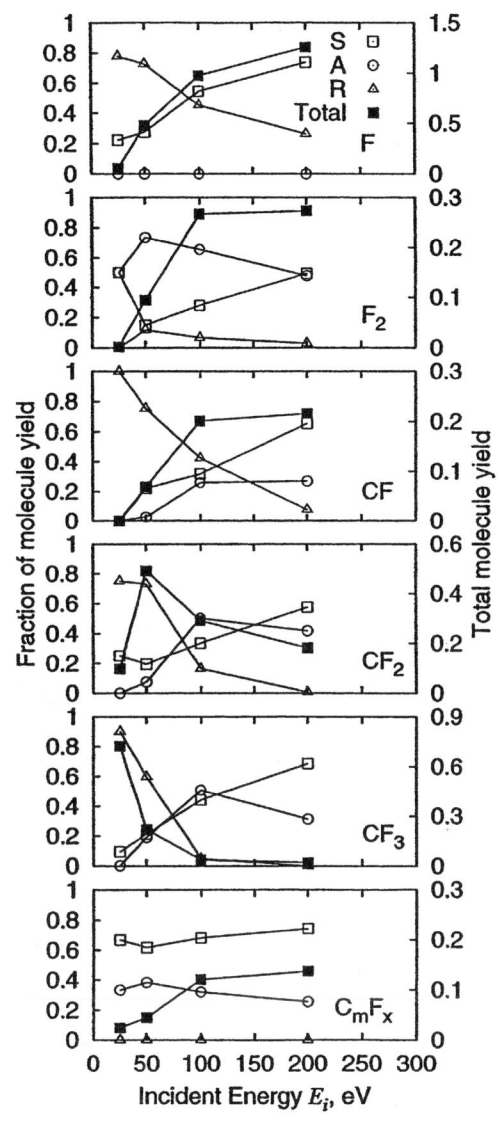

FIG. 27. Yields of major products F, F$_2$, CF, CF$_2$, CF$_3$, and C$_m$F$_x$ vs E_i. Filled symbols are the total yield (molecules/ion), and open symbols are the fraction of the total yield resulting from each mechanism: sputtering (S), abstraction (A), and reflection (R). (From Abrams and Graves, 2000c.)

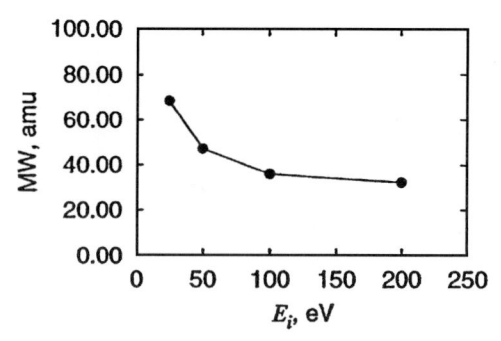

FIG. 28. Average product molecular weight vs E. The standard deviations are smaller than the symbol size. (From Abrams and Graves, 2000c.)

conditions, on average 1 C and 3 F must leave the surface per incident CF_3. By 200 eV, the main ejected species is F.

Figure 28 is a plot of the average product molecular weight as a function of ion energy. Obviously, as energy increases, the average size of the ejected species decreases. However, a careful examination of Table II indicates that the higher molecular weight products increase as the energy increases as well, and these relatively few but heavy species can contain a significant fraction of the C and F emitted from the surface.

From the point of view of etched feature profile evolution, two other characteristics of the ejected species are important: the energy and angle of ejection of products. The product kinetic energy distributions for the set of cases we have examined in this section are shown in Fig. 29. As energy increases from 25 to 200 eV, the primary ejection species changes from CF_3 to F. The solid line corresponds to linear cascade theory (LCT), and as ion impact energy increases, the total kinetic energy distribution gets closer to this model. The average kinetic energy of the ejected species as a function of ejection angle is shown in Fig. 30. An angle greater than 90° corresponds to species moving into the surface at the end of the trajectory. Note that most species are ejected with an average energy of 1 eV and at an angle that peaks around 50° from normal. Species that leave the bottom of an etched feature are more likely to impact the sidewalls than they are simply to move straight up the feature. In addition, a species leaving with 1 eV corresponds to a temperature of about 10^4 K and, therefore, is capable of scaling a significant activation energy barrier for a subsequent chemical reaction.

3. CF^+ and CF_2^+ Deposition on Silicon

As noted above, under some conditions, fluorocarbon ions will deposit as a film rather than etch the underlying silicon. Since C is the depositing species and F the etching species, by reducing the F/C ratio in the ion, one

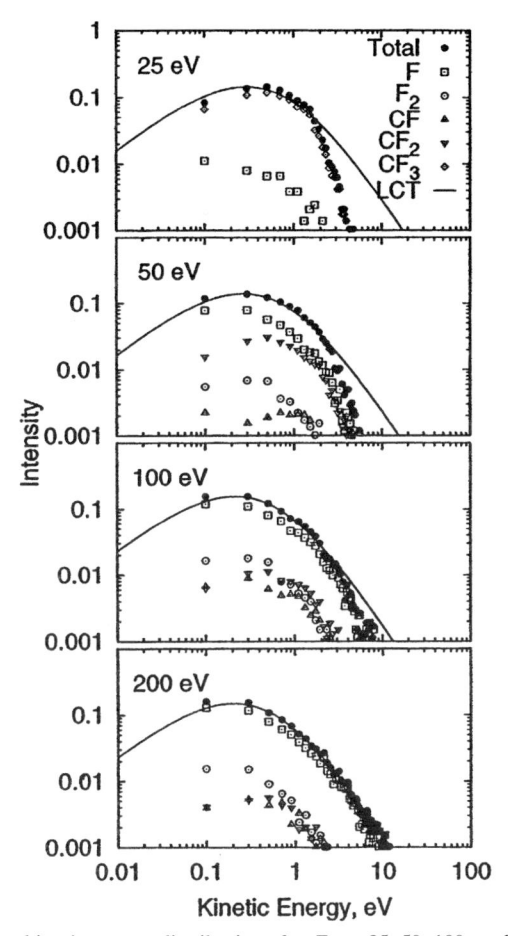

FIG. 29. Product kinetic energy distributions for $E_i = 25, 50, 100,$ and 200 eV. The solid line is a best fit to the "total" distribution using linear cascade theory (Sigmund, 1969). (From Abrams and Graves, 2000c.)

moves in the direction of net deposition: hence the choice of CF^+ and CF_2^+ in this section, which results in net deposition on Si in the energy and angle of impact range we have studied here.

The transition between etching and deposition depends on the etch yield of the material impacting the surface and the underlying material. Steady state corresponds to a constant composition depth profile, after translating the origin with either the net etching or the deposition rate. Therefore, if steady-state etching is to occur, the underlying material will be mixed into the region near the surface with some of the incoming ion material. Steady state can be reached only if this mixing occurs. In contrast, under depositing

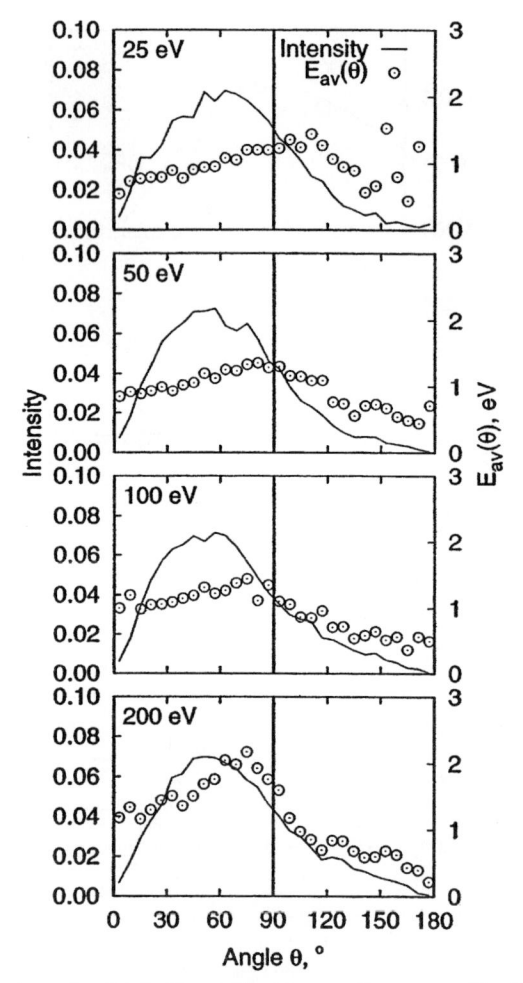

FIG. 30. Product angular distributions, with corresponding average kinetic energy as a function of angle, for E's of 50, 100, and 200 eV. Products for which $\Theta > 90°$ are "direct" and products for which $\Theta > 90°$ are "indirect." (From Abrams and Graves, 2000c.)

conditions, the material from the impacting ion cannot be removed faster than it deposits, and a film builds up. Of course, the composition of the film is not, in general, the same as the composition of the incoming ion. Some preferential sputtering of one component relative to another will generally occur.

In Fig. 31, the evolution of the composition of the surface as a function of ion fluence is shown for CF^+ impact (normal incidence) at 50, 100, and 200 eV. Note that the C and F content increases continuously, representing steady deposition. A small amount of Si is lost at the early stages for the

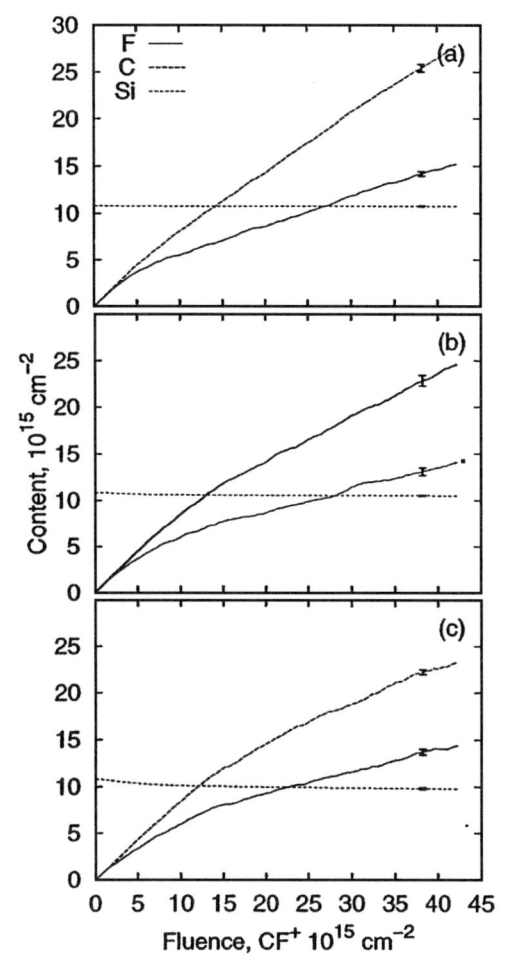

FIG. 31. Uptake of F and C as a function of fluence of CF$^+$ for $E_i =$ (a) 50 eV, (b) 100 eV, and (c) 200 eV. Data shown are the average of three independent runs for each E_i, and error bars show standard deviations averaged over the fluence for each element content. (From Abrams and Graves, 2000a.)

higher energies. Note also that at the early stages of impact, the slopes of the lines are nearly unity, indicating near-unity sticking coefficients on bare Si. The slopes decrease until reaching a steady value. This occurs after about 5×10^{15} cm^{-2} fluence at 50 eV but not until about 15×10^{15} cm^{-2} at 200 eV. The corresponding plot for CF$_2^+$ is shown in Fig. 32. Note that after an initial period of deposition for these cases, the coverages (equivalently, surface content) of C and F appear to stop increasing significantly. CF$_2^+$ appears to result in a finite film thickness that stops increasing after it reaches a

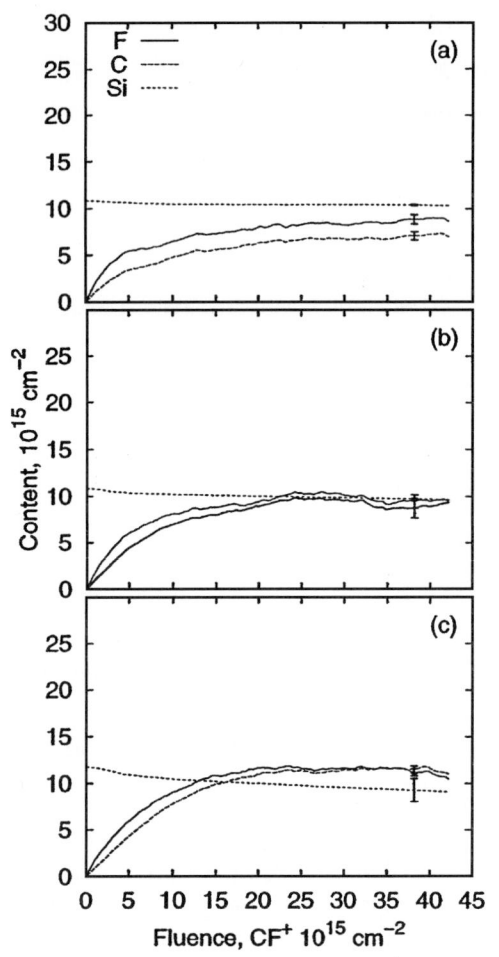

FIG. 32. Uptake of F and C as a function of the fluence of CF_2^+ for $E_i =$ (a) 50 eV, (b) 100 eV, and (c) 200 eV. Data shown are the average of three independent runs for each E_i, and error bars show standard deviations averaged over the fluence for each element content. (From Abrams and Graves, 2000a.)

certain value. This is an example of a case in which the film reaches a certain thickness and stops growing. Of course, it needs to be kept in mind that given the approximate nature of the interatomic potentials, the behavior we observe in the simulations may or may not represent quantitatively what would be observed in the lab under the same conditions. However, tests of the sensitivity of the results to the choice of form of interatomic potential and to parameter values within a form of potential (Abrams, 2000) indicate that qualtitative results are trustworthy.

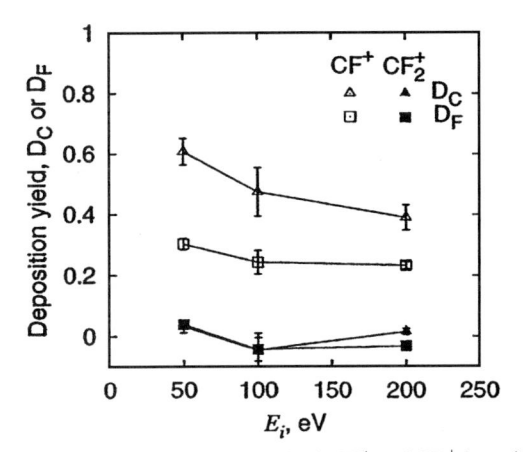

FIG. 33. Deposition yield of F and C vs E_i for both CF$^+$ and CF$_2^+$ ions. (From Abrams and Graves, 2000a.)

Figure 33 is a plot of the (steady-state) deposition yields of C and F for the two ions at 50, 100, and 200 eV. The deposition yield is defined as the slopes of the uptake curves in Figs. 31 and 32, that is, they are the net probability of deposition. The values near zero for CF$_2^+$ indicates that steady-state deposition probabilities are near zero. For CF$^+$, the ratio of C/F in the film stays near 2, but decreases slightly as energy increases. The sticking coefficients for these cases are plotted in Fig. 34. The sticking coefficients are defined in terms of the C and F in the impacting ions alone, ignoring

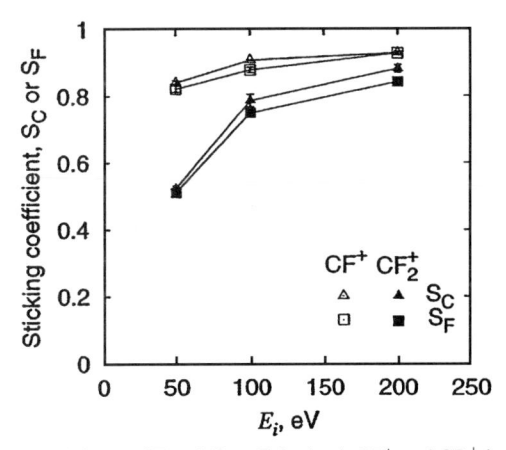

FIG. 34. Sticking coefficients of F and C vs E_i for both CF$^+$ and CF$_2^+$ ions. The magnitudes of the standard deviations for each data point are generally smaller than the symbol size. (From Abrams and Graves, 2000a.)

simultaneous removal of material already on the surface. The ion fragments more on impact at higher energy, and the fragments penetrate more deeply, resulting in a higher probability of sticking. The sticking coefficients for both C and F are above 0.8 for both ions at 200 eV.

The depth-resolved composition profiles as a function of the ion fluence for CF$^+$ at 100 eV are shown in Figs. 35a–d. Note that these profiles are the spatial analogues of the integrated content curves shown in Fig. 31. At a fluence of 2×10^{15} cm^{-2}, the C and F profiles are nearly identical, but by 10^{14} cm^{-2}, the C content is nearly twice the F content. From this point, it is clear that depositon will predominate, and to first order, the film simply grows

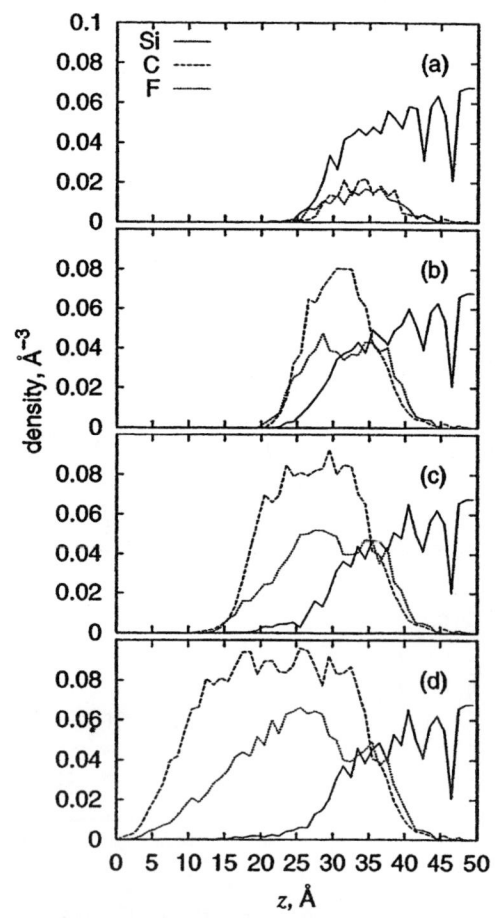

FIG. 35. Evolution of the C, F, and Si depth profiles for the surface layer grown by 100-eV CF$^+$ bombardment, at fluences of (a) 2.0, (b) 10.0, (c) 20.0, and (d) 40.0×10^{15} cm^{-2}. (From Abrams and Graves, 2000a.)

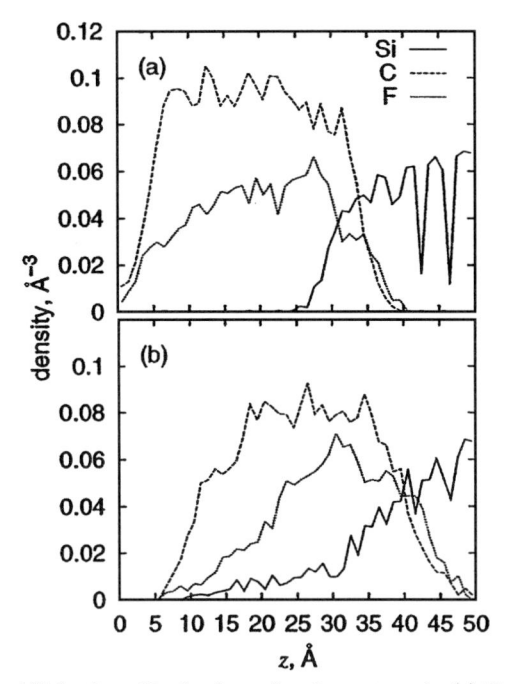

FIG. 36. C, F, and Si depth profiles for the surface layer grown by (a) 50-eV and (b) 200-eV CF$^+$ bombardment, at a fluence of 40×10^{15} cm^{-2}. (From Abrams and Graves, 2000a.)

at a constant composition with $C/F \sim 2$. Figures 36a and b are snapshots of the depth profiles after a fluence of 4×10^{14} cm^{-2} for 50- and 200-eV ions, respectively. The main point here is that the higher-energy ions have a lower deposition rate, and the Si has been more strongly mixed into the fluorocarbon layer. The C/F ratio is about 2 for both ion energies. Obviously, the film composition is not the same as the impacting ion: more C sticks than F, probably because F is attached to the surface with a single bond, whereas C can bond much more strongly. Finally, the approximate thickness of the layer in which C and Si overlap—the "interfacial thickness"—is plotted as a function of the ion energy for the two ions in Fig. 37. This value increases from about 6 to above 10 Å as the energy is increased from 50 to 200 eV.

The results of these simulations have shown that the transition from etching to deposition depends critically on the amount of F that remains in the film, especially in the first 1 or 2 ML of fluence. At 50 eV, the underlying Si is scarcely disturbed and the Si profile remains mostly unchanged, but at 200 eV, the Si mixes over 10–15 Å or more. These results have implications for ion-assisted film deposition and adhesion promotion and the character of the film–substrate interface. Also, in etch tools, it is very common to observe

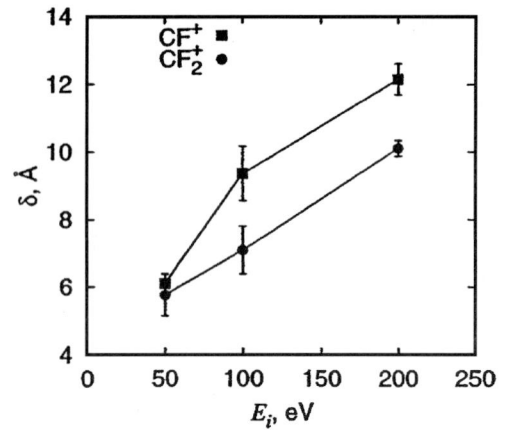

FIG. 37. Interfacial $Si_xC_yF_z$ layer thickness d at steady state vs E_i for CF^+ and CF_2^+ ions. The error bars show the standard deviation among the three runs averaged for each point. (From Abrams and Graves, 2000a.)

film deposition on walls and even on parts of the wafer surface while etching is occurring. Both tool-scale and feature-scale models require reasonably quantitative models of these processes, and the hope is that the atomic-scale insights provided by MD simulations will lead to reliable phenomenological models suitable for those purposes.

IV. Concluding Remarks

MD simulations of plasma–surface, especially ion– or fast neutral–surface interactions, have clearly generated considerable insight into some of the processes that are important in surface modification by plasmas. Equally apparent is that the work has only begun truly to understand and control plasma–surface processes. MD works very well for a few species interacting strongly over a few tens of angstroms and several picoseconds to perhaps a nanosecond. Even with these restrictions, only a handful of the many possible compounds have had many body potentials developed for this use. The potentials that have been applied are clearly approximations, and more work is needed to make them more quantitative. The role of charged species and excited states may be important in some cases, and this has been largely unexplored. Diffusion, reaction, and other processes that take place over time scales from microseconds to seconds under typical conditions need to be incorporated into the simulation scheme, since room-temperature radicals and stable molecules are known to play important roles in surface modification. The focus in this article has been on etching, and deposition is treated elsewhere in this volume. There are other applications involving replacement or alteration of surface functional groups by plasmas that have not been studied to any significant extent using atomistic simulations.

It can be stated with confidence that developing improved control of atomic structure and configuration at surfaces over macroscopic length scales (such as a 300-mm-diameter wafer) will continue to increase in importance. Evolutionary improvements in electronic devices, photonic devices, and magnetic storage and display devices, among others, will continue to drive this need. Simulations of the processes that are used to manufacture such devices will likely become crucial aids to the interpretation of experiment and to the development of the associated manufacturing technology.

ACKNOWLEDGMENTS

The authors relied heavily on the work of Maria Barone and Bryan Helmer, and their many insights and contributions are gratefully acknowledged. Junichi Tanaka and Koji Satake contributed many insights as well, and we are grateful to them. The authors acknowledge support from the University of California SMART, Lam Research Corporation, National Science Foundation, and Semiconductor Research Corporation.

REFERENCES

Abrams, C. F., *Molecular Dynamics Simulations of Plasma-Surface Chemistry*, Ph.D. dissertation, University of California, Berkeley, 2000.
Abrams, C. F., and Graves, D. B., *J. Vac. Sci. Technol. A*, vol. 16, no. 5, pp. 3006–3019 (1998).
Abrams, C. F., and Graves, D. B., *J. Appl. Phys.* **86**(11), 5938–5948 (1999).
Abrams, C. F., and Graves, D. B., *J. Vac. Sci. Techol. A*, **in press** (2000a).
Abrams, C. F., and Graves, D. B., *J. Appl. Phys.* **88**(6), 3734–3738 (2000b).
Abrams, C. F., and Graves, D. B., *Thin Solid Films* **374**(2), 150–156 (2000c).
Abrams, C. F., and Graves, D. B., *J. Vac. Sci. Technol. A* **18**(2), 411–416 (2000d).
Allen, M. P., and Tildesley, D. J., "Computer Simulation of Liquids." Oxford University Press, New York, 1987.
Athavale, S. D., and Economou, D. J., *J. Vac. Sci. Technol. A* **13**(3), 966–971 (1995).
Barone, M. E., Atomistic Simulations of Plasma-Surface Interactions, Ph.D. dissertation, University of California, Berkeley, 1995.
Barone, M. E., and Graves, D. B., *J. Appl. Phys.* **77**(3), 1263–1274 (1995a).

Barone, M. E., and Graves, D. B., *J. Appl. Phys.* **78**(11), 6604–6615 (1995b).
Barone, M. E., and Graves, D. B., *Plasma Sources Sci. Technol.* **5**, 1–6 (1996).
Beardmore, K., and Smith, R., *Phil. Mag. A* **74**(6), 1439–1466 (1996).
Brenner, D. W., *Phys. Rev. B* **42**(15), 9458–9471 (1990).
Brenner, D. W., *Phys. Rev. B* **46**(3), 1948 (1992).
Chen, F. F., "Introduction to Plasma Physics and Controlled Fusion, 2nd ed." Vol. 1. *Plasma Physics*, Plenum Press, New York, 1984.
Coburn, J. W., and Winters, H. F., *J. Appl. Phys.* **50**, 3189 (1979).
Coburn, J. W., Winters, H. F., and Chuang, T. J., *J. Appl. Phys.* **48**(8), 3532–3540 (1977).
Coronell, D. G., Hanson, D. E., Voter, A. F., Liu, C.-L., Liu, X.-Y., and Kress, J. D., *Appl. Phys. Lett.* **73**(26), 3860–3862 (1998).
Feil, H., Dieleman, J., and Garrison, B. J., *J. Appl. Phys.* **74**(2), 1303–1309 (1993).
Galijatovic, A., Darcy, A., Acree, B., Fullbright, G., McCormac, R., Green, B., Krantzman, K. D., and Schoolcraft, T. A., *J. Phys. Chem.* **100**, 9471–9479 (1996).
Garrison, B. J., Reimann, C. T., Winograd, N., and Harrison Jr., D. E., *Phys. Rev. B* **36**(7), 3516–3521 (1987).
Haile, J. M., *Molecular Dynamics Simulation*, John Wiley and Sons, New York, 1992.
Hanson, D. E., Voter, A. F., and Kress, J. D., *J. Appl. Phys.* **82**(7), 3552–3559 (1997).
Hanson, D. E., Kress, J. D., and Voter, A. F., *J. Chem. Phys.* **110**(12), 5983–5988 (1998).
Harrison, D. E., *Crit. Rev. Solid State Mater. Sci.* **14**(Suppl. 1), S1 (1988).
Harrison, D. E. Jr., Levy, N. S., Johnson, J. P., and Effron, H. M., *J. Appl. Phys.* **39**(8), 3742–3761 (1968).
Harrison, D. E. Jr., Kelly, P. W., Garrison, B. J., and Winograd, N., *Surf. Sci.* **76**, 311–322 (1978).
Helmer, B. A., *Computer Simulations of Plasma-Surface Chemistry*, Ph.D. dissertation, University of California, Berkeley, 1998.
Helmer, B. A., and Graves, D. B., *J. Vac. Sci. Technol. A* **15**, 2252 (1997).
Helmer, B. A., and Graves, D. B., *J. Vac. Sci. Technol. A* **16**(6), 3502–3514 (1998).
Kress, J. D., Hanson, D. E., Voter, A. F., Liu, C. L., Liu, X.-Y., and Coronell, D. G., *J. Vac. Sci. Technol. A* **17**(5), 2819–2825 (1999).
Kubota, N. A., Economou, D. J., and Plimpton, S. J., *J. Appl. Phys.* **83**(8), 4055–4063 (1998).
Layadi, N., Donnelly, V. M., and Lee, J. T. C., *J. Appl. Phys.* **81**, 6738 (1997).
Lieberman, M. A., and Lichtenberg, A. J., *Principles of Plasma Discharges and Materials Processing*, John Wiley and Sons, 1994.
Oostra, D., Haring, A., de Vries, A., Sanders, F., and van Veen, G., *Nucl. Instrum. Methods Phys. Res.* **13**, 556 (1986).
Post, D. E., and Behrisch, R., (eds.), *Physics of Plasma-Wall Interactions in Controlled Fusion*, NATO ASI Series B, Plenum Press, New York, 1986.
Rapaport, D. C., *The Art of Molecular Dynamics Simulation*, Cambridge University Press, Cambridge, 1995.
Sigmund, P., *Phys. Rev.* **184**(2), 383–416 (1969).
Sikola, T., Armour, D. G., and den Berg, J. A. V., *J. Vac. Sci. Technol. A* **14**, 3156–3163 (1996).
Smith, R., Harrison, D. E. Jr., and Garrison, B. J., *Phys. Rev. B* **40**(1), 93–101 (1989).
Smith, R., Harrison, D. E. Jr., and Garrison, B. J., *Nucl. Instrum. Methods Phys. Res. B* **46**, 1–11 (1990).
Stillinger, F. H., and Weber, T. A., *Phys. Rev. Lett.* **62**(18), 2144–2147 (1989).
Tanaka, J., Abrams, C. F., and Graves, D. B., *J. Vac. Sci. Technol. A* **18**(3), 938–945 (2000).
Tersoff, J., *Phys. Rev. B* **38**(14), 9902–9905 (1988).
Tersoff, J., *Phys. Rev. B* **39**(8), 5566–5568 (1989).
Thompson, D. J., and Helms, C. R., *Surf. Sci.* **236**, 41–47 (1990).
Tully, J. C., *Annu. Rev. Phys. Chem.* **31**, 319–343 (1980).

Vyvoda, M. A., Abrams, C. F., and Graves, D. B., *IEEE Trans. Plas. Sci.* **27**(5), 1433–1440 (1999).
Vyvoda, M. A., *Surface Evolution during Integrated Circuit Processing,* Ph.D. thesis, University of California, Berkeley, 1999.
Winters, H. F., and Coburn, J. W., *Surf. Sci. Rep.* **14**(4–6), 161–269 (1992).
Wolf, S., and Tauber, R. N., *Silicon Processing for the VLSI Era, Vol. 1. Process Technology,* 2nd ed., Lattice Press, Sunset Beach, CA, 2000.

CHARACTERIZATION OF POROUS MATERIALS USING MOLECULAR THEORY AND SIMULATION

Christian M. Lastoskie

Department of Chemical Engineering, Michigan State University, East Lansing, Michigan 48824

Keith E. Gubbins

Department of Chemical Engineering, North Carolina State University, Raleigh, North Carolina 27695

I. Introduction	203
II. Disordered Structure Models	206
A. Porous Glasses	206
B. Microporous Carbons	209
C. Xerogels	213
D. Templated Porous Materials	216
III. Simple Geometric Pore Structure Models	218
A. Molecular Simulation Adsorption Models	222
B. Density Functional Theory Adsorption Models	225
C. Semiempirical Adsorption Models	231
D. Classical Adsorption Models	239
IV. Conclusions	244
References	246

I. Introduction

Characterization of the pore structure of amorphous adsorbents and disordered porous catalysts remains an important chemical engineering research problem. Pore structure characterization requires both an effective experimental probe of the porous solid and an appropriate theoretical or numerical model to interpret the experimental measurement. Gas adsorption porosimetry [1] is the principal experimental technique used to probe the structure of the porous material, although various experimental alternatives have been proposed including immersion calorimetry [2–4], positron annihilation [5], transmission electron microscopy [6, 7], and small-angle X-ray [7, 8] or neutron [9] scattering. Consequently, most pore structure analysis methods attempt to address the question of how properly to interpret gas sorption uptake measurements on porous solids to obtain an accurate picture of the porous material structure.

Two approaches may be taken in devising a pore structure model to apply to experimental measurements. In one approach, a *disordered model microstructure* is assembled that explicitly incorporates the amorphous nature of the adsorbent into the theoretical model. For example, an activated carbon might be represented as a randomly oriented assemblage of graphitic disks of uniform size and composition [10]. In the other approach, a *simple geometric model* of the adsorbent pore structure is utilized, and the amorphous character of the porous solid is implicitly represented through structure distribution functions. In this case, the pore volume of the aforementioned activated carbon might be modeled as an array of slit-shaped pores with a distribution of slit widths [11]. Not surprisingly, the substitution of a simplified geometric model for the actual structure of a disordered adsorbent invokes major assumptions regarding pore shape, connectivity, and chemical composition. At best, these assumptions will introduce a degree of uncertainty into the characterization results, and in some instances, the simple pore geometries may be altogether unsuitable for describing the adsorbent structure. However, analysis methods that are based upon simple geometric models generally lend themselves to convenient and efficient numerical solution, whereas fitting the structure parameters of a model disordered adsorbent can be quite difficult and computationally intensive to carry out. For this reason, a great majority of the thermodynamic models for interpreting experimental gas adsorption isotherms are based upon simple geometric representations of pore structure that, for the most part, do not directly incorporate the effects of pore shape variation, pore connectivity, or chemical heterogeneity.

The thermodynamic models used to represent adsorption in either the disordered or the simple geometric description of the pore volume may be classified into four categories, presented in decreasing order of computational intensiveness.

1. *Molecular simulation* calculations of the theoretical adsorption isotherm for a model pore structure under the experimental conditions: This involves principally either grand canonical Monte Carlo (GCMC) [12] or Gibbs ensemble Monte Carlo (GEMC) [13] simulation of adsorption in regular pore geometries but also includes reverse Monte Carlo (RMC) [14] simulation methods for the interpretation of pore morphologies from scattering data.

2. *Statistical thermodynamic theories* such as atomistic density functional theory (DFT) for the computation of adsorption isotherms in simple pore geometries such as slits [15] or cylindrical capillaries [16]: This category also includes integral equation methods for porous matrices [17] and templated porous materials [18].
3. *Semiempirical models* such as the Horvath–Kawazoe (HK) method [19] and the Dubinin model [20] and their derivatives: These models generally make specific assumptions regarding the shape of the pores and/or the distribution function that describes the pore sizes within the adsorbent. To varying degrees, the semiempirical methods incorporate adsorbate–adsorbent interaction energies into the calculation of the theoretical isotherm.
4. *Classical thermodynamic models* of adsorption based upon the Kelvin equation [21] and its modified forms: These models are constructed from a balance of mechanical forces at the interface between the liquid and the vapor phases in a pore filled with condensate and, again, presume a specific pore shape. The Kelvin-derived analysis methods generate model isotherms from a continuum-level interpretation of the adsorbate surface tension, rather than from the atomistic-level calculations of molecular interaction energies that are predominantly utilized in the other categories.

As a general rule, the more sophisticated and computationally demanding thermodynamic models yield a more realistic description of the adsorption process and, consequently, a more accurate characterization of the pore structure when used to analyze experimental adsorption data. However, even numerically intensive analysis methods such as GCMC simulation and DFT will yield erroneous pore structure results if the model is not properly posed. For example, large errors may result in the pore size distribution interpreted from nitrogen adsorption of an activated carbon using DFT if unrealistic values are selected for the nitrogen–carbon potential parameters [22]. Thus, accurate characterization of the pore volume of an adsorbent requires both a realistic adsorption model, obtained from one of the four previously noted categories, and a method of fitting the key adjustable parameters of the adsorption model (in most cases, the intermolecular potential parameters). A third complicating factor, one that may introduce additional uncertainty into the characterization results, is the numerical technique used to interpret the experimental isotherm in terms of the designated adsorption model. Regularization methods, for example, may have a profound effect on the computed physical properties of a porous adsorbent [23].

In the remainder of this paper, adsorption models for adsorbent characterization based upon both amorphous (disordered) and idealized (simple geometric) pore structures are reviewed. A description of the principal structure analysis methods in each of the four thermodynamic model categories is also provided, and applications of these methods are highlighted. In Section II, amorphous porous microstructure models are discussed; in Section III, idealized pore geometric models are surveyed. Finally, in Section IV a summary of the state of the art in porous material characterization is presented, with an assessment of the relative merits of the currently available analysis methods and a set of recommendations for future improvements to these adsorption models.

II. Disordered Structure Models

Pore structure analysis methods based upon realistic disordered microstructures may be classified into two types. In one approach, the experimental procedures used to fabricate the material are reproduced, to the greatest extent possible, via molecular simulation, and the resulting amorphous material structure is then statistically analyzed to obtain the desired structural information. In the other approach, adsorbent structural data (e.g., small-angle neutron scattering) is used to construct a model disordered porous structure that is statistically consistent with the experimental measurements. As in the first approach, molecular simulations can then be carried out using the derived model structure to obtain the structural characteristics of the original adsorbent.

Several examples follow of recent efforts to describe explicitly porous adsorbent materials using disordered structure models.

A. Porous Glasses

An illustration of the first type of amorphous material modeling is the use of quench molecular dynamics (MD) methods to mimic the spinodal decomposition of a liquid mixture of oxides, producing model porous silica glasses that are topologically similar to controlled pore glasses (CPGs) or Vycor glasses [24, 25]. In the quench MD simulations, a homogeneous binary mixture of a large number (10^5 to 10^6) of spherical atomic particles at an elevated temperature is subjected to a sudden decrease in temperature, causing phase segregation to occur as shown in Fig. 1. As the quenching time increases, the extent of phase segregation also increases. By removing the atoms associated with the discontinuous phase (i.e., the atomic species with the smaller mole fraction), an irregularly shaped, highly interconnected

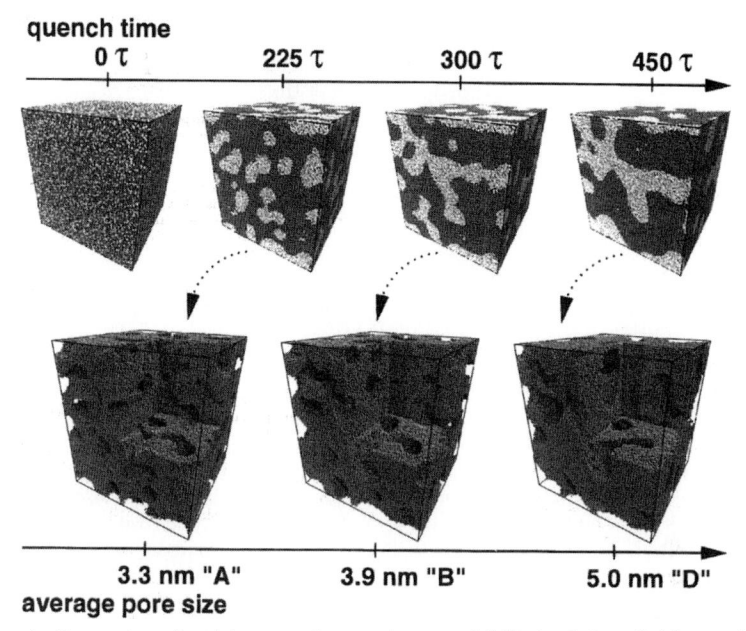

FIG. 1. Generation of model porous glasses using quench MD simulation of a binary mixture with a mole fraction of 0.70. Quenching produces a series of phase-separated structures, which may be converted at any time into a porous network (shown in cutaway view at the bottom) by removing the component with the smaller mole fraction [25].

network of pores is fashioned. For longer quenching times, larger phase segregated domains are formed as the binary mixture progresses toward the equilibrium phase-separated condition. This results in a larger average pore size in the porous network when the discontinuous phase is removed. By changing the quench time, the pore structure of the model adsorbent can therefore be tailored to some extent.

The quench MD procedure simulates the preparation of CPGs and Vycor glasses [26], in which the near-critical phase separation of a mixture of SiO_2, Na_2O, and B_2O_3, or a similar oxide mixture, is carried out, followed by etching to remove the borosilicate phase. The etching treatment produces a CPG silica matrix with a porosity ranging between 50 and 75% and an average pore size that is adjustable anywhere between 4.5 and 400 nm by varying the duration of the quenching stage. Vycor glasses prepared by a similar procedure have a porosity near 28% and an average pore diameter between 4 and 7 nm [27].

Once the model porous glass structures have been assembled using quench MD simulation, their geometric pore size distributions can be determined by sampling the pore volume accessible to probe molecules

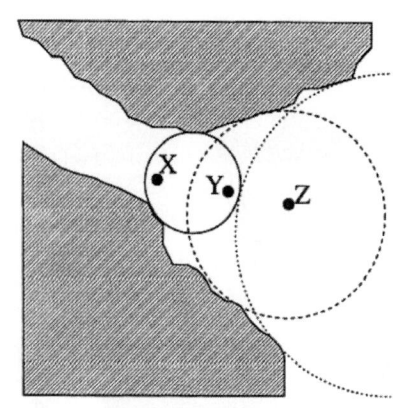

FIG. 2. Two-dimensional illustration of the geometric definition of the pore size distribution [25]. Point Z may be overlapped by all three circles of differing radii, whereas point Y is accessible only to the two smaller circles and point X is excluded from all but the smallest circle. The geometric pore size distribution is obtained by determining the size of the largest circle that can overlap each point in the pore volume. (Reproduced with permission from S. Ramalingam, D. Maroudas, and E. S. Aydil. Interactions of SiH radicals with silicon surfaces: An atomic-scale simulation study. *Journal of Applied Physics*, 1998;84:3895–3911. Copyright © 1998, American Institute of Physics.)

of different radii, as illustrated in Fig. 2. For each point in the pore volume, the radius of the largest spherical molecule is found which overlaps the given point but does not overlap any atoms of the solid matrix. For spherical probe molecules of radius r, there is a pore volume $V(r)$ that is accessible to the molecules. The pore volume function $V(r)$ is a monotonically decreasing function of r and is directly analogous to the cumulative pore volume curves commonly reported from interpretation of experimental adsorption isotherms [28]. The derivative $-dV(r)/dr$ is the fraction of the pore volume accessible to spheres of radius r but not accessible to spheres of radius $r + dr$. This derivative is a direct definition of the pore size distribution function and it can be calculated by Monte Carlo volume integration [29].

Using this technique, it has been shown that the model disordered adsorbents created using quench MD simulation have pore size distributions, porosities, and specific surface areas that closely resemble those of actual porous glasses. An advantage of the geometric pore size definition, as presented in Fig. 2, is that for irregularly shaped pore volumes this definition is fully applicable. In contrast, most analytic methods for obtaining the cumulative pore volume distribution from the experimental isotherm are based upon simple geometric pore models that bear little resemblance to the complex porous glass structures shown in Fig. 1.

To develop amorphous structure model for the direct interpretation of adsorption isotherms measured on CPGs and porous glasses, one could in principle assemble a sequential array of disordered structures from quench MD using different quenching times for each model structure. By carrying out GCMC molecular simulations (described in Section III) to construct the adsorption isotherm of each candidate structure, a representative porous glass model structure could then be selected as the one that gives the best agreement with the experimentally measured probe gas isotherm. This procedure may be computationally expensive and time-consuming, however, because of the large number of atoms required to generate realistic pore structures from the quench MD simulations. An alternative method that has been suggested for interpreting porous glass structures is to use off-lattice reconstruction methods [30, 31] to build realistic models of porous glasses. In this method, a model microstructure is identified based upon the volume autocorrelation function obtained from TEM data. A method similar in theme is the reverse Monte Carlo (RMC) technique of matching the structures of model disordered porous carbons to data obtained from small-angle X-ray or neutron scattering [32]. This method is an example of the second approach to molecular modeling of amorphous solids, and it is described in more detail in the next section.

B. MICROPOROUS CARBONS

Experimental radial distribution functions (RDFs) for the atoms in a porous solid may be obtained from the Fourier transform of the structure factor measured using small-angle X-ray or neutron scattering [33]. The RDF for carbon–carbon atomic pairs in a microporous carbon provides the experimental input for developing a model disordered carbon microstructure using RMC simulation. The RMC method follows the procedure of traditional Monte Carlo methods in that an initial atomic configuration is systematically changed through a well-defined stochastic procedure. The principal distinction between various Monte Carlo techniques is in the criteria used to accept or reject trial configurations. Rather than basing configuration acceptance on ensemble probability functions, as in the well-known Metropolis algorithm [34], new configurations are accepted in the RMC method based upon whether they improve the agreement of simulated structure correlation functions with experimental inputs.

A morphological model for microporous carbons has recently been reported [32] in which rigid aromatic sheets of sp^2 bonded carbon are randomly placed in a three-dimensional cubic simulation cell with periodic boundaries. A typical carbon plate has the structure shown in Fig. 3a. The plates are roughly aligned in the simulation cell, as illustrated in Fig. 3b, but with random variations in their angles of tilt. RMC simulation is carried out by sampling three types of changes to the carbon structure: (i) translation and

(a)   (b)

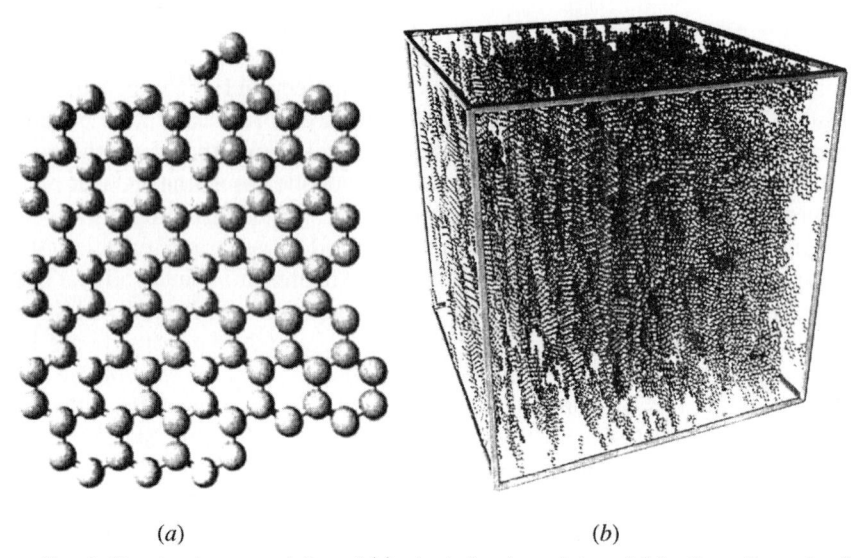

FIG. 3. Structural representation of (a) a typical carbon plate and (b) a three-dimensional disordered carbon microstructure obtained from RMC simulation of a collection of carbon plates [32].

rotation of the plate centers of mass, (ii) addition or removal of aromatic rings from the edges of the carbon plates, and (iii) creation or annihilation of whole carbon plates. Sampled configurations are accepted if they bring the carbon–carbon radial distribution function $g(r)$ of the simulated structure into closer agreement with the experimental target distribution $g_{\exp}(r)$. The objective acceptance criterion χ^2 is given as

$$\chi^2 = \sum_{i=1}^{n} [g(r_i) - g_{\exp}(r_i)]^2, \tag{1}$$

where the range of separation distances over which RDF matching is to be carried out has been divided into n equally spaced intervals. A trial structure is accepted if $\chi^2_{\text{new}} < \chi^2_{\text{old}}$.

The RMC procedure has been used to study the structural morphology of microporous carbons. Porous carbons such as activated carbons, carbon fibers, carbon blacks, and vitreous carbons are often envisioned as disordered arrangements of defective crystallites of graphite [35]. The size and shape of the graphite crystallites are known to vary according to the ability of the material to transform to graphite under severe heat treatment. Domains of orientational order are usually observed in which microcrystallites align in similar direction [36]. A variety of models has been suggested to try to account for the deviations of activated carbon morphologies from

CHARACTERIZATION OF POROUS MATERIALS 211

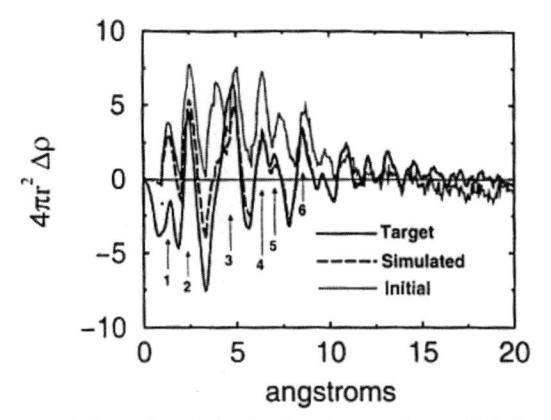

FIG. 4. Experimental (target) and simulated carbon–carbon radial distribution function for an activated mesocarbon microbead adsorbent [32]. The RDFs are reported in the form $4\pi r^2 \rho[g(r) - 1]$, where ρ is the average carbon density and r is the carbon pair separation distance noted on the horizontal axis. The simulated RDFs for the initial and converged (dashed line) model structures are shown, with numbered peaks referenced in the text.

ideal slit-shaped pore geometries [10, 37–40]; none of these models, however, provides a method for constructing the model carbon microstructure from the experimental isotherm or scattering data. The RMC method, on the other hand, generates disordered structures that realistically simulate the structure correlation functions of carbon adsorbents.

Figure 4 shows a comparison between the simulated and the experimentally measured carbon–carbon RDFs for an activated mesocarbon microbead (MCMB) [32]. Using the RMC algorithm, the RDF of the initial adsorbent structure is brought into nearly exact agreement with the experimental RDF of the MCMB, determined from small-angle X-ray scattering (SAXS), for C—C distances greater than 7 Å. The differences between the best-fit simulated RDF and the experimental RDF below 5 Å are attributed to inaccuracies in the SAXS resolution of the structure factor. The RDFs in Fig. 4 combine both the intraplate and the interplate contributions to the C—C radial distribution function, the former comprising carbon atom pairs on the same plate and the latter from pairs located on different plates. The RDF for separation distances less than 5 Å corresponds almost exclusively to intraplate contributions, and SAXS experiment tends to undercount the carbon density in this portion of the measurement. The first RDF peak (labeled 1) represents the direct intraplate C—C distance along sp^2 bonds. Peaks 2 and 3 correspond to C—C distances for atom pairs in the same aromatic ring and in neighboring rings. Peak 3 also contains a contribution due to interplate separations between neighboring plates that are aligned

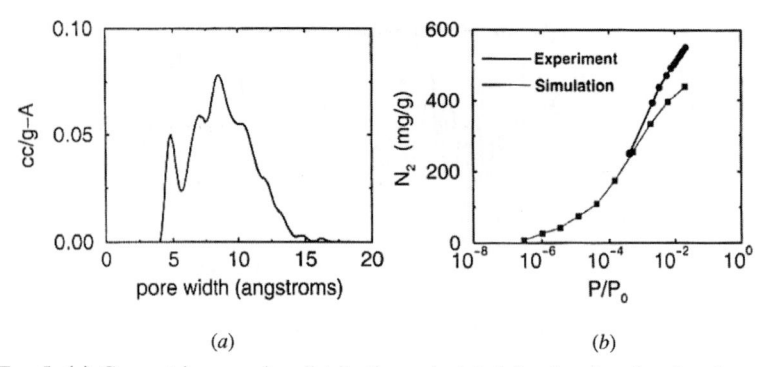

FIG. 5. (a) Geometric pore size distribution calculated for the disordered carbon plate model of an activated MCMB adsorbent. (b) Comparison of experimental (circles) and simulated (squares) nitrogen adsorption isotherms at 77 K on activated MCMB [32].

nearly in parallel (i.e., stacked layers of graphitic crystallites). Peaks at larger separation distances correspond to specific interplate C—C distances; e.g., Peak 6, at 8.2 Å, is due primarily to second-nearest-neighbor interplate distances.

The geometric pore size distribution (PSD) can be found for the disordered porous carbon model structures using the same procedure as described for CPGs in Section II.A. The PSD so obtained for the converged simulated MCMB structure is shown in Fig. 5a. There is a direct correspondence between certain peaks in the PSD and peaks in the adsorbent RDF. For example, the PSD peak at ~5 Å arises from a staggered configuration of three stacked plates, in which the two outer plates extend beyond the end of the middle plate. This plate configuration yields a pore space approximately 5 Å in width and is observed throughout the MCMB structure shown in Fig. 3b.

The porosity, specific surface area, and micropore size distributions of the simulated model carbon structures are in good agreement with the physical properties of the target structures fitted using RMC simulation [32]. Another test of the RMC-derived structure is to calculate the nitrogen adsorption isotherm of the model structure at 77 K using grand canonical Monte Carlo simulation [41] (see Section III.A) and compare it with the experimental nitrogen isotherm. The result for the MCMB adsorbent is shown in Fig. 5b. It can be seen that the experimental and simulated isotherms are in excellent agreement for relative pressure P/P_0 below $\sim 10^{-3.3}$. At higher relative pressures, the simulated isotherm predicts a lower nitrogen uptake than the experimental result. This is due to the absence of mesopores in the simulated model structure, which condense nitrogen at relative pressures $P/P_0 > 10^{-3}$. There is no direct means of incorporating mesoporous structures into the RMC model because the structure correlation function $g(r)$ of the adsorbent material has a weak sensitivity to long-range correlations (see Fig. 4 for $r > 15$Å). One possible route for incorporate mesopore-sized void regions in the disordered model structure would be to include correlation functions in the RMC optimization that are specifically sensitive to long-range structure, such as those obtained from TEM micrographs. It should also be recognized that the optimized structure for MCMB shown in Fig. 3b, while consistent with the experimental structure and adsorption measurements, is by no means a unique solution to the adsorbent structure. Other carbon microstructures may be generated using the RMC technique, and these alternative structures will be equally consistent with the modeling constraints. In the case of purely graphitic crystalline models, variations among the plausible microstructures may not significantly effect the calculated adsorption properties. However, this may not necessarily be the case for more sophisticated models that incorporate plate defects or that include heteroatoms, such as oxygen-, nitrogen-, sulfur-containing surface functional groups that bridge between carbon plates. In the latter case, these factors may be expected to have significant effects on the pore morphology and adsorption properties.

C. Xerogels

Capillary condensation and hysteresis have been investigated in disordered porous materials that resemble silica xerogels using GCMC molecular simulation [42–44]. In GCMC simulation of adsorption, configurational sampling of the adsorbate is carried out at a fixed chemical potential, temperature, and volume using three types of perturbations: (i) displacement of molecules to new positions, (ii) insertion of new molecules into the pore volume, and (iii) deletion of molecules from the pore volume. The displacement trial moves provide thermal equilibration of the fluid and are accepted with a probability p_{dis} given as

$$p_{dis} = \min\left[1, \exp\left(\frac{-\Delta U}{kT}\right)\right], \quad (2)$$

where k is the Boltzmann constant and T is the temperature. The configurational energy change ΔU associated with the molecular displacement includes both adsorbate–adsorbate and adsorbate–adsorbent potential interactions. The insertion and deletion trial configurations provide material equilibration at imposed chemical potential μ and are accepted with respective probabilities p_{ins} and p_{del} that depend on the current number of molecules N in the simulated pore volume according to the equations

$$p_{\text{ins}} = \min\left[1, \left(\frac{a}{N+1}\right)\exp\left(\frac{-\Delta U}{kT}\right)\right], \tag{3}$$

$$p_{\text{del}} = \min\left[1, \left(\frac{N}{a}\right)\exp\left(\frac{-\Delta U}{kT}\right)\right], \tag{4}$$

where

$$a = \left(\frac{h^2}{2\pi mkT}\right)\exp\left(\frac{\mu}{kT}\right), \tag{5}$$

and h and m are the Planck constant and the adsorbate molecular weight, respectively. The adsorption isotherm is constructed from GCMC simulation by statistically sampling the density of the equilibrated confined fluid for a sequence of chemical potential values, which can be related to the relative pressure through the appropriate bulk fluid equation of state (e.g., Ref. 45). By increasing the chemical potential in increments and computing the adsorbate density at each potential value, the adsorption branch of the isotherm is obtained; similarly, the desorption branch of the isotherm is recovered by computing the adsorbate density for a decreasing series of input chemical potential values.

The GCMC simulation method has been applied to the study of adsorption in model silica gels, in which the material is represented as a disordered collection of solid spherical particles [46, 47]. Figure 6a shows visualizations of the coexisting liquid and vapor phases of a Lennard–Jones fluid in a disordered array of spherical particles of uniform size [44]. The disordered model silica structure was obtained by equilibrating an ordered face-centered cubic array of spherical particles with the desired porosity (shown at the right in Fig. 6a) using molecular dynamics simulation. The effects of disorder may be discerned by comparing the isotherms obtained for the ordered and the amorphous structures. Figure 6b shows subcritical methane adsorption and desorption isotherms for two values of the methane–silica potential interaction strength. A hysteresis loop is seen that bears a resemblance to the category H2 loop in the IUPAC classification of type IV and type V isotherms observed for mesoporous solids [48]. This hysteresis is purely thermodynamic in nature; another manifestation of sorption hysteresis, arising from pore blocking effects in porous networks, may be investigated using the grand canonical molecular dynamics simulation technique [44, 49, 50].

The GCMC adsorption simulation method, when applied to model disordered assemblies of spherical silica particles, yields isotherms that are similar to the experimental isotherms of silica xerogels [51]. In principle,

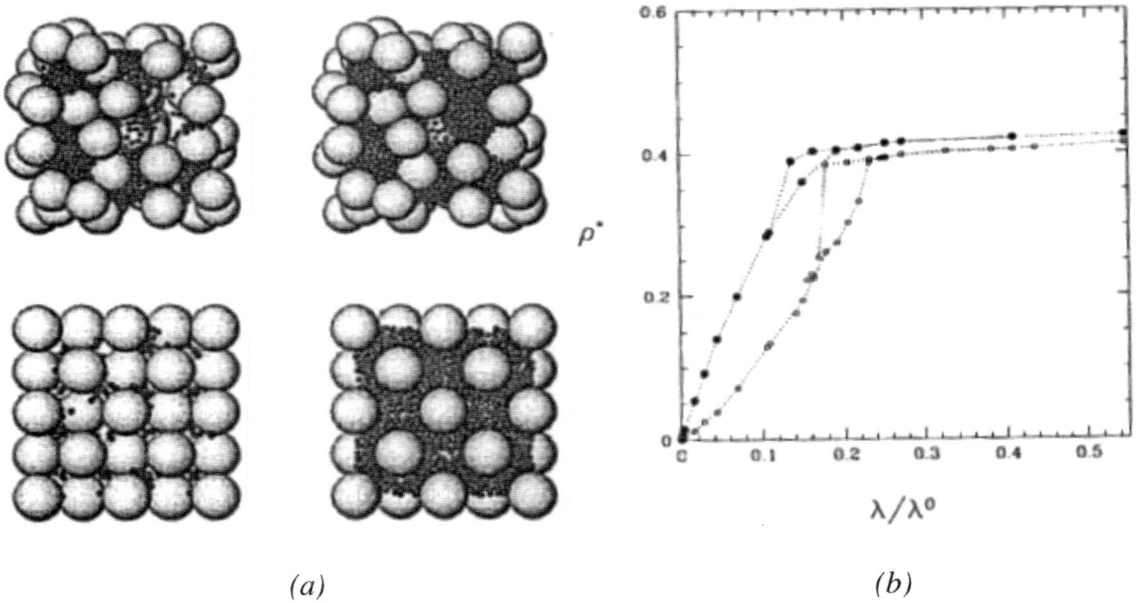

FIG. 6. (a) Visualizations of configurations from GCMC simulations of confined fluids in model silica xerogels [44]. The small spheres are fluid molecules, and the large spheres are immobile silica particles. The top visualizations are for a disordered material and the bottom visualizations are for an ordered material of the same porosity. The visualizations on the left are for the saturated vapor state, and those on the right are for the corresponding saturated liquid state. (b) Simulated adsorption and desorption isotherms for Lennard-Jones methane in a silica xerogel at reduced temperature $kT/\varepsilon_{ff} = 0.7$. The reduced adsorbate density $\rho^* = \rho\sigma^3$ is plotted vs the relative pressure λ/λ_0 for methane–silica/methane–methane well depth ratios $\varepsilon_{sf}/\varepsilon_{ff} = 1.5$ (open circles) and 1.8 (filled circles) [44]. (Reproduced with permission from S. Ramalingam, D. Maroudas, and E. S. Aydil. Atomistic simulation study of the interactions of SiH$_3$ radicals with silicon surfaces. *Journal of Applied Physics*, 1999;86:2872–2888. Copyright © 1999, American Institute of Physics.)

structure correlation data from scattering experiments could be used in conjunction with the adsorption isotherm to construct model disordered structures for specific silica gel adsorbents, using the methodologies described in Section II.B for activated microporous carbons.

D. Templated Porous Materials

Another variation on the theme of quenched disordered structures noted in Section II.A is to employ a removable template such as an organic molecule, colloid, or metal ion during the synthesis of a porous material [52–54]. Following formation of the quenched material structure, as shown in Fig. 7, the template is removed, leaving behind a matrix of particles with a pore space that mimics, to some extent, the original template. Because templates of diverse size and shape are available, templating offers the prospect of designing porous materials whose architectures are tailored for specific applications.

Model templated structures can be assembled from Monte Carlo simulations of binary mixtures of matrix and template particles [55–57]. Upon removal of the template from the quenched equilibrated structure, a porous matrix is recovered with an enhanced accessible void volume for adsorption. GCMC simulation studies have established that the largest enhancement of adsorption uptake occurs when the template particles used to fashion the porous matrix are the same size as the adsorbate molecules for which the adsorbent is intended [55]. The enhanced adsorption capacity of the templated material relative to a nontemplated matrix is noticeable even for modest template particle densities [55].

A adsorption model for a templated porous material may be posed in terms of seven replica Ornstein–Zernicke (ROZ) integral equations [58–60] that relate the direct and total correlation functions, $c_{ij}(r)$ and $h_{ij}(r)$, respectively, of the matrix–adsorbate system

$$h_{00} = c_{00} + \rho_0 c_{00} \otimes h_{00} + \rho_{0'} c_{00'} \otimes h_{00'} \tag{6}$$

$$h_{00'} = c_{00'} + \rho_0 c_{00} \otimes h_{00'} + \rho_{0'} c_{00'} \otimes h_{0'0'} \tag{7}$$

$$h_{0'0'} = c_{0'0'} + \rho_0 c_{00'} \otimes h_{00'} + \rho_{0'} c_{0'0'} \otimes h_{0'0'} \tag{8}$$

$$h_{01} = c_{01} + \rho_0 c_{00} \otimes h_{01} + \rho_{0'} c_{00'} \otimes h_{0'1} + \rho_1 c_{01} \otimes h_c \tag{9}$$

$$h_{0'1} = c_{0'1} + \rho_0 c_{00'} \otimes h_{01} + \rho_{0'} c_{0'0'} \otimes h_{0'1} + \rho_1 c_{0'1} \otimes h_c \tag{10}$$

$$h_{11} = c_{11} + \rho_0 c_{01} \otimes h_{01} + \rho_{0'} c_{0'1} \otimes h_{0'1} + \rho_1 c_c \otimes h_{11} + \rho_1 c_b \otimes h_c \tag{11}$$

$$h_c = c_c + \rho_1 c_c \otimes h_c. \tag{12}$$

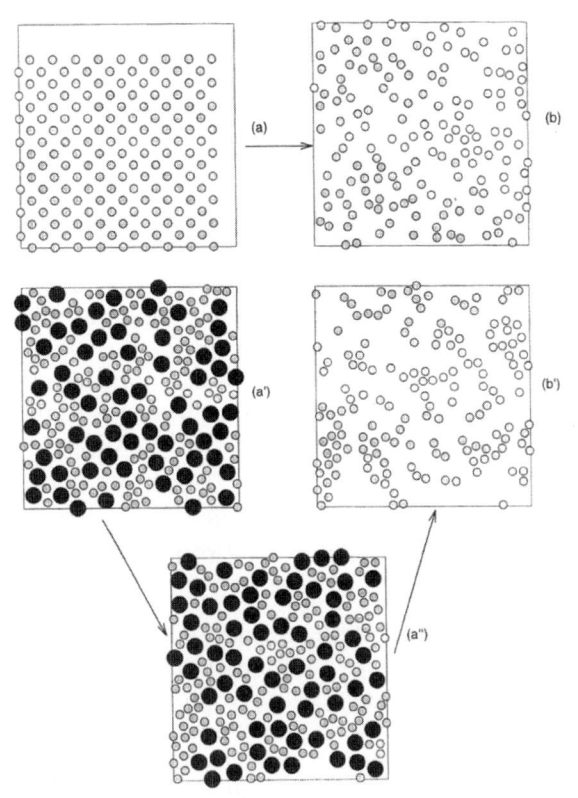

FIG. 7. Illustration of porous matrix formed via templating. The initial configuration of particles shown in a is equilibrated at a high temperature without a template and then quenched to yield the structure shown in b. The initial configuration in a' has the same density of matrix particles (small circles) as in a, but template particles (large circles) are also present in this system. The template particles are removed from the quenched equilibrated matrix + template system (a'') to yield the structure shown in b'. It is clear to the eye that the structure in b' possesses a more "open" pore structure with more available void volume than the structure in b [55]. (Reproduced with permission from S. Ramalingam, D. Maroudas, and E. S. Aydil. Atomistic simulation study of the interactions of SiH$_3$ radicals with silicon surfaces. *Journal of Applied Physics,* 1999;86:2872–2888. Copyright © 1999, American Institute of Physics.)

In Eqs. (6)–(12), ρ_i is the density of component i; $\otimes$ denotes a convolution in r-space; and the subscripts 0, 0', and 1 refer to the matrix, template, and adsorbate, respectively. The subscript c denotes the correlation function between a pair of fluid particles whose graphical expansion results in diagrams that all possess at least one path involving only fluid particles. The subscript b, in contrast, refers to diagrams containing only paths that pass through at least one matrix particle. Solving the integral equations requires specification of

(i) a pairwise additive potential $u_{ij}(r)$, e.g., the hard-sphere potential, and (ii) a set of closures, such as the well-known Percus–Yevick closure [61],

$$c_{ij}(r) = [1 - \exp(u_{ij}(r)/kT)][h_{ij}(r) + 1]. \qquad (13)$$

Once the correlation functions have been solved, adsorption isotherms can be obtained from the Fourier transform of the direct correlation function $c_c(r)$ [55]. The ROZ integral equation approach is noteworthy in that it yields model adsorption isotherms for disordered porous materials that have irregular pore geometries without resort to molecular simulation. In contrast, most other disordered structural models of porous solids implement GCMC or other simulation techniques to compute the adsorption isotherm. However, no method has yet been demonstrated for determining the pore structure of model disordered or templated structures from experimental isotherm measurements using integral equation theory.

III. Simple Geometric Pore Structure Models

The severe computational burden associated with assembling and carrying out adsorption calculations on disordered model microstructures for porous solids, such as those discussed in Sections II.A and II.B, has until recently limited the development of pore volume characterization methods in this direction. While the realism of these models is highly appealing, their application to experimental isotherm or scattering data for interpretation of adsorbent pore structure remains cumbersome due to the structural complexity of the models and the computational resources that must be brought to bear in their utilization. Consequently, approximate pore structure models, based upon simple pore shapes such as slits or cylinders, have been retained in popular use for pore volume characterization.

The most commonly used approximate model for pore topology is to represent the pore volume of the adsorbent as an array of independent, chemical homogeneous, noninterconnected pores of some simple geometry; usually, these are slit-shaped for activated carbons and cylindrical-shaped for glasses, silicas, and other porous oxides. Usually, the heterogeneity is approximated by a distribution of pore sizes, it being implicitly assumed that all pores have the same shape and the same surface chemistry. In this case, the excess adsorption, $\Gamma(P)$, at bulk gas pressure P can be represented by the adsorption integral equation

$$\Gamma(P) = \int_{H_{\min}}^{H_{\max}} \Gamma(P, H) f(H) dH, \qquad (14)$$

where the local isotherm $\Gamma(P, H)$ is the specific excess adsorption for an adsorbent in which all the pores are of size H; and $f(H)$ is the pore size distribution (PSD), such that $f(H)\,dH$ is the fraction of pores with sizes between H and $H + dH$. The integration is taken over all pore sizes between the minimum and the maximum pore sizes, H_{min} and H_{max}, present in the adsorbent. Equation (14) assumes either that geometric and chemical heterogeneity are entirely absent or that they can be treated as effectively equivalent to pore size heterogeneity with regard to adsorption. An alternative approach [62] is to approximate the heterogeneity as due exclusively to chemical heterogeneity, with a distribution of adsorbate–adsorbent interaction energies $f(\varepsilon)$ such that Eq. (14) is replaced by

$$\Gamma(P) = \int_{\varepsilon_{min}}^{\varepsilon_{max}} \Gamma(P, \varepsilon) f(\varepsilon) d\varepsilon, \qquad (15)$$

where $\Gamma(P, \varepsilon)$ is the local isotherm for pores with a uniform surface interaction energy ε.

At the level of approximation invoked by the simple geometric model, the mathematical problem becomes one of inverting Eq. (14), a linear Fredholm integral equation of the first kind, to obtain the PSD. The kernel $\Gamma(P, \varepsilon)$ represents the thermodynamic adsorption model, $\Gamma(P)$ is the experimental function, and the pore size distribution $f(H)$ is the unknown function. The usual method of determining $f(H)$ is to solve Eq. (14) numerically via discretization into a system of linear equations,

$$\sum_{j=1}^{m} \Gamma(P_i, H_j) \Delta H_j\, f(H_j) = \Gamma(P_i), \qquad (16)$$

where the interval of pore sizes between H_{min} and H_{max} has been partitioned into m intervals, and $f(H_j)$ is a histogram coefficient that represents the fraction of pores with sizes ranging between $H_j - \Delta H_j/2$ and $H_j + \Delta H_j/2$. If the experimental isotherm is fitted at n different bulk gas pressures, then Eq. (16) yields a system of n linear equations in m unknowns. Such a system can be solved by minimizing the residual error, e.g., by a least-squares match to the experimental isotherm,

$$E = \sum_{i=1}^{n} \left[\Gamma(P_i) - \sum_{j=1}^{m} \Gamma(P_i, H_j) \Delta H_j\, f(H_j) \right]^2, \qquad (17)$$

with the solution vector of coefficients $f(H_j)$ sought that minimizes the value of E. Alternatively, one may assign a nonnegative functional form to the PSD, such as a Gaussian distribution or a multimodal Γ distribution, and then fit the shape Γ parameters of the distribution so as to minimize the residual error given by Eq. (17). A variety of numerical minimization techniques may be used to carry out this task [63].

Fredholm integral equations are by nature numerically ill posed [64]. This means that experimental errors in the measurement of $\Gamma(P)$ may be transmitted to the unknown result $f(H)$ in such a way that the PSD is completely distorted. Application of the least-squares minimization method of Eq. (17) may in some cases lead to strong oscillations in the function $f(H)$, obscuring the true adsorbent PSD. The ill-posed nature of Eq. (14) is of mathematical origin and is not due to incorrect formulation of the physical problem. To circumvent the problems caused by the ill-posedness of the Fredholm integral equation, one may either (i) postulate an analytical form of the unknown function $f(H)$, as noted earlier, and find its parameter values via least-squares fitting [15]; (ii) fit an analytical equation to the experimental isotherm $\Gamma(P)$ that gives an analytical solution of integral equation and fit the coefficients of $f(H)$ via least squares; or (iii) use a regularization method to introduce an additional set of constraints into the solution of the adsorption integral equation, thus stabilizing the PSD with respect to perturbations in the experimental data [65–67]. The latter method involves the selection of constraints that are physically consistent with the expected form of the PSD. The most commonly adopted regularization procedure [68, 69] is to modify the objective function of Eq. (17) by adding a term that is proportional to the square of the second derivative of the pore size distribution function $f(H)$,

$$E = \sum_{i=1}^{n} \left[\Gamma(P_i) - \sum_{j=1}^{m} \Gamma(P_i, H_j) \Delta H_j \, f(H_j) \right]^2 + \alpha \sum_{j=1}^{m} \Delta H_j \, [f''(H_j)]^2, \quad (18)$$

where the proportionality coefficient $\alpha > 0$ is referred to as the smoothing parameter. Essentially, regularization forces a slightly worse fit to the experimental isotherm $\Gamma(P)$ to generate a smoother PSD, under the presumption that the adsorbent is most likely to exhibit a relatively smooth distribution of pore sizes, centered around a few dominant pore sizes [68]. Figure 8 shows the effect of regularization for a range of smoothing parameter values employed in fitting the PSD of activated carbon to methane adsorption data using the objective function given by Eq. (18) [69]. It can be seen that as the calculated PSD becomes smoother, as expected, as the value of the smoothing parameter increases. The PSD progresses from a tetramodal distribution for $\alpha = 1$ to a unimodal distribution for $\alpha \geq 100$. To select which PSD is the "correct" PSD, one must consider the error bounds associated with the experimental measurements. If the experimental error is small, the tetramodal PSD shown in Fig. 8 may indeed be the correct PSD. However, if there is a greater uncertainty in the measured values of the experimental isotherm, then it is equally plausible that the PSD is unimodal, bimodal (e.g., $\alpha = 10$) or tetramodal in shape. One criterion that has been suggested for selecting

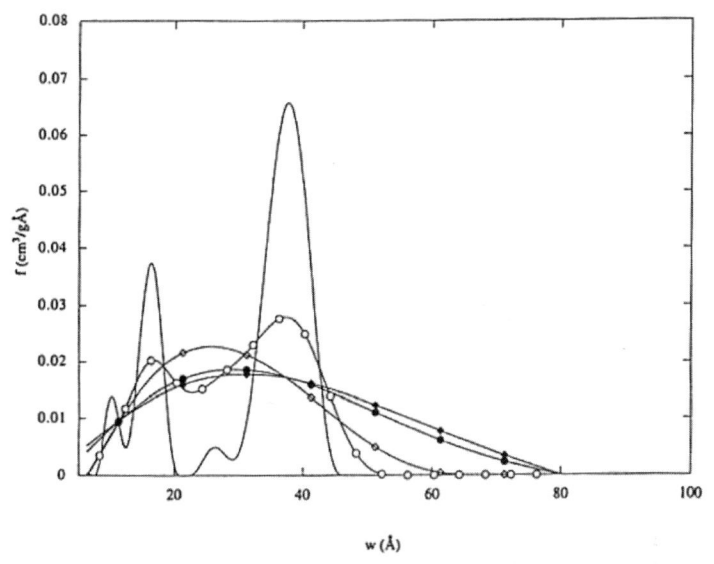

FIG. 8. PSDs obtained for methane adsorption in square model carbon pores using molecular simulation to interpret an activated carbon isotherm. PSD results are shown for regularization smoothing parameter values of 1 (solid line), 10 (open circles), 100 (open diamonds), 600 (filled circles), and 800 (filled diamonds) [69].

the proper PSD is to accept the smoothest distribution (i.e., largest value of α) to within the experimental error of the isotherm measurement [69].

A consequence of the ill-posed nature of Eq. (14), therefore, is that different PSD results can be obtained for the same material if different methods are applied to solve the adsorption integral equation, even if the same experimental data and adsorption model are used in both cases. A standard protocol has not yet been agreed upon for the use of regularization in pore size characterization. To avoid confusion in comparing PSD results, therefore, the numerical method employed to solve for the PSD and the type of regularization, if any, implemented to smooth the PSD should both be clearly identified.

Duly recognizing the potential pitfalls associated with inversion of the adsorption integral equation, the determination of the PSD reduces to the problem of selecting an adsorption model that represents the uptake of adsorbate in pores of different sizes that have the same shape and surface composition. This is a critical choice because the PSD results obtained from the inversion of Eq. (14) are highly sensitive to the selected adsorption model. As noted in Section I, adsorption models for simple geometric pore shapes can be categorized into four classes, based on the computational method used to generate the local adsorption isotherm for the model pore geometry. In the remainder of this section, selected examples from the four principal types of adsorption models are reviewed and discussed.

A. Molecular Simulation Adsorption Models

GCMC simulation has been used to generate model isotherms for the adsorption of methane at supercritical temperatures [70] and the adsorption of carbon dioxide at subcritical temperatures [71] in slit-shaped carbon pores. The methane adsorbate is typically represented as a spherical Lennard–Jones molecule, whereas a three-center Lennard–Jones model is frequently used to represent the potential interactions of carbon dioxide [71, 72], with electrostatic point charges on the atoms to account for the CO_2 quadrupole. Adsorbate interactions with the carbon slit pore surfaces are most often modeled using the 10–4–3 potential in which the adsorbent is treated as stacked graphitic planes of Lennard–Jones atoms [73]. To determine activated carbon pore size distributions, GCMC model isotherms must usually be calculated for a set of one or two dozen slit pore widths spanning the micropore and mesopore size range between 7 and 50 Å. At the cryogenic temperatures used for nitrogen or argon porosimetry (77 or 87 K), adsorbate diffusion may be slow in carbon micropores, and long equilibration times may thus be required to measure the experimental adsorption isotherm. In such cases, the use of carbon dioxide or methane model isotherms at elevated temperatures (e.g., 195.5 or 273 K for CO_2, 308 K for methane) is an effective means of reducing mass transfer resistances in the adsorbent microstructure. For supercritical methane, however, the GCMC model can be used to determine only micropore PSDs, because the excess isotherms for methane adsorption are indistinguishable for pores larger than 20 Å [70].

A method of using GCMC simulation in conjunction with percolation theory [74, 75] has been suggested for simultaneous determination of the PSD and network connectivity of a porous solid [76]. In this method, isotherms are measured for a battery of adsorbate probe molecules of different sizes, e.g., CH_4, CF_4, and SF_6. As illustrated in Fig. 9a, the smaller probe molecules are able to access regions of the pore volume that exclude the larger adsorbates. Consequently, each adsorbate samples a different portion of the adsorbent PSD, as shown in Fig. 9b. By combining the PSD results for the individual probe gases with a percolation model, an estimate of the mean connectivity number of the network can be obtained [76].

An alternative molecular simulation technique for obtaining model adsorption isotherms for simple geometric pore structures is the Gibbs ensemble Monte Carlo (GEMC) simulation method [13, 22]. In GEMC simulation, configurational sampling is carried out between two simulation cells, (I) and (II), that represent regions of the pore volume. A schematic of the GEMC methodology for a slit-shaped pore geometry is shown in Fig. 10. Three types of sampling moves are performed in GEMC simulation: (a) displacement of molecules in each pore region; (b) interchange of

CHARACTERIZATION OF POROUS MATERIALS 223

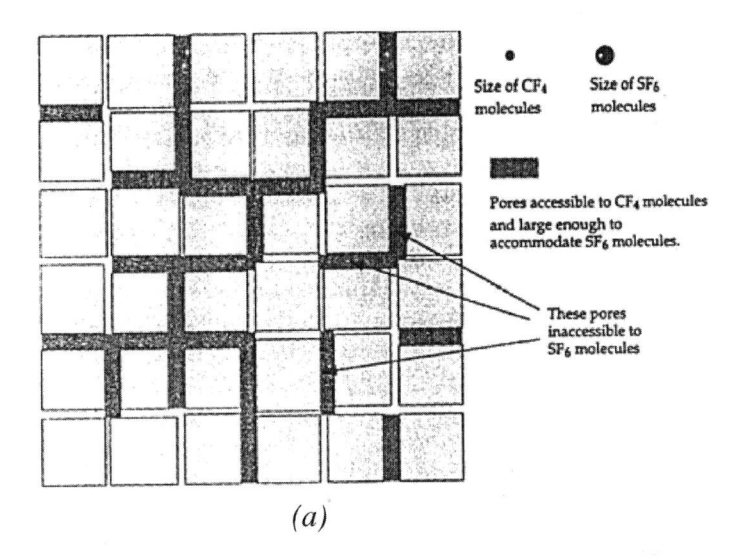

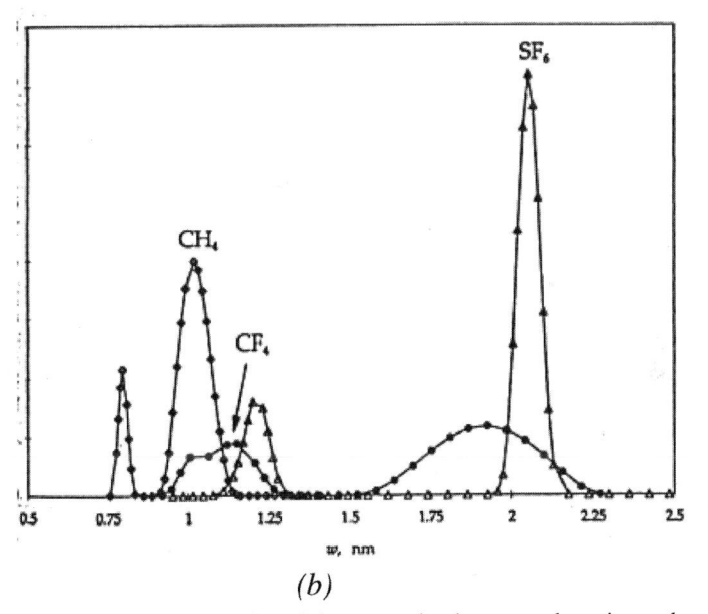

Fig. 9. (a) Schematic illustration of the connection between adsorption and percolation in a porous network. Some pores large enough to accommodate large adsorbate molecules (e.g., SF_6) remain unfilled because they cannot be accessed through smaller connecting pore channels. (b) PSD results obtained using GCMC simulation to interpret CH_4, CF_4, and SF_6 isotherms measured on an activated carbon at 296 K. The smaller probe molecules sample regions of the pore volume that are inaccessible to the large adsorbate molecules [76].

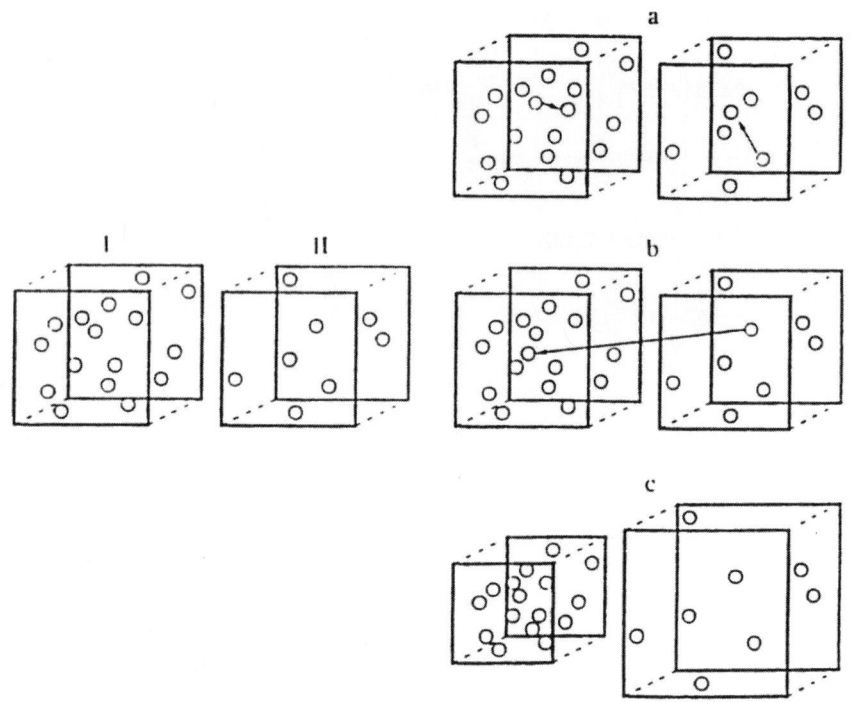

FIG. 10. Schematic of the Gibbs ensemble Monte Carlo simulation method for calculation of phase equilibria of confined fluids [22].

molecules between the two regions; and (c) exchange of pore volume between the two regions, with the total pore volume and the slit pore width held constant. The acceptance probabilities for the sampling moves are given as

$$p = \min\left[1, \exp\left(-\frac{\Delta U}{kT}\right)\right], \quad (19)$$

where the change in total energy ΔU depends on the type of perturbation. For molecular displacements,

$$\Delta U_{\text{dis}} = \Delta U_{\text{I}} + \Delta U_{\text{II}}, \quad (20)$$

where ΔU_i is the energy change that occurs in simulation cell i. For interchange steps,

$$\Delta U_{\text{int}} = \Delta U_{\text{I}} + \Delta U_{\text{II}} + kT \ln\left[\frac{(N_{\text{I}}+1)V_{\text{II}}}{N_{\text{II}}V_{\text{I}}}\right], \quad (21)$$

where N_i and V_i are the number of molecules and volume, respectively, of simulation cell i. For volume exchange steps,

$$\Delta U_{\text{vol}} = \Delta U_\text{I} + \Delta U_\text{II} - N_\text{I} kT \ln\left(\frac{V_\text{I} - \Delta V}{V_\text{I}}\right) - N_\text{II} kT \ln\left(\frac{V_\text{II} + \Delta V}{V_\text{II}}\right), \tag{22}$$

where ΔV is the amount of volume exchanged between the two simulation cells. Configurational sampling brings the two regions into thermal, material, and mechanical equilibrium, thus yielding the densities of the coexisting liquid and vapor states in the pore. GEMC simulations can also be carried out to obtain coexistence properties for an adsorbed fluid in equilibrium with a bulk vapor [13, 22]. By performing a series of GEMC simulations starting from different initial densities in the two simulation cells, the isotherm for a pore of prescribed width can be constructed. A distinct advantage of GEMC over GCMC simulation is that in the GEMC method, the chemical potential of the bulk vapor phase need not be specified as an input to the computer simulation. When the GEMC simulations are performed correctly, the pressures and chemical potentials of regions I and II will be the same at equilibrium. Thus, in GEMC the adsorption isotherm can be calculated without direct knowledge of the chemical potential. This is particularly useful for situations in which the bulk fluid equation of state is not known.

GEMC model isotherms have been constructed for nitrogen adsorption in slit-shaped carbon pores [22] and cylindrical oxide pores [11]. The filling pressure at which capillary condensation is expected to occur in a pore of given size can be extracted in a straightforward manner from GEMC calculations and the correlation is shown for the nitrogen–carbon system in Fig. 11. The pore filling correlations predicted by several other adsorption models for slit-shaped pores are also shown in Fig. 11. These models, and the reasons for the differences evident in Fig. 11, are discussed in the sections that follow.

B. Density Functional Theory Adsorption Models

The DFT method [15, 77–79] is derived from statistical thermodynamics and offers an efficient means of computing model isotherms for simple pore geometries. The accuracy of the DFT model isotherms rivals that of the isotherms obtained from molecular simulation, but the computational time required by DFT is typically about 1% of the CPU time needed to complete GCMC or GEMC isotherm calculations for a comparable system. The DFT method retains its computational advantage over molecular simulation only for pore shapes of low dimensionality, such as slits, spheres, or

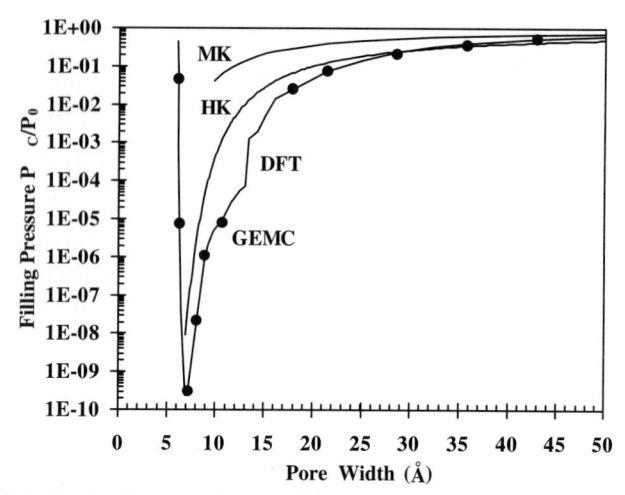

FIG. 11. Relationship between the pore filling pressure and the pore width predicted by the modified Kelvin equation (MK), the Horvath–Kawazoe method (HK), density functional theory (DFT), and Gibbs ensemble Monte Carlo simulation (points) for nitrogen adsorption in carbon slit pores at 77 K [11].

cylinders, so DFT is not applicable to realistic disordered structures such as those considered in Section II.

In the DFT method [80], each individual pore has a fixed geometry and is in open contact with the bulk vapor-phase adsorbate at a fixed temperature. For this system, the grand canonical ensemble provides the appropriate description of the thermodynamics. In this ensemble, the chemical potential μ, temperature T, and pore volume V are specified. In the presence of a spatially varying external potential V_{ext}, the grand potential functional Ω of the fluid is [15]

$$\Omega[\rho(\mathbf{r})] = F[\rho(\mathbf{r})] - \int d\mathbf{r}\rho(\mathbf{r})[\mu - V_{\text{ext}}(\mathbf{r})], \qquad (23)$$

where F is the intrinsic Helmholtz free energy functional, $\rho(\mathbf{r})$ is the local fluid density at position $\mathbf{r}$, and the integration is carried out over the entire pore volume. The free energy functional is expanded to first order about a reference system of hard-sphere adsorbate molecules, with the temperature-dependent diameter d given by the Barker–Henderson prescription [81]

$$F[\rho(\mathbf{r})] = F_{\text{h}}[\rho(\mathbf{r}); d] + \frac{1}{2} \int\int d\mathbf{r}\, d\mathbf{r}'\, \rho(\mathbf{r})\rho(\mathbf{r}')\phi_{\text{att}}(|\mathbf{r} - \mathbf{r}'|), \qquad (24)$$

where F_{h} is the hard-sphere Helmholtz free energy functional and ϕ_{att} is the attractive part of the adsorbate–adsorbate potential. Equation (24) invokes the mean field approximation, wherein pair correlations between molecules due to attractive forces are neglected. The attractive part of the adsorbate pair potential is represented using the WCA division of the Lennard–Jones potential [82],

$$\phi_{\text{att}}(|\mathbf{r}-\mathbf{r}'|) = \phi_{\text{ff}}(|\mathbf{r}-\mathbf{r}'|), \quad |\mathbf{r}-\mathbf{r}'| > r_{\text{m}}$$
$$-\varepsilon_{\text{ff}}, \quad |\mathbf{r}-\mathbf{r}'| < r_{\text{m}}, \quad (25)$$

where $r_{\text{m}} = 2^{1/6}\sigma_{\text{ff}}$ is the location of the minimum of the Lennard–Jones potential ϕ_{ff}, and ε_{ff} and σ_{ff} are the Lennard–Jones well depth and molecular diameter, respectively. The hard-sphere free energy functional F_{h} can be written as the sum of two terms,

$$F_{\text{h}}[\rho(\mathbf{r}); d] = kT \int d\mathbf{r}\rho(\mathbf{r})[\ln(\Lambda^3\rho(\mathbf{r})) - 1] + kT \int d\mathbf{r}\rho(\mathbf{r})f_{\text{ex}}[\bar{\rho}(\mathbf{r}); d], \quad (26)$$

where $\Lambda = h/(2\pi mkT)^{1/2}$ is the thermal de Broglie wavelength, m is the molecular mass, h and k are the Planck and Boltzmann constants, respectively, and f_{ex} is the excess molar Helmholtz free energy, i.e., the total molar free energy less the ideal gas contribution. The excess molar free energy of the hard-sphere fluid is calculated from the Carnahan–Starling equation of state [83]. The first term on the right-hand side of Eq. (26) is the ideal-gas contribution, which is exactly local; i.e., its value at $\mathbf{r}$ depends only on $\rho(\mathbf{r})$. The second term on the right-hand side of Eq. (26) is the excess contribution, which is nonlocal and is calculated using a smoothed density, $\bar{\rho}(\mathbf{r})$, that represents a suitable weighted average of the local density $\rho(\mathbf{r})$

$$\bar{\rho}(\mathbf{r}) = \int d\mathbf{r}'\rho(\mathbf{r}')w[|\mathbf{r}-\mathbf{r}'|; \bar{\rho}(\mathbf{r})]. \quad (27)$$

The choice of the weighting function w depends on the version of DFT used. If the δ function is assigned for w, the smoothed density reverts to the local density and the so-called local DFT method is obtained [77]. The local DFT model correctly describes many of the structural features of inhomogeneous fluids but becomes inaccurate for fluids confined in small pores because of the strong short-ranged correlations present in such fluids. For highly inhomogeneous confined fluids, a smoothed or nonlocal density approximation is needed that gives a good description of the direct correlation function for the hard-sphere fluid over a wide range of fluid densities. Among the several weighting function sets that have been proposed [84–88], Tarazona's model [86–88] is the one most commonly used for the nonlocal version of DFT. This model has been shown to give very good agreement with simulation results for the density profile and surface tension of Lennard–Jones fluids near attractive surfaces. The Tarazona prescription for the weighting functions uses a power series expansion in the smoothed density, truncating at the second-order term

$$w[|\mathbf{r} - \mathbf{r}'|; \bar{\rho}(\mathbf{r})] = \sum_{i=0}^{2} w_j(|\mathbf{r} - \mathbf{r}'|)\bar{\rho}(\mathbf{r})^i. \tag{28}$$

The effect of the weighting is to flatten the sharp oscillations of the local density profile into a smoothed density profile that is input into the equation of state for the excess hard-sphere free energy. The nonlocal DFT method thus avoids difficulties associated with applying the equation of state to unrealistically large values of the local density, as is the case with the local DFT method.

The equilibrium density profile is determined by minimizing the grand potential functional with respect to the local density,

$$\frac{\delta \Omega[\rho(\mathbf{r})]}{\delta \rho(\mathbf{r})} = 0 \quad \text{at} \quad \rho = \rho_{\text{eq}}. \tag{29}$$

An iteration scheme is used to numerically solve this minimization condition to obtain $\rho_{\text{eq}}(\mathbf{r})$ at the selected temperature, pore width, and chemical potential. For simple geometric pore shapes such as slits or cylinders, the local density is a function of one spatial coordinate only (the coordinate normal to the adsorbent surface) and an efficient solution of Eq. (29) is possible. The adsorption and desorption branches of the isotherm can be constructed in a manner analogous to that used for GCMC simulation. The chemical potential is increased or decreased sequentially, and the solution for the local density profile at previous value of μ is used as the initial guess for the density profile at the next value of μ. The chemical potential at which the equilibrium phase transition occurs is identified as the value of μ for which the liquid and vapor states have the same grand potential.

The DFT excess isotherm is found by taking the volume average of the local adsorbate density profile, e.g., for a slit-shaped pore of width H,

$$\Gamma(P, H) = \frac{1}{H} \int_0^H [\rho(z) - \rho_b] dz, \tag{30}$$

where z is the spatial coordinate normal to the pore walls and ρ_b is the bulk vapor phase density at pressure P. The pressure may be obtained from the chemical potential using the bulk fluid equation of state [89].

DFT model isotherms have been reported for adsorption of nitrogen [15, 22, 77, 79, 90–93], argon [90–92], methane [93, 94], helium [93], and carbon dioxide [72, 92, 95] in slit-shaped graphitic carbon pores and adsorption of nitrogen [8, 11, 78, 96] and argon [8, 78] in cylindrical capillaries with oxide surfaces. Most of the DFT models are developed at cryogenic temperatures,

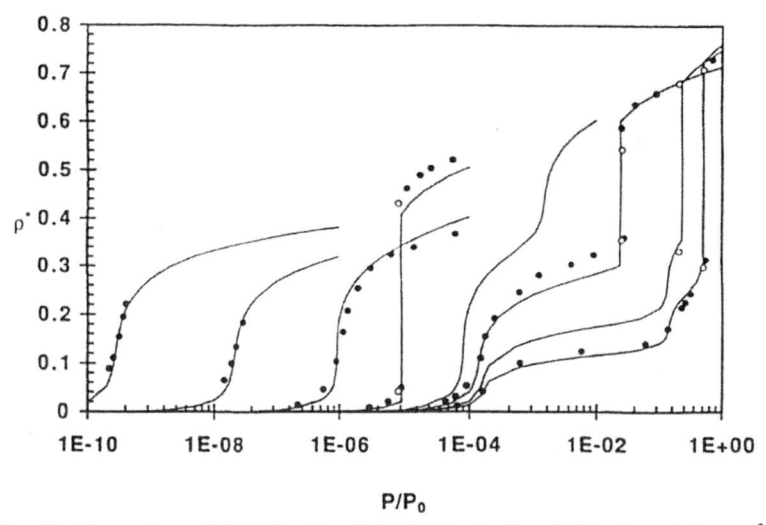

FIG. 12. Comparison of DFT (lines) and GEMC (points) model isotherms $\rho^* = \rho\sigma_{\text{ff}}^3$ obtained for nitrogen adsorption at 77 K in slit-shaped carbon pores of width (reading from left to right) $H = 7.1, 8.0, 8.9, 10.7, 13.4, 17.9, 28.6,$ and 42.9 Å [22].

where capillary condensation provides a distinct experimental "fingerprint" of the PSD. As noted in Section III.A, isotherm databases may also be constructed to interpret uptake measurements at elevated temperatures in cases where slow mass transfer is a concern.

Theoretical isotherms obtained using DFT method rival the accuracy of model isotherms constructed from molecular simulation, as shown in the comparison in Fig. 12 for N_2 adsorption results in model graphitic slit pores at 77 K [22]. The DFT model correctly predicts the capillary condensation pressure, the most prominent feature of the subcritical isotherm, relative to the exact computer simulation results shown in Fig. 11. DFT also provides a good description of the secondary structure of the mesopore isotherm (e.g., $H = 42.9$ Å in Fig. 12), in which capillary condensation may be preceded by one or more wetting transitions.

A test of the robustness of the DFT method is the consistency of the adsorbent PSDs calculated from adsorption experiments using different adsorbate molecules to probe the pore volume. Figure 13 shows a comparison of porosimetry results obtained using nitrogen and argon to probe the pore structures of activated carbons [90]. The specific pore volume, mean pore diameter, and PSD maximum obtained from nitrogen and argon porosimetry were all found to agree to within 8% for the two probe gases using DFT to interpret the adsorption isotherms measured for both species. The differences, where noted, may be ascribed in part to the heterogeneity of the carbon samples, which all possess an appreciable mass

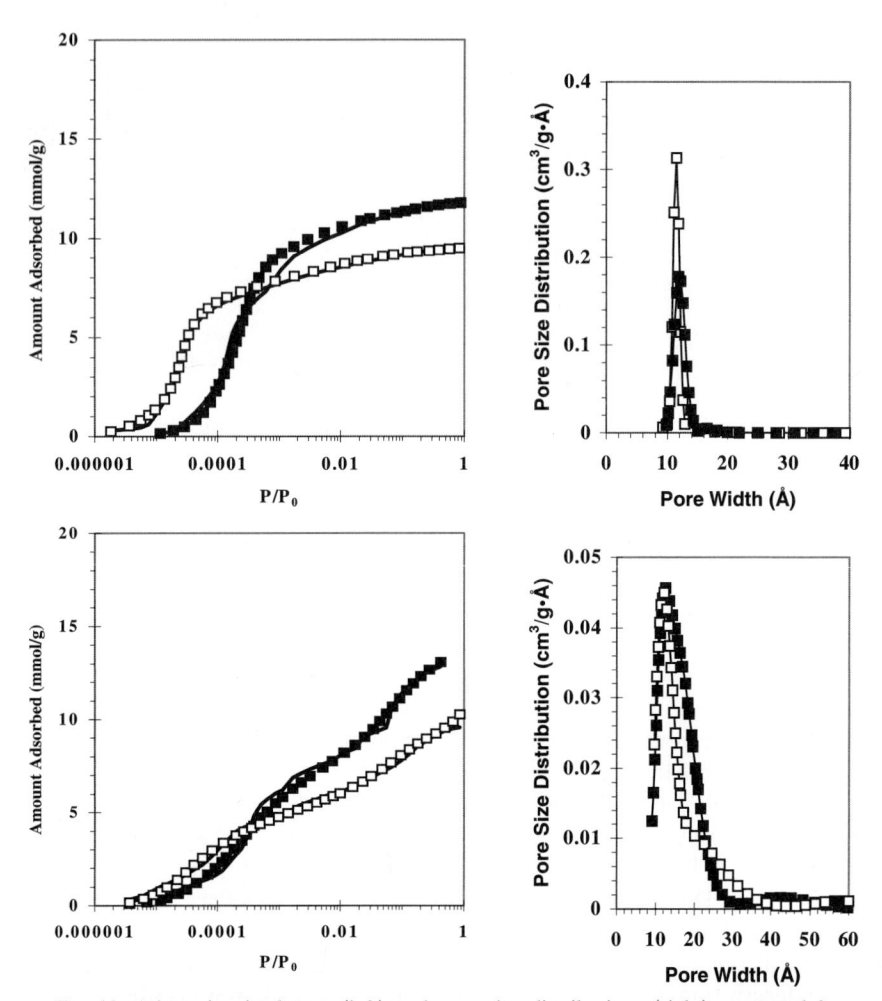

FIG. 13. Adsorption isotherms (left) and pore size distributions (right) computed from nitrogen (open squares) and argon (filled squares) adsorption at 77 K on a porous Saran char (top) and a granular activated carbon (bottom) [90]. (Reproduced with permission from S. Ramalingam. Plasma-surface interactions in deposition of silicon thin films: An atomic-scale analysis. PhD Thesis, University of California, Santa Barbara; 2000.)

fraction (9 to 13%) of hydrogen-, nitrogen-, oxygen-, and sulfur-containing surface functional groups [90]. These heteroatoms can interact with the nitrogen quadrupole and adsorb nitrogen at pressures lower than for N_2 adsorption on bare graphitic surfaces. Another study [94] reported similar findings for PSD differences between quadrupolar nitrogen and nonpolar methane on microporous carbons. For adsorbents that exhibit substantial chemical heterogeneity, the use of a probe molecule (e.g., argon) that has no permanent electrostatic moments is therefore recommended.

DFT-based pore size distribution results have also been reported for the family of templated siliceous mesoporous molecular sieves first synthesized by Mobil researchers [97] and designated M41S. The most well-known members of this family are MCM-41, which has a hexagonal array of unidirectional pores [98], and MCM-48, which has a cubic pore system [99]. The pore diameters of these adsorbents typically range between 30 and 40 Å. The PSDs of MCM-41 and MCM-48 adsorbents have been calculated to a high degree of consistency using a cylindrical oxide pore DFT model to interpret experimental nitrogen and argon isotherms, as shown in Fig. 14a for a MCM-41-type adsorbent [8]. By subtracting the DFT result for the internal pore diameter from X- ray diffraction measurement of the pore spacing, the pore wall thickness of an MCM-class adsorbent can be estimated. Figure 14b shows the wall thickness calculated for five MCM-41 adsorbents using the diffraction data in combination with the adsorption isotherm. The wall thicknesses, were found to be between 6 and 8 Å for four of the five adsorbents, consistent with other independent estimates of the wall thickness of MCM-41 materials obtained from transmission electron microscopy [100].

C. SEMIEMPIRICAL ADSORPTION MODELS

Two principal semiempirical adsorption models have enjoyed widespread use for adsorbent PSD characterization: the Horvath-Kawazoe (HK) method [19] and its derivatives, and approaches based upon the ideas of Dubinin [20] for modeling micropore distributions. Each of these methodologies is considered in turn.

1. The Horvath–Kawazoe Method

In the original HK method [19], an analytic pore filling correlation is obtained by calculating the mean free energy change of adsorption $\bar{\phi}$ that occurs when an adsorbate molecule is transferred from the bulk vapor phase to the condensed phase in a slit pore of width H

$$\ln(P_c/P_0) = \bar{\phi}(H)/kT. \qquad (31)$$

In Eq. (31), P_0 is the saturation pressure and P_c is the pore condensation pressure. The assumed exponential dependence of the condensation pressure on the adsorption free energy change is similar in basis to the Polanyi potential theory [101] and the Frenkel–Halsey–Hill (FHH) theory [102–104]. In the HK method, the mean free energy change due to adsorption is calculated

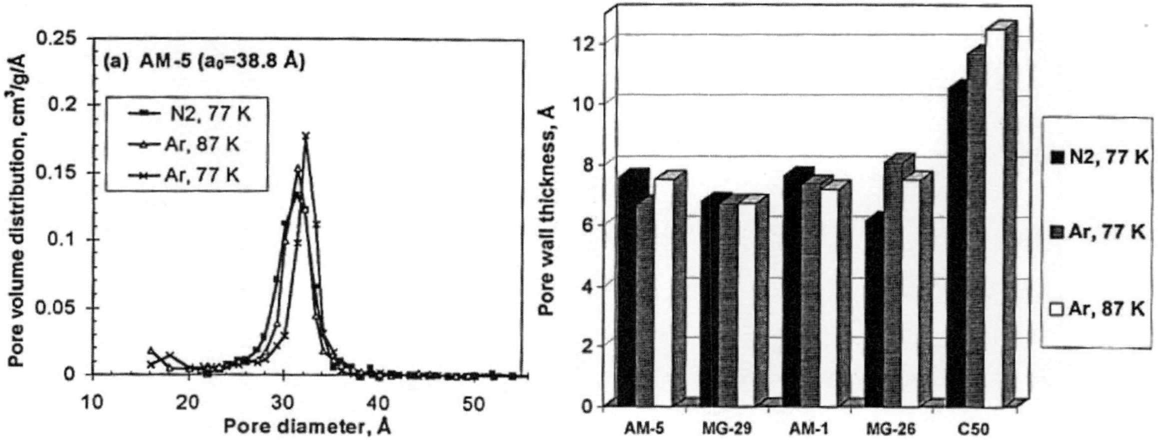

Fig. 14. (a) PSDs obtained for MCM-41-type adsorbents from nitrogen and argon DFT porosimetry. (b) Pore wall thickness calculated for five MCM-41 adsorbents using a combination of X-ray diffraction data and N_2/Ar adsorption measurements [8].

from an unweighted average of the adsorbate interaction potential ϕ, measured over the slit pore volume between $z = \sigma_{sf}$ and $z = H - \sigma_{sf}$

$$\bar{\phi} = \frac{\int_{\sigma_{sf}}^{H-\sigma_{sf}} \phi(z)dz}{\int_{\sigma_{sf}}^{H-\sigma_{sf}} dz} = \frac{\int_{\sigma_{sf}}^{H-\sigma_{sf}} \phi(z)dz}{H - 2\sigma_{sf}}, \qquad (32)$$

where z is the distance of the adsorbate center of mass from the surface layer plane of atomic nuclei, and σ_{sf} is the arithmetic mean of the adsorbate and adsorbent diameters. The interaction energy of the adsorbate molecule and the pore wall is represented by the sum of two 10–4 potentials, one that represents the adsorbate–adsorbent potential ϕ_{sf} and the other representing an effective adsorbate–adsorbate interaction ϕ_{ff}

$$\phi(z) = \phi_{sf}(z) + \phi_{ff}(z) = \frac{N_s A_{sf} + N_f A_{ff}}{2d^4}\left[\left(\frac{d}{z}\right)^{10} - \left(\frac{d}{z}\right)^4\right.$$

$$\left. + \left(\frac{d}{H-z}\right)^{10} - \left(\frac{d}{H-z}\right)^4\right]. \qquad (33)$$

In Eq. (33), N is the molecular density (i.e., molecules per unit surface area); A is the dispersion coefficient; the subscripts s and f refer to the adsorbent and adsorbate, respectively; and $d = (2/5)^{1/6}\sigma_{sf}$. The 10–4 potential is obtained by integrating the Lennard–Jones potential over an unbounded planar surface [105]. The dispersion coefficients are calculated using the Kirkwood–Muller equations (as reported in Ref. 19)

$$A_{sf} = \frac{6mc^2 \alpha_s \alpha_f}{(\alpha_s/\chi_s) + (\alpha_f/\chi_f)} \qquad (34)$$

$$A_{ff} = \frac{3}{2}mc^2 \alpha_f \chi_f. \qquad (35)$$

In Eqs. (34) and (35), m is the mass of an electron, c is the velocity of light, and α and χ are the polarizability and the magnetic susceptibility, respectively. Substitution of Eqs. (32) and (33) into Eq. (31) and integration yields the HK correlation between the pore condensation pressure and the slit pore width

$$\ln\left(\frac{P_c}{P_0}\right) = \frac{N_s A_{sf} + N_f A_{ff}}{kTd^3(H - 2\sigma_{sf})}\left[\frac{1}{9}\left(\frac{d}{\sigma_{sf}}\right)^9 - \frac{1}{3}\left(\frac{d}{\sigma_{sf}}\right)^3\right.$$

$$\left. - \frac{1}{9}\left(\frac{d}{H-\sigma_{sf}}\right)^9 + \frac{1}{3}\left(\frac{d}{H-\sigma_{sf}}\right)^3\right]. \qquad (36)$$

Correlations similar to Eq. (36) have been derived for cylindrical [106] and spherical [107] pore geometries and for adsorbates other than nitrogen, such as argon [108] and methyl chloride [109]. The analytic HK correlation may be conveniently applied to adsorbent PSD analysis using a simple spreadsheet, whereas a computer algorithm is required to obtain theoretical isotherms via DFT or molecular simulation.

In Fig. 11, the pore filling correlation predicted by Eq. (36) is shown for nitrogen adsorption in carbon slit pores at 77 K, using dispersion parameters reported in the original HK paper [19]. The HK method correctly predicts a sharp decrease in the pore filling pressure as the pore width decreases into the micropore range. This dramatic decrease occurs because of enhancement of the gas–solid potential in narrow slit pores, and the HK method qualitatively accounts for this effect. It is evident from Fig. 11, however, that the HK adsorption model, as originally posed, does not agree with the pore filling correlations obtained from DFT or GEMC simulation for the nitrogen–carbon system. The differences in the model results can be attributed to three causes.

(i) The HK model uses the 10–4 potential in the calculation of the adsorption free energy change, whereas DFT and molecular simulation models routinely represent the gas–solid potential using the 10–4–3 potential [73], which contains an additional attractive term for adsorbate interactions with the subsurface layers of adsorbent atoms.

(ii) The adsorbate–adsorbate dispersion parameter calculated using Eq. (34) in the original HK paper is considerably smaller than the comparable parameter obtained in the DFT and simulation models by fitting the model isotherm to a nonporous graphitized carbon reference adsorbent [22].

(iii) The adsorbate potential representation given by Eq. (33) is posed incorrectly. As noted in two recent publications [110, 111], in the FHH formulation, the adsorbate–adsorbate interaction is subtracted from the adsorbate–adsorbent interaction, rather than added to it as shown in Eq. (33). The theoretical construct of FHH theory is that a slab of adsorbate is removed and replaced with a slab of adsorbent atoms [102]. In the original HK treatment, the adsorbate–adsorbate interactions are, without justification, directly superimposed onto the adsorbate–adsorbent interactions in a physically implausible manner [110, 111]. Additionally, a factor of two error appears in the Kirkwood–Muller formula for the adsorbate dispersion coefficient A_{ff} reported in the original HK method [19]. As noted by Muller [112], the leading coefficient on the right-hand side of Eq. (34) should be 3 rather than 3/2.

It has been shown that much better agreement between the HK and the DFT pore filling pressure correlations is obtained if the same form of the gas–solid potential and the same potential parameter values are used in comparing the two models [111].

2. Modified Horvath–Kawazoe Models

Several modifications of the original HK method have been suggested to bring the semiempirical model into better agreement with DFT and GEMC results [110, 111, 113, 114]. One approach is to use weighting functions (e.g., a multimodal Gaussian distribution) to represent the spatial variations in the local density due to fluid layering near the pore walls [111, 113]. In this representation, the mean free energy change of adsorption is calculated as

$$\bar{\phi} = \frac{\int_d^{H-d} \rho(z)\phi(z)dz}{\int_d^{H-d} \rho(z)dz}, \qquad (37)$$

where the integration is now weighted with respect to the local density $\rho(z)$. This approach yields lower pore filling pressures than are obtained from Eq. (32) because of the additional weighting toward the regions of the pore volume where the gas–solid potential is the most attractive (i.e., the monolayer well adjacent to each pore wall).

Another modification that has been suggested is a geometric approximation [110] in which different potentials are used to calculate the interaction energy of the adsorbate layer adjacent to the pore wall (molecules marked "A" in Fig. 15a) and adsorbate layers in the interior of the pore (molecules marked "B"). Type A molecules are assumed to interact with the surface adsorbent atoms on one side and with adsorbate molecules on the other side, whereas Type B molecules interact only with the two adjacent layers of adsorbate. This method of averaging does away with the integration of Eq. (32) and replaces it with a summation of the energies of molecules sited at discrete positions. The PSD results obtained from this modified HK model for molecular sieve carbons, aluminophosphates and zeolites were in better agreement with crystallographic data for these adsorbents than PSD results obtained from the original HK model.

A major limitation of the original HK method is that it does not account for monolayer formation and film growth in mesopores prior to the capillary condensation transition (see Fig. 12). The model isotherm given by the original HK model is essentially a step-function isotherm in which the pore is either completely empty (if the pressure is below the pore condensation pressure) or completely filled (if $P > P_c$). The wetting of mesopore surfaces can constitute a substantial contribution to the overall experimental isotherm at low pressures, and hence models that omit such contributions will generate erroneous PSD results. To remedy this deficiency, a two-stage HK adsorption model [114] has been proposed in which the capillary condensation step in the mesopore isotherm is preceded by a step that corresponds to the formation of a monolayer on the pore surface, as shown in Fig. 15b. The two-stage HK model provides a better fit than the original HK model to experimental isotherm data for mesoporous adsorbents.

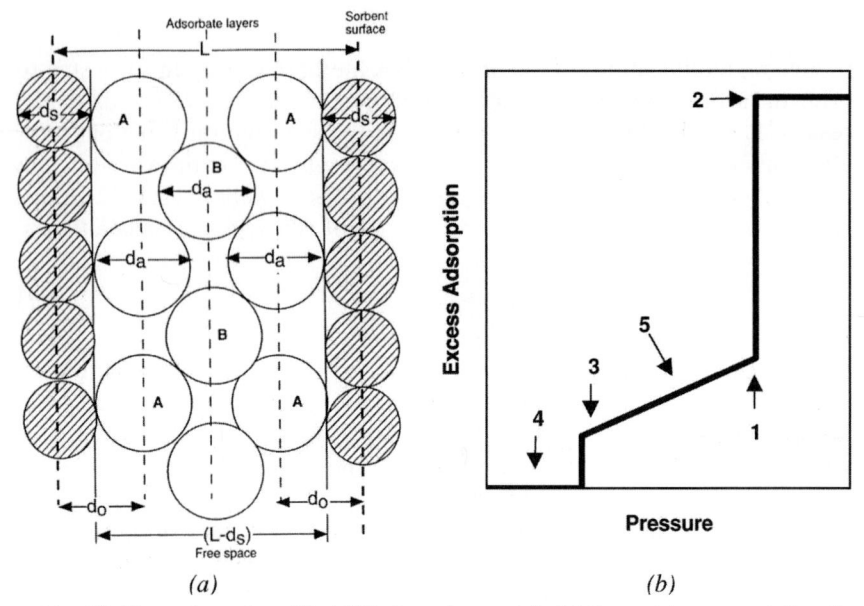

FIG. 15. Illustrations of modified HK adsorption models. (a) Geometric representation slit pore filled with adsorbate [110]. (b) Two-stage HK mesopore isotherm model [114] in which capillary condensation (1) to the filled state (2) is preceded by a wetting transition (3) from an empty state (4) to an intermediate condition characterized by film growth on the pore walls (5). (Reproduced with permission from S. Ramalingam, E. S. Aydil, and D. Maroudas. Molecular dynamics study of the interactions of small thermal and energetic silicon clusters with crystalline and amorphous silicon surfaces. *Journal of Vacuum Science and Technology B*, 2000;19:634–644. Copyright © 2001, AVS.)

3. Dubinin Adsorption Models

In the Dubinin–Radushkevitch (DR) equation [115], an adsorption model derived from a concept of Dubinin [20] based on Polanyi potential theory, the fluid volume V adsorbed in micropores at pressure P is represented empirically as

$$V/V_0 = \exp\{-[RT \ln(P_0/P)]^2/(\beta E_0)^2\} = \exp[-(A/E)^2], \quad (38)$$

where V_0 is the micropore volume, and E_0 and β are characteristic parameters of the adsorbent surface energy and the adsorbate affinity, respectively. A plot of $\ln(V)$ vs $\ln^2(P_0/P)$ is thus expected to yield a linear relationship. This has been found to be the case over a wide range of relative pressures for some microporous carbons [116], but for many other adsorbents the linear DR relationship does not hold [1]. If a linear plot is obtained, the micropore volume can be determined from the intercept at $P = P_0$.

Dubinin and Astakhov [117] put forward a more general form of Eq. (38), termed the DA equation, in which the square exponent is replaced with an empirical constant with a value between 2 and 6. No physical basis was identified, however, for selecting the value of the exponent. To cast the DR equation into a form more suitable for PSD analysis of heterogeneous microporous solids, Stoeckli [118] suggested an integral form of Eq. (38) involving a structure distribution function $J(B)$ for the micropore PSD,

$$V/V_0 = \int_0^\infty \exp[-B(A/\beta)^2] J(B) dB. \quad (39)$$

In the Dubinin–Stoeckli (DS) method, a Gaussian pore size distribution is assumed for $J(B)$ in Eq. (39), based on the premise that for heterogeneous carbons, the original DR equation holds only for those carbons that have a narrow distribution of micropore sizes. This assumption enables Eq. (39) to be integrated into an analytical form involving the error function [119] that relates the structure parameter B to the relative pressure $A = -RT \ln(P/P_0)$. The structure parameter B is proportional to the square of the pore half-width, for carbon adsorbents that have slit-shaped micropores.

In recent work along similar lines [120], the DS approach was used to calculate PSDs for a set of activated carbons, with a Γ distribution function used to represent the PSD in Eq. (39). The activated carbons were subjected to different activation times, so that the microporosity was more fully developed in the samples with longer activation periods. The pore size results obtained from the DS, HK, and DFT methods were compared. It was found that the HK and DFT methods correctly predicted an increase in the micropore PSD for samples prepared with longer activation times, whereas the DS method did not predict this trend correctly. In another evaluation of the adsorption models, a mock isotherm was generated via GCMC simulation for a hypothetical graphitic carbon with a Gaussian pore size distribution (Fig. 16). The DS, HK, and DFT models were then applied to this hypothetical isotherm to see if the original PSD could be recovered. In general, the shape of the original Gaussian PSD was not reproduced by any of the analysis routines. However, the large amount of "noise" in the unregularized DFT pore size distributions in Fig. 16 strongly suggests that experimental errors have been transmitted into these PSD results and that additional smoothing constraints are needed to interpret the experimental data properly. The DFT model did perform the best in recovering the PSD maxima for hypothetical Gaussian distributions that were centered in the micropore range [120].

Dubinin adsorption models have been used to calculate carbon micropore distributions from experimental isotherm measurements of a number of adsorbates, including nitrogen [120–122], carbon dioxide [122, 123], methane [123], and several other organic molecules [124]. It is has generally been

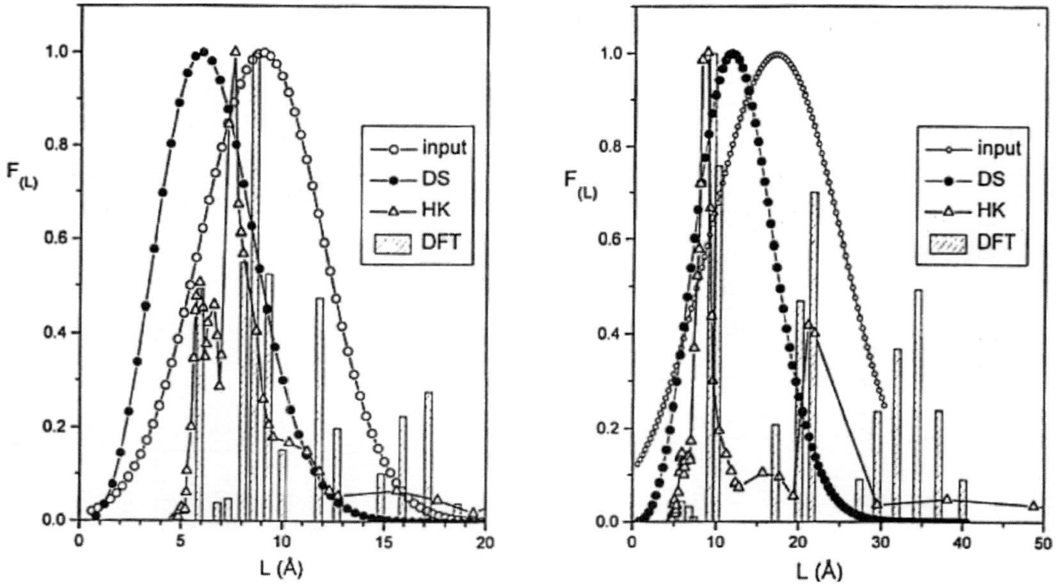

Fig. 16. Comparison of PSDs obtained using the Dubinin–Stoeckli (DS), Horvath–Kawazoe (HK), and density functional theory (DFT) methods to interpret an isotherm generated from molecular simulation of nitrogen adsorption in a model carbon that has a Gaussian distribution of slit pore widths [120]. Results are shown for mean pore widths of 8.9 Å (left) and 16.9 Å (right).

observed that the Dubinin-class adsorption models have a tendency to overconstrain the shape of the PSD, so that the true PSD function is not obtained in many cases. Also, the values of the empirical parameters E_0 and β that appear in the DR and DS equations vary according to the adsorbent–adsorbate pair, but a consistent method for selecting the parameters has not been reported.

D. CLASSICAL ADSORPTION MODELS

The classical model for describing adsorption in simple geometric pores is based on the Kelvin equation [125], which is derived from the condition of mechanical equilibrium for a curved interface between coexisting vapor and liquid phases in a pore. If the adsorbed liquid completely wets the pore walls, as shown in Fig. 17a, and the vapor phase is assumed to be an ideal gas, then mechanical equilibrium requires that

$$\ln(P_c/P_0) = -2\gamma_1/RT\rho_1(r_k - t), \qquad (40)$$

where γ_1 is the surface tension, ρ_1 is the liquid density, and t is the equilibrium thickness of the film adsorbed on the pore wall. Equation (40) is actually a modified Kelvin (MK) equation in which the pore radius, r_p, is calculated as the sum of the core radius of the vapor–liquid interface, r_k, and the adsorbed film thickness t. The latter quantity is obtained from an

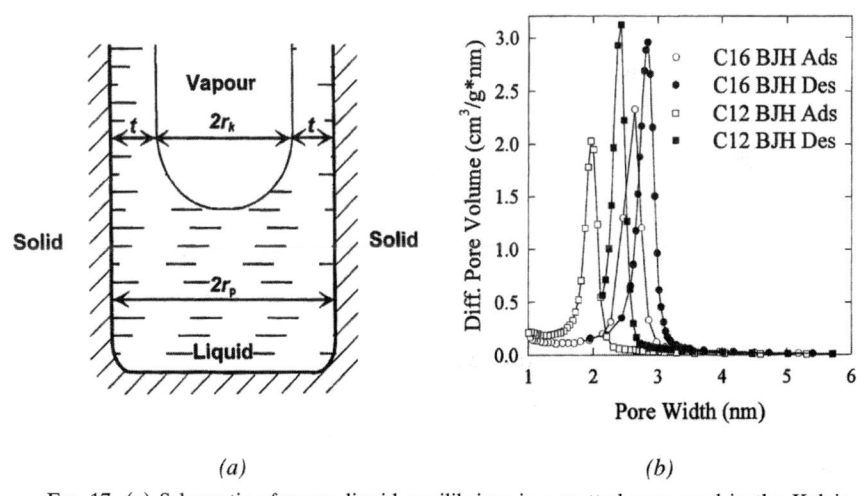

(a)   (b)

FIG. 17. (a) Schematic of vapor–liquid equilibrium in a wetted pore used in the Kelvin equation model [2]. (b) Comparison of PSD results obtained using the BJH Kelvin equation model to interpret the PSDs of two mesoporous MCM-41 structures, C16 (circles) and C12 (squares) [130]. The open and filled symbols denote PSD results obtained from the adsorption and desorption branches of the isotherm, respectively.

experimental isotherm measured upon a nonporous substrate of the same material as the porous solid [126]. The analytic form of the MK equation can be applied to PSD analysis of mesoporous solids; various prescriptions have been suggested for interpreting the experimental isotherm using this model [28, 127, 128]. For mesoporous materials that exhibit adsorption/desorption hysteresis, the PSD that is obtained will depend on whether the adsorption or desorption branch of the experimental isotherm is selected for the PSD analysis. For capillary condensation in cylindrical pores, the meniscus of the adsorbate is usually assumed to be cylindrical in shape during adsorption but hemispherical during desorption [129]. This results in forms of Eq. (40) for adsorption and desorption that differ by a factor of two on the right-hand side. In general, smaller mean pore sizes are obtained when the adsorption branch is used to interpret the PSD rather than the desorption branch, as shown in Fig. 17b [130] for the PSDs of two MCM-41 samples calculated using the Barrett–Joyner–Halenda (BJH) method [28], the most popular Kelvin-based analysis method.

It has been known for about two decades that the classical Kelvin and MK equation methods underestimate the mean pore sizes of adsorbents that have pores smaller than 75 Å [131, 132]. The error becomes very large for microporous adsorbents, as shown in Fig. 11 [11]. The continuum description of a vapor–liquid interface in a filled capillary breaks down for micropores, in which the adsorbed fluid has a coarse-grained structure and the construct of the surface tension cannot be applied. Also, the Kelvin equation altogether neglects interactions between adsorbate molecules and the pore surface, a significant omission since these interactions are greatly enhanced in micropores. Various studies have established that traditional Kelvin-based adsorption models consistently underestimate the mean pore sizes and PSD maxima of microporous solids. Figure 18, for example, demonstrates that the mean pore sizes obtained for model controlled pore glasses (see Fig. 1) using the BJH method [28] are found to be approximately 1 nm smaller [25] than the pore sizes obtained from the exact geometric calculation as noted in Fig. 2.

The shortcomings of the classical Kelvin-based adsorption models have spurred efforts to modify the Kelvin equation method further so that it may be applied to the characterization of microporous adsorbents. In recent studies of MCM-41 adsorbents [129, 133, 134], it was noted that the MK pore filling correlation can be brought into close agreement with experimental measurements of capillary condensation (Fig. 19) by subtracting a constant value from the $(r_k - t)$ term on the right-hand side of Eq. (40). The value of the adjustment factor was found to be 0.3 nm for nitrogen adsorption at 77 K [129] and 0.438 nm for argon adsorption at 87 K in MCM-41 samples [134]. Although the corrections reproduce the experimental pore filling pressures

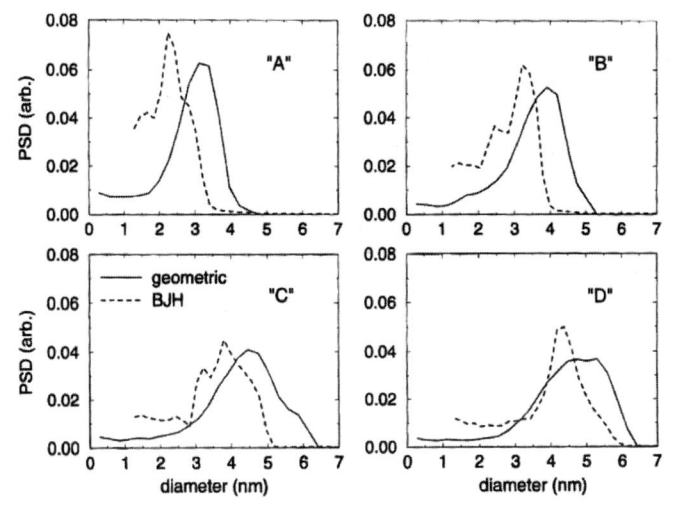

FIG. 18. PSDs for model porous silica glasses [25]. A, B, C, and D are sample glasses prepared by quench MD; the samples differ in mean pore size and porosity. The solid curves are exact geometric PSD results for the model adsorbents; the dashed lines are the PSDs predicted from BJH pore size analysis of simulated nitrogen isotherms for the model porous glasses.

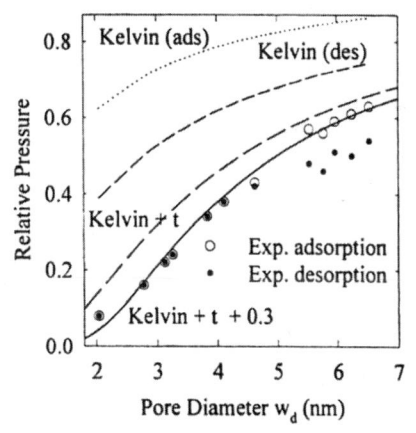

FIG. 19. Comparison of pore filling correlations developed to interpret experimental measurements of the nitrogen condensation pressure at 77 K in MCM-41 samples that have different pore diameters [129]. The dotted and short-dashed lines denote the results for the original Kelvin equation, i.e., $t = 0$ in Eq. (40), with a cylindrical and hemispherical meniscus assumed respectively for the adsorption and desorption cases. The long-dashed line shows the pore filling correlation for the MK model of Eq. (40), and the solid line shows the result when the pore width is adjusted by an additional factor of 0.3 nm.

of MCM-41 materials with reasonable accuracy, no physical justification has been put forth to rationalize this modified form of the Kelvin equation. Consequently, there is no a priori method to predict the correction factor needed for other probe gases and other temperatures, nor is it known whether the empirical correlation will hold for materials other than MCM-41.

Another pore filling model based upon capillary equilibrium in cylindrical pores has recently been proposed in which the condition of thermodynamic equilibrium is modified to include the effects of surface layering and adsorbate–adsorbent interactions [135–137]. Assuming that the vapor–liquid interface is represented by a cylindrical meniscus during adsorption and by a hemispherical meniscus during desorption, and invoking the Defay–Prigogene expression for a curvature-dependent surface tension [21], the equilibrium condition for capillary coexistence in a cylindrical pore is obtained as

$$\frac{d(\Delta G)}{dN} = 0 = \int_{P_g}^{P_0} v_g dP + \widetilde{\phi}(t, R) - \frac{v_1 \gamma_\infty (R-t)}{(R-t-\lambda/2)^2}$$

$$= \int_{P_g}^{P_0} v_g dP + \widetilde{\phi}(R, R) - \frac{2 v_1 \gamma_\infty (R-t)^2}{[(R-t)(R-t-\lambda) + \lambda \sigma_{\mathrm{ff}}/4](R-t-\lambda)},$$

(41)

where $d(\Delta G)/dN$ is the Gibbs free energy change per mole adsorbed, v_g and v_1 are the respective gas and liquid molar volumes at the saturation pressure P_0, γ is the surface tension of the vapor–condensate interface, σ_{ff} is the adsorbate molecular diameter, R is the pore radius, and $\lambda = (\frac{2}{3})^{1/2} \sigma_{\mathrm{ff}}$ [135]. The potential field $\widetilde{\phi}(r, R)$ represents the net interaction of an adsorbate molecule at radial coordinate r with the surrounding condensate and solid. This field is given as the difference between the adsorbate–adsorbent and the adsorbate–adsorbate potentials, and it is represented using the 9–3 potential for an adsorbate confined within a cylindrical pore of a structureless Lennard–Jones solid. The adsorbed layer thickness t can be determined for a given pore radius and potential field by solving the equality given by Eq. (41). The capillary coexistence curve can then be constructed by calculating the condensation pressure P_g from Eq. (41) using the known values of film thickness and pore radius. This method was shown to give good agreement with the capillary coexistence curve and pore criticality predicted by DFT for adsorption of nitrogen [135–137] and other probe gases [136] on MCM-41 materials. The method also qualitatively predicts some wetting features of the isotherm below the capillary condensation pressure [135, 136].

Another suggested MK method incorporates an improved model of the statistical adsorbed film thickness into Eq. (40) [138, 139]. In this approach,

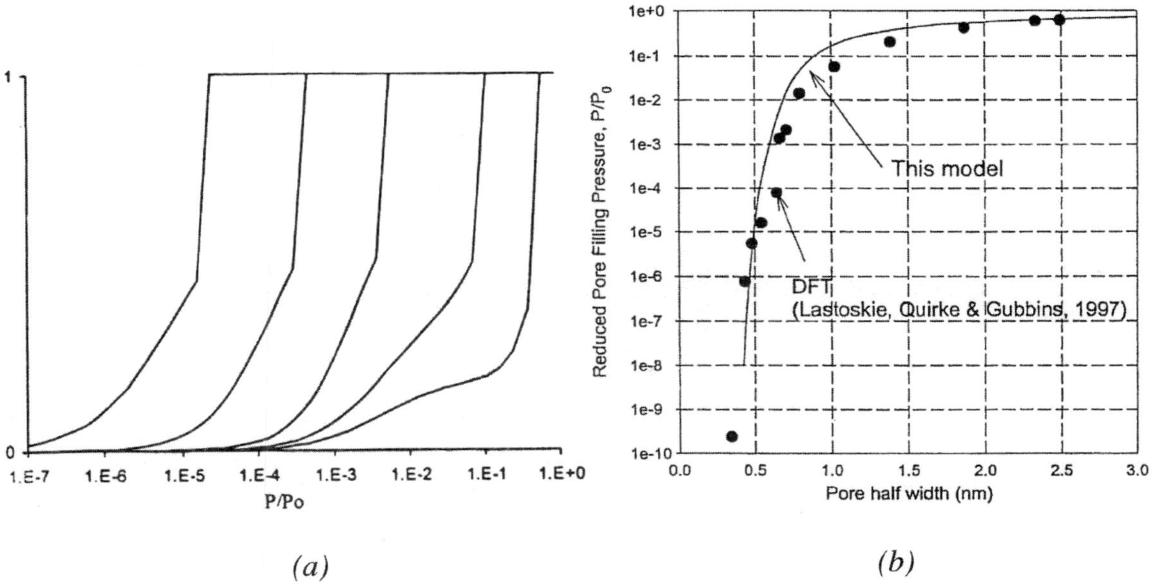

FIG. 20. (a) Model nitrogen adsorption isotherms at 77 K calculated using a modified Kelvin-BET method for carbon slit pores of physical width (reading from left to right) 10.0, 11.4, 14.3, 21.4 and 42.9 Å [138]. (b) Comparison of pore filling pressure correlations for DFT (points) and the MK-BET method (line) for nitrogen adsorption in carbon slit pores at 77 K [139].

the film thickness t is calculated using a modified BET equation to account for molecular layering in pores of finite dimension, with the BET shape parameter calculated from the heat of adsorption of the adsorbate molecule in the micropore. The modified BET equation is also used to represent the amount adsorbed below the capillary condensation pressure, as shown in the model isotherms in Fig. 20a. The MK-BET method is computationally efficient, and as shown in Fig. 20b, it yields a nitrogen pore filling correlation that is in good agreement with DFT results [138, 139]. The MK-BET method assumes that the surface tension and molar volume of the confined fluid do not depend on the pore size. Although there is no physical justification in support of this assumption, the empirical combination of Eq. (40) with a modified BET model for the film thickness gives a surprisingly accurate pore filling correlation for the case of subcritical nitrogen adsorption in slit-shaped carbon pores.

IV. Conclusions

At the present time, the pore filling models that are most frequently used to interpret pore structure from gas adsorption measurements are those based upon simple geometric pore models: density functional theory, the Horvath–Kawazoe method, and variations of the Kelvin equation. It should be recognized that these models are all limited by similar assumptions regarding the pore shape, the chemical homogeneity of the adsorbent surface, and the neglect of pore connectivity. Among the computationally efficient adsorption models that assume a simple pore geometry, DFT is the best model for the determination of the adsorbent pore size distribution [140]. Validation against molecular simulation results has established that DFT provides a realistic model of pore filling for chemically homogeneous porous solids that have simple pore geometries. Molecular simulation methods such as GCMC and GEMC can also be directly employed in simple geometric pore models for PSD characterization, but the molecular simulation approach is considerably more computer-intensive than the analytic pore filling models (i.e., HK, Kelvin equation) or DFT.

DFT and molecular simulation adsorption models give good results over a wide range of temperatures, both subcritical and supercritical, and over the full range of pressures sampled in the experimental adsorption isotherm. The DFT and molecular simulation methods may be used for PSD analysis in both the micropore and the mesopore size range. In contrast, the Kelvin and HK pore filling models are computationally convenient but are more limited in their range of application, although recent modifications of these methods have significantly extended their capabilities. For microporous solids, it is particularly important that the adsorption model includes the effects of gas–solid interactions in the calculation of the theoretical isotherms. This is best accomplished using DFT or a molecular simulation technique.

The accuracy and realism of the pore filling model are the principal consideration in obtaining the correct PSD from inversion of Eq. (14). It should be noted, however, that experimental and numerical considerations also factor into the range and accuracy of the calculated PSD results. Using commercially available gas porosimeters, for example, one can measure the experimental isotherm of the adsorbent to a minimum pressure of approximately 10^{-6} atm. For microporous carbon adsorbents with slit-shaped pores, theoretical results indicate that nitrogen condensation at 77 K occurs at pressures below 10^{-6} atm in pores narrower than 9 Å (Fig. 11). Thus, conventional nitrogen porosimetry cannot access the ultramicropore size distribution and the PSDs of materials such as molecular sieve carbons must be determined by other methods, e.g., size exclusion porosimetry. Another potential source of uncertainty in the PSD result, as noted in Section III, is the extent to which regularization is applied for the inversion of the ill-posed integral of Eq. (14). There is presently no consensus on the form of the regularization constraint that should be used in obtaining the PSD, nor is there an established procedure for selecting the proper value of the smoothing parameter that appears in the objective function. It is important to recognize that regularization can have a profound effect on the shape of the computed pore size distribution and that highly dissimilar PSD results can be obtained from the same experimental isotherm and the same theoretical adsorption model if different regularization techniques are applied.

Although DFT is computationally more efficient than molecular simulation for simple geometric pore models, molecular simulation methods will become progressively more desirable for carrying out pore structure analysis as computer hardware capabilities steadily improve. Molecular simulation has the advantage that it can be readily applied to more complicated adsorbate molecules, and particularly to more complex pore topologies such as the disordered porous model structures described in Section II. This is not the case for DFT. The use of molecular simulation is particularly attractive for developing realistic disordered model porous structures by mimicking the processing of adsorbent materials, as discussed in Section II.A for quench MD simulation of controlled pore glasses. This simulation method has the advantage that it gives a unique pore structure. However, a different simulation approach is required for each new class of materials. Another application of molecular simulation that shows great promise is to use reverse Monte Carlo sampling to generate model amorphous microstructures that are in statistical agreement with adsorbent structural data obtained from scattering experiments. The RMC method described for microporous carbons in Section II.B can be applied to other types of materials, but in general it does not yield a unique structure for the porous solid. How important this feature of nonuniqueness is for pore structure characterization and for prediction of adsorption properties remains to be determined.

It may be possible to alleviate the nonuniqueness problem by using more than one experimental probe technique in the pore structure analysis. Some preliminary work (e.g., Fig. 14) has shown that structure factor measurements from X-ray or neutron diffraction can be effectively combined with sorption porosimetry and/or adsorption calorimetry measurements to characterize semiamorphous porous solids. Sophisticated molecular modeling methods for disordered porous solids are still in an early stage of development and are highly computer-intensive. However, with further improvements in both computing speeds and modeling algorithms, a new set of powerful simulation-based characterization methods for amorphous materials may be expected in the next decade. These simulated model structures offer the tangible prospect of incorporating important features such as pore connectivity, pore shape variations, and surface chemical heterogeneity directly into the adsorbent characterization procedure.

ACKNOWLEDGMENTS

We thank the National Science Foundation (Grant CTS-9733086) and the Department of Energy (Grant DE-FG02-98ER14847) for support of this research.

REFERENCES

1. S. J. Gregg and K. S. W. Sing, *Adsorption, Surface Area and Porosity*, Academic Press, London, 1982, p. 37.
2. F. Rouquerol, J. Rouquerol, and K. Sing, *Adsorption by Powders and Porous Solids*, Academic Press, London, 1999, p. 227.
3. C. G. Salazar, A. Sepulveda-Escribano, and F. Rodriguez-Reinoso, *Stud. Surf. Sci. Catal.* **128**, 303 (2000).
4. A. Albiniak, E. Broniek, M. Jasienko-Halat, A. Jankowska, J. Kaczmarczyk, T. Siemieniewska, R. Manso, and J. A. Pajares, *Stud. Surf. Sci. Catal.* **128**, 653 (2000).
5. T. Goworek, B. Jasinska, J. Wawryszczuk, K. Ciesielski, and J. Goworek, *Stud. Surf. Sci. Catal.* **128**, 557 (2000).
6. H. Kanda, M. Miyahara, T. Yoshioka, and M. Okazaki, *Langmuir* **16**, 6622 (2000).

7. M. Kruk, M. Jaroniec, Y. Sakamoto, O. Terasaki, R. Ryoo, and C. H. Ko, *J. Phys. Chem. B* **104**, 292 (2000).
8. A. V. Neimark, P. I. Ravikovitch, M. Grun, F. Schuth, and K. K. Unger, *J. Colloid Interface Sci.* **207**, 159 (1998).
9. J. D. F. Ramsay, S. Kallus, and E. Hoinkis, *Stud. Surf. Sci. Catal.* **128**, 439 (2000).
10. E. I. Segarra and E. D. Glandt, *Chem. Eng. Sci.* **49**, 2953 (1994).
11. C. M. Lastoskie, N. Quirke, and K. E. Gubbins, *Stud. Surf. Sci. Catal.* **104**, 745 (1997).
12. D. Nicholson and N. G. Parsonage, *Computer Simulation and the Statistical Mechanics of Adsorption*, Academic Press, London, 1982.
13. A. Z. Panagiotopoulos, *Mol. Phys.* **62**, 701 (1987).
14. R. L. McGreevy and L. Putszai, *Mol. Simul.* **1**, 359 (1988).
15. C. M. Lastoskie, K. E. Gubbins, and N. Quirke, *J. Phys. Chem.* **97**, 4786 (1993).
16. P. I. Ravikovitch, G. L. Haller, and A. V. Neimark, *Stud. Surf. Sci. Catal.* **117**, 77 (1998).
17. W. G. Madden and E. D. Glandt, *J. Stat. Phys.* **51**, 537 (1988).
18. L. Zhang and P. R. Van Tassel, *J. Chem. Phys.* **112**, 3006 (2000).
19. G. Horvath and K. Kawazoe, *J. Chem. Eng. Japan* **16**, 474 (1983).
20. M. M. Dubinin and L. V. Radushkevich, *Proc. Acad. Sci. USSR* **55**, 331 (1947).
21. R. Defay and I. Prigogine, *Surface Tension and Adsorption*, Longmans, Green, Bristol, 1966.
22. C. M. Lastoskie, K. E. Gubbins, and N. Quirke, *Langmuir* **9**, 2693 (1993).
23. G. M. Davies, N. A. Seaton, and V. S. Vassiliadis, *Langmuir* **15**, 8235 (1999).
24. L. D. Gelb and K. E. Gubbins, *Langmuir* **14**, 2097 (1998).
25. L. D. Gelb and K. E. Gubbins, *Langmuir* **15**, 305 (1999).
26. W. Haller, *Nature* **206**, 693 (1965).
27. P. Levitz, G. Ehret, S. K. Sinha, and J. M. Drake, *J. Chem. Phys.* **95**, 6151 (1991).
28. E. P. Barrett, L. G. Joyner, and P. P. Halenda, *J. Am. Chem. Soc.* **73**, 373 (1951).
29. P. Pfeifer, G. P. Johnston, R. Deshpande, D. M. Smith, and A. J. Hurd, *Langmuir* **7**, 2833 (1991).
30. P. Levitz, *Adv. Colloid Interface Sci.* **76**, 71 (1998).
31. R. J. M. Pellenq, A. Delville, H. van Damme, and P. Levitz, *Stud. Surf. Sci. Catal.* **128**, 1 (2000).
32. K. T. Thomson and K. E. Gubbins, *Langmuir* **16**, 5761 (2000).
33. B. E. Warren, *X-Ray Diffraction*, Dover, New York, 1990, p. 116.
34. N. Metropolis and S. Ulam, *J. Am. Stat. Assoc.* **44**, 335 (1949).
35. B. McEnaney, *Carbon* **26**, 267 (1988).
36. H. March and P. L. Walker, Jr., in *Chemistry and Physics of Carbon*, Marcel Dekker, New York, 1979, Vol. 15, Chap. 3.
37. B. McEnaney, T. J. Mays, and X. Chen, *Fuel* **77**, 557 (1998).
38. J. Rodriguez, F. Ruette, and J. Laine, *Carbon* **32**, 1536 (1994).
39. J. R. Dahn, W. Xing, and Y. Gao, *Carbon* **35**, 825 (1997).
40. N. A. Seaton, S. P. Friedman, J. M. D. MacElroy, and B. J. Murphy, *Langmuir* **13**, 1199 (1997).
41. M. P. Allen and D. J. Tildesley, *Computer Simulaton of Liquids*, Clarendon Press, Oxford, 1987.
42. K. S. Page and P. A. Monson, *Phys. Rev. E* **54**, R29 (1996).
43. K. S. Page and P. A. Monson, *Phys. Rev. E* **54**, 6557 (1996).
44. L. Sarkisov and P. A. Monson, *Stud. Surf. Sci. Catal.* **128**, 21 (2000).
45. J. K. Johnson, J. A. Zollweg, and K. E. Gubbins, *Mol. Phys.* **78**, 591 (1993).
46. R. D. Kaminsky and P. A. Monson, *J. Chem. Phys.* **95**, 2936 (1991).
47. J. M. D. MacElroy and K. Raghavan, *J. Chem. Phys.* **93**, 2068 (1990).

48. K. S. W. Sing, D. H. Everett, R. A. W. Haul, L. Moscou, R. A. Pierotti, J. Rouquerol, and T. Siemieniewska, *Pure Appl. Chem.* **57,** 603 (1985).
49. M. W. Maddox, C. M. Lastoskie, N. Quirke, and K. E. Gubbins, in *Fundamentals of Adsorption 5* (M. D. LeVan, ed.), Kluwer Academic, Boston, 1996, p. 571.
50. G. S. Heffelfinger and F. van Swol, *J. Chem. Phys.* **100,** 7548 (1994).
51. W. D. Machin and P. D. Golding, *J. Chem. Soc. Faraday Trans.* **86,** 175 (1990).
52. Y. Lu, G. Cao, R. P. Kale, S. Prabakar, G. P. Lopez, and C. J. Brinker, *Chem. Mater.* **11,** 1223 (1999).
53. O. D. Velev, T. A. Jede, R. F. Lobo, and A. M. Lenhoff, *Nature* **389,** 447 (1997).
54. C. T. Kresge, M. E. Leoniwicz, W. J. Roth, J. C. Vartuli, and J. S. Beck, *Nature* **359,** 710 (1992).
55. L. Zhang and P. R. Van Tassel, *Mol. Phys.* **98,** 1521 (2000).
56. P. R. Van Tassel, *J. Chem. Phys.* **107,** 9530 (1997).
57. P. R. Van Tassel, *Phys. Rev. E* **60,** R25 (1999).
58. J. A. Given and G. Stell, *J. Chem. Phys.* **97,** 4573 (1992).
59. M. L. Rosinberg, G. Tarjus, and G. Stell, *J. Chem. Phys.* **100,** 5172 (1994).
60. P. R. Van Tassel, J. Talbot, P. Viot, and G. Tarjus, *Phys. Rev. E* **56,** R1229 (1997).
61. J. K. Percus and G. J. Yevick, *Phys. Rev.* **110,** 1 (1958).
62. J. P. Olivier, in *Fundamentals of Adsorption 6* (F. Meunier, ed.), Elsevier, Paris, 1998, p. 207.
63. W. H. Press, *Numerical Recipes: The Art of Scientific Computing,* Cambridge University Press, New York, 1986.
64. M. v. Szombathely, P. Brauer, and M. Jaroniec, *J. Comput. Chem.* **13,** 17 (1992).
65. W. A. House, *J. Colloid Interface Sci.* **67,** 166 (1978).
66. P. H. Merz, *J. Comp. Phys.* **38,** 64 (1980).
67. C. H. W. Vos and L. K. Koopal, *J. Colloid Interface Sci.* **105,** 183 (1985).
68. G. M. Davies and N. A. Seaton, *Langmuir* **15,** 6263 (1999).
69. G. M. Davies and N. A. Seaton, *Carbon* **36,** 1473 (1998).
70. V. Y. Gusev, J. A. O'Brien, and N. A. Seaton, *Langmuir* **13,** 2815 (1997).
71. S. Samios, A. K. Stubos, N. K. Kanellopoulos, R. F. Cracknell, G. K. Papadopoulos, and D. Nicholson, *Langmuir* **13,** 2795 (1997).
72. A. Vishnyakov, P. Ravikovitch, and A. Neimark, *Langmuir* **15,** 8736 (1999).
73. W. A. Steele, *Surf. Sci.* **36,** 317 (1973).
74. H. Liu, L. Zhang, and N. A. Seaton, *Chem. Eng. Sci.* **47,** 4393 (1992).
75. H. Liu and N. A. Seaton, *Chem. Eng. Sci.* **49,** 1869 (1994).
76. M. Lopez-Ramon, J. Jagiello, T. Bandosz, and N. Seaton, *Langmuir* **13,** 4435 (1997).
77. N. A. Seaton, J. R. P. B. Walton, and N. Quirke, *Carbon* **27,** 853 (1989).
78. P. I. Ravikovitch, G. L. Haller, and A. V. Neimark, *Adv. Colloid Interface Sci.* **76,** 203 (1998).
79. R. Nilson and S. K. Griffiths, *J. Chem. Phys.* **111,** 4281 (1999).
80. R. Evans, in *Inhomogeneous Fluids* (D. Henderson, ed.), Marcel Dekker, New York, 1992, Chap. 5.
81. J. A. Barker and D. Henderson, *J. Chem. Phys.* **47,** 4714 (1967).
82. J. D. Weeks, D. Chandler, and H. C. Andersen, *J. Chem. Phys.* **54,** 5237 (1971).
83. N. F. Carnahan and K. E. Starling, *J. Chem. Phys.* **51,** 635 (1969).
84. Y. Rosenfeld, *Phys. Rev. Lett.* **63,** 980 (1989).
85. E. Kierlik and M. L. Rosinberg, *Phys. Rev. A* **42,** 3382 (1990).
86. P. Tarazona, *Phys. Rev. A* **31,** 2672 (1985).
87. P. Tarazona, *Phys. Rev. A* **32,** 3148 (1985).
88. P. Tarazona, U. Marini Bettolo Marconi, and R. Evans, *Mol. Phys.* **60,** 573 (1987).

89. T. M. Reed and K. E. Gubbins, *Applied Statistical Mechanics,* Butterworth–Heinemann, Stoneham, MA, 1973.
90. R. Dombrowski, D. Hyduke, and C. M. Lastoskie, *Langmuir* **16,** 5041 (2000).
91. J. P. Olivier, *J. Porous Mater.* **2,** 9 (1995).
92. P. Ravikovitch, A. Vishnyakov, R. Russo, and A. Neimark, *Langmuir* **16,** 2311 (2000).
93. A. V. Neimark and P. I. Ravikovitch, *Langmuir* **13,** 5148 (1997).
94. N. Quirke and S. R. R. Tennison, *Carbon* **34,** 1281 (1996).
95. S. Scaife, P. Kluson, and N. Quirke, *J. Phys. Chem. B* **104,** 313 (2000).
96. P. I. Ravikovitch, S. C. O. Domhnaill, A. V. Neimark, F. Schuth, and K. K. Unger, *Langmuir* **11,** 4765 (1995).
97. C. T. Kresge, M. E. Leonowicz, W. J. Roth, J. C. Vartuli, and J. S. Beck, *Nature* **359,** 710 (1992).
98. P. Ravikovitch, D. Wei, W. T. Chueh, G. L. Haller, and A. Neimark, *J. Phys. Chem. B* **101,** 3671 (1997).
99. K. Schumacher, P. I. Ravikovitch, A. Du Chesne, A. Neimark, and K. K. Unger, *Langmuir* **16,** 4648 (2000).
100. C. Y. Chen, S. Q. Xiao, and M. E. Davis, *Micropor. Mater.* **4,** 1 (1995).
101. R. A. Pierotti and H. E. Thomas, *Surf. Colloid Sci.* **4,** 241 (1971).
102. J. Frenkel, *Kinetic Theory of Liquids,* Clarendon Press, Oxford, 1946.
103. G. D. Halsey, *J. Am. Chem. Soc.* **74,** 1082 (1952).
104. T. L. Hill, *J. Chem. Phys.* **17,** 668, (1949).
105. D. H. Everett and J. C. Powl, *J. Chem. Soc. Faraday Trans. 1* **72,** 619 (1976).
106. A. Saito and H. C. Foley, *AIChE J.* **37,** 429 (1991).
107. L. S. Cheng and R. T. Yang, *Chem. Eng. Sci.* **49,** 2599 (1994).
108. L. S. Cheng and R. T. Yang, *Adsorption* **1,** 187 (1995).
109. R. K. Mariwala and H. C. Foley, *Ind. Eng. Chem. Res.* **33,** 2314 (1994).
110. S. U. Rege and R. T. Yang, *AIChE J.* **46,** 734 (2000).
111. R. J. Dombrowski, C. M. Lastoskie, and D. R. Hyduke, *Colloid Surf. A* (2001), in press.
112. A. Muller, *Proc. Roy. Soc. London* **A154,** 624 (1936).
113. C. M. Lastoskie, *Stud. Surf. Sci. Catal.* **128,** 475 (2000).
114. R. J. Dombrowski and C. M. Lastoskie, submitted for publication (2001).
115. M. M. Dubinin, *Chem. Rev.* **60,** 235 (1960).
116. M. M. Dubinin, in *Chemistry and Physics of Carbon* (P. L. Walker, ed.), Marcel Dekker, New York, 1966, p. 51.
117. M. M. Dubinin and V. A. Astakhov, *Adv. Chem. Ser.* **102,** 69 (1970).
118. H. F. Stoeckli, *J. Colloid Interface Sci.* **59,** 184 (1977).
119. R. C. Bansal, J. B. Donnet, and H. F. Stoeckli, *Active Carbon,* Marcel Dekker, New York, 1988.
120. D. L. Valladares, F. Rodriguez-Reinoso, and G. Zgrablich, *Carbon* **36,** 1491 (1998).
121. M. el-Merraoui, M. Aoshima, and K. Kaneko, *Langmuir* **16,** 4300 (2000).
122. D. Cazorla-Amoros, J. Alcaniz-Monge, M. A. de la Casa-Lillo, and A. Linares-Solano, *Langmuir* **14,** 4589 (1998).
123. F. Stoeckli, A. Guillot, D. Hugi-Cleary, and A. M. Slasli, *Carbon* **38,** 938 (2000).
124. M. Domingo-Garcia, I. Fernandez-Morales, F. J. Lopez-Garzon, and C. Moreno-Castilla, *Langmuir* **13,** 1218 (1997).
125. W. T. Thomson, *Phil. Mag.* **42,** 448 (1871).
126. M. Jaroniec, M. Kruk, and J. P. Olivier, *Langmuir* **15,** 5410 (1999).
127. R. W. Cranston and F. A. Inkley, in *Advances in Catalysis,* Vol. 9, Academic Press, New York, 1957, p. 143.
128. D. Dollimore and G. R. Heal, *J. Appl. Chem.* **14,** 109 (1964).

129. M. Kruk, M. Jaroniec, and A. Sayari, *Langmuir* **13,** 6267 (1997).
130. M. Kruk, M. Jaroniec, and A. Sayari, *J. Phys. Chem. B* **101,** 583 (1997).
131. J. R. Fisher and J. N. Israelachvili, *J. Colloid Interfac. Sci.* **80,** 528 (1981).
132. J. P. R. B. Walton and N. Quirke, *Mol. Simul.* **2,** 361 (1989).
133. M. Kruk, M. Jaroniec, J. M. Kim, and A. R. Ryoo, *Langmuir* **15,** 5279 (1999).
134. M. Kruk and M. Jaroniec, *Chem. Mater.* **12,** 222 (2000).
135. C. G. Sonwane and S. K. Bhatia, *Chem. Eng. Sci.* **53,** 3143 (1998).
136. C. G. Sonwane, S. K. Bhatia, and N. Calos, *Ind. Eng. Chem. Res.* **37,** 2271 (1998).
137. S. K. Bhatia and C. G. Sonwane, *Langmuir* **14,** 1521 (1998).
138. C. Nguyen and D. D. Do, *Langmuir* **15,** 3608 (1999).
139. D. D. Do, C. Nguyen, and H. D. Do, *Colloid Surf. A* in press, (2001).
140. C. M. Lastoskie and K. E. Gubbins, *Stud. Surf. Sci. Catal.* **128,** 41 (2000).

MODELING OF RADICAL–SURFACE INTERACTIONS IN THE PLASMA-ENHANCED CHEMICAL VAPOR DEPOSITION OF SILICON THIN FILMS

Dimitrios Maroudas

Department of Chemical Engineering, University of California, Santa Barbara, Santa Barbara, California 93106-5080

I. Introduction	252
II. Computational Methodology	254
A. The Hierarchical Approach	255
B. Density-Functional Theory	257
C. Empirical Description of Interatomic Interactions	258
D. Methods of Surface Preparation	260
E. Methods of Surface Characterization and Reaction Analysis	263
III. Surface Chemical Reactivity with SiH_x Radicals	264
A. Structure of Crystalline and Amorphous Silicon Surfaces	265
B. Interactions of SiH_x Radicals with Crystalline Silicon Surfaces	266
C. Interactions of SiH_x Radicals with Surfaces of Amorphous Silicon Films	270
IV. Plasma–Surface Interactions during Silicon Film Growth	273
A. Surface Chemical Reactions during Film Growth	274
B. Mechanism of Amorphous Silicon Film Growth	280
C. Surface Evolution and Film Structural Characterization	281
D. Film Surface Composition and Comparison with Experiment	283
E. The Role of the Dominant Deposition Precursor	284
F. The Role of Chemically Reactive Minority Species	286
V. Summary	290
References	291

I. Introduction

Growth of thin films by plasma-enhanced chemical vapor deposition (PECVD) enables a wide range of technologies for manufacturing electronic, optoelectronic, and photovoltaic devices (Reif, 1990; Crowley, 1992; Sinke, 1993). The use of highly reactive gas plasmas in PECVD allows for growth of thin films at temperatures that are much lower than those needed for other (CVD) processes (Randhawa, 1991; Graves, 1994; Smith, 1996). This ability to deposit films at low temperatures is due to the highly nonequilibrium environment of the gas discharge that leads to generation of ions and chemically reactive radicals through, e.g., electron impact dissociation. In spite of the widespread use of PECVD, process optimization and reactor design rely on completely empirical procedures that are not transferable among different reactor types or chemical systems. However, advances in plasma processing applications to surpass existing technological limits require a fundamental understanding of plasma physics and homogeneous and heterogeneous plasma chemistry, as well as species transport in plasma reactors (Graves *et al.*, 1996).

Among the least understood aspects of plasma processing technologies are plasma–surface interactions, i.e., phenomena occurring on the growth surface upon impingement of chemically reactive radicals and energetic species from the plasma (Graves *et al.*, 1996). The absence of a fundamentally based framework of surface reaction kinetics and its integration with species transport and reaction kinetics in the gas phase is a long-standing problem that limits the predictive capabilities of plasma reactor models (Proud *et al.*, 1991; Graves *et al.*, 1996). Typically, plasma modeling studies treat surface reactions in a phenomenological manner, accounting only for the presence of the surface and its effects on the gas-phase species concentrations through lumped surface reaction kinetics. However, development of predictive plasma reactor models requires detailed information on the reactions of species impinging on the surface from the plasma (Maroudas and Shankar, 1996).

Systematic investigations of plasma–surface interactions under realistic PECVD conditions require knowledge of fluxes and energies of species incident onto the surface from the plasma, structure and composition of the deposition surface, and elementary reaction processes that produce the surface species from the radicals and ions that impinge on the surface. Although the surface composition and species fluxes and energies can be determined experimentally through simultaneous use of plasma and surface diagnostics, surface reaction processes cannot be observed directly during deposition. This problem, however, can be addressed by atomic-scale simulation methods, which provide ideal means for detailed nanoscopic modeling of the interactions of radicals and molecular fragments from the plasma with surfaces, as well as determination of surface reaction mechanisms and kinetics.

PECVD from silane (SiH_4) and hydrogen (H_2) containing discharges is used widely for the production of hydrogenated amorphous silicon (a-Si:H) and nanocrystalline silicon (nc-Si:H). Thin films of these materials have numerous industrial applications ranging from solar cells to the manufacturing of thin-film transistors for flat panel displays (Crowley, 1992; Sinke, 1993). The identities and the fluxes of chemically reactive species that originate in the plasma and impinge onto the deposition surface determine the structural quality and, therefore, the electronic properties of the deposited films. Experimental research on silicon plasma deposition has provided empirical ranges of processing conditions, such as substrate temperatures and feed gas composition, for improving the structural quality of the PECVD-grown films (see, e.g., Abelson, 1993). However, even after decades of research into the growth of these silicon thin films, the reaction mechanisms between the species from the gas phase and the deposition surface, as well as the kinetics of these surface reactions remain largely unknown; even the identities of the deposition precursors are still under debate (Abelson, 1993).

This article presents an integrated atomic-scale computational methodology for the fundamental mechanistic and quantitative understanding of plasma–surface interactions that govern the plasma-assisted deposition of thin films. The article focuses on the growth of silicon thin films and, in particular, a-Si:H films. PECVD of Si films is viewed as an ideal prototypical system: in spite of its own technological importance, the study of this system can be used to examine and develop modeling tools and strategies that can be transferable to a wide class of materials processing systems. Among the various types of species generated in the discharge, this article places special emphasis on chemically reactive radicals. Recent progress in modeling radical–surface interactions in the plasma deposition of a-Si:H films is reviewed: radical–surface interactions are identified, reaction mechanisms are elucidated, reaction energetics are analyzed, and computationally deposited a-Si:H films are characterized. In spite of the focus of this article on radicals, the important roles of the ions impinging on the surface in modifying the near-surface region is realized: energetic ions can cause surface sputtering and induce various surface reactions, such as desorption reactions. The role of such ions in the surface structural and chemical characteristics of the computationally deposited films also is discussed briefly. Given that even the computational generation of a realistic amorphous network remains an open research subject, addressing the above problems rigorously at the atomic scale is a very challenging task. Integration of the resulting knowledge of plasma–surface chemistry into a reactor-scale description with predictive capabilities for process optimization is even more challenging. Chemical engineering contributions in this area can have major impact on microelectronics and other electronic device fabrication technologies.

II. Computational Methodology

Atomic-scale theoretical studies can contribute significantly to our fundamental and quantitative understanding of plasma–surface interactions and the development of an accurate chemical reaction database that is critical for progress in plasma process modeling. Quantitative accuracy in the analysis of plasma–surface interactions is guaranteed only by first-principles quantum mechanical (QM) calculations, i.e., *ab initio* calculations of the electronic structure for given atomic positions and the forces on the atoms when displaced from their equilibrium configuration (Car and Parrinello, 1985; Payne *et al.*, 1992). Such accurate calculations are limited to small clusters of atoms, or small periodic supercells that contain up to a hundred (or a few hundreds of) atoms, and integration of the equations of atomic motion over a few picoseconds. Plasma–surface interactions, however, are characterized by a much broader range of length and time scales. Thus, addressing the problem theoretically based solely on accurate QM calculations is computationally impractical. In spite of the continuous increase in computational power that will enable QM calculations over larger length scales and longer time scales, parametric studies at this level of theory will remain impractical. On the other hand, targeted accurate QM analysis of surface reactions is computationally feasible (Ricca and Musgrave, 1999; Ramalingam *et al.*, 1998c) for the development of databases of energetic and kinetic reaction parameters.

Atomistic simulation methods based on classical force fields provide computationally efficient means for the investigation of nanoscopic mechanisms that determine the interactions between surfaces and chemically reactive species from a plasma. These methods include molecular-statics (MS), Monte Carlo (MC), and molecular-dynamics (MD) techniques (Abraham, 1986; Ciccotti *et al.*, 1987; Allen and Tildesley, 1990; Maroudas and Brown, 1993; Rapaport, 1998). Combination of these techniques allows for monitoring of dynamical phenomena, analysis of reaction paths, and relaxation of the surface structure. Fundamental mechanistic understanding gained by such simulations can provide the necessary insights to guide efficiently parametric studies and design of experimental protocols. These simulations, however, are limited by the accuracy of the classical interatomic potential used to calculate the total energy and the resulting interatomic forces. Fully empirical parametrization of classical force fields is not transferable to widely different physicochemical environments. The transferability of such force fields can be improved substantially by systematic parametrization schemes based on accurate QM-calculated energy surfaces, including energy landscapes for radical–surface interactions.

Regardless of the accuracy and transferability of a classical force field, dynamical simulations based on MD techniques are limited to time scales of the order of nanoseconds. However, these time scales that can be captured directly by MD are shorter by orders of magnitude than the actual time scales characteristic, e.g., of surface relaxation phenomena during a surface growth process. In spite of recent progress in accelerating the MD simulation of infrequent events (Voter, 1997a,b; 1998; Pal and Fichthorn, 1999), kinetic MC (KMC) simulation methods provide the most practical approach to overcome time-scale limitations in complex chemically reactive systems. Such KMC methods can extend the propagation of the dynamical system to time scales longer by several orders of magnitude than those that can be captured directly by MD within an atomic-scale description of the physicochemical process (Limoge and Bocquet, 1988; Kang and Weinberg, 1989; Fichthorn and Weinberg, 1991). The inherent limitation of KMC methods is that their accuracy depends fully on the completeness and accuracy of the database of transition probabilities that is used as input to KMC simulations: the development of such an accurate database for a complex physicochemical process is a particularly challenging task.

Over the past decade, atomic-scale simulation based on MD and KMC techniques has been applied extensively to study deposition processes, most notably semiconductor epitaxial growth (Srivastava and Garrison, 1990, 1991; Crowley *et al.*, 1993) and diamond film growth (Dawnkaski *et al.*, 1996; Battaile *et al.*, 1997; 1998). In the area of plasma–surface interactions, atomic-scale simulations have focused mostly on understanding sputtering and etching processes on silicon surfaces exposed to various discharges (Schoolcraft and Garrison, 1991; Weakliem and Carter, 1993; Barone and Graves, 1995a,b; Helmer and Graves, 1997, 1998, 1999; Kubota *et al.*, 1998; Kubota and Economou, 1999; Kim *et al.*, 1999; Hanson *et al.*, 1999; Srivastava *et al.*, 1999; Halicioglu and Srivastava, 1999). In addition, substantial simulation effort has been devoted recently to the understanding and analysis of interactions between Si surfaces and chemically reactive species from silane-containing plasmas (Ramalingam *et al.*, 1998a–c, 1999a–c, 2000).

A. THE HIERARCHICAL APPROACH

The advantages of accuracy and computational efficiency of the different levels of theory outlined above can be realized through a hierarchical computational approach that links QM calculations with MS/MC/MD and KMC simulations. Such a hierarchical approach is illustrated in Fig. 1. The

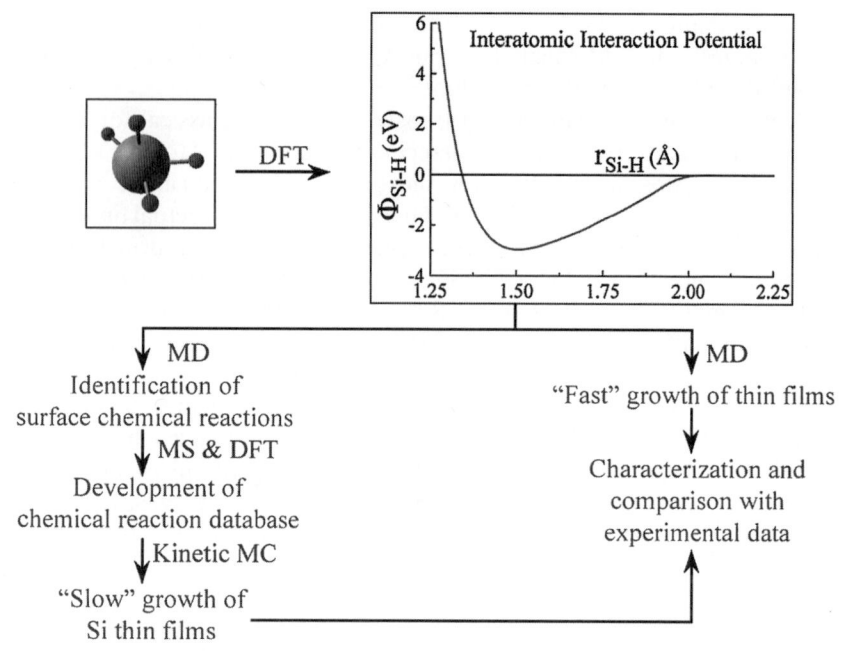

FIG. 1. Diagrammatic outline of the hierarchical approach to modeling plasma–surface interactions in the plasma-enhanced chemical vapor deposition of silicon thin films.

starting point is QM calculation within the framework of density-functional theory (DFT) (Hohenberg and Kohn, 1964; Kohn and Sham, 1965; Payne *et al.*, 1992). DFT-based energy calculations can be used to evaluate the parameters of classical interatomic interaction potentials, which can be used to perform MS, MC, and MD simulations; such *ab initio* potential parametrization is a key to improving the transferability of the classical force field. In Fig. 1, an interatomic potential energy function for Si–H interactions is given as an example of such a parametrization (Ohira *et al.*, 1995).

The classical interatomic potential can be used to carry out MD simulations of "fast" film growth on a substrate. Although the MD growth rates are several orders of magnitude faster than the experimental rates, the MD-deposited films and their surfaces can be characterized in detail and compared with experimental measurements. The main aim of such MD simulations is a fundamental mechanistic understanding and comprehensive identification of chemical reactions that occur on the deposition surfaces, as well as analysis of surface diffusion and relaxation mechanisms. Reaction identification is a very important part of the computational hierarchy: it is the key to interpretation of various experimental observations and construction of the list of reactions needed for KMC simulation of film growth. The identified set of reactions can be analyzed further to contribute to the process chemical reaction database. Upon identification, the reaction energetics and kinetics can be computed accurately based on DFT calculations that employ a model of the actual surface where the reaction occurs.

KMC modeling of the growth process over coarser time scales requires as input the rates of the surface reactions that have been identified by the MD simulations. These reaction rates can be predicted within the framework of variational rate theory (Voter and Doll, 1985; Hanggi et al., 1990; Makarov and Metiu, 1997) using the computed energy surfaces along the reaction paths. The computed rates for the identified surface reaction processes depend strongly on the local structure and composition of the deposition surface: for Si film growth, the degree of crystallinity and the H coverage of the surface are very important rate determining factors. Simple lattice KMC methods are not appropriate for the study of amorphous or nanocrystalline film growth. In such cases, hybrid off-lattice KMC simulations should be implemented. These MC simulations combine a KMC propagator of the dynamical system over long time scales with an equilibrium MC method for surface relaxation between infrequent events. Such a hybrid KMC approach was implemented recently in the simulation of diamond film growth (Clark et al., 1996a,b).

Executing schedules of film growth dynamical simulations, as outlined in Fig. 1, enables parametric studies toward development of optimal growth strategies. Specification of system parameters that govern plasma–surface interactions may require additional links to gas-phase plasma models and/or plasma diagnostics experiments. Such important parameters include the substrate temperature, the chemical identities and fluxes of species impinging on the deposition surface from the gas phase, the kinetic energies of the impinging species, and the angles of incidence and molecular orientations of the impinging species with respect to the surface.

B. DENSITY-FUNCTIONAL THEORY

Computationally efficient *ab initio* quantum mechanical calculations within the framework of DFT play a significant role in the study of plasma–surface interactions. First, they are used to parametrize classical force fields for MD simulations. Second, they provide the quantitative accuracy needed in the development of a chemical reaction database for KMC simulations over long time scales: upon identification of a surface chemical reaction through MD simulation, DFT can be used to address in quantitative detail the reaction energetics and kinetics. Third, DFT-based chemical reaction analysis and comparison with the corresponding predictions of the empirical interatomic potential used in the MD simulations provides further validation (or insights for further improvement) of the interatomic potential by generating accurate data beyond the data set employed in the potential parametrization.

The choice of the "level of theory," i.e., the type of approximations employed within DFT that determines the level of accuracy of the DFT results, is essential for computationally efficient quantum mechanical analysis of plasma-surface interactions. Such an acceptable level of theory, within a cluster model representation of the surface, has been outlined by Ramalingam et al. (1998c). The B3LYP functional (Becke, 1993; Stephens et al., 1994) is used with Gaussian basis sets for the representation of the electronic wavefunctions (Frisch et al., 1984). An effective core potential is employed to represent the $Si1s^22s^22p^6$ core (Stevens et al., 1984). The cluster that represents the surface includes terminating H atoms that mimic the actual local bonding environment, i.e., Si–Si bonds with bulk Si atoms are replaced by Si–H bonds. The geometry of the cluster atoms is relaxed based on local energy minimization techniques (Payne et al., 1992) and the cluster size effects on the computed energies should be examined by appropriate convergence tests. For analysis of reactions on the Si(001) surface, clusters with four or five atomic layers parallel to the surface plane can be used: the smallest surface model is a four-layer Si_9H_{12} cluster, while a typical size is a five-layer $Si_{23}H_{24}$ cluster. For DFT studies of radical interactions with amorphous surfaces, the necessary cluster model can be constructed using initial configurations taken from the surface region of a classical MD supercell in the vicinity of the reaction site followed by DFT-based structural relaxation of the cluster atoms.

C. EMPIRICAL DESCRIPTION OF INTERATOMIC INTERACTIONS

Several empirical and semiempirical interatomic potentials have been developed for the Si:H system based on extensions and modifications of well-known potentials for Si including up to three-body interactions (Stillinger and Weber, 1985; Biswas and Hamann, 1985; Biswas et al., 1987; Mousseau and Lewis, 1991; Baskes, 1992). Recent atomic-scale simulation work of plasma-surface interactions in the PECVD of Si thin films has been based on an empirical description of interatomic interactions in the Si:H system according to Tersoff's (1986, 1988, 1989) potential for Si, as extended by Ohira and co-workers (1994, 1995, 1996) to incorporate Si–H, H–H, and the corresponding three-body interactions. The extension of the potential to include the presence of hydrogen adopted the Tersoff parametrization to fit results of ab initio calculations for the structure and energetics of SiH_x, $x \leq 4$, species in the gas phase (Ohira et al., 1994, 1995, 1996). A similar form of potential energy functions also was developed by Murty and Atwater (1995). The extended Tersoff potential, U, has the form of a two-body potential with three-body interactions implicitly built in, given by

$$U = \sum_i \sum_{j>i} [a_{ij} V_r(r_{ij}) + b_{ij} V_a(r_{ij})] f_c(r_{ij}), \quad (1)$$

where

$$V_r(r_{ij}) = A_{ij} \exp(-\lambda_{ij} r_{ij}), \qquad V_a(r_{ij}) = -B_{ij} \exp(-\mu_{ij} r_{ij}) \quad (2)$$

$$a_{ij} = \varepsilon_{ij} (1 + \beta_i^{n_i} \tau_{ij}^{n_i})^{-1/2n_i}, \qquad b_{ij} = \chi_{ij} (1 + \beta_i^{n_i} \xi_{ij}^{n_i})^{-m_i/2n_i} \quad (3)$$

$$\tau_{ij} = \sum_{k \neq i,j} f_c(r_{ik}) \delta_{ik} g(\theta_{ijk}), \qquad \xi_{ij} = \sum_{k \neq i,j} f_c(r_{ik}) \omega_{ik} g(\theta_{ijk})$$

$$\times \exp[\sigma_{ik}(r_{ij} - r_{ik})] \quad (4)$$

$$g(\theta_{ijk}) = 1 + \frac{c_i^2}{d_i^2} - \frac{c_i^2}{[c_i^2 + (h_i - \cos\theta_{ijk})^2]}, \quad (5)$$

and

$$f_c(r_{ij}) = \begin{cases} 1, & r_{ij} < R_{ij} \\ \frac{1}{2} + \frac{1}{2} \cos\left(\frac{\pi(r_{ij} - R_{ij})}{S_{ij} - R_{ij}}\right), & R_{ij} < r_{ij} < S_{ij}, \\ 0, & r_{ij} > R_{ij}. \end{cases} \quad (6)$$

In Eqs. (1)–(6), r_{ij} denotes the interatomic distance between atom i and atom j and θ_{ijk} denotes the angle between the distance vectors $\mathbf{r}_{ij}$ and $\mathbf{r}_{jk}$ subtended at vertex j. Repulsive and attractive interactions are represented in Eqs. (1) and (2) by the potential functions V_r and V_a, respectively. The smooth cutoff function, f_c, limits the range of the potential and reduces the computational cost for energy calculations. The potential is adapted to changes in the local bonding environment by the factors a_{ij} and b_{ij} which multiply in Eq. (1) the repulsive and attractive pair interactions, $V_r(r_{ij})$ and $V_a(r_{ij})$, respectively. These factors represent a measure of bond order and have the appropriate form, Eqs. (3)–(6), to describe the dependence of the local bonding environment on atomic coordination and structure. A complete list of the potential parameters is given by Ohira et al. (1996) and Ramalingam et al. (1998b).

In spite of the parametrization effort involved in the construction of an empirical interatomic potential, capturing the complexity of plasma–surface interactions poses severe demands on the accuracy and transferability of such a model of interatomic interactions. Thus, careful testing is required prior to using any empirical model for the study of plasma–surface chemistry. The quantitative accuracy of the interatomic potential of Eqs. (1)–(6) for the study of radical–surface interactions was tested exhaustively through comparisons of its predictions with experimental measurements and results of *ab initio* calculations. Such successful comparisons included the structure and energetics of gas-phase radicals SiH_x ($x = 1, 2, 3$) and clusters Si_nH_m (n > 1), crystalline and amorphous Si bulk phases and surfaces, and mechanisms and energetics of reactions of SiH_x radicals ($x = 1, 2, 3$) with Si surfaces (Ramalingam, 2000; Ramalingam *et al.*, 1998a–c, 1999b, 2000). These comparisons also indicate that the interatomic potential of Eqs. (1)–(6) is capable of quantitative descriptions of radical surface diffusion processes on hydrogen-covered crystalline and amorphous Si surfaces.

D. METHODS OF SURFACE PREPARATION

Careful preparation of the deposition surface is an important step in the study of surface interactions with chemically reactive species from the gas plasma phase. Two types of silicon surfaces are of particular interest: crystalline and amorphous surfaces that are exposed to the chemically reactive plasma species during the PECVD process. Crystalline surfaces, especially the Si(001) surface, are representative of commonly used crystalline substrates (Si wafers) in various experimental studies and practical applications. Amorphous surfaces are representative growth surfaces at later stages of the PECVD process, after an a-Si:H film has been deposited on the substrate.

The crystalline Si(001) surface is prepared in its equilibrium configuration, i.e., a dimerized surface layer according to the (2 × 1) reconstruction (Yin and Cohen, 1981; Dabrowski and Scheffler, 1992). A bulk crystalline silicon supercell is constructed and a free Si(001) surface is created by removing the periodic boundary conditions in the [001] direction. The atoms on the resulting unreconstructed free surface are given initial displacements toward dimerization of the surface. The system is then mapped onto its minimum-energy configuration using a local nonlinear optimization method, such as a conjugate-gradient algorithm (Gill *et al.*, 1981). Subsequently, the system with the dimerized surface is equilibrated by MD at the temperature of interest to obtain a relaxed Si(001)-(2 × 1) surface for analysis of plasma–surface interactions. H-terminated Si(001)-(2 × 1) surfaces (Northrup, 1991) are constructed by introducing a number of hydrogen atoms corresponding to the surface coverage of interest within interaction distance of the Si dangling bonds of the Si(001)-(2 × 1) surface. The system is relaxed first using a conjugate gradient to minimize its energy and, subsequently, equilibrated at the temperature of interest through MD for several picoseconds.

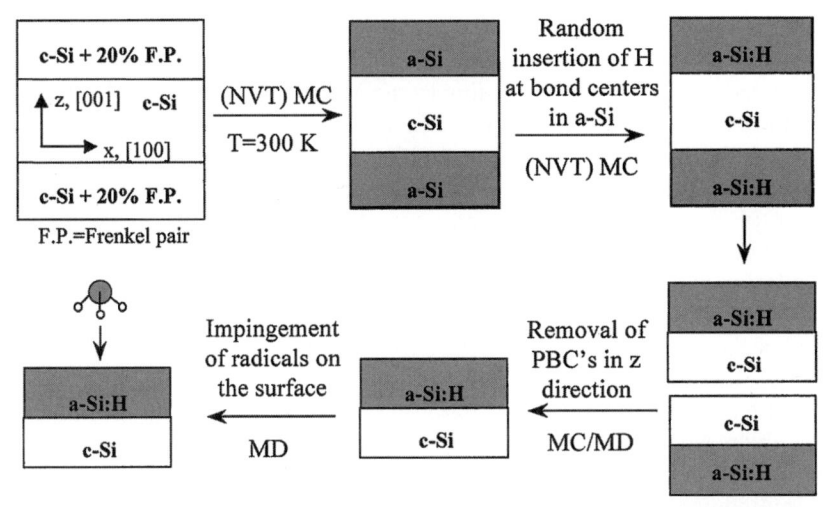

FIG. 2. Schematic illustration of a computational procedure employed to generate a-Si:H/c-Si film/substrate systems through a sequence of steps that involve MC and MD simulations.

The structures of amorphous films and their surfaces vary according to the experimental conditions of film growth. In similar manner, various computational methods can be employed to prepare amorphous films on substrates. One such method (Ramalingam et al., 1998b), based on insights drawn from solid-state amorphization upon irradiation (Barone and Maroudas, 1997), follows the sequence of steps outlined in Fig. 2. First, the volume of a bulk simulation supercell is fixed according to the lattice parameter of crystalline Si (c-Si) at the temperature of interest, as computed through an isothermal–isobaric (NpT) MC simulation. The top and bottom quarters of the c-Si supercell are "damaged" by introduction of Frenkel-pair defects, i.e., pairs of vacancies and self-interstitials removed beyond their recombination range, at concentrations above the critical concentrations for amorphization (Barone and Maroudas, 1997). The resulting structure is equilibrated using canonical (NVT) MC, which yields a layered a-Si/c-Si superstructure with c-Si sandwiched by amorphous silicon (a-Si) layers. Subsequently, hydrogen atoms at a given concentration are introduced randomly at Si–Si bond centers in the amorphized regions of the sample and the system is equilibrated using canonical MC. The MC equilibration consists of a repeated sequence of chemical relaxation and structural relaxation stages. Chemical relaxation involves MC sweeps over the H atoms only, consisting of trial moves of the H atoms for given positions of Si atoms in the a-Si matrix; these moves are accepted or rejected according to the Metropolis criterion (Allen and

Tildesley, 1990). Next, structural relaxation is performed based on MC trial moves of all the atoms in the supercell and acceptance or rejection of the generated configurations according to the Metropolis criterion. This hydrogenation of the a-Si layers results in a layered a-Si:H/c-Si superstructure with structurally stable interfaces. The simulation cell is then cut in half and the periodic boundary conditions (PBCs) are removed in the direction normal to the plane of the a-Si:H/c-Si interface. Additional c-Si atomic layers are placed at the appropriate interlayer distance at the bottom of the resulting simulation cell and are held fixed to simulate contact with an infinite rigid substrate. MC equilibration is used to distribute the hydrogen in the presence of the free surface. Finally, the surface is relaxed further by MD at the temperature of interest.

More commonly, amorphous films on substrates are generated computationally by MD through repeated impingement of species from a gas phase on the growth surface. Such an MD method was used to generate a-Si:H films by impingement of SiH_x radicals ($x = 1, 2, 3$) on an initially H-terminated Si(001)-(2 × 1) surface for studies of plasma species interactions with the amorphous deposition surface (Ramalingam, 2000; Ramalingam et al., 1998c, 1999b,c). In the MD simulations, the classical equations of motion are integrated using a fifth-order predictor-corrector algorithm, rescaling of the atomic velocities for substrate temperature control, and an integration time step typically of the order of 10^{-16} s; the fine time discretization is dictated mainly by the light mass of the H atoms in the dynamical system. The simulation cells typically include 20 dynamical (001) planes from the initial substrate surface in contact with fixed layers at the bottom of the cell. Typical supercell surface areas are $4a \times 4a$ or larger, where a denotes the Si lattice parameter at the temperature of interest and zero pressure. The impinging radicals are placed initially just beyond the range of interaction with the surface atoms and they are directed normal to the surface, toward randomly chosen surface locations, at random molecular orientations with respect to the surface, and with kinetic energies that correspond to the substrate temperature. For the range of impingement conditions used in these MD simulation studies, the implemented temperature control does not affect either the sequence or the mechanisms of the observed surface chemical reactions. Between successive radical impingement events the system is equilibrated through MD for several picoseconds; a period of 4 ps has been shown to be sufficient to achieve surface relaxation when SiH_3 is used as the sole deposition precursor (Ohira et al., 1995; Ramalingam et al., 1999b). Such MD simulations of film growth require several thousands of impingement events, i.e., a long overall trajectory with a duration of several nanoseconds, and they are particularly demanding computationally. The unique advantage of this procedure is that interactions of radicals can be studied throughout the simulated growth process with various growth surfaces that are representative of various stages of the actual growth process.

E. Methods of Surface Characterization and Reaction Analysis

Chemical characterization of the computationally generated surfaces and its correlation with the surface structure and composition is an important step in our understanding of the deposition mechanism governed by radical–surface interactions. The surface "reactivity map" introduced by Ramalingam *et al.* (1998a–c, 1999a–c) is an important characterization tool of the surface chemical reactivity with the impinging radical as a function of the location of impingement on the surface. Such maps are surface specific, depend on the molecular orientation of the radical with respect to the surface plane, and are constructed based on a sequence of energy minimization (MS) calculations. Specifically, the total potential energy of the interacting atoms in the simulation supercell is minimized with respect to the atomic coordinates and under the appropriate constraints based on an iterative optimization scheme, such as a conjugate-gradient algorithm (Ramalingam *et al.*, 1998b). The energy gain, ΔE, upon interaction of the surface with the radical approaching the surface in a given (constrained) molecular orientation is calculated over a grid of locations (x, y) on the surface. ΔE provides a measure of the radical reactivity at each surface location: a high energy gain corresponds to a high reactivity of this surface location. This energy gain is expressed with respect to a reference state comprising the surface in the absence of the radical and the free radical. It is defined by

$$\Delta E(x, y) \equiv E_{S-R}(x, y) - [E_S + E_R], \tag{7}$$

where $E_{S-R}(x, y)$ is the minimized potential energy of the system with the radical impinged at (x, y) in a specified molecular orientation, E_s is the potential energy of the sample with the radical removed beyond interaction range with the surface, and E_R is the potential energy of the free radical. The reactivity map is the resulting contour plot of $\Delta E(x, y)$ over the surface. For amorphous film surfaces, systematic analysis of the surface reactivity map also requires detailed surface structural characterization through, e.g., the distribution on the surface of the atomic coordination as a measure of the surface dangling bond density distribution. It should be mentioned that such reactivity maps can be constructed, as a postprocessing characterization step, for a sequence of growth surfaces generated by the MD simulation procedure described in Section II.D.

Upon identification of a reaction as part of an MD trajectory, a similar MS-based procedure is used to derive the energetic progress of the reaction, the reaction energy landscape, along the corresponding reaction path (Ramalingam et al., 1998c, 1999b). The energy of the atomic configurations along the reaction path is minimized following a local energy minization technique. The computed energies of the reactants, reaction products, and saddle-point configurations are used to obtain reaction energies and thermal activation energy barriers. Such energetic analysis of surface reactions can be performed based on either DFT or an empirical description of interatomic interactions; the DFT analysis is aided significantly by the reaction path identified through MD.

III. Surface Chemical Reactivity with SiH_x Radicals

Plasma-assisted deposition occurs through chemical reactions of species from the plasma with the deposition surface, which may lead to attachment of such species to reactive sites on the surface. In SiH_4-containing plasmas, electron impact dissociation of SiH_4 generates various chemically reactive radicals, such as SiH, SiH_2, and SiH_3, which play key roles in the PECVD process and determine the structure and properties of the deposited films. The structure and chemical composition of a deposition surface exposed to the chemically reactive gas discharge also play an important role in determining the radical–surface reaction mechanisms. The interactions of a radical with a particular surface can be viewed as a set of reactions involving two reactants: the radical and the surface. For a given radical and substrate temperature, the reaction mechanism and products depend on the local structure and chemical composition of the surface in the vicinity of the surface location where the radical impinges, as well as the radical's kinetic energy and molecular orientation with respect to the surface. It should be mentioned that this view of radical–surface interactions also can be useful for lumping reactions into categories for more efficient incorporation of such surface reactions into macroscopic reactor-scale modeling.

A first step toward a fundamental and quantitative understanding of radical–surface interactions is the detailed characterization of well-defined surfaces with respect to their reactivity with isolated radicals impinging from the gas phase at thermal energies. This is a necessary step for understanding the elementary surface chemical mechanisms responsible for film deposition, prior to any analysis of more complex physicochemical processes that take place on a structurally and compositionally evolving surface during PECVD. Among crystalline substrate surfaces, the Si(001) surface is of particular importance for such targeted studies of radical reactivity because of its very frequent use in experimental studies and in engineering applications. Among all possible orientations of SiH$_x$, $x = 1, 2, 3$, radicals with respect to the growth surface at normal incidence, two are of particular interest because they represent two limiting cases of high and low reactivity, respectively. In the first one, the Si atom of the radical is closer to the surface, with its dangling bond(s) directed toward the surface: this is termed the Si-down radical configuration. In the second one, the radical impinges on the surface, with its H atom(s) pointing toward the surface, i.e., the Si dangling bond(s) of the radical points away from the surface: this is termed the H-down configuration.

The reactivity of a surface with impinging radicals can be characterized in detail through the reactivity maps discussed above, which are constructed according to Eq. (7) for a given description of interatomic interactions. The computational cost of constructing a reactivity map for a surface is comparable to that for obtaining a short MD trajectory: the cost of the corresponding MS iterative computation for 10^2 grid points on the surface cell and 10^2 iterations for each grid point is comparable to generating an MD trajectory of 10^4 time steps. Although these maps are useful for understanding the reactivity of specific surface locations, reaching conclusions about the overall reactivity of the surface and the reaction mechanisms requires a schedule of MD simulations where the radical is impinged on many different surface locations in different molecular orientations. The reactivity maps can be used as a complementary tool that aids in understanding the trajectory of a radical after it comes within interaction range with the surface. Careful examination of the energy landscapes expressed by the reactivity maps can lead to predictions of even detailed features of the radical dynamics on the surface and can provide insights for designing MD simulation protocols.

A. STRUCTURE OF CRYSTALLINE AND AMORPHOUS SILICON SURFACES

The reactivity of a growth surface depends strongly on the distribution of atomic coordination on the surface, which indicates the location of surface atoms with dangling bonds available for reaction. Thus, knowledge of surface structure can yield immediate conclusions on surface reactivity. For example, a pristine Si(001)-(2 × 1) surface is expected to be highly reactive due to the high density of dangling bonds associated with the dimer atoms: all the atoms of the pristine surface are threefold coordinated, i.e., undercoordinated. The structure of the pristine Si(001)-(2 × 1) surface (top view) is shown in Fig. 3a (see color insert). Locations 1–6 denote high-symmetry points in the unit cell of the dimerized surface: sites 1 and 2 are between dimers of the same dimer row, 3 is at the center of the Si–Si dimer bond, 4 is right at a dimer atom, and 5 and 6 are sites in the trough between neighboring dimer rows. On the other hand, a Si(001)-(2 × 1) surface terminated by one monolayer of H atoms, which is denoted henceforth H:Si(001)-(2 × 1), is expected to be less reactive: all of the Si atoms on this surface are fully (fourfold) coordinated. The structure of the H:Si(001)-(2 × 1) surface (top view) is shown in Fig. 3b (see color insert).

The distribution of atomic coordination on amorphous surfaces is more complicated due to the complete disorder and increased corrugation of the surface. For the surfaces of the a-Si:H films generated as outlined in Fig. 2, surface atomic coordination distributions are shown for films with low (16%) and high (50%) hydrogen atomic fractions in Figs. 3c and d (see color insert), respectively. These coordination distributions were computed by calculating the number of bonds formed by the Si atoms at the a-Si:H film surface within a slice of thickness corresponding to two layers of the crystalline substrate; the Si–Si and Si–H bond lengths were used in determining the coordination of the Si surface atoms (Ramalingam et al., 1998b). Figures 3c and d show clearly that the surface of the sample with the lower H concentration is characterized by a higher density of undercoordinated Si atoms than the surface of the film with the higher H concentration; the latter surface is characterized mainly by fourfold coordinated and overcoordinated Si atoms. In general, Figs. 3c and d suggest that the surface of the film with the lower hydrogen content is expected to be overall more reactive than that of the film with the high hydrogen content.

B. Interactions of SiH$_x$ Radicals with Crystalline Silicon Surfaces

Reactivity maps for the interactions of the pristine Si(001)-(2 × 1) surface with the SiH, SiH$_2$, and SiH$_3$ radicals impinging on the surface in the Si-down configuration are shown in Figs. 4a–c (see color insert) respectively. The pristine crystalline surface is indeed very reactive. For all of the three radicals, the most reactive regions lie between the Si dimers in the same row (from site 1 to site 2) and along the Si–Si dimer bond (from site 3 to site 4). The least reactive locations lie in the trough between neighboring dimer rows, i.e., along the line connecting sites 5 and 6. Changes in the molecular orientation of the radicals with respect to the surface plane affect the reactivity of the radicals with the surface. Detailed examination of surface reactivity with the radicals impinging in the H-down configuration has shown that the overall reactivity of the surface is reduced significantly (Ramalingam et al., 1998b, 1999b,c). In these configurations, the dangling bond(s) of the Si atom of the radicals points away from the surface and the presence of the H atom(s) of the radicals reduces or inhibits the Si–Si interaction. As a result, it is observed frequently in MD simulations of radical impingement on the surface in the H-down configuration that the radicals rotate to point their dangling bond(s) toward the surface and react with the surface.

Compared to the pristine Si(001)-(2 × 1) surface, the H:Si(001)-(2 × 1) surface has a significantly reduced overall reactivity with the radicals. This is due to the presence of the one monolayer of H atoms on this H-terminated surface that passivates the Si dangling bonds. The corresponding reactivity maps for interactions with the SiH, SiH$_2$, and SiH$_3$ radicals impinging on the surface in the Si-down configuration are shown in Figs. 4d,e, and f (see color insert), respectively. In all cases, the center of the Si–Si dimer bond (site 3) is reactive, which suggests the possibility of radical insertion reactions that may involve breaking of the Si–Si dimer bond. For the SiH and SiH$_3$ radicals, the vicinity of the surface Si–H bond (above site 4) also is reactive: this indicates that H abstraction may be possible through breaking of this Si–H bond upon radical impingement in this vicinity. For SiH$_2$, regions parallel to the dimers (along the line connecting sites 2, 1, and 6) also exhibit some reactivity. For all of the three radicals, regions between neighboring dimer rows are the least reactive. Again, the surface reactivity with the radicals is reduced significantly when the radicals are directed toward the surface in the H-down configuration (Ramalingam *et al.*, 1998b, 1999b,c).

The detailed interactions between the SiH$_x$ radicals and silicon surfaces can be captured by MD simulations of radical impingement at a grid of locations on the surface with varying radical molecular orientation with respect to the surface. The reactions of SiH with the pristine Si(001)-(2 × 1) surface can be classified broadly into two classes (Ramalingam *et al.*, 1998b). The radical either adsorbs dissociatively onto the surface or penetrates below the top surface layer into the substrate also resulting in dissociation. The H atom that is released upon radical dissociation becomes an interstitial impurity of the substrate Si lattice and it can migrate rapidly or channel deeper into the substrate. These reactions can be represented as

$$\text{SiH}_{(g)} + \text{surface} \rightarrow \text{Si}_{(s)} + \text{H}_{(i)} \tag{8}$$

and

$$\text{SiH}_{(g)} + \text{surface} \rightarrow \text{Si}_{(b)} + \text{H}_{(i)}, \tag{9}$$

where the subscripts (g), (s), (b), and (i) refer to gas, surface, bulk, and interstitial species, respectively. When SiH attaches to the H:Si(001)-(2 × 1) surface, the reaction mechanism can be classified into one of two categories (Ramalingam *et al.*, 1998b). The radical either stays intact or adsorbs dissociatively, releasing its H atom into the gas phase. These classes of reactions can be represented as

$$\text{SiH}_{(g)} + \text{surface} \rightarrow \text{SiH}_{(s)} \tag{10}$$

and

$$\text{SiH}_{(g)} + \text{surface} \rightarrow \text{Si}_{(s)} + \text{H}_{(g)}. \qquad (11)$$

The SiH$_2$ radical attaches to the pristine Si(001)-(2 × 1) surface at one of five possible adsorption sites (Ramalingam et al., 1999c). These reactions can be represented as

$$\text{SiH}_{2(g)} + \_{(s)} \rightarrow \text{SiH}_{2(s)}, \qquad (12)$$

or

$$\text{SiH}_{2(g)} + 2\_{(s)} \rightarrow \text{SiH}_{2(s)}, \qquad (13)$$

or

$$\text{SiH}_{2(g)} + 2\text{Si}_{(b)} \rightarrow \text{Si-SiH}_2\text{-Si}_{(s)}, \qquad (14)$$

where the additional subscript "_" is used to denote a Si dangling bond on the pristine surface. Reaction (12) leads to a threefold coordinated Si atom in the original SiH$_2$ radical by attachment to the dangling bond of one of the surface dimer atoms. Reaction (13) leads to fourfold coordination of the Si atom of the radical, as well as two Si atoms on the pristine surface; thus, in general, this reaction is preferred energetically. Reaction (13) represents three mechanisms: the radical may attach to both atoms of a Si–Si dimer with or without breaking the dimer bond, or it may form Si–Si bonds with two Si atoms of adjacent dimer rows and bridge the dimer rows across the trough that separates them. Reaction (14) represents attachment of the radical to two second-layer Si atoms in the trough between dimer rows, which leads to formation of overcoordination defects. However, the energy gain associated with passivating both of the dangling bonds of the radical outweighs the defect formation energy. The reactions of the SiH$_2$ radical with the H:Si(001)-(2 × 1) surface can be classified into four categories (Ramalingam et al., 1999c) represented by

$$\text{SiH}_{2(g)} + \text{HSi-SiH}_{(s)} \rightarrow \text{HSi-SiH}_2\text{-SiH}_{(s)}, \qquad (15)$$

or

$$\text{SiH}_{2(g)} + \text{HSi}_{(s)} \rightarrow \text{HSi-SiH}_{2(s)}, \qquad (16)$$

or

$$\text{SiH}_{2(g)} + 2\text{HSi}_{(s)} \rightarrow \text{HSi-SiH}_2\text{-SiH}_{(s)}, \qquad (17)$$

or

$$\text{SiH}_{2(g)} + 2\text{Si}_{(b)} \rightarrow \text{Si} - \text{SiH}_2\text{-Si}_{(s)}. \qquad (18)$$

Reaction (15) is an insertion reaction and corresponds to breaking of the Si–Si dimer bond and subsequent bonding of the radical to both atoms of the dimer. Reactions (16) and (17) correspond to attachment of the Si atom of the radical to one and to two surface Si atoms, respectively, and do not result in the breaking of any existing surface Si bonds; in the latter reaction, a second Si–Si bond is formed with a Si atom of an adjacent dimer in the same dimer row. Reaction (18) is similar to Reaction (14) with the pristine surface and represents attachment of the radical to two second-layer Si atoms in the trough between dimer rows.

The SiH_3 radical attaches to the pristine Si(001)-(2 × 1) surface, more commonly, in one of four absorbed configurations (Ramalingam et al., 1999b). The corresponding reactions can be represented as

$$SiH_{3(g)} + -_{(s)} \rightarrow SiH_{3(s)}, \tag{19}$$

or

$$SiH_{3(g)} + 2-_{(s)} \rightarrow SiH_{3(s)}. \tag{20}$$

Reaction (19) leads to a fourfold coordinated Si atom in the original silyl radical by formation of a Si–Si bond between the radical and one surface atom. Reaction (20) leads to fivefold coordination of the Si atom of the radical and represents three mechanisms: the radical forms two Si–Si bonds by either inserting into a surface dimer through breaking of the dimer bond, or bridging two Si atoms of adjacent dimers in the same dimer row, or forming a bridge between two Si atoms in adjacent dimer rows. Although the Si atom of the radical that adsorbs on the surface through reaction (20) is overcoordinated, the resulting energy gain is substantial due to the passivation of two surface dangling bonds. Two types of reactions may occur upon interaction of the SiH_3 radical with the H:Si(001)-(2 × 1) surface (Ramalingam et al., 1999b). First, the radical may insert into the surface dimer (at site 3) by breaking of the Si–Si dimer bond and forming two Si–Si bonds with the atoms of the dimer. Subsequent relaxation of the adsorbed configuration results in transfer of an H atom from the radical to one of the Si atoms of the original surface dimer. This dissociative adsorption reaction can be represented as

$$SiH_{3(g)} + HSi\text{–}SiH_{(s)} \rightarrow HSi\text{–}SiH_2\text{–}SiH_{2(s)}. \tag{21}$$

In addition, the silyl radical may abstract an H atom from the surface, generate a surface Si dangling bond, and return into the gas phase as a silane molecule. This abstraction reaction can be represented as

$$SiH_{3(g)} + HSi_{(s)} \rightarrow SiH_{4(g)} + Si-_{(s)}. \tag{22}$$

Reactions (21) and (22) are analyzed in some detail in Section IV.

C. Interactions of SiH$_x$ Radicals with Surfaces of Amorphous Silicon Films

The chemical reactivity of a-Si:H film surfaces with impinging radicals from the gas phase is correlated strongly with the local composition of the more complex, completely disordered, corrugated, amorphous surfaces. Reactivity maps for radical–surface interactions are shown in Fig. 5 (see color insert), with the radicals impinging in the Si-down configuration on the surfaces of a-Si:H films that were generated according to the procedure outlined in Fig. 2. Figures 5a,b, and c show the reactivity maps for interactions of the SiH, SiH$_2$, and SiH$_3$ radicals, respectively, with the surface of such a film with a 16% H concentration. Clearly, SiH is the most reactive of the three radicals. The correlation between surface atomic coordination (Fig. 3c) and surface reactivity is evident for all three types of interactions: the reactivity, measured as energy gain upon interaction, is higher for surface regions with a higher density of undercoordinated Si atoms that correspond to surface Si dangling bonds. In addition, in a manner similar to the interactions with the crystalline surfaces, the overall reactivity of the amorphous surface is reduced significantly if the radicals impinge on the surface in the H-down configuration with the Si dangling bond(s) of the radicals pointing away from the surface (Ramalingam et al., 1998b, 1999b,c).

Reactivity maps for interactions of the SiH, SiH$_2$, and SiH$_3$ radicals with the surface of an a-Si:H film with a 50% H concentration is shown in Figs. 5d, e, and f, respectively. Comparisons with the reactivity maps in Figs. 5a,b, and c, respectively, confirms that the major factor in determining the chemical reactivity of the amorphous surface is its H coverage. Increasing the H concentration in the a-Si:H film leads to an increased H coverage of its surface and, consequently, a decreased density of undercoordinated surface Si atoms due to increased passivation of surface dangling bonds. Again, the reactivity of SiH with this H-rich surface is clearly the highest of the three radicals and the correlation between surface atomic coordination (Fig. 3d) and surface reactivity remains evident. Finally, changing the molecular orientation of the radicals with respect to the surface to the corresponding H-down configurations results in a further significant reduction of the surface reactivity (Ramalingam et al., 1998b, 1999b,c).

In general, the interactions of SiH$_x$ radicals ($x = 1, 2, 3$) with the surfaces of the a-Si:H films are determined strongly by the availability of surface Si dangling bonds and can be classified into three categories, as identified by MD simulations (Ramalingam et al., 1998b, 1999b,c). If the radicals impinge in the vicinity of surface dangling bonds, they adsorb immediately onto the surface, forming strong Si–Si bonds with surface Si atoms. This is frequently the case on surfaces of a-Si:H films with a low H concentration, since the low H coverage of such surfaces leads to increased density of unpassivated surface dangling bonds. If the radicals impinge on the surface at locations that are completely terminated by H atoms, either they are reflected back into the gas phase or they migrate on the surface until they find energetically favorable surface sites to attach to. Reflection of radicals back into the gas phase is observed frequently for radical impingement on almost completely passivated surfaces of a-Si:H films with a high H concentration.

The above interaction mechanisms are illustrated in Figs. 6a and b, which show two-dimensional projections of MD trajectories of the SiH$_3$ radical on the surfaces of amorphous films with low (16%) and high (50%) H concentrations, respectively (Ramalingam et al., 1999b). The trajectories are superimposed on the corresponding atomic coordination maps for these surfaces (Figs. 3c and d), for better illustration of the mechanism dependence on H coverage and surface dangling bond density. In Figs. 6a and b, the magnitude of the radical's displacement on the surface from its initial impingement location is indicative of the radical's mobility on the surface. In Fig. 6a, a trajectory is shown of a SiH$_3$ radical that was impinged near a Si dangling bond on the surface of the film with the low H concentration. Practically immediately (within 1 ps), the radical reacts and adsorbs onto the surface. On the other hand, Fig. 6b shows the trajectory of a SiH$_3$ radical impinged at a less reactive location on the surface of the film with the high H concentration. The radical migrates on the surface for over 3 ps, until it attaches to an undercoordinated surface Si atom.

■ Undercoordinated Si   O 4-fold coordinated Si   ■ Overcoordinated Si

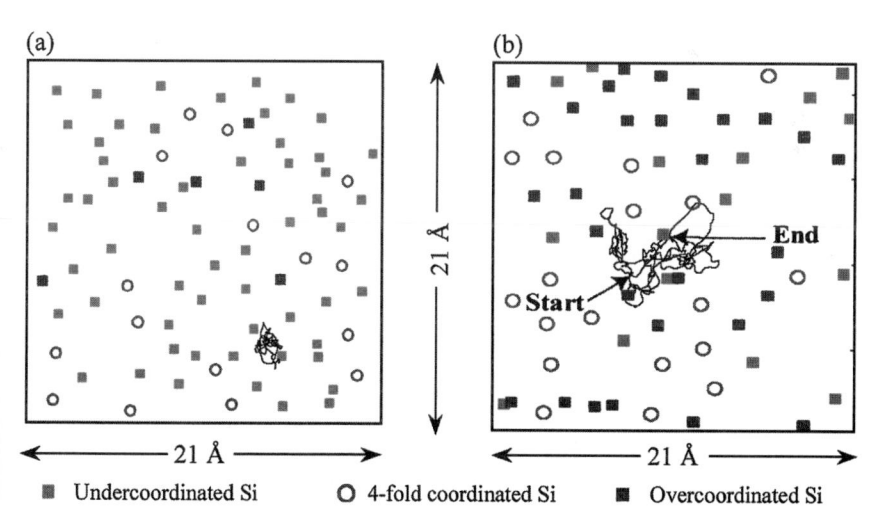

FIG. 6. Two-dimensional projections of SiH$_3$ center-of-mass trajectories superimposed on the coordination maps of surfaces of a-Si:H films with H concentrations of (a) 16% and (b) 50% (from Ramalingam et al., 1999b).

In general, SiH is the least mobile radical on the surfaces of these a-Si:H films due to its extremely high reactivity with these surfaces that are characterized primarily by surface monohydride species (Ramalingam et al., 1998b). SiH has three Si dangling bonds; as a result, it can often form two Si–Si bonds with the amorphous surface without overly straining any one of its bonds. It is not energetically favorable to break these two Si–Si bonds with simultaneous formation of a Si–Si bond between the surface and the remaining dangling bond of the radical, which limits the mobility of SiH on the surface. On the contrary, the SiH$_2$ and SiH$_3$ radicals are very mobile on these amorphous surfaces (Ramalingam et al., 1999b,c). SiH$_3$ can migrate particularly easily on a-Si:H surfaces with a high H coverage, where most of the surface dangling bonds are passivated by H atoms. On amorphous surfaces with a low H coverage, the SiH$_2$ radical is generally more mobile than the SiH$_3$ radical. Although SiH$_3$ can form a single Si–Si bond with the surface, it must break this bond and form another one with the surface to migrate, hopping from dangling bond to dangling bond on the surface; a relatively high activation energy barrier is associated with this process. In contrast, SiH$_2$ has two Si dangling bonds, and thus, it can move on the surface by breaking one Si–Si bond with the surface while simultaneously forming another, which is characterized by a lower migration energy barrier.

MD simulations of radical impingement on the surfaces of these a-Si:H films can be used to provide estimates of the overall sticking probability of a radical with these model amorphous surfaces. These overall probabilities correspond to the percentage of reactive events observed in a finite sample of MD trajectories; in the MD simulations, the radicals are directed normal to the surface, in either the Si-down or the H-down configuration, and impinge on the nodes of a grid on the amorphous surface for a-Si:H films of various H concentrations (Ramalingam et al., 1998b, 1999b,c). An alternative approach to calculating sticking probabilities is discussed in Section IV. The estimated overall sticking probabilities from the MD simulations are 96, 70, and 50% for the SiH, SiH$_2$, and SiH$_3$ radicals, respectively. SiH is the only silane fragment whose sticking probability on a-Si:H film surfaces has been measured directly in a well-defined experiment; this measured sticking probability is in excess of 94% (Ho et al., 1989), i.e., the computational result is in excellent agreement with the experimental measurement. Due to the specific nature of the films generated by the procedure in Fig. 2 and the relatively small size of the sample employed in the estimation of the above sticking probabilities, detailed comparisons with experimental measurements are not discussed here; more systematic comparisons are discussed below, in Section IV.

IV. Plasma–Surface Interactions during Silicon Film Growth

Our current knowledge of film growth mechanisms by PECVD is inferred largely from macroscopic observations, such as the radical concentrations in the plasma during deposition and the variation of the deposition rates and film properties with the deposition conditions. For example, it is believed widely that the silyl (SiH$_3$) radical is the dominant precursor in the plasma deposition of a-Si:H films (Robertson and Gallagher, 1986; Matsuda *et al.*, 1990; Abelson, 1993; Perrin *et al.*, 1998). This belief is based largely on the observation that SiH$_3$ is the most abundant radical in silane plasmas under typical deposition conditions (Abelson, 1993). Dissentions from this view exist and the identity of the primary deposition precursor remains under debate; in general, the deposition precursors depend on the deposition conditions. Several SiH$_3$–surface reaction mechanisms have been proposed in the experimental literature as important elementary processes that occur during plasma deposition of a-Si:H films. These include abstraction of H atoms from surface sites by impinging SiH$_3$ radicals, diffusion of SiH$_3$ radicals on the deposition surface, and reaction between two SiH$_3$ radicals adsorbed on the surface to form Si$_2$H$_6$ (Robertson and Gallagher, 1986; Perrin, 1991; Abelson, 1993; Matsuda, 1998). These radical–surface reaction mechanisms can be examined in detail through atomic-scale simulation of film growth by radical impingement on the growth surface.

Intensive nanosecond-time-scale MD simulations of a-Si:H film growth were carried out recently based on the interatomic potential of Eqs. (1)–(6) (Ramalingam, 2000). These amorphous film growth simulations followed the methodology outlined in Section II. Specifically, the initial configuration consisted of a crystalline silicon substrate with an H:Si(001)-(2 × 1) surface and the substrate temperature was varied over the range from 500 to 773 K. Motivated by the common belief on the primary importance of the SiH$_3$ radical as a deposition precursor, special emphasis was placed on the analysis of the MD trajectories of SiH$_3$ impingement on the growth surface. The MD trajectories for the growth simulations were longer by about three orders of magnitude compared to those analyzed in Section III to characterize the reactivity of well-defined surfaces; thus, these simulations provide a massive amount of atomic-scale information about the film growth mechanisms. In these MD simulations, the surface structure and composition evolve governed by reaction mechanisms between the impinging radicals and the evolving growth surface, which complicates the analysis of these interactions as the growth proceeds.

A. Surface Chemical Reactions during Film Growth

MD simulations of a-Si:H film growth from thermal SiH$_3$ precursors have revealed a wide range of elementary surface reaction mechanisms that occur on the growth surface (Ramalingam, 2000). These mechanisms can be divided into broad classes that include

- Si incorporation into the grown film either by insertion into surface Si–Si bonds or by direct attachment to surface Si dangling bonds;
- H removal from the growth surface through various abstraction and desorption mechanisms;
- Si–Si bond formation processes by adsorbed species on the surface;
- Si removal from the growth surface, i.e., etching mechanisms that may compete with Si incorporation mechanisms; and
- migration mechanisms of adsorbed mobile surface species that lead to Si transport on the growth surface.

Some of the most important elementary surface chemical reactions are analyzed below, prior to their integration in understanding the complex film deposition mechanism.

One mechanism of Si incorporation into the growing film is dissociative adsorption that involves insertion of the SiH$_3$ radical into Si–Si bonds (Ramalingam, 2000). At the very initial stages of growth, where the radical impinges on the crystalline H:Si(001)-(2 × 1) surface, such a mechanism can be represented by reaction (21). The progress of this dissociative adsorption mechanism is illustrated in Fig. 7 through a sequence of snapshots, A, B, C, and D, from MD simulation. In A, the radical is shown impinging in the Si-down configuration on the H:Si(001)-(2 × 1) surface at the center of the Si–Si surface dimer bond. In B, the radical inserts into the dimer bond: the Si–Si dimer bond is broken and the radical forms Si–Si bonds with both atoms of the original surface dimer. In this configuration, the radical's Si atom is overcoordinated; as a result, one of the Si–Si bonds formed in B is broken, as shown in snapshot C. The MD trajectory reveals that the Si atom of the radical forms bonds alternately, with each of the dimer atoms having its bond with the other broken; this occurs through a hopping process that involves a transition state where the radical is bonded with both of the dimer atoms (Ramalingam *et al.*, 1999b). Upon surface relaxation and at a state where the radical's Si is bonded to only one of the dimer atoms, a hydrogen atom is transferred from the radical to the other atom of the original surface dimer, leading to the formation of a surface dihydride species. The resulting configuration is shown in snapshot D in Fig. 7. This dissociative adsorption mechanism, predicted using the interatomic potential of Eqs. (1)–(6), is fully consistent with DFT-based *ab initio* calculations

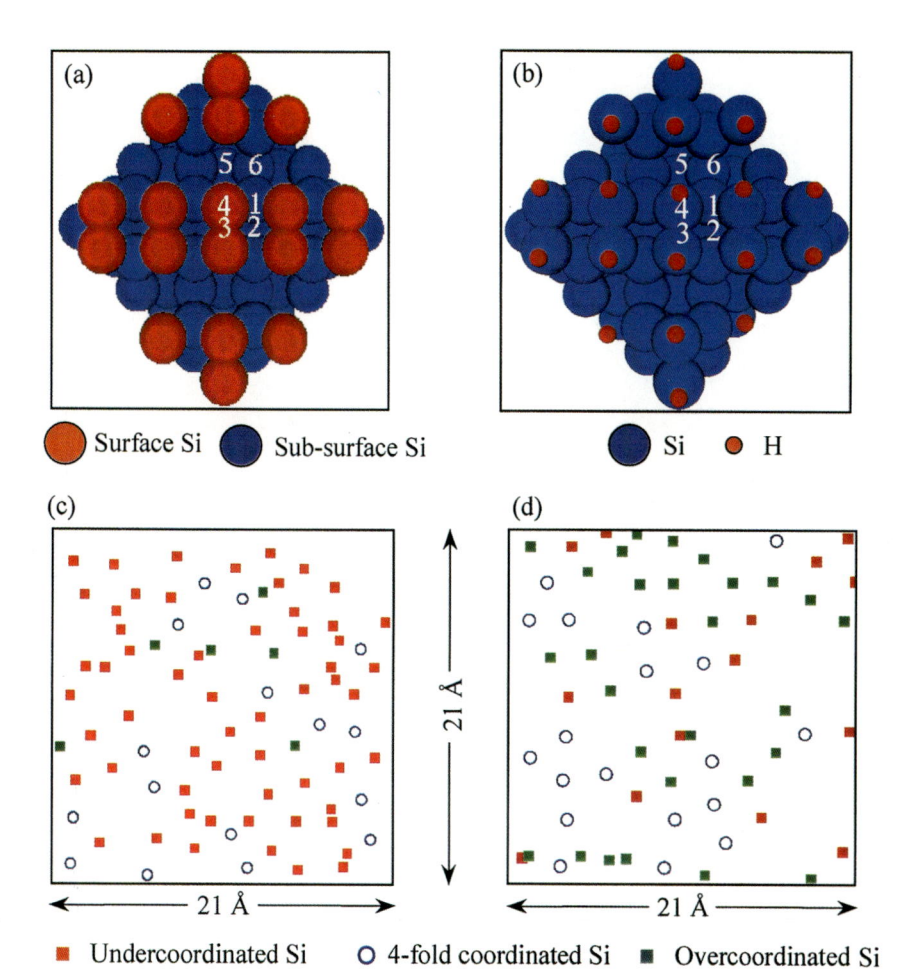

FIG. 3. Structure of silicon surfaces (from Ramalingam *et al.,* 1998b). (a) Pristine Si(001)-(2×1) surface, (b) Si(001)-(2×1) surface terminated by one monolayer of H atoms, and surfaces of a-Si:H films with (c) 16% and (d) 50% H atomic fractions in the amorphous film. In a and b, locations 1–6 denote high-symmetry points in the unit cell of the dimerized surface. In c and d, surface structure is expressed by the distribution of atomic coordination on the amorphous surface.

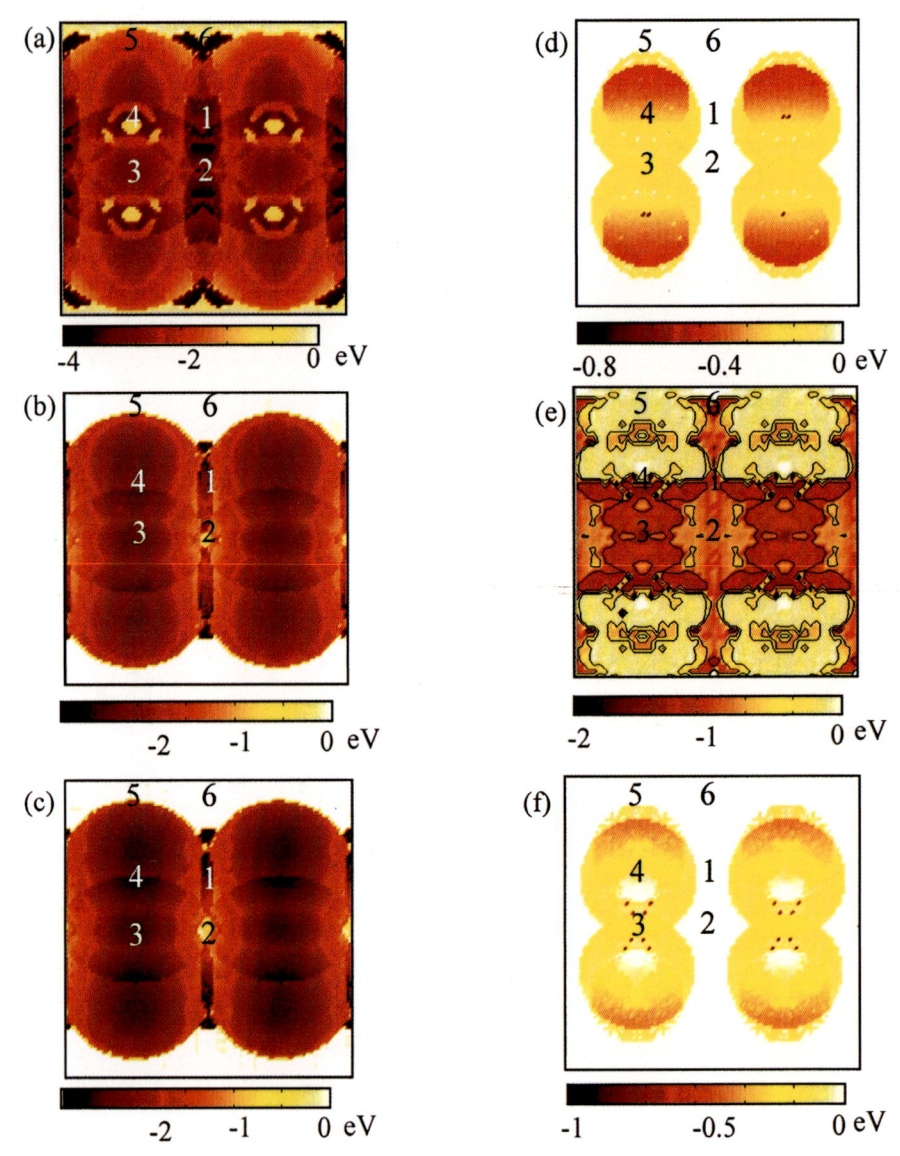

FIG. 4. Reactivity maps for SiH$_x$ radicals ($x=1,2,3$) impinging on crystalline silicon surfaces in the Si-down configuration (from Ramalingam et al., 1998b, 1999a–c). (a) SiH, (b) SiH$_2$, and (c) SiH$_3$ on the pristine Si(001)-(2×1) surface. (d) SiH, (e) SiH$_2$, and (f) SiH$_3$ on the H:Si(001)-(2×1) surface. In all cases, the surface area shown corresponds to two (2×1) surface unit cells and the dimer rows on the surface are oriented horizontally. The numbers on the colored scale indicate the energy difference (eV) associated with the interaction of the radical with the surface, Eq. (7).

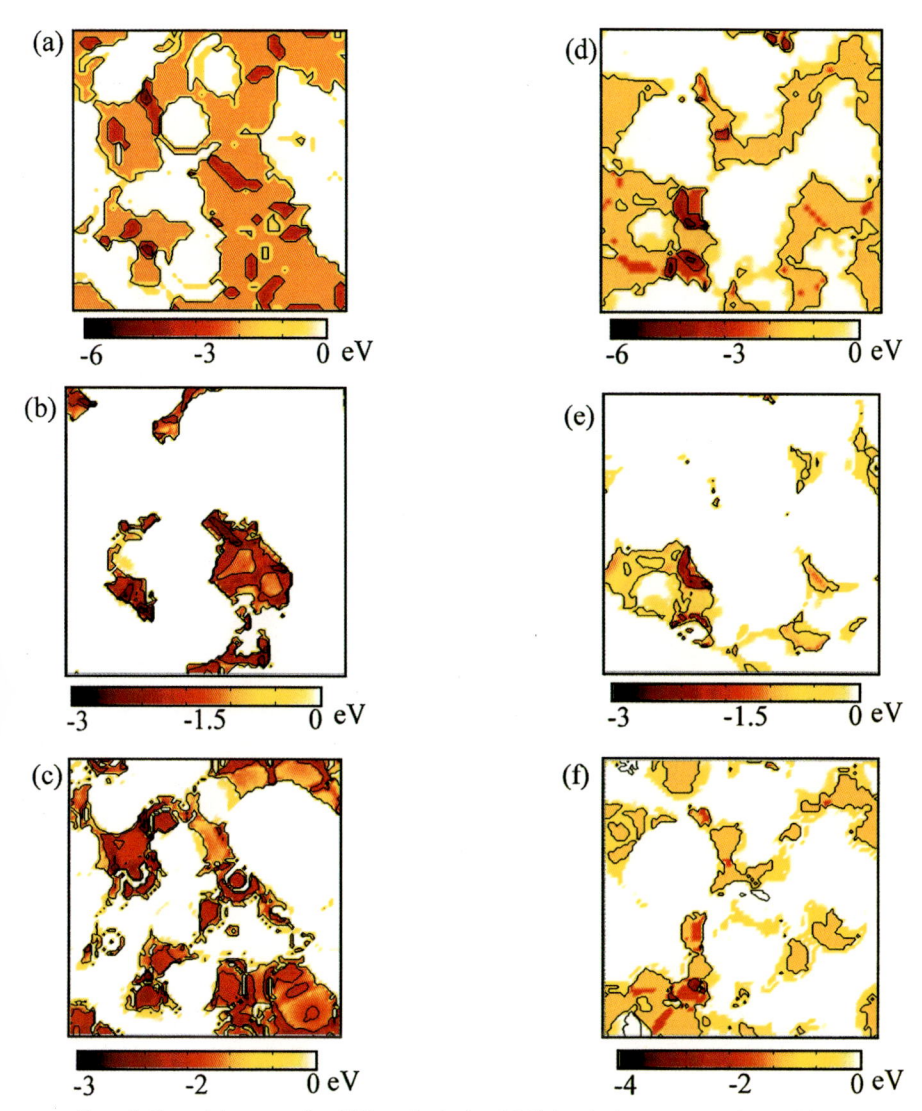

FIG. 5. Reactivity maps for SiH$_x$ radicals (x=1,2,3) impinging on amorphous silicon surfaces in the Si-down configuration (from Ramalingam *et al.*, 1998b, 1999a–c). (a) SiH, (b) SiH$_2$, and (c) SiH$_3$ on the surface of an a-Si:H film with a H concentration of 16%. (d) SiH, (e) SiH$_2$, and (f) SiH$_3$ on the surface of an a-Si:H film with a H concentration of 50%. In all cases, the surface area shown is 21×21 Å^2. The numbers on the colored scale indicate the energy difference (eV) associated with the interaction of the radical with the surface, Eq. (7).

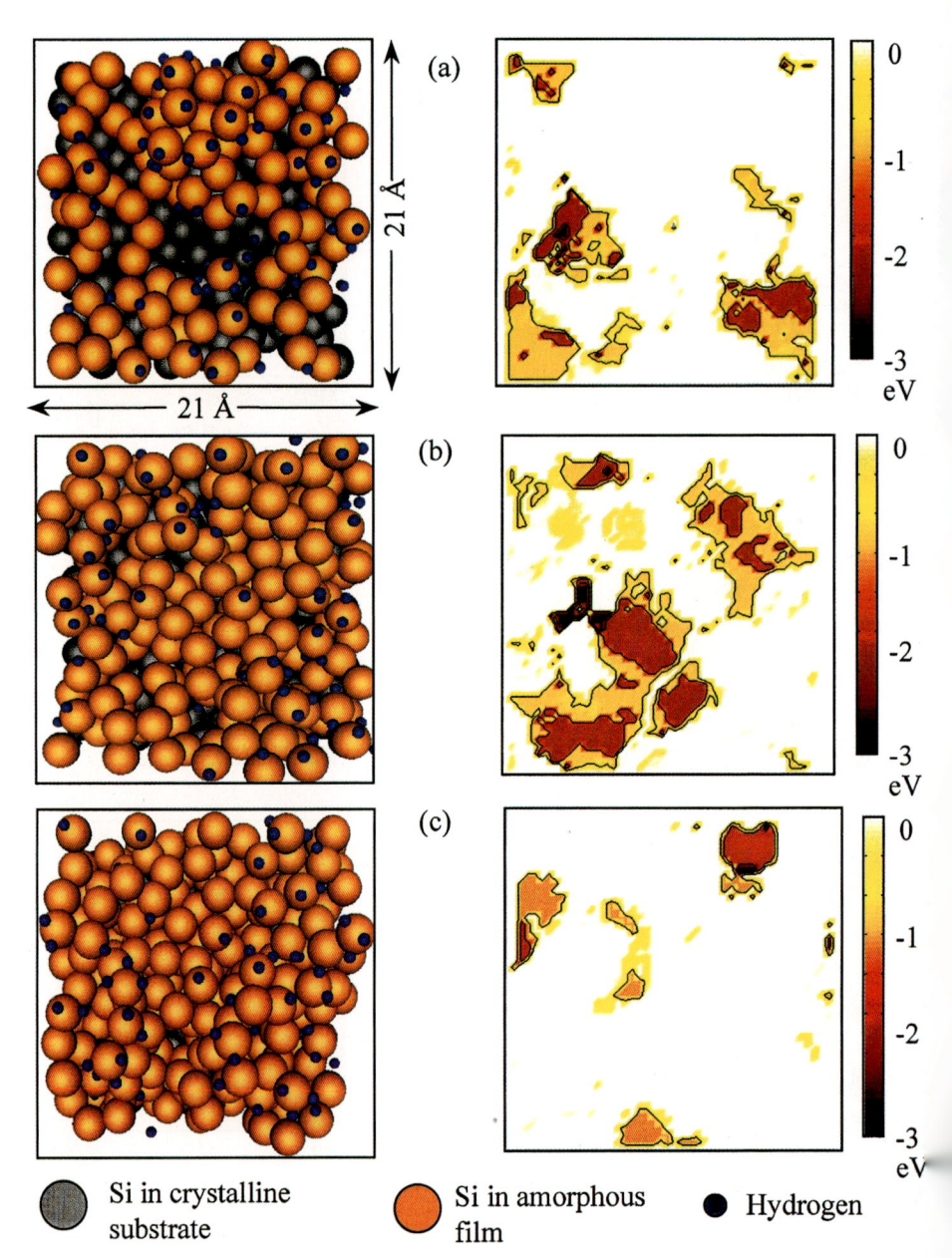

FIG. 11. Evolution of the surface structure (left-hand frame) and reactivity with the SiH$_3$ radical impinging in the Si-down configuration (right-hand frame) of an a-Si:H film during its deposition by MD simulation. The reactivity is expressed by the energy gain distribution upon radical–surface interaction, $\Delta E(x,y)$, according to Eq. (7). The corresponding duration of simulated film growth is (a) 1 ns, (b) 2 ns, and (c) 5 ns.

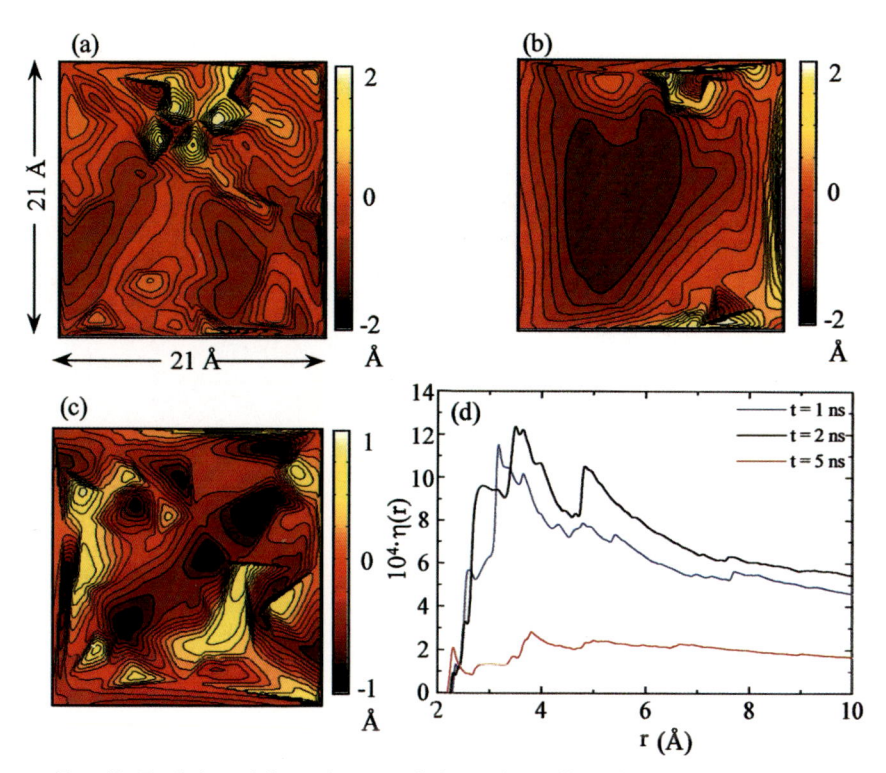

FIG. 12. Evolution of the surface morphology of an a-Si:H film during its deposition by MD simulation. The distribution of the surface height, $h(x,y)$, is shown after (a) 1 ns, (b) 2 ns, and (c) 5 ns of simulated film growth; $h = 0$ is the average location of the plane of the surface. The corresponding evolution of the film surface roughness is shown in (d), as expressed by the height–height correlation function, $\eta(r)$.

MODELING OF RADICAL–SURFACE INTERACTIONS 275

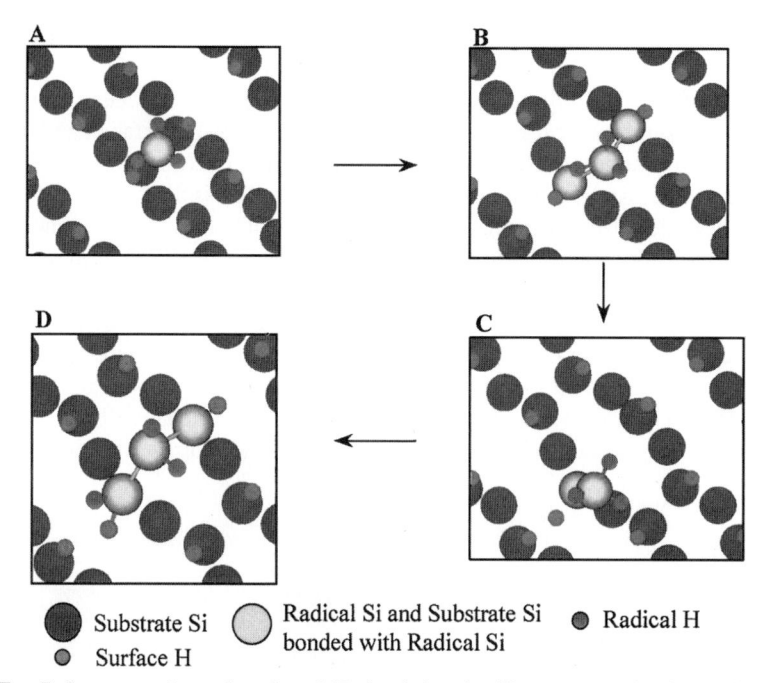

FIG. 7. Sequence of snapshots from MD simulation that illustrate a mechanism of dissociative adsorption of SiH$_3$ on an H-terminated Si(001)-(2 × 1) surface (from Ramalingam et al., 1999b).

in terms of its energetic progress along the reaction path and the geometry of the corresponding atomic configurations (Walch et al., 2000).

Several SiH$_3$ insertion reactions and dissociative adsorption events, where an insertion reaction is followed by H transfer from the radical to the surface, have been observed during MD simulations of a-Si:H film growth (Ramalingam, 2000). This reaction has important implications for the PECVD of Si films. The impinging SiH$_3$ radical leads to the formation of two SiH$_2$ species on the deposition surface. In conjunction with other reactions that can produce dihydride species on the surface of a-Si:H films, this SiH$_3$ dissociative adsorption mechanism can explain the large proportion of surface dihydrides measured on plasma-deposited films (Marra et al., 1998a,b). These MD simulation results also are consistent with the proposal of von Keudell and Abelson (1999) based on their investigation of SiH$_3$ interactions with a-Si:H film surfaces using infrared spectroscopy. These authors proposed that SiH$_3$ inserts into the dimer bonds of the crystalline surface during the initial stages of growth and into strained Si–Si bonds on the amorphous growth surface at later stages of deposition.

A very common mechanism of hydrogen removal from the deposition surface during MD simulation of Si film growth is through surface H abstraction by the impinging SiH$_3$ radical. In the initial stages of growth, such a hydrogen abstraction from the H:Si(001)-(2 × 1) surface can be represented by reaction (22). This abstraction reaction occurs through an Eley–Rideal mechanism (Ramalingam *et al.*, 1999b), which is illustrated in Fig. 8a. Three MD-generated atomic configurations, A, TS, and B, are shown that capture the radical and the surface at three stages of the abstraction reaction. At A, the SiH$_3$ radical is impinging on the H:Si(001)-(2 × 1) surface, but it is still outside interaction range with the surface. TS is the transition-state configuration, where the surface H atom that is being abstracted is bonded both to the Si atom of the impinging silyl radical and to the surface Si atom. At configuration B in Fig. 8a, the Si–H bond of the H atom with the surface is broken and the H atom is attached only to the Si atom of the radical: the

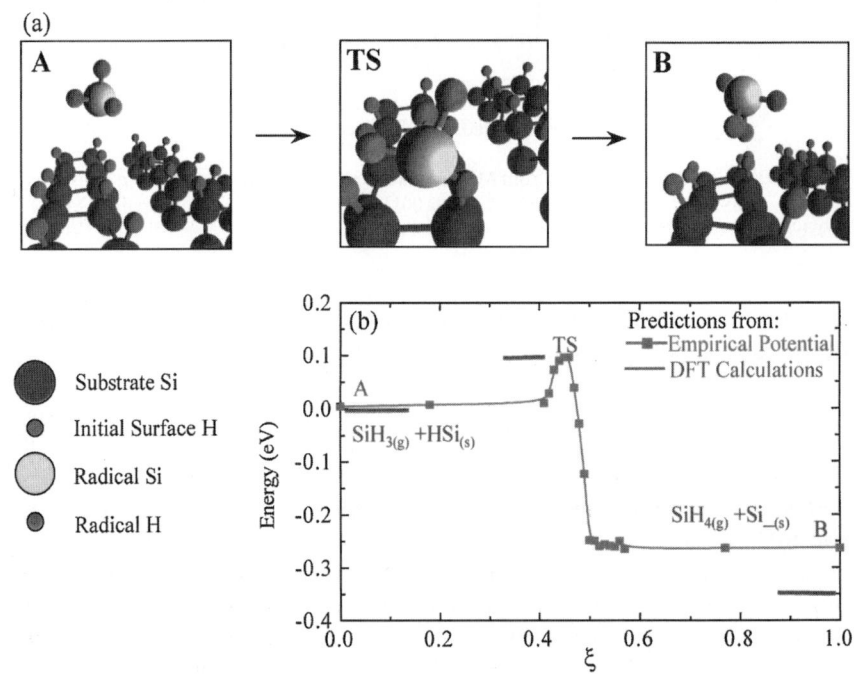

FIG. 8. Hydrogen abstraction by SiH$_3$ from an H-terminated Si(001)-(2 × 1) surface (from Ramalingam *et al.*, 1998c, 1999b). (a) Three atomic configurations from MD simulation showing the radical and the surface at different stages of the abstraction reaction; A, TS, and B correspond to the reactant, transition-state, and reaction product configurations. (b) Energetic progress of the reaction along its path as a function of a reaction coordinate, ξ, which expresses a dimensionless Si–H distance between the Si of the radical and the abstracted H.

SiH₄ molecule produced returns to the gas phase. The energetic progress of the Eley–Rideal abstraction along the reaction path is shown in Fig. 8b as a function of a reaction coordinate, ξ. the MD trajectory is followed along the reaction path and the sequence of atomic configurations is mapped onto their local minimum-energy configurations, as discussed in Section II.E. In Fig. 8b, the reactant state, configuration A, is chosen to set the level of zero energy. The reaction coordinate is defined by

$$\xi \equiv \begin{cases} 0, & d_{\text{Si-H}} > d_{\max} \\ \dfrac{d_{\max} - d_{\text{Si-H}}}{d_{\max} - d_{\text{Si-H}}^{\text{eq}}}, & d_{\text{Si-H}}^{\text{eq}} < d_{\text{Si-H}} \leq d_{\max}, \\ 1, & d_{\text{Si-H}} \leq d_{\text{Si-H}}^{\text{eq}} \end{cases} \qquad (23)$$

where $d_{\text{Si-H}}$ is the distance in the optimized configurations between the Si atom of the impinging radical and the H atom that is abstracted from the surface, $d_{\max}$ is the maximum Si–H distance at which the two atoms come within interaction range, and $d_{\text{Si-H}}^{\text{eq}}$ is the equilibrium Si–H bond length in the SiH₄ molecule. The transition-state configuration TS in Fig. 8a is formed at $\xi \approx 0.45$. The reaction energetics in Fig. 8b, predicted according to the interatomic potential of Eqs. (1)–(6), yields an activation energy barrier of 0.095 eV and an exothermic reaction energy of 0.26 eV. These results are in very good agreement with DFT-based pseudopotential cluster calculations, which predict an activation barrier of 0.09 eV and an exothermicity of 0.35 eV (Ramalingam *et al.*, 1998c).

This Eley–Rideal H abstraction reaction occurs not only on the crystalline H:Si(001)-(2 × 1) surface but also on the deposited a-Si:H film surfaces at later stages of the growth process and can be represented, in general, as

$$\text{SiH}_{3(g)} + \text{H}_{(s)} \rightarrow \text{SiH}_{4(g)} + -_{(s)}. \qquad (24)$$

The computed activation energy barriers and exothermicities for reaction (24) on a-Si:H surfaces generated during MD simulation of film growth over the temperature range from 500 to 773 K are in excellent agreement with the energy values computed for the abstraction reaction on the H-terminated crystalline surface (Ramalingam, 2000). This abstraction reaction is very important for the PECVD process, because it provides a mechanism of hydrogen removal from an H-passivated surface. This mechanism generates surface dangling bonds which act as reaction sites on the surface for subsequently impinging SiH₃ radicals.

MD simulations of Si film growth also have revealed a Langmuir–Hinselwood mechanism of hydrogen removal from the growth surface (Ramalingam, 2000). This mechanism involves reaction of an adsorbed SiH₃ species on the surface with surface H to form a weakly adsorbed SiH$_4$ molecule, which desorbs subsequently into the gas phase. This reaction can be represented as

$$\text{SiH}_{3_{(s)}} + \text{H}_{(s)} \rightarrow \text{SiH}_{4_{(s)}} \rightarrow \text{SiH}_{4_{(g)}} + -_{(s)}. \tag{25}$$

In reaction (25), the Si atom of the adsorbed silane intermediate is overcoordinated; it is actually highly (sixfold) overcoordinated if the SiH$_3$ radical has adsorbed through an insertion reaction into a Si–Si surface bond. H transfer is particularly favorable energetically if the H atom is bonded to an overcoordinated surface atom; such coordination defect formation is very common during film growth. The empirical potential of Eqs. (1)–(6) predicts that the H transfer step of reaction (25), from an overcoordinated Si to an inserted SiH$_3$ radical, occurs with an activation energy barrier of 0.6 eV and leads to an energy gain of 0.45 eV, while the subsequent silane desorption step is endothermic with an energy requirement of 1.4 eV. Reaction (25) is an important surface hydrogen removal mechanism; it occurs on both crystalline and amorphous surfaces, i.e., during both early and later stages of the amorphous film growth process, and it is accelerated substantially by increasing the growth temperature. Other mechanisms of surface H removal that have been studied through MD simulation (Ramalingam, 2000) include abstraction by impinging H atoms on the growth surface and desorption of surface H induced by impingement of energetic species on the surface. Of course, these possible mechanisms of H removal are not active during MD simulation of film growth solely from a thermal SiH$_3$ precursor.

Hydrogen removal, through either an Eley–Rideal or a Langmuir–Hinselwood mechanism, provides an additional path for Si incorporation into the growing film. As a result, impinging radicals can be incorporated into the growing film through two types of reactions: insertion into a surface Si–Si bond and direct attachment to a surface dangling bond. The energetics of these two reaction classes on a-Si:H films are comparable to those of the corresponding Si adsorption mechanisms on the crystalline Si(001)-(2 × 1) surface. In all cases, the energy gain upon SiH$_3$ adsorption is 2.0 eV or higher, as computed by the empirical potential of Eqs. (1)–(6); the energetics of adsorption on the crystalline surface are in agreement with predictions of DFT calculations (Walch et al., 2001). For both reaction mechanisms, the energy gain upon radical adsorption on the amorphous surface is lower by 0.2–0.3 eV than that for radical adsorption on the crystalline surface.

Two adjacent SiH$_3$ radicals adsorbed on the growth surface react commonly with each other to form a surface disilane (Si$_2$H$_6$) species, where one or both of the Si atoms of the radical are overcoordinated (Maroudas et al.,

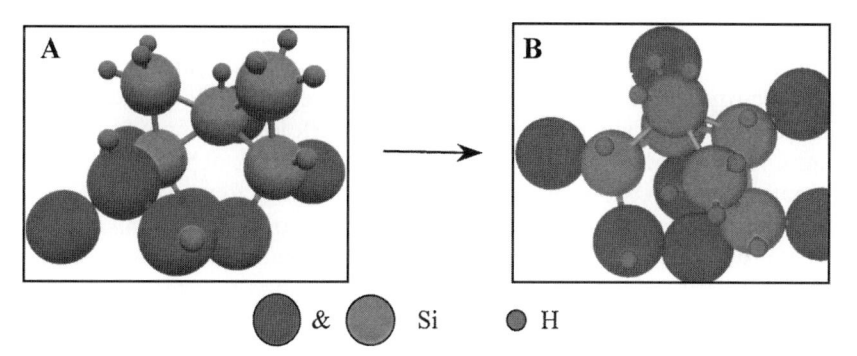

FIG. 9. Surface disilane formation reaction (from Ramalingam, 2000). Two atomic configurations are shown from MD simulation corresponding to the reactants (A) and reaction products (B), respectively.

1999; Ramalingam, 2000). This surface reaction can be represented by

$$SiH_{3(s)} + SiH_{3(s)} \rightarrow Si_2H_{6(s)}. \qquad (26)$$

Typically, the two reactant SiH_3 species form a Si–Si bond with a common surface Si atom before they react with each other. The mechanism of such a disilane formation reaction during the initial stages of growth is illustrated in Fig. 9. Configuration A in Fig. 9 shows two adsorbed neighboring SiH_3 radicals with Si–Si bonds with a common surface atom coming within interaction range with each other. Configuration B shows the formation of a Si–Si bond between the two adsorbed radicals producing a $Si_2H_{6(s)}$ species. The Si atoms of $Si_2H_{6(s)}$ and on the surface in the vicinity of the adsorbed species are overcoordinated. This particular mechanism of $Si_2H_{6(s)}$ formation is athermal, i.e., the reaction in Fig. 9 is barrierless, and has an exothermicity of 0.5 eV according to the empirical potential of Eqs. (1)–(6). The high density of coordination defects associated with $Si_2H_{6(s)}$ species favors Si–Si bond breaking processes and leads to easy migration of disilane on the surface and/or desorption from the surface; these thermally activated processes are accelerated at high growth temperatures. Note that desorption of disilane constitutes an etching mechanism, which leads to removal of Si from the surface and begins to compete with Si incorporation as the growth temperature increases. Finally, analysis of MD trajectories has revealed that $Si_2H_{6(s)}$ dissociation, i.e., the reverse of reaction (26), also is possible after its migration on the surface (Ramalingam, 2000).

An important factor in determining the morphology of the amorphous film grown through MD simulations is the observed mobility of the SiH_3 radical on the amorphous growth surface as the deposition proceeds (Ramalingam, 2000). The dynamics of the radical on the surface is qualitatively similar to those illustrated in Fig. 6 for isolated radical–surface interactions. If the radical impinges on a location of low surface reactivity, its trajectory resembles that in Fig. 6b; surface regions of low reactivity are characterized typically by a high density of fourfold coordinated surface Si atoms. The radical migrates on the surface by a sequence of hops that involve formation and breaking of "weak" Si–Si bonds with fourfold coordinated surface Si atoms; the activation energy barriers associated with this hopping mechanism are very low on the amorphous surface. This sequence of hops continues until the radical finds a reactive site, such as a surface Si dangling bond, and attaches to the surface at this site by forming a "strong" Si–Si bond. Here, the bond strength refers strictly to the bonding energetics, i.e., stronger bonds correspond to greater energy gain upon bond formation. The migration of other species that are weakly adsorbed on the surface, such as $Si_2H_{6(s)}$, also is an important mechanism of Si transport on the growth surface and can affect the morphology of the amorphous film surface.

B. Mechanism of Amorphous Silicon Film Growth

As the MD simulation of a-Si:H film growth proceeds, detailed visualization of the generated MD trajectories is required to analyze the Si bonding patterns and elucidate the complex mechanism of the amorphous network formation process (Ramalingam et al., 2000). Initially, the SiH_3 radicals adsorb on high-symmetry locations of the crystalline substrate surface. When a fractional coverage of the bare crystalline surface is achieved, Si–Si bonds start forming between adjacently adsorbed silyl radicals. Such Si–Si bond formation results in overcoordination of the Si atoms that have attached to the surface, and these newly formed bonds become strained. In addition, the fraction of a monolayer that covers the original growth surface retains largely the crystallinity of the substrate.

As the deposition process continues through radical impingement on the surface, the bonded pairs that were formed during the initial stage of growth evolve to networks of small clusters that consist of three or more adsorbed species bonded with each other. These small clusters form through surface reactions that include Si–Si bond formation, H transfer between adsorbed species, and desorption of species attached to the surface. Cross-linking between the small clusters leads to the formation of larger clusters on the deposition surface. These larger clusters, however, are unstable because their Si–Si bonds are highly strained. As a result, the clusters dissociate through desorption processes accompanied by surface migration of species detached from the clusters. In addition, the high strain in the clusters induces local crystalline-to-amorphous transitions of the bonding network. Some of the

Si–Si bonds on the crystalline substrate surface also break to aid in the accommodation of strain in the clusters of deposited material, resulting in partial loss of the surface structural order. The coordination defects of the Si atoms in the partially amorphized surface are passivated by transport of available H atoms that are bonded to overcoordinated Si atoms in the deposited clusters. These processes determine the structure and composition of the interface that forms between the deposited material and the crystalline substrate. The above relaxation mechanisms stabilize the amorphous pockets on the deposition surface, which grow and coalesce as more radicals arrive from the gas phase to form an amorphous film on the crystalline substrate.

C. Surface Evolution and Film Structural Characterization

The structural evolution of an a-Si:H film deposited by MD at 773 K with SiH_3 as the sole deposition precursor is shown in Fig. 10 (Ramalingam, 2000). Figure 10a shows the profile evolution in the growth direction, z, of

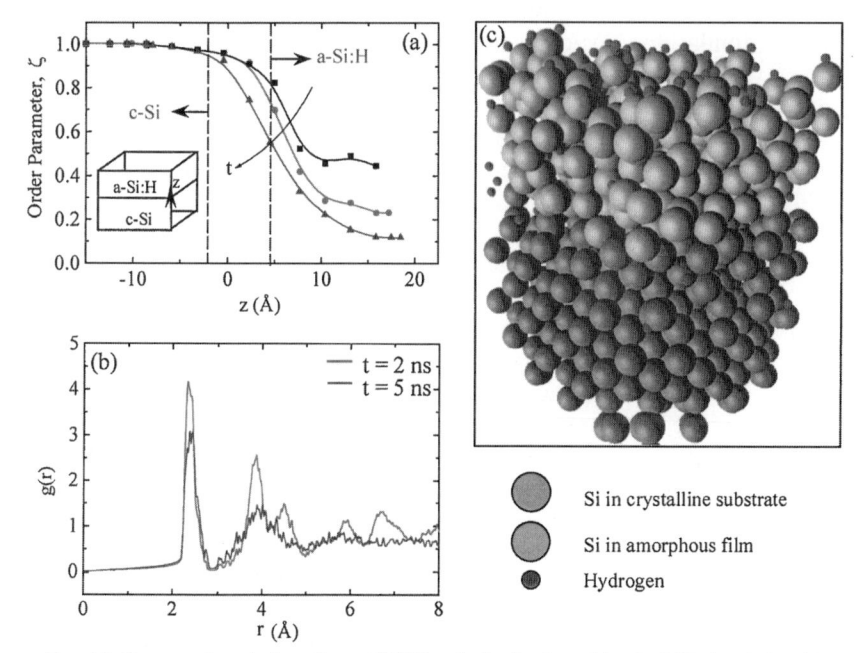

FIG. 10. Structural evolution of an a-Si:H film during its deposition by MD simulation (from Ramalingam, 2000). (a) Profile evolution of the order parameter, ζ, in the direction of growth, z; profiles are shown after 1, 2, and 5 ns, respectively. (b) Si–Si radial distribution function, $g(r)$, after 2 and 5 ns. (c) a-Si:H/c-Si film/substrate configuration after 5 ns of film growth simulation.

an order parameter, ζ, computed over a number of equal thickness slices of the film/substrate system; $z = 0$ corresponds to the initial location of the crystalline substrate surface. The order parameter is defined based on the static structure factor as by Barone and Maroudas (1997); $\zeta = 1$ and $\zeta = 0$ express perfect crystallinity and complete disorder, respectively. Consistently with the film growth mechanism described above, some crystallinity is observed in the deposited film at an early stage of the growth process, which is eventually lost ($\zeta \to 0$) as the growth proceeds resulting in the deposition of an amorphous film. These structural characteristics of the deposited film are elucidated further in Fig. 10b by monitoring the evolution of the Si–Si radial distribution function, $g(r)$, in the film. At early stages of growth, the $g(r)$ peaks indicate that the main crystalline character of the substrate is retained in the deposited film. At later stages of growth, however, the film's short-range order is lost, as expressed by the disappearance of the peaks corresponding to third and higher coordination shells of crystalline (diamond-cubic) Si. A representative a-Si:H/c-Si film/substrate configuration is shown in Fig. 10c; this configuration is obtained after 5 ns of MD simulation of film growth.

Detailed characterization of the deposition surfaces during the film growth simulation elucidate the significant factors that determine the surface reactivity with the impinging radicals (Ramalingam, 2000). Figures 11a,b, and c (see color insert) show the evolution of the growth surface reactivity after MD simulated film growth for 1, 2, and 5 ns, respectively. The reactivity is expressed by the energy gain distribution, as defined by Eq. (7), upon radical–surface interaction with the radical impinging in the Si-down configuration. The evolution of surface reactivity is monitored in conjunction with the evolution of the surface structure and compositional distribution. There is a strong correlation between the surface reactivity distribution and the distribution of H on the surface. Specifically, surface areas which are passivated by H atoms appear as less reactive regions on the reactivity maps. The regions of the surface which are not covered with H atoms are the most reactive: these regions are characterized by the highest densities of surface Si dangling bonds and the lowest density of surface overcoordination defects.

The evolution of the surface roughness during the same MD simulation of film growth is shown in Fig. 12 (see color insert) (Ramalingam, 2000). Figures 12a–c show the evolution of the film height distribution during the simulated growth process. Comparison of Figs. 12a–c with Figs. 11a–c indicates a clear correlation between surface roughness and surface reactivity. Specifically, valleys in the surface topography are preferable regions for concentration of Si dangling bonds, which can be explained by H diffusion away from these valleys to passivate dangling bonds created at the hills of the surface through, e.g., H abstraction reactions. Such H migration away from the surface valleys is observed within MD time scales, due to the locally very strong driving force for migration toward H-deficient sites on surface hills. In addition, the depth of such valleys on the surface is determined by the presence of mobile species on the surface, which can migrate, "fill" such valleys, and passivate the surface dangling bonds. The surface roughness can be measured by the height–height correlation function, $\eta(r)$ (Barabási and Stanley, 1995). The evolution of $\eta(r)$ during the MD simulation of a-Si:H film growth is shown in Fig. 12c. The local roughness is seen to increase during the initial stages of growth when the deposited film is semicrystalline. At later growth stages, however, the deposited film becomes amorphous, its surface roughness decreases substantially, and $\eta(r)$ is an almost-uniform distribution. Therefore, deposition with SiH_3 as the sole deposition precursor results in a-Si:H films with a remarkably smooth surface. In addition, these films are characterized by a very homogeneous density distribution and a negligible void fraction.

D. FILM SURFACE COMPOSITION AND COMPARISON WITH EXPERIMENT

Although MD simulation of film growth from a single thermal precursor does not take into account various surface phenomena that affect the structure and composition of the deposited films, it is worth comparing the MD simulation results with experimental measurements on plasma deposited a-Si:H films. Such comparisons provide insights about the validity of certain assumptions made in the computer simulations.

In the case of the a-Si:H films generated by the MD simulations discussed above (Ramalingam, 2000), comparisons were made of the computed film surface composition with the experimentally measured one in amorphous films deposited by PECVD from SiH_4 containing plasmas without any H_2 dilution; the measurements were based on ATR-FTIR spectroscopy as described by Marra *et al.* (1998a,b). These comparisons are summarized in Fig. 13 for the surface hydride composition at low and high growth temperatures; the SiH_x ($1 \le x \le 4$) concentrations are time averaged over several picoseconds after the end of the film growth simulation and are expressed as percentages of the total number of surface silicon hydrides. In Fig. 13a, the growth temperatures in the simulation and the experiment were identical and equal to 500 K. In Fig. 13b the film growth temperature was higher in the simulation (773 K) than in the experiment (640 K); the higher temperature in the simulation was used to accelerate thermally activated surface reactions and speed up the simulated film growth process further. Both experiments and simulations show that SiH_2 is the most abundant surface hydride species

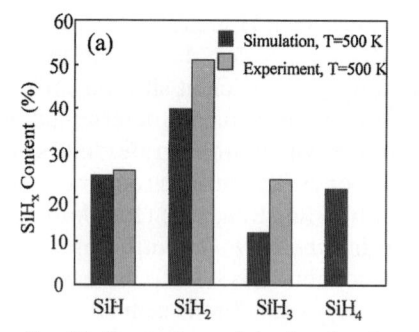

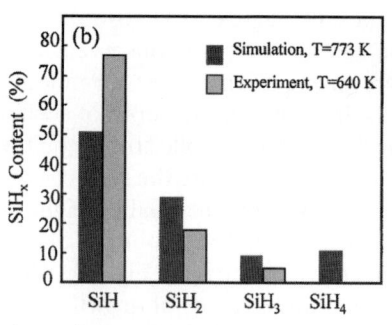

FIG. 13. Comparison of simulated and experimentally measured silicon hydride composition of a-Si:H film surfaces at (a) a low temperature (500 K) and (b) a high temperature (experiment at 640 K, simulation at 773 K).

at 500 K. In contrast, at the higher temperature, the most abundant surface species is SiH: this trend is captured correctly by the MD simulations. The simulations show that SiH is produced through successive thermally activated dissociation of SiH_3 and SiH_2. Furthermore, there is quantitative agreement between the computed surface coverage and the measured coverages at temperatures higher than 430 K (Ramalingam, 2000).

The overall agreement between experimental measurements and MD simulation results indicates that the MD simulations of film growth solely from a thermal SiH_3 precursor have captured several of the key surface reaction mechanisms during a-Si:H film growth and their temperature dependence. However, differences between simulation and experimental results are worth taking into account to improve future simulation strategies. For example, SiH_4 species are not detected in the experiments but they are present on the simulated film surfaces at both temperatures. Such discrepancies are most likely due to unaccounted effects in the MD simulations of energetic ion impingement on the surface and longer surface relaxation between successive radical impingement events. Indeed, silane is weakly bonded to the surface and it is likely to desorb over longer time scales or by ions impinging on the growth surface at kinetic energies higher than thermal.

E. The Role of the Dominant Deposition Precursor

Assuming a dominant precursor for deposition simplifies tremendously the study of plasma–surface interactions. In spite of the oversimplification, it is worth examining the effects on the film structure and composition of growth simulations solely from a given chemically reactive radical. First, direct comparison with experimental data provides an assessment for the possible role of a given precursor that can be used for more realistic simulation studies even if precise knowledge of impinging species fluxes is not available. More importantly, such studies identify the chemical reactions of this precursor with the surface and elucidate the corresponding mechanisms. Knowledge of such surface chemistry is important prior to examining more complex plasma–surface interactions that involve several chemically reactive species impinging on the surface, both in terms of understanding more complex surface chemistry and in terms of constructing and completing a chemical reaction database.

Several experimental studies have reported reaction probabilities for SiH_3 with Si surfaces during deposition over the range from 5 to 30% (Matsuda et al., 1990; Jasinski, 1993; Nuruddin et al., 1994; Perrin et al., 1998). MD simulations of surface reactivity and film growth have elucidated the importance of the detailed surface structure and chemical composition in determining the corresponding reaction probabilities (Ramalingam et al., 1999b; Ramalingam, 2000). In addition, such atomistic simulation studies can be used to clarify the dependences of reaction probabilities on processing parameters, such as substrate temperature and fluxes and kinetic energies of impinging species on the surface. MD simulations of a-Si:H growth solely from an SiH_3 precursor determined a total reaction probability of 15% from over 1200 impinging radicals on the deposition surface; this probability includes both abstraction and adsorption reactions (Ramalingam, 2000). The reaction probability was found to be practically independent of substrate temperature over the range from 500 to 773 K, which is consistent with experimental data discussed in Section IV.D on the temperature dependence of surface H coverage. These MD simulations do not take into account ion bombardment of the surface, which results in surface dangling bond production and is expected to increase substantially the radical adsorption probability. Indeed MD simulations of energetic particle (25-eV Si atoms) bombardment of films deposited solely from an SiH_3 precursor have shown that such bombardment of the film surface induces hydrogen desorption and generation of surface dangling bonds. The SiH_3 sticking probability was found to be closer to 30% under conditions that lead to high fluxes of energetic ions on the deposition surface (Matsuda et al., 1990; Perrin et al., 1998). However, in the downstream region of a silane discharge, where ion bombardment of the film surfaces is expected to be negligible, the corresponding measured sticking coefficients were less than 10% (Jasinski, 1993). Thus, the experimental observations are fully consistent with the physical understanding obtained from the MD simulations.

For further understanding of the role of SiH_3 as a deposition precursor, a-Si:H films were deposited on crystalline silicon substrates by MD simulation solely from thermal SiH_2 or SiH precursors (Ramalingam, 2000).

Film growth through repeated impingement of SiH_2 on an H:Si(001)-(2 × 1) surface resulted in a surface that is covered with silicon mono-, di-, and trihydride species, in reasonable agreement with the experimental measurements discussed in Section IV.D; the corresponding total reaction probability of the radical was found to be approximately 40%. On the other hand, when SiH was used as the sole precursor for deposition on the same substrate surface, the resulting surface of the deposited amorphous film contained predominantly silicon monohydrides and some dihydrides; the percentage of the surface dihydride species is too low compared to the experimental value. The computed total reaction probability for SiH was found to be 95% for over 250 radical impingement events on the deposition surface, in excellent agreement with the experimentally measured value (Ho et al., 1989). The surfaces of the films grown from SiH_2 and SiH precursors were rough and the density of structural defects in the films was high: in both cases, the MD deposited films are characterized by a highly inhomogeneous density distribution and a substantial void fraction.

In conclusion, the MD simulations of a-Si:H film growth from the SiH_3, SiH_2, and SiH radicals as the sole deposition precursors support that SiH_3 is the dominant deposition precursor and have elucidated its role in determining the detailed film structure and film surface morphology and composition (Ramalingam, 2000).

F. THE ROLE OF CHEMICALLY REACTIVE MINORITY SPECIES

In spite of the role of the dominant deposition precursor in determining the structure and composition of the deposited film, it is also worth examining the possible roles of minority species that are present at low concentrations in the plasma and are not expected to have significant quantitative contribution to the film growth process. Of particular interest are species that are very reactive with the growth surfaces, such as SiH radicals and small Si_nH_m clusters with $n > 1$. Indeed, such species may have both beneficial and detrimental effects on the structural quality of the deposited film.

Chemically reactive radicals attach to the surface of a-Si:H films either immediately, if they impinge at a chemically reactive location on the surface such as a Si dangling bond, or after some migration on the surface until they find a reactive surface site to attach to. For the SiH radical, detailed analysis of MD trajectories identified a third mechanism of attachment in the case of films with a high hydrogen concentration (Ramalingam et al., 1998a,b). This mechanism is illustrated in Fig. 14a through the evolution of the height coordinate of the radical's center of mass. In trajectories A and B, the radical was impinged on highly reactive locations on an a-Si:H surface with a low

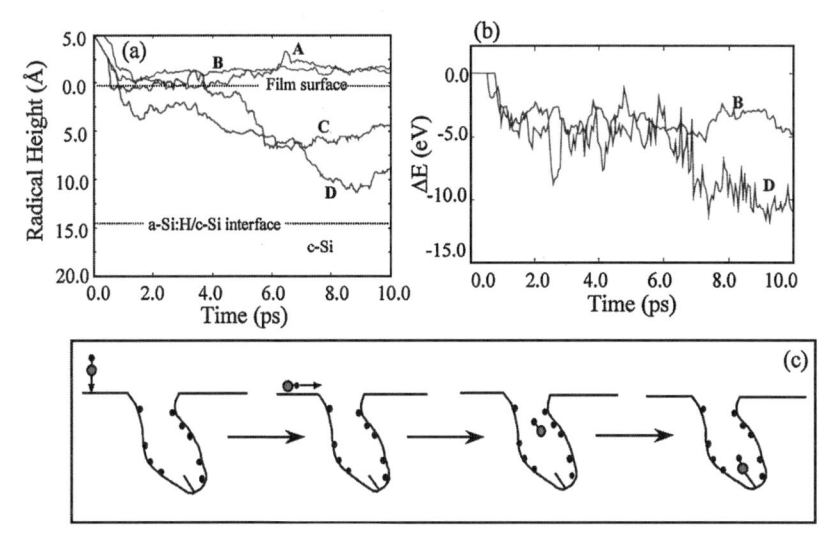

FIG. 14. SiH penetration into a-Si:H films (from Ramalingam et al., 1998a,b). (a) Evolution of the height coordinate of the radical's center of mass for four MD trajectories, A, B, C, and D, of radical impingement on a-Si:H surfaces: in C and D, the radical penetrates into the amorphous film. (b) Energy evolution, $\Delta E(t)$, along trajectories B and D shown in a. (c) Schematic illustration of a common mechanism of radical penetration through an H-passivated nanocavity emanating from the surface of the film.

(16%) and a high (50%) hydrogen concentration in the amorphous film, respectively. In both cases, the radical impinged in the vicinity of Si dangling bonds on the surface, formed strong Si–Si bonds with the corresponding surface atoms and remained attached to the surface for the entire duration of the simulation. In trajectories C and D, however, the radical was directed toward less reactive surface locations of the film with the high hydrogen content; these locations correspond to sites within the lightly colored regions of the reactivity map shown in Fig. 5d, occupied mainly by fourfold and overcoordinated Si atoms. It is evident that in these cases, the radical penetrated into the a-Si:H film. Figure 14b elucidates the energetics of the observed SiH penetration into the a-Si:H film through the evolution of the energy during both MD trajectories, C and D: penetration of the radical into the film is energetically preferable because it leads to Si–Si bond formation due to the availability of Si dangling bonds in the a-Si:H film.

A common mechanism of SiH penetration into hydrogen-rich a-Si:H films is outlined schematically in Fig. 14c (Ramalingam et al., 1998b). This mechanism involves the participation of nanocavities that may be present in the film emanating from the film surface. The presence of such nanocavities is typical of low-quality amorphous films and depend on the deposition processing conditions; computationally, such nanocavity formation has been observed in various methods of amorphous network generation ranging from defect-induced amorphization to deposition through direct radical impingement. After impingement on an unreactive surface location close to such a nanovoid and short migration on the surface, the SiH radical may enter the nanoscopic cavity; in general, the details of the entrance process depend on the size of the radical, the hydrogen coverage, and the resulting interactions between the nanocavity walls that control the size of the cavity. The walls of the nanocavities are passivated by hydrogen in the form of monohydride, as a result of the high hydrogen content in the film. The radical does not form any Si–Si bonds immediately, but channels through the cavity into the bulk of the film. Typically, the radical penetrates until it reaches the bottom of the cavity, which is characterized by a high Si dangling bond density. As a result, SiH reacts and attaches itself to the bottom surface of the cavity. Contrary to the immediate attachment mechanism, the film penetration mechanism has potentially beneficial effects on the structural quality of the deposited film: it leads to passivation of coordination defects in the film and also contributes to healing larger-scale defects, through filling void space in film nanocavities.

Small neutral and cationic clusters containing several Si atoms also have been shown to be present in silane-containing plasmas (Theil and Powell, 1994; Kessels *et al.*, 1998, 1999a,b;) the cluster size can range up to 9 or 10 Si atoms. Various experimental studies have indicated that clusters containing six Si atoms are particularly stable (Brown *et al.*,1987; Honea *et al.*,1993; Watanabe *et al.*,1997, 1998; Kessels *et al.*,1998, 1999a,b). Approximately 5–10% of the Si incorporated into the film impinges on the surface in the form of such clusters (Kessels *et al.*, 1998, 1999a,b); this percentage is high enough to affect significantly the structure and electronic properties of the entire film. In addition, DFT calculations indicate that $Si_6H_{13}^+$ is the most stable among the $Si_6H_m^+$ clusters (Miyazaki *et al.*, 1996). A significant fraction of cationic clusters is neutralized through collisions with Auger electrons; it should be mentioned, however, that charged clusters have the same framework symmetries as the corresponding neutral clusters (Ragavachari and Logovinsky, 1985). The computed structure of the Si_6H_{13} cluster according to the extended Tersoff interatomic potential, Eqs. (1)–(6), is shown in Fig. 15a; this structure is in excellent agreement with that predicted for $Si_6H_{13}^+$ by *ab initio* calculations (Watanabe *et al.*, 1998). The structure consists of a ring made up of six SiH_2 species connected together by a twofold coordinated hydrogen atom; the presence of this coordination defect suggests the possibility of breaking the ring at the overcoordinated H atom upon interaction of the cluster with the surface. In this discussion, Si_6H_{13} is chosen also as a representative H-rich cluster; the cluster H content is expected to affect the reactivity of the cluster with the surface and the resulting film surface

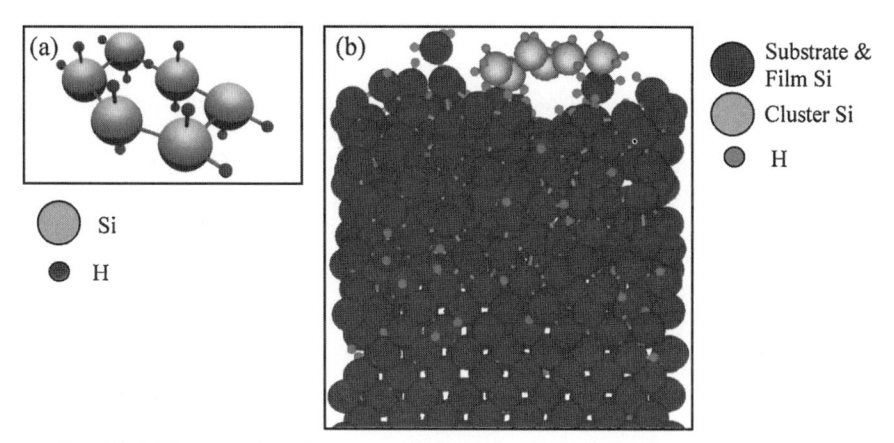

FIG. 15. (a) Computed stable structure of the Si$_6$H$_{13}$ cluster (from Ramalingam *et al.*, 2000). (b) Representative configuration of an Si$_6$H$_{13}$ cluster adsorbed on an a-Si:H film surface (from Ramalingam *et al.*, 2000).

structure upon reaction. The qualitative results presented here are expected to apply to both charged and neutral clusters. Furthermore, charged clusters are accelerated toward the surface by strong electric fields in the plasma region; in MD simulations, this is taken into account by impinging the clusters on the surface with appropriately specified kinetic energies.

MD simulations of the interactions of a thermal Si$_6$H$_{13}$ cluster with a-Si:H film surfaces revealed a high cluster reactivity and cluster-mediated defect formation in the amorphous film. The amorphous films used in the study of the interactions between the cluster and the a-Si:H surfaces were generated by successive impingement (every 4 ps) of thermal SiH$_3$ radicals on an originally H-terminated Si(001)-(2 × 1) surface for a substrate temperature of 773 K. The cluster reacts with unit probability and the details of the reaction mechanism depend on the cluster molecular orientation and location of impingement on the surface. In general, the cluster maximizes the number of Si–Si bonds it forms with the surface upon reaction by breaking at the twofold coordinated H atom and undergoing subsequent structural rearrangement (Ramalingam *et al.*, 2001). A possible origin, through cluster–surface interaction, of a deposition-related void defect in the amorphous film is illustrated in Fig. 15b (Ramalingam *et al.*, 2001); this atomic configuration is stable after surface relaxation for over 15 ps. The cluster breaks at the twofold coordinated hydrogen atom forming a five-member ring, attaches to the surface through the formation of three Si–Si bonds with surface Si atoms, and generates a bridge-like structure over a hydrogen passivated trough that already exists on the film surface. This attached cluster structure results in an enclosed void and provides a possible cluster-mediated mechanism of defect formation in the film.

Furthermore, impact of energetic (up to 100-eV) Si_6H_{13} clusters on the a-Si:H film surface can generate additional defects in the film and hydride species on the film surface (Ramalingam *et al.*, 2001). Upon impingement on the film surface, the energetic clusters dissociate. Some of the cluster atoms break Si–Si surface bonds, penetrate into subsurface layers generating overcoordination defects, and form mono-, di-, and trihydride surface species. Therefore, formation of surface hydride species also is possible due to interaction with the deposition surface of minority species, such as Si_nH_m clusters ($n > 1$); traditionally, the presence of surface hydride species in plasma-deposited a-Si:H films has been attributed solely to impinging free SiH_x radicals from the plasma discharge.

V. Summary

Significant progress has been made recently in the fundamental mechanistic and quantitative understanding of radical–surface interactions during plasma deposition processes of Si thin films based on a hierarchical atomic-scale simulation approach. Structurally and chemically well-defined crystalline and amorphous silicon surfaces have been characterized in detail with respect to their reactivity with radicals that are present in silane-containing plasmas. A broad class of chemical reactions that occur on the growth surface during deposition has been identified and analyzed. In addition, model film structures grown on crystalline substrates have been generated computationally and characterized. Furthermore, more realistic input of growth simulation parameters has been identified as an important need to improve the present status of the simulation predictive capabilities.

In spite of these recent advances in the modeling of radical–surface interactions, many important questions on the film growth mechanisms during PECVD remain unanswered and the corresponding chemical reaction database for accurate long-time scale dynamical simulation remains largely incomplete. Future directions for modeling in this area must follow hierarchical simulation approaches, establish direct links to accurate experimental plasma and surface diagnostics, and incorporate additional complexity regarding the identities, fluxes, and energies of radicals impinging on the deposition surface. Addressing these particularly challenging tasks and making use of the ever-increasing capabilities of high-performance computing will enable fully quantitative modeling of plasma deposition processes.

Plasma deposition of semiconductor thin films is just one case of a fundamentally interesting, complex process, where multiscale modeling

(Maroudas and Shankar, 1996; Jensen et al., 1998; Maroudas, 2000) offers a powerful tool toward innovative design for materials property optimization. Advances in the development of multiscale modeling approaches and their implementation for addressing a broad class of problems in electronic materials thin-film processing and reliability presents a unique challenge and opportunity for chemical engineering.

ACKNOWLEDGMENTS

It is a pleasure to acknowledge my long-standing collaboration with Professor Eray S. Aydil at UCSB and the valuable contributions of our graduate student, Shyam Ramalingam, in the area of plasma deposition of silicon thin films. Additional contributions to our research in this area by our students Saravanapriyan Sriraman, Pushpa Mahalingam, Denise Marra, and Sumit Agarwal and our collaborator, Dr. Stephen P. Walch, at NASA Ames Research Center also are gratefully acknowledged. This work was supported by the National Science Foundation, the NSF/DoE Partnership for Basic Plasma Science and Engineering (Award Nos. DMR-9713280 and ECS-0078711), the University of California Energy Institute (Award No. 08960648), the Universities Space Research Association through Cooperative Agreement No. NCC 2-1006 (Award Nos. 8006-001-02, 8008-001-003-001, and 8008-003-005-001), and the Camille & Henry Dreyfus Foundation through a Camille Dreyfus Teacher-Scholar Award.

REFERENCES

Abelson, J. R., Plasma deposition of hydrogenated amorphous silicon: Studies of the growth surface. *Appl. Phys. A Solids Surf.* **56,** 493–512 (1993).
Abraham, F. F., Computational statistical mechanics: Methodology, applications, and supercomputing. *Adv. Phys.* **35,** 1–111 (1986).
Allen, M. P., and Tildesley, D. J., "Computer Simulation of Liquids." Oxford University Press, Oxford, 1990.
Barabási, A.-L., and Stanley, H. E., "Fractal Concepts in Surface Growth." Cambridge University Press, Cambridge, 1995.
Barone, M. E., and Graves, D. B., Chemical and physical sputtering of fluorinated silicon. *J. Appl. Phys.* **77,** 1263–1274 (1995a).
Barone, M. E., and Graves, D. B., Molecular dynamics simulations of direct reactive ion etching of silicon by fluorine and chlorine. *J. Appl. Phys.* **78,** 6604–6615 (1995b).
Barone, M. E., and Maroudas, D., Defect-induced amorphization of crystalline silicon as a mechanism of disordered-region formation during ion implantation. *J. Comp.-Aid. Mater. Des.* **4,** 63–73 (1997).
Baskes, M. I., Modified embedded-atom potentials for cubic materials and impurities. *Phys. Rev. B* **46,** 2727–2742 (1992).

Battaile, C. C., Srolovitz, D. J., and Butler, J. E., Kinetic Monte Carlo method for the atomic-scale simulation of chemical vapor deposition: Application to diamond. *J. Appl. Phys.* **82,** 6293–6300 (1997).
Battaile, C. C., Srolovitz, D. J., and Butler, J. E., Atomic-scale simulations of chemical vapor deposition on flat and vicinal diamond substrates. *J. Crystal Growth* **194,** 353–368 (1998).
Becke, A. D., Density-functional thermochemistry. III. The role of exact exchange. *J. Chem. Phys.* **98,** 5648–5652 (1993).
Biswas, R., and Hamann, D. R., Interatomic potentials for silicon structural energies. *Phys. Rev. Lett.* **55,** 2001–2004 (1985).
Biswas, R., Grest, G. S., and Soukoulis, C. M., Generation of amorphous silicon structures with use of molecular-dynamics simulations. *Phys. Rev. B* **36,** 7437–7441 (1987).
Brown, W. L., Freeman, R. R., Raghavachari, K., and Schluter, M., Covalent group IV atomic clusters. *Science* **235,** 860–865 (1987).
Car, R., and Parrinello, M., Unified approach for molecular dynamics and density-functional theory. *Phys. Rev. Lett.* **55,** 2471–2474 (1985).
Ciccotti, C., Frenkel, D., and McDonald, I. R. (eds.), "Simulations of Liquids and Solids: Molecular Dynamics and Monte Carlo Methods in Statistical Mechanics." North-Holland, Amsterdam, 1987.
Clark, M. M., Raff, L. M., and Scott, H. L., Kinetic Monte Carlo studies of early surface morphology in diamond film growth by chemical vapor deposition of methyl radical. *Phys. Rev. B* **54,** 5914–5919 (1996a).
Clark, M. M., Raff, L. M., and Scott, H. L., Hybrid Monte Carlo method for off-lattice simulation of processes involving steps with widely varying rates. *Comput. Phys.* **10,** 584–590 (1996b).
Crowley, L., Plasma enhanced chemical vapor deposition for flat panel displays. *Solid State Technol.* **35,** 94–97 (1992).
Crowley, M. F., Srivastava, D., and Garrison, B. J., Molecular-dynamics investigation of the MBE growth of Si on Si(100). *Surf. Sci.* **284,** 91–102 (1993).
Dabrowski, J., and Scheffler, M., Self-consistent study of the electronic and structural properties of the clean Si(001)-(2 × 1) surface. *Appl. Surf. Sci.* **56–58,** 15–19 (1992).
Dawnkaski, E. J., Srivastava, D., and Garrison, B. J., Growth of diamond films on a diamond (001)-(2 × 1):H surface by time dependent Monte Carlo simulations. *J. Chem. Phys.* **104,** 5997–6008 (1996).
Fichthorn, K. A., and Weinberg, W. H., Theoretical foundations of dynamical Monte Carlo simulations. *J. Chem. Phys.* **95,** 1090–1096 (1991).
Frisch, M. J., Pople, J. A., and Binkley, J. S., Self-consistent molecular orbital methods. XXV. Supplementary functions for Gaussian basis sets. *J. Chem. Phys.* **80,** 3265–3269 (1984).
Gill, P. E., Murray, W., and Wright, M. H., "Practical Optimization." Academic Press, London, 1981.
Graves, D. B., Plasma processing. *IEEE Trans. Plasma Sci.* **22,** 31–42 (1994).
Graves, D. B., Kushner, M. J., Gallagher, J. W., Garscadden, A., Oehrlein, G. S., and Phelps, A. V., "Database Needs for Modeling and Simulation of Plasma Processing." National Research Council, Panel on Database Needs in Plasma Processing, National Academy Press, Washington, DC, 1996.
Halicioglu, T., and Srivastava, D., Energetics for bonding and detachment steps in etching of Si by Cl. *Surf. Sci.* **437,** L773–L778 (1999).
Hanggi, P., Talkner, P., and Borkovec, M., Reaction rate theory—50 years after Kramers. *Rev. Mod. Phys.* **62,** 251–341 (1990).
Hanson, D. E., Kress, J. D., and Voter, A. F., Reactive ion etching of Si by Cl and Cl_2 ions: Molecular dynamics simulations with comparisons to experiment. *J. Vac. Sci. Technol. A* **17,** 1510–1513 (1999).

Helmer, B. A., and Graves, D. B., Molecular-dynamics simulations of fluorosilyl species impacting fluorinated silicon surfaces with energies from 0.1 to 100 eV. *J. Vac. Sci. Technol. A* **15**, 2252–2261 (1997).

Helmer, B. A., and Graves, D. B., Molecular-dynamics simulations of Ar$^+$ and Cl$^+$ impacts onto silicon surfaces: Distributions of reflected energies and angles. *J. Vac. Sci. Technol. A* **16**, 3502–3514 (1998).

Helmer, B. A., and Graves, D. B., Molecular-dynamics simulations of Cl$_2^+$ impacts onto a chlorinated silicon surface: Energies and angles of the reflected Cl$_2$ and Cl fragments. *J. Vac. Sci. Technol. A* **17**, 2759–2770 (1999).

Ho, P., Breiland, G., and Buss, R. J., Laser studies of the reactivity of SiH with the surface of a depositing film. *J. Chem. Phys.* **91**, 2627–2634 (1989).

Hohenberg, P., and Kohn, W., Inhomogeneous electron gas. *Phys. Rev.* **136**, B864–B871 (1964).

Honea, E. C., Ogura, A., Murray, C. A., Raghavachari, K., Sprenger, W. O., Jarrold, M. F., and Brown, W. L., Raman spectra of size-selected silicon clusters and comparison with calculated structures. *Nature* **366**, 42–44 (1993).

Jasinski, J. M., Surface loss coefficients for the silyl radical. *J. Phys. Chem.* **97**, 7385–7387 (1993).

Jensen, K. F., Rodgers, S. T., and Venkataramani, R., Multiscale modeling of thin film growth. *Curr. Opin. Solid State Mater. Sci.* **3**, 562–569 (1998).

Kang, H. C., and Weinberg, W. H., Dynamic Monte Carlo with a proper energy barrier: Surface diffusion and two-dimensional domain ordering. *J. Chem. Phys.* **90**, 2824–2829 (1989).

Kessels, W. M. M., van de Sanden, M. C. M., and Schram, D. C., Hydrogen-poor cationic silicon clusters in an expanding argon-hydrogen-silane plasma. *Appl. Phys. Lett.* **72**, 2397–2399 (1998).

Kessels, W. M. M., Leewis, C. M., Leroux, A., van de Sanden, M. C. M., and Schram, D. C., Formation of large positive silicon-cluster ions in a remote silane plasma. *J. Vac. Sci. Technol. A* **17**, 1531–1535 (1999a).

Kessels, W. M. M., Leewis, C. M., van de Sanden, M. C. M., and Schram, D. C., Formation of cationic silicon clusters in a remote silane plasma and their contribution to hydrogenated amorphous silicon film growth. *J. Appl. Phys.* **86**, 4029–4039 (1999b).

Kim, C. K., Kubota, A., and Economou, D. J., Molecular-dynamics simulation of silicon surface smoothing by low-energy argon cluster impact. *J. Appl. Phys.* **86**, 6758–6762 (1999).

Kohn, W., and Sham, L. J., Self-consistent equations including exchange and correlation effects. *Phys. Rev.* **140**, A1133–A1138 (1965).

Kubota, A., and Economou, D. J., A molecular-dynamics simulation of ultrathin oxide films on silicon: Growth by thermal O atoms and sputtering by 100 eV Ar$^+$ ions. *IEEE Trans. Plasma Sci.* **27**, 1416–1425 (1999).

Kubota, N. A., Economou, D. J., and Plimpton, S. J., Molecular-dynamics simulations of low-energy (25–200 eV) argon ion interactions with silicon surfaces: Sputter yields and product formation pathways. *J. Appl. Phys.* **83**, 4055–4063 (1998).

Limoge, Y., and Bocquet, L., Monte Carlo simulation in diffusion studies: Time scale problems. *Acta Metall.* **36**, 1717–1725 (1988).

Makarov, D. E., and Metiu, H., The reaction rate constant in a system with localized trajectories in the transition region: Classical and quantum dynamics. *J. Chem. Phys.* **107**, 7787–7799 (1997).

Maroudas, D., Multi-scale modeling of hard materials: Challenges and opportunities for chemical engineering. *AIChE J.* **46**, 878–882 (2000).

Maroudas, D., and Brown, R. A., Calculation of thermodynamic and transport properties of intrinsic point defects in silicon. *Phys. Rev. B* **47**, 15562–15577 (1993).

Maroudas, D., and Shankar, S., Electronic materials process modeling. *J. Comp.-Aid. Mater. Des.* **3**, 36–48 (1996).

Maroudas, D., Ramalingam, S., and Aydil, E. S., Atomic-scale modeling of plasma-surface interactions in the PECVD of silicon. In "Fundamental Gas-Phase and Surface Chemistry of Vapor-Phase Materials Synthesis," (Allendorf, M. D., Zachariah, M. R., Mountziaris, T. J., and McDaniel, A. H., Eds.), Electrochemical Society Proceedings Series, Vol. 98-23, Electrochemical Society, Pennington, NJ, 1999, pp. 179–190.

Marra, D. C., Edelberg, E. A., Nanone, R. L., and Aydil, E. S., Silicon hydride composition of plasma-deposited hydrogenated amorphous and nanocrystalline silicon films and surfaces. *J. Vac. Sci. Technol. A* **16,** 3199–3210 (1998a).

Marra, D. C., Edelberg, E. A., Nanone, R. L., and Aydil, E. S., Effect of H_2 dilution on the surface composition of plasma deposited silicon films from SiH_4. *Appl. Surf. Sci.* **133,** 148–151 (1998b).

Matsuda, A., Plasma and surface reactions for obtaining low defect density amorphous silicon at high growth rates. *J. Vac. Sci. Technol. A* **16,** 365–368 (1998).

Matsuda, A., Nomoto, K., Takeuchi, Y., Suzuki, A., Yuuki, A., and Perrin, J., Temperature dependence of the sticking and loss probabilities of silyl radicals on hydrogenated amorphous silicon. *Surf. Sci.* **227,** 50–56 (1990).

Miyazaki, T., Uda, T., Stich, I., and Terakura, K., Theoretical study of the structural evolution of small hydrogenated silicon clusters: Si_6H_x. *Chem. Phys. Lett.* **261,** 346–352 (1996).

Mousseau, N., and Lewis, L. J., Dynamical models of hydrogenated amorphous silicon. *Phys. Rev. B* **43,** 9810–9817 (1991).

Murty, M. V. R., and Atwater, H. A., Empirical interatomic potential for Si-H interactions. *Phys. Rev. B* **51,** 4889–4893 (1995).

Northrup, J. E., Structure of Si(001)H: Dependence on the H chemical potential. *Phys. Rev. B* **44,** 1419–1422 (1991).

Nuruddin, A., Doyle, J. R., and Abelson, J. R., Surface reaction probability in hydrogenated amorphous silicon growth. *J. Appl. Phys.* **76,** 3123–3129 (1994).

Ohira, T., Inamura, T., and Adachi, T., Molecular dynamics simulation of hydrogenated amorphous silicon with Tersoff potential. *Mater. Res. Soc. Symp. Proc.* **336,** 177–182 (1994).

Ohira, T., Ukai, O., Adachi, T., Takeuchi, Y., and Murata, M., Molecular-dynamics simulations of SiH_3 radical deposition on hydrogen-terminated Si(100) surfaces. *Phys. Rev. B* **52,** 8283–8287 (1995).

Ohira, T., Ukai, O., Noda, M., Takeuchi, Y., and Murata, M., Molecular-dynamics simulations of hydrogenated amorphous silicon thin-film growth. *Mater. Res. Soc. Symp. Proc.* **408,** 445–450 (1996).

Pal, S., and Fichthorn, K. A., Accelerated molecular dynamics of infrequent events. *Chem. Eng. J.* **74,** 77–83 (1999).

Payne, M. C., Teter, M. P., Allan, D. C., Arias, T. A., and Joannopoulos, J. D., Iterative minimization techniques for ab initio total-energy calculations: Molecular dynamics and conjugate gradients. *Rev. Mod. Phys.* **64,** 1045–1097 (1992).

Perrin, J., Plasma and surface reactions during a-Si:H film growth. *J. Non-Cryst. Solids* **137–138,** 639–644 (1991).

Perrin, J., Shiratani, M., Kae-Nune, P., Videlot, H., Jolly, J., and Guillon, J., Surface reaction probabilities and kinetics of H, SiH_3, Si_2H_5, CH_3, and C_2H_5 during deposition of a-Si:H and a-C:H from H_2, SiH_4, and CH_4 disharges. *J. Vac. Sci. Technol. A* **16,** 278–289 (1998).

Proud, J. M., Gottscho, R. A., Bondur, J., Garscadden, A., Heberlein, J., Herb, G. K., Kushner, M. J., Lawler, J., Lieberman, M., Mayer, T. M., Phelps, A. V., Roman, W., Sawin, H. H., Winters, H., Perepezko, J., Hazi, A. U., Kennel, C. F., and Gerardo, J., "Plasma Processing of Materials: Scientific Opportunities and Technological Challenges." National Research Council, Report on Processing, National Academy Press, Washington, DC, 1991.

Raghavachari, K., and Logovinsky, V., Structure and bonding in small silicon clusters. *Phys. Rev. Lett.* **55,** 2853–2856 (1985).

Ramalingam, S., "Plasma-Surface Interactions in Deposition of Silicon Thin Films: An Atomic-Scale Analysis," Ph.D. thesis, University of California, Santa Barbara, 2000.
Ramalingam, S., Maroudas, D., and Aydil, E. S., Atomistic simulation of SiH interactions with silicon surfaces during deposition from silane containing plasmas. *Appl. Phys. Lett.* **72**, 578–580 (1998a).
Ramalingam, S., Maroudas, D., and Aydil, E. S., Interactions of SiH radicals with silicon surfaces: An atomic-scale simulation study. *J. Appl. Phys.* **84**, 3895–3911 (1998b).
Ramalingam, S., Maroudas, D., Aydil, E. S., and Walch, S. P., Abstraction of hydrogen by SiH_3 from hydrogen-terminated Si(001)-(2 × 1) surfaces. *Surf. Sci.* **418**, L8–L13 (1998c).
Ramalingam, S., Maroudas, D., and Aydil, E. S., Visualizing radical-surface interactions in plasma deposition processes: Reactivity of SiH_3 radicals with Si surfaces. *IEEE Trans. Plasma Sci.* **27**, 104–105 (1999a).
Ramalingam, S., Maroudas, D., and Aydil, E. S., Atomistic simulation study of the interactions of SiH_3 radicals with silicon surfaces. *J. Appl. Phys.* **86**, 2872–2888 (1999b).
Ramalingam, S., Mahalingam, P., Aydil, E. S., and Maroudas, D., Theoretical study of the interactions of SiH_2 radicals with silicon surfaces. *J. Appl. Phys.* **86**, 5497–5508 (1999c).
Ramalingam, S., Aydil, E. S., and Maroudas, D., Molecular-dynamics study of the interactions of small thermal and energetic silicon clusters with crystalline and amorphous silicon surfaces. *J. Vac. Sci. Technol. B* **19**, 634–644 (2001).
Randhawa, H., Review of plasma-assisted deposition processes. *Thin Solid Films* **196**, 329–349 (1991).
Rapaport, D. C., "The Art of Molecular Dynamics Simulation." Cambridge University Press, Cambridge, 1998.
Reif, R., Plasma enhanced chemical vapor deposition of thin films for microelectronics processing. In "Handbook of Plasma Processing Technology: Fundamentals, Etching, Deposition, and Surface Interactions," (Rossnagel, S. M., Cuomo, J. J., and Westwood, W. D., Noyes, Eds.), Park Ridge, NJ, 1990.
Ricca, A., and Musgrave, C. B., Theoretical study of the Cl-passivated Si(111) surface. *Surf. Sci.* **430**, 116–125 (1999).
Robertson, R., and Gallagher, A., Mono- and di-silicon radicals in silane-argon DC discharges. *J. Appl. Phys.* **59**, 3402–3411 (1986).
Schoolcraft, T. A., and Garrison, B. J., Initial stages of etching of the Si(001)-(2 × 1) surface by 3.0-eV normal incident fluorine atoms: A molecular-dynamics study. *J. Am. Chem. Soc.* **113**, 8221–8228 (1991).
Sinke, W. C., The photovoltaic challenge. *MRS Bull.* **18**(10), 18–20 (1993).
Smith, D. L., "Thin Film Deposition." McGraw–Hill, New York, 1996.
Srivastava, D., and Garrison, B. J., Growth mechanisms of Si and Ge epitaxial films on the dimer reconstructed Si(100) surface via molecular dynamics. *J. Vac. Sci. Technol. A* **8**, 3506–3511 (1990).
Srivastava, D., and Garrison, B. J., Modeling the growth of semiconductor epitaxial films via nanosecond time scale molecular-dynamics simulations. *Langmuir* **7**, 683–692 (1991).
Srivastava, D., Halicioglu, T., and Schoolcraft, T. A., Fluorination of Si(001)-(2 × 1) surface near step edges: A mechanism for surface defect induced etching. *J. Vac. Sci. Technol. A* **17**, 657–661 (1999).
Stephens, P. J., Devlin, F. J., Chabalowski, C. F., and Frisch, M. J., Ab initio calculation of vibrational absorption and circular dichroism spectra using density functional force fields. *J. Phys. Chem.* **98**, 11623–11627 (1994).
Stevens, W. J., Basch, H., and Krauss, M., Compact effective potentials and efficient shared exponent basis sets for the first- and second-row atoms. *J. Chem. Phys.* **81**, 6026–6033 (1984).

Stillinger, F. H., and Weber, T. A., Computer simulation of local order in condensed matter phases of silicon. *Phys. Rev. B* **31,** 5262–5271 (1985).
Tersoff, J., New empirical model for the structural properties of silicon. *Phys. Rev. Lett.* **56,** 632–635 (1986).
Tersoff, J., New empirical approach for the structure and energy of covalent systems. *Phys. Rev. B* **37,** 6991–7000 (1988).
Tersoff, J., Modeling solid state chemistry: Interatomic potentials for multicomponent systems. *Phys. Rev. B* **39,** 5566–5568 (1989).
Theil, J. A., and Powell, G., The effects of He plasma interactions with SiH_4 in remote plasma enhanced chemical vapor deposition. *J. Appl. Phys.* **75,** 2652–2666 (1994).
Von Keudell, A., and Abelson, J. R., Direct insertion of SiH_3 radicals into strained Si–Si surface bonds during plasma deposition of hydrogenated amorphous silicon films. *Phys. Rev. B* **59,** 5791–5798 (1999).
Voter, A. F., A method for accelerating the molecular-dynamics simulation of infrequent events. *J. Chem. Phys.* **106,** 4665–4677 (1997a).
Voter, A. F., Hyperdynamics: Accelerated molecular dynamics of infrequent events. *Phys. Rev. Lett.* **78,** 3908–3911 (1997b).
Voter, A. F., Parallel replica method for dynamics of infrequent events. *Phys. Rev. B* **57,** R13985–R13988 (1998).
Voter, A. F., and Doll, J. D., Dynamical corrections to transition state theory for multistate systems: Surface cell diffusion in the rare event regime. *J. Chem. Phys.* **82,** 80–92 (1985).
Walch, S. P., Ramalingam, S., Aydil, E. S., and Maroudas, D., Mechanisms and energetics of dissociative adsorption of SiH_3 on the hydrogen-terminated Si(001)-(2 × 1) surface. *Chem. Phys. Lett.* **329,** 304–310 (2000).
Walch, S. P., Ramalingam, S., Sriraman, S., Aydil, E. S., and Maroudas, D., Mechanisms and energetics of SiH_3 adsorption on the pristine Si(001)-(2 × 1) surface. *Chem. Phys. Lett.,* in press (2001).
Watanabe, M. O., Murakami, H., Miyazaki, T., and Kanayama, T., Three types of stable structures of hydrogenated silicon clusters. *Appl. Phys. Lett.* **71,** 1207–1209 (1997).
Watanabe, M. O., Kawashima, N., and Kanayama, T., Ambient gas dependence of hydrogenated silicon clusters grown in an ion trap. *J. Phys. D Appl. Phys.* **31,** L63–L66 (1998).
Weakliem, P. C., and Carter, E. A., Surface chemical reactions studied via ab initio-derived molecular-dynamics simulations: Fluorine etching of Si(100). *J. Chem. Phys.* **98,** 737–745 (1993).
Yin, M. T., and Cohen, M. L., Theoretical determination of surface atomic geometry: Si(001)-(2 × 1). *Phys. Rev. B* **24,** 2303–2306 (1981).

NANOSTRUCTURE FORMATION AND PHASE SEPARATION IN SURFACTANT SOLUTIONS

Sanat K. Kumar

Department of Materials Science and Engineering, Pennsylvania State University, University Park, Pennsylvania 16802

M. Antonio Floriano

Department of Chemistry, Università della Calabria, 87036 Arcavacata di Rende (CS), Italy

Athanassios Z. Panagiotopoulos*

Department of Chemical Engineering, Princeton University, Princeton, New Jersey 08544

I. Introduction	298
II. Simulation Details	300
A. Models and Methods	300
B. Some Methodological Issues	301
III. Results	302
A. Homopolymer Chains	302
B. Role of Different Interaction Sets	304
IV. Discussion	308
V. Conclusions	310
References	310

We have studied the phase and micellization behavior of a series of model surfactant systems using Monte Carlo simulations on cubic lattices of coordination number $z = 26$. The phase behavior and thermodynamic properties were studied through the use of histogram reweighting methods, and the nanostructure formation was studied through examination of the behavior of the osmotic pressure as a function of composition and through analysis of configurations.

---

*To whom correspondence should be addressed. E-mail: azp@princeton.edu.

Our results show that only phase separation occurs for the surfactants with short head groups. Nanostructure formation then occurs for long head groups, or for short heads, where these moieties repel each other strongly. The transition from phase separation to nanostructure formation is thus quasi-continuous and can be readily tuned by varying surfactant architecture and interactions. The results obtained from these simulations, which can be qualitatively understood through theories valid for block copolymers in selective solvents, provide the first benchmark studies against which analytical theories can be compared. © 2001 Academic Press.

I. Introduction

Surfactant molecules are well known to assemble spontaneously into many different nanostructures, e.g., micelles, bilayers, and vesicles (Israelachvili, 1992; Ruckenstein and Nagarajan, 1981; Tanford, 1980; Szleifer, 1985). The competition between the propensity of the head group(s) for the solvent and the tendency of hydrocarbon tails to minimize solvent contact dictates the equilibrium structure. Previous *analytical* work in this area has stressed the importance of various microscopic parameters, including the sizes of the hydrocarbon tails and the head groups, as well as the interaction energies between these different species. Israelachvili *et al.* (1972) derived criteria for the shapes assumed by nanostructures based on packing arguments for the hydrocarbon tails in the surfactant assembly. The central parameter in this development is the dimensionless ratio:

$$\Delta \equiv \frac{v}{a_0 l_c}, \quad (1)$$

where v is the hydrocarbon volume, and l_c the maximum extension of the hydrocarbon tail. a_0 is the optimal surface area per head, which is generally different from its geometric area a. For $\Delta \leq 1/3$, spherical micelles are predicted to form, cylindrical micelles are predicted for $1/3 < \Delta \leq 1/2$, and bilayers for $\Delta > 1/2$. While these relationships are quite useful, unfortunately, they require knowledge of the optimal area per head group, which is normally known only by examining the resulting structure. Consequently, it has been common practice to replace a_0 by a, the geometric area of the head group. As we have shown in previous work (Floriano *et al.*, 1999), and as we show below, this often results in qualitatively incorrect predictions for the nanostructures adopted by these surfactants in solution.

Typical time scales for formation of micellar aggregates in solution are tens to hundreds of microseconds (Kositza et al., 1999). Because these time scales are much longer than those that can be covered in a simulation of an atomistically detailed model, simulation studies of self-assembly are performed using simplified models, either in the continuum (Smit et al., 1990) or, more commonly, on lattices (Nelson et al., 1997; Xing and Mattice, 1997; Viduna et al., 1998; Girardi and Figueiredo, 2000).

In past work, we have began to investigate systematically the factors controlling the self-assembly of amphiphilic molecules utilizing Monte Carlo simulations on a lattice (Floriano et al., 1999; Chatterjee and Panagiotopoulos, 1999). We have shown that thermodynamic properties obtained from relatively small systems can pinpoint the critical micelle concentration. This fact circumvents another problem encountered in typical simulations, where a large number of molecules have to be considered to observe the formation of enough micelles to obtain accurate averages. In our notation H denotes the "water soluble" (hydrophilic) head moieties, while T denotes the tail (hydrophobic) groups. The phase behavior and self-assembly of the symmetric surfactants H_3T_3 and H_4T_4 were studied by grand canonical Monte Carlo simulations which were able to determine the solubility limits of the surfactants as well as their aggregation behavior. The critical micelle concentration was shown to be accompanied by a change in slope of a plot of the osmotic pressure vs the surfactant concentration, thus suggesting that micellization was accompanied by changes in thermodynamic properties. However, micellization is not a true thermodynamic transition, due to the finite size of the structures which result.

We have reanalyzed the data presented by Floriano et al. (1999) and found that, for H_4T_4, Δ changes from a value of 0.25 at $T^* = 6$ to a value of 0.19 at $T^* = 8$, which is close to the upper micellization temperature. These results are consistent with the observations that these surfactants yield spherical nanostructures. However, note that this parameter alone is insufficient to anticipate the fact that for $T^* > 8.5$ no coherent micelles form.

In this context we note that the formation of nanostructures by amphiphilic molecules represents a careful balance between the "solvophobic" and the "solvophilic" interactions. A dominance of either can lead to either phase separation or complete solubility of the amphiphile in the solvent. While the subtle balance between these forces is recognized in the current literature, the crossover from one behavior to the next remains to be understood in a quantitative fashion.

To gain more insights into the micellization processes, and the competition between phase separation and self-assembly, here we report Monte Carlo results for surfactant molecules with asymmetric chemical structures. We start with pure T molecules, which phase separate from solvent but show no self-assembly and progressively increase the length of the H block. Self-assembly occurs before phase separation for longer H blocks and we find, progressively, the occurrence of worm-like micelles and spherical micelles as the H length is increased.

II. Simulation Details

A. MODELS AND METHODS

Short diblock and triblock surfactants of structure H_xT_y and $T_xH_yT_x$ were studied on cubic lattices of coordination number $z = 26$. This lattice model for surfactants was originally proposed by Larson (Larson et al., 1985; Larson, 1988; Larson, 1996) and used by us previously. The lattice consists of surfactant and holes, a system which is isomorphic with an incompressible solution. Nearest-neighbor pairs of chain monomers have interaction energies as follows: ϵ_{HH} and ϵ_{HT} and ϵ_{TT}, where the subscript H stands for the head group, while T denotes the tail monomer. The energy of interaction of any species with a hole is identically zero. We utilized two interaction sets (I1) = (0, 0, −2) and (I2) = (+2, +1, −2), where the interactions are given in the order (HH, HT, TT). The I2 interaction set (Chatterjee and Panagiotopoulos, 1999) approximates more closely energetic interactions in nonionic surfactant solutions of the ethoxylated alkyl series. The surfactant molecules and the interaction sets studied are summarized in Table I. The reduced temperature, T^*, is defined with respect to the unit of interaction energies. An interesting point to note is that the exchange energy parameters, χ_{ij}, defined as

$$\chi_{ij} = \frac{1}{kT}\left[\epsilon_{ij} - \frac{\epsilon_{ii} + \epsilon_{jj}}{2}\right] \qquad (2)$$

TABLE I
SURFACTANT ARCHITECTURES AND INTERACTION ENERGY SETS STUDIED

Molecule	I1	I2
T_x ($x = 1–4$)	Phase separates	Phase separates
H_1T_4	Phase separates	Cylindrical micelles
H_2T_4	Worm-like micelles	Spherical micelles
H_4T_4	Spherical micelles [6]	Spherical micelles
$T_2H_4T_2$	Phase separates	Spherical micelles
$T_2H_8T_2$	Spherical micelles	

are the same for the HT and Th [h is an empty lattice position] in the two interaction sets. However, the χ_{Hh} is more negative for the (I2) interaction set relative to (I1). In other words, although tails are repelled from holes and heads in the same manner in (I1) and (I2), (I2) has the additional factor that the heads are more strongly solubilized. Thus, these molecules will show a higher propensity to create nanostructures, rather than macroscopically phase separate.

We determined the phase coexistence of these solutions by the histogram reweighting method as described previously (Floriano et al., 1999). The micellization of these surfactant solutions was determined by monitoring the dependence of the osmotic pressure on the composition. A relatively sudden decrease in the slope of this plot indicates either the formation of aggregates or phase separation as has been discussed previously (Floriano et al., 1999). For macroscopic phase separation, the slope after the break is a function of the system size, decreasing to zero in the thermodynamic limit. In contrast, for aggregating systems, the slope is independent of the overall system size. Snapshots of the molecular positions, which are often used in simulation studies of aggregation, can be misleading, indicating aggregation for systems that do not have stable, long-lived aggregates. Analysis of molecular configurations, however, provides information on the sizes and shapes of the aggregates.

B. Some Methodological Issues

Given any reasonable definition for assigning two molecules to a single cluster, e.g., by asking that they are separated by less than a certain cutoff distance, one will always find aggregates in any solution. The question of how one determines whether such aggregates should be called micelles deserves some additional discussion. The size of a cluster is not a good indicator. After all, any system near a critical point for a first-order phase transition shows the formation of large clusters due to order parameter fluctuations. However, these clusters are weakly bound. Cluster size distributions show broad peaks, as there is no clearly preferred aggregation number. These distributions are also strong functions of the overall concentration. Some recent simulation studies of aggregation (Girardi and Figueiredo, 2000) have been performed in this regime, because the low thermodynamic stability of the aggregates facilitates sampling of phase space.

In contrast, systems that form stable micellar aggregates that can be compared to experimentally observed micelles have large free energy barriers separating the aggregated and free states. In our previous work (Floriano et al., 1999) we have shown that for the surfactants H_3T_3 and H_4T_4, we have "true" micelles only at low temperatures. As the temperature is increased, the aggregates that form are less and less stable. We have, somewhat arbitrarily, set an upper limit for the slope of the osmotic pressure curves to separate "true" micellization from weak aggregation.

Finally, aggregation to large, stable clusters can be easily mistaken for a first-order transition in simulation studies of relatively small systems, as it is accompanied by hysteresis. However, the true thermodynamic character of a transition is easily revealed by performing an analysis of the system size dependence of the transition location. In particular, except near critical points, the effect of system size on the location of a first-order transition is only minor. In contrast, for systems forming finite aggregates, there is a strong dependence of the apparent location of a transition. This can be understood as follows. Consider a surfactant system that preferentially forms aggregates of $N = 100$ molecules at very low concentrations of free molecules. Simulation in a system of volume $V = 1000$ will show a "phase" of mole fraction $N/V = 0.1$, corresponding to a single aggregate forming in the simulation cell. Doubling the system volume would not change the preferred aggregation number and thus the apparent transition will move to half the previous mole fraction. Thus, by examining the system size dependence of the location of (apparently) first-order phase transition boundaries, we can reliably distinguish between micellization and macroscopic phase separation.

III. Results

A. Homopolymer Chains

Figure 1 shows phase diagrams for T chains of length 1 to 5 on a $z = 26$ lattice. As expected, we find only liquid–vapor coexistence [or, isomorphically, coexistence between a T-rich and a T-lean phase]. Note that for chains of length $x = 4$, $T_c \approx 18$. To map the lattice chain lengths onto realistic alkanes in Fig. 2, the volume fractions of T_x in the dilute phase are plotted as a function of the inverse reduced temperature. Note that there are two unknowns at this juncture: these are the mapping of a lattice monomer to the corresponding number of monomers on a realistic chain and the energy of interaction between two lattice monomers. The mole fractions are compared to experimental data for the solubility of normal alkanes in water from the correlations of Tsonopoulos (1999), which are based on a critical evaluation of experimental measurements available for these systems. We find that utilizing $\epsilon = 60$ K and mapping two real monomers into a lattice model permits remarkably good agreement for the slope of the solubility curves at high temperatures and, also, for the length dependence of the solubility.

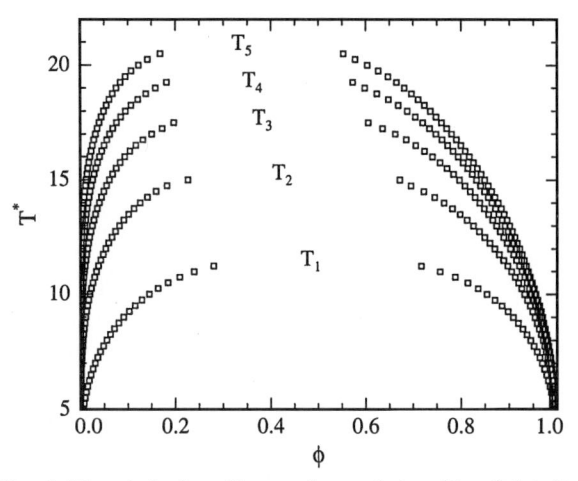

FIG. 1. Phase behavior of homopolymer chains of length 1 to 5.

Note that the experimental curves in Fig. 2 were scaled only along the temperature axis and that the scaling of chain lengths then becomes apparent on visual inspection of the resulting figure. While this agreement is satisfactory, the deficiencies of the lattice model become apparent at lower temperatures, for which the solubility of hydrocarbons passes through a minimum.

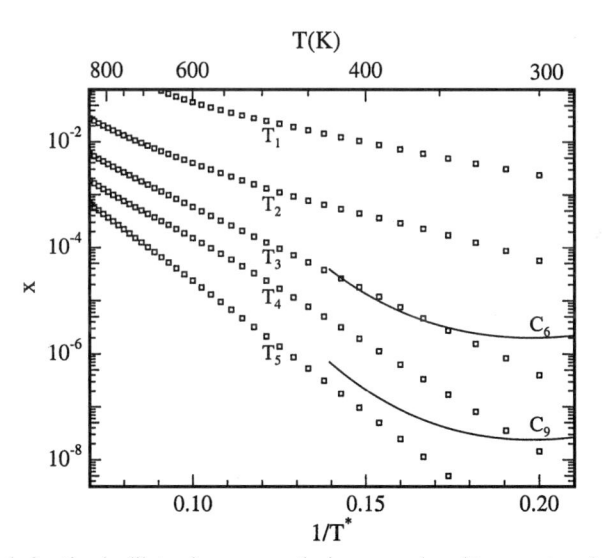

FIG. 2. Mole fraction in dilute phase versus the inverse reduced temperature for homopolymer chains of length 1 to 5 (points; top curve is $r = 1$). The top scale gives the corresponding real temperature, assuming that $\epsilon = 60$ K. Experimental data for n-hexane and n-nonane are shown as solid lines (top curve is for n-hexane).

This solubility minimum is thought to be related to changes in the hydrogen-bonding structure of water. This fact cannot be captured by a solvent model with temperature-independent interaction energies. The resulting real temperature scale is shown at the top of the graph in Fig. 2.

B. Role of Different Interaction Sets

We now proceed to examine the behavior of the surfactants and consider the (I1) interaction set. Phase separation is found for the molecules with the shortest heads, i.e., $H_1T_4(I1)$ and $T_2H_4T_2(I1)$. Figure 3 shows phase diagrams for $H_1T_4(I1)$ and $T_2H_4T_2(I1)$. Two system sizes, $L=15$ (open symbols) and $L=10$ (filled symbols), are shown in the graph. The phase diagrams in Fig. 3 do not depend on size; thus these are true phase transitions.

We now compare the behavior of the (I1) and (I2) interaction sets. We find that increasing the chain length of the H group or changing the interaction set from (I1) to (I2) has the consequence of facilitating nanostructure formation before phase separation [see Table I and Figs. 4 and 5]. This is possible since either of these variations results in a dramatic lowering of the critical temperature, as expected. Note that, in Figs. 4 and 5, the nanostructures which form are worm-like or cylindrical, suggesting that the Δ values lies between 1/2 and 1/3, which implies a large optimal area per head group, a_0. Increasing the H length for (I2) to two groups, then yields spherical micelles, which are characterized by an even smaller value of Δ (see Fig. 6). These

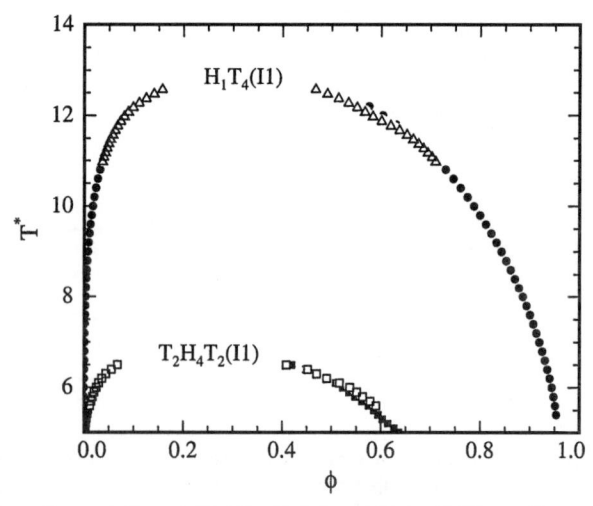

Fig. 3. Phase diagrams for amphiphiles that do not form micelles on the $z=26$ lattice.

NANOSTRUCTURE FORMATION AND PHASE SEPARATION 305

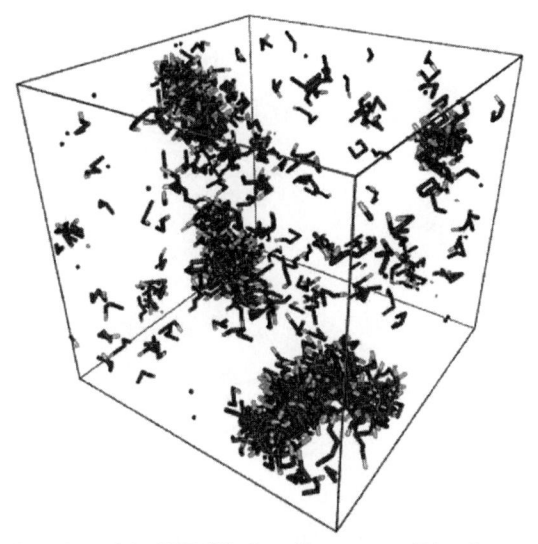

Fig. 4. Snapshot of the $H_1T_4(I2)$, $L = 40$, system at $T^* = 8$, $\mu = -47.8$.

results, in combination, suggest that as the H group length is increased, the Δ value decreases systematically. The transition from phase separation to worm-like micelles to spherical micelles happens in a quasi-continuous manner, stressing that the nanostructure formation is a subtle balance of solvophilic and solvophobic forces as noted above.

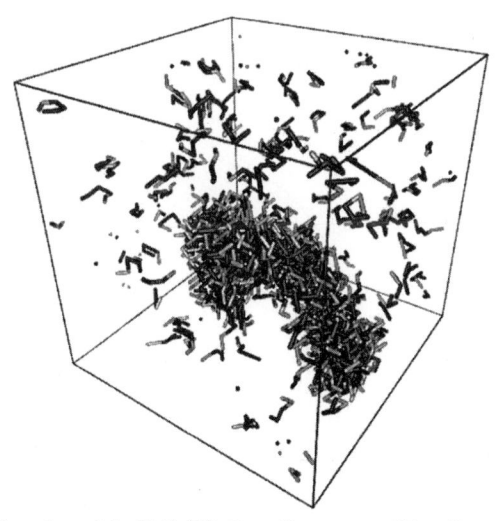

Fig. 5. Snapshot of the $H_2T_4(I1)$, $L = 40$, system at $T^* = 8$, $\mu = -51.4$.

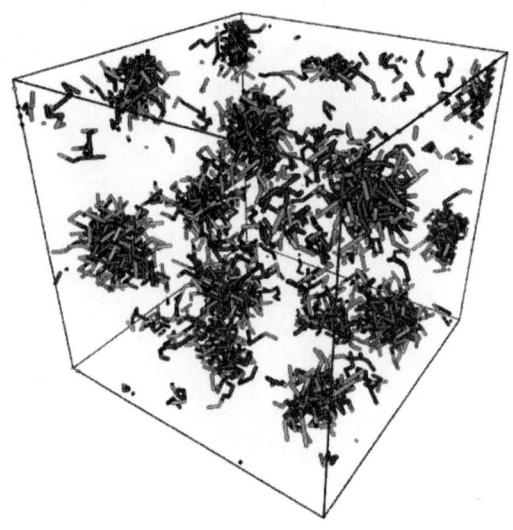

FIG. 6. Snapshot of the H_2T_4(I2), $L = 40$, system at $T^* = 7$, $\mu = -40.3$.

The critical micelle concentrations (cmc's) were determined with a slightly modified methodology of Floriano et al. (1999). Figure 7 illustrates the concept. The reduced osmotic pressure is obtained from the logarithm of the grand partition function, a quantity which can in turn be determined from the simulations to within an additive constant (Floriano et al., 1999). For the system H_2T_4 (I2), $L = 10$, we obtain the two curves shown in Fig. 7 at the two temperatures indicated there. There is a clear break, indicating the

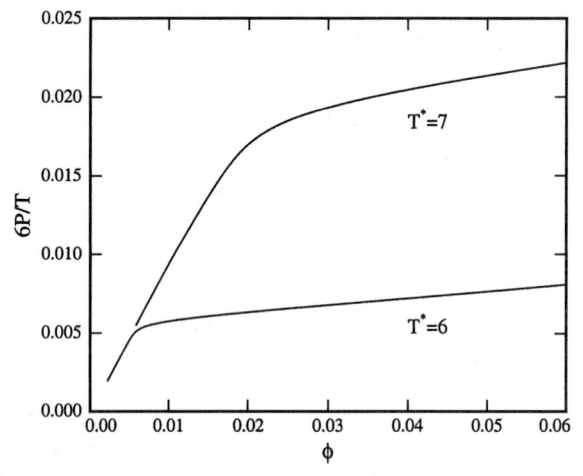

FIG. 7. Reduced osmotic pressure, $6P^*/T^*$, versus volume fraction of surfactant for the H_2T_4 (I2) system, $L = 10$.

formation of aggregates. Instead of fitting the parts of the curve before and after the break to straight lines following Floriano *et al.* (1999), we have chosen to identify the cmc as the volume fraction of surfactant for which the second derivative of this curve is a maximum. The two definitions are essentially equivalent for the range of temperatures over which micellization is observed. In Floriano *et al.* (1999), it was demonstrated that this definition results in cmc's consistent with the macroscopic interpretation, namely, as the concentration of surfactant for which the first well-defined micellar aggregate appears in solution. As the temperature is increased, the change in slope of the osmotic pressure curves becomes more gradual and the break becomes less well defined.

Figure 8 gives the cmc's, determined as the point at which the osmotic pressure curves have the maximum second derivative, for all surfactants studied that do not macroscopically phase separate. As the H chain length increases, the cmc is increased, in agreement with experimental observations for polyoxyethylenic nonionic surfactants (Rosen, 1989). Surfactants of the same architecture have generally higher cmc's for the (I2) interaction set, consistent with the observation earlier that the (I2) interaction set results in more hydrophilic interactions of the head groups. The triblock surfactants aggregate to micelles at much lower temperatures than the corresponding diblocks. This can be understood as a consequence of the much greater entropy loss for the two opposite hydrophobic blocks to come together.

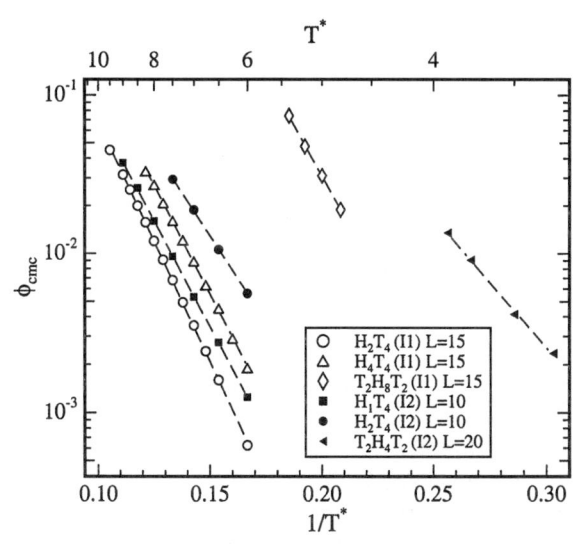

FIG. 8. The critical micelle concentration, ϕ_{cmc}, plotted against the inverse temperature for all systems that form micelles.

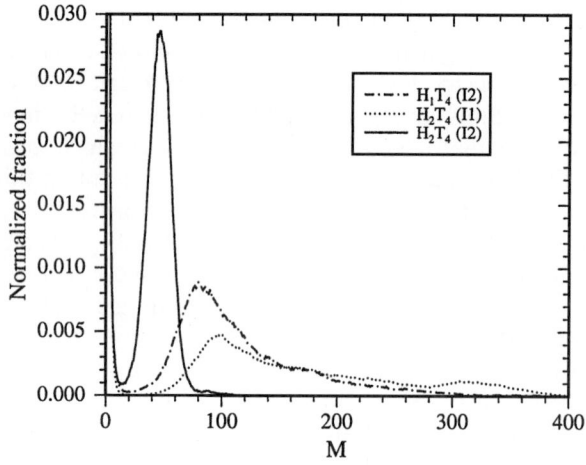

FIG. 9. Normalized cluster probability distributions corresponding to Figs. 4, 5, and 6, respectively.

Figure 9 shows the normalized cluster size distributions for the systems corresponding to Figs. 4, 5 and 6. For the H_2T_4 (I2) system, the distribution is almost perfectly Gaussian and the corresponding aggregates nearly spherical. For the H_1T_4 (I2) system, the distribution is much broader and has a long tail characteristic of cylindrical micelles. For the H_2T_4 (I1) system, there is an even broader distribution, extending to aggregation numbers near 400. It is interesting to note that the next step in this progression is macroscopic phase separation. These distributions are qualitatively similar to those of Nelson et al. (1997), who have interpreted them as resulting from a mixture of spherical and cylindrical micelles. The system of Nelson et al. (1997) was H_2T_2 surfactants on a face centered cubic lattice ($z = 12$) using the (I1) interaction set.

IV. Discussion

The key result of the present work is the progression from macroscopic phase separation to formation of micellar aggregates at low concentrations for amphiphilic molecules as the surfactant architecture and interactions are varied. This progression is analogous to that seen experimentally, for example, in the polyoxyethylene glycol monoether series, C_iE_j, with increasing hydrophilic segment length j. Our results show that only macroscopic phase separation occurs for surfactants in the case of short heads where the H group is athermal to the solvent. Situations where the heads are repelled appear not to phase separate for the T lengths examined. The propensity for microphase separation then reflects the opposite trend.

To try and place these results in perspective, we evoke the current state of the understanding of the related situation of block copolymer phase behavior, especially when the molecule is dissolved in a preferential solvent (Bates and Fredrickson, 1990). There are two relevant enthalpic parameters in this situation, $\chi_{HT}zN$ and $\chi_{sh}zN$. The first dimensionless group is relevant to the microphase separation of the surfactant, which occurs when this quantity increases above a threshold number. For long block copolymers, mean field theory suggests that this threshold value is 10.5. The second quantity, where χ_{sh} is the effective interaction parameter between a surfactant molecule and the holes, triggers phase separation. The Flory lattice model dictates that the critical value of this quantity is 2 for mixtures of long-chain molecules. Note that both of these quantities, which will suggest the point where enthalpic interaction cause microphase or macrophase separation, scale linearly with z and N, the chain length of the surfactant.

Based on these ideas, then, we understand the complex behavior displayed by these surfactant molecules as follows. The pure T molecules phase separate only from solvent as expected. As one adds H moieties the propensity for phase mixing increases, since χ_{sh} decreases. When one goes from the (I1) to the (I2) set, the χ_{sh} becomes smaller, although the χ_{HT} stays unchanged. Thus, (I2) systems show a lower propensity to phase separate and, hence, a higher tendency to form micelles. These results are in qualitative agreement with simulation but await quantitative treatment through the use of mean field theories, such as the single-chain theory of Szleifer *et al.* (1985).

A final point we note is the progression of structures for the (I2) interaction set with increasing length of the H block. For the shortest H block the system forms cylindrical micelles, while for longer H blocks they form spheres. These results may also be understood by seeking guidance from the nanostructures formed by block copolymers in selective solvents. The fact that the H_1T_4 (I2) molecule forms cylinders can be understood by realizing that this head block is strongly solubilized in solvent, since it strongly repels other H moieties. Thus, this solubilized block is effectively much longer, justifying its cylindrical micellar geometry. A logical progression of these ideas suggests that the longer blocks can form either lamellae (at high overall volume fractions) or spheres (at low volume fractions). Consequently, all longer H blocks must result in spheres near the cmc, as is found.

In the present paper, we have focused on the solution behavior at low overall concentrations of surfactant. We have not investigated the formation of liquid crystalline phases at higher concentrations. The formation of lamellar phases has been used experimentally as a probe of the onset of order in solution as the lengths of the head and tail groups are varied (Koehler *et al.*, 1994).

V. Conclusions

We have found that the balance between phase separation and nanostructure formation is delicate in these surfactant molecules. Increasing the effective "dislike" between a surfactant molecule and the solvent always favors the formation of phase separated structures. The inverse correlation is found for micellization. The Monte Carlo simulations results presented here clearly illustrate that there is a gradual transition between macroscopic phase separation and micellization. Small structural changes can result in dramatic variations in the resulting microstructures. As the length of the H groups increases we find a progression of nanostructures from worm-like micelles to cylinders and, finally, spheres. These ideas are qualitatively consistent with notions derived from the block copolymer literature. These simulation results represent the first systematic efforts to understand these complex morphological transitions which occur at low concentrations for short-chain surfactant systems.

ACKNOWLEDGMENTS

This work was initiated at the University of Maryland, where funding was provided by the Department of Energy, Office of Basic Energy Sciences (Grant DE-FG02-98ER14858), and by the Petroleum Research Fund administered by the American Chemical Society (Grant 34164AC9). At Penn State, this research was funded by the National Science Foundation (Grant CTS-9975625).

REFERENCES

Bates, F. S., and Fredrickson, G. H., *Annu. Rev. Phys. Chem.* **41,** 525 (1990).
Chatterjee, A. P., and Panagiotopoulos, A. Z., in *Computer Simulation Studies in Condensed Matter Physics,* Springer Proceeding in Physics, Vol. 85 (D. P. Landau, Ed.), Springer, New York, 1999, pp. 211–222.
Floriano, M. A., Caponetti, E., and Panagiotopoulos, A. Z., *Langmuir* **15,** 3143 (1999).
Girardi, M., and Figueiredo, W., *J. Chem. Phys.* **112,** 4833 (2000).
Israelachvili, J., *Intermolecular and Surface Forces.* Academic Press, London, 1992.
Israelachvili, J., Mitchell, D., and Ninham, B. W., *J. Chem. Soc. Faraday Trans.* **72,** 1525 (1972).
Koehler, R. D., Schubert, K.-V., Strey, R., and Kaler, E. W., *J. Chem. Phys.* **101,** 10843 (1994).
Kositza, M. J., Bohne, C., Alexandridis, P., Hatton, T. A., and Holzwarth, J. F., *Macromolecules* **32,** 5539 (1999).
Larson, R. G., Scriven, L. E., and Davis, H. T., *J. Chem. Phys.* **83,** 2411 (1985).
Larson, R. G., *J. Chem. Phys.* **89,** 1642 (1988).

Larson, R. G., *J. Phys. II France* **6,** 1441 (1996).
Nelson, P. H., Rutledge, G. C., and Hatton, T. A., *J. Chem. Phys.* **107,** 10777 (1997).
Rosen, M. J., *Surfactants and Interfacial Phenomena,* 2nd ed., John Wiley and Sons, New York, 1989.
Ruckenstein, E., and Nagarajan, R., *J. Phys. Chem.* **85,** 3010 (1981).
Smit, B., Hilbers, P., Esselink, K., Rupert, L., van Os, N. M., and Schlijper, A. G., *Nature* **348,** 624 (1990).
Szleifer, I., Ben-Shaul, A., and Gelbart, W. M., *J. Chem. Phys.* **83,** 2612 (1985).
Tanford, C., *The Hydrophobic Effect,* Wiley, New York, 1980.
Tsonopoulos, C., *Fluid Phase Equil.* **156,** 21 (1999).
Viduna, D., Milchev, A., and Binder, K., *Macromol. Theor. Simul.* **7,** 649 (1998).
Xing, L., and Mattice, W. L., *Macromolecules* **30,** 1711 (1997).

SOME CHEMICAL ENGINEERING APPLICATIONS OF QUANTUM CHEMICAL CALCULATIONS

Stanley I. Sandler,* Amadeu K. Sum, and Shiang-Tai Lin

Center for Molecular and Engineering Thermodynamics, Department of Chemical Engineering, University of Delaware, Newark, Delaware 19716

I.	Introduction	314
II.	*Ab Initio* Interaction Potentials and Molecular Simulations	315
III.	Infinite Dilution Activity Coefficients and Partition Coefficients from Quantum Mechanical Continuum Solvation Models	325
IV.	Use of Computational Quantum Mechanics to Improve Thermodynamic Property Predictions from Group Contribution Methods	335
V.	Use of *ab Initio* Energy Calculations for Phase Equilibrium Predictions	341
VI.	Conclusions	347
	References	348

Quantum chemical calculations have long been used in the chemical process industry as a method of computing ideal gas thermodynamic and spectroscopic properties, analyzing reaction pathways, and the calculation of heats of formation and related properties. However, it is only recently, as a result of improvements in computational hardware and software, that accurate interaction energies can be computed that can be used directly or combined with molecular simulation to predict the thermodynamic properties of nonideal systems of interest in the practice of chemical engineering. It is this application of quantum chemical calculations that is reviewed in this chapter. © 2001 Academic Press.

*To whom correspondence should be addressed.

I. Introduction

Computational quantum chemistry has been used in many ways in the chemical industry. The simplest of such calculations is for an isolated molecule; this provides information on the equilibrium molecular geometry, electronic energy, and vibrational frequencies of a molecule. From such information the dissociation energy at 0 K is obtained, and using ideal gas statistical mechanics, the entropy and other thermodynamic properties at other temperatures in the ideal gas state can be computed. Such calculations have provided information on heats of formation of compounds and, when used with transition state theory, on reaction pathways and reaction selectivity. As these applications are well documented in the literature, they are not discussed here.

Instead, our focus is on several new applications of *ab initio* quantum chemical methods to chemical engineering. We do not attempt to describe the basic theory of quantum chemistry calculations or their details (e.g., level of theory and choice of basis sets) but, instead, refer the interested reader to several recent textbooks (Levine, 1991; Szabo and Ostlund, 1996; Lowe, 1993). Of special interest is a recent compilation of computational chemistry theory, concepts, and techniques (Schleyer, 1998).

Most processes in chemical processing occur in mixtures, frequently in liquid solutions. Our interest in this review is in the use of quantum chemical methods for such systems, and we focus on three computational methods related to thermodynamic properties and phase behavior. The first is a conceptually straightforward application in which one uses interaction energies between molecules obtained from quantum chemistry in Monte Carlo or molecular dynamics simulation to compute bulk thermodynamic properties. Historically, this has been done as a two-step process. First, the interactions between molecules, usually only two-body interactions, are computed from *ab initio* quantum chemistry. These energies are fit with a model potential function and then used in simulation: in the past, assuming pairwise additivity and, more recently, by including multibody interactions. We discuss such calculations here. More recently, *ab initio* simulations have started to appear. In this case, at each stage of the simulation, a quantum mechanical calculation of the energy for the assembly of molecules is done. This rigorous method is gaining in importance as computing power increases. At present, however, it is so computer intensive that its use is restricted to very simple molecules.

Because of the computer-intensive nature of calculating the interaction energy of a large assembly of molecules, as for a solute in a solvent, several other methods have been developed. We discuss two here. One is based on treating a solvent as a polarizable continuum. In this model the energy, especially the free energy, change of adding a solute molecule to a solvent is computed based on the reaction field of a continuum with the same dielectric constant as the solvent. This is a far less time-consuming calculation than considering all the solvent molecules explicitly. The second method is based on the observation that the adjustable parameters in most activity coefficient (or excess Gibbs free energy) models used by chemical engineers are interpreted as being average energies of interaction (actually, usually differences in energies of interaction between like and unlike molecules). Because of molecular packing in a dense phase, these energies are different than in the ideal gas state. In this method, the average interaction energies in small assemblies of molecules are computed and then used as the parameters in activity coefficient models to predict mixture phase behavior, thereby avoiding complex statistical mechanical calculations.

II. *Ab Initio* Interaction Potentials and Molecular Simulations

Molecular simulation is an area of active research in the chemical, physical, and biological sciences and in engineering. Molecular simulations are used as a tool to understand phenomena at the microscopic level and, in certain instances, as a replacement to experiment. Indeed, the term "computer experiment" has been often used in the literature to refer to molecular simulations, especially in cases where simulations were performed for systems or conditions inaccessible by conventional experimental methods.

The first molecular simulations were performed almost five decades ago by Metropolis *et al.* (1953) on a system of hard disks by the Monte Carlo (MC) method. Soon after, hard spheres (Rosenbluth and Rosenbluth, 1954) and Lennard-Jones (Wood and Parker, 1957) particles were also studied by both MC and molecular dynamics (MD). Over the years, the simulation techniques have evolved to deal with more complex systems by introducing different sampling or computational algorithms. Molecular simulation studies have been made of molecules ranging from simple systems (e.g., noble gases, small organic molecules) to complex molecules (e.g., polymers, biomolecules).

Monte Carlo computer simulation methods require the energy of an assembly of molecules to determine whether a trial move is accepted or rejected, while in MD methods the forces on molecules along their trajectories are needed. Simple analytic forms, such as the Lennard–Jones potential, are commonly used to describe interaction potentials in which all atoms are treated explicitly with distinct parameters or a small collection of atoms is treated as a single interaction site (united-atom approximation). Most potentials used are empirical in that the parameters have been fitted to reproduce some experimental data (e.g., density, heat capacity, heat of vaporization, atom–atom or site–site pair correlation functions derived from X-ray or neutron scattering data) but fail for the prediction of other properties. Examples of such models include OPLS (Jorgensen and Tirado-Rives, 1988), AMBER (Cornell *et al.*, 1995), CHARMM (Brooks *et al.*, 1983), and others specifically for certain compounds, such as SPC (Berendsen *et al.*, 1981), SPC/E (Berendsen *et al.*, 1987), and TIP4P (Jorgensen *et al.*, 1983) for water. However, there are a number of problems with such potentials. For example, while they may result in a reasonable correlation of the data over the range of conditions to which they were fit, predictions for other properties or for states outside the fitting range may be inaccurate. In particular, "effective" two-body potentials fitted to some liquid state data result in inaccurate second virial coefficients.

The empirical determination of potential parameters has usually been performed assuming pairwise additivity for the molecular interactions, that is, that the total interaction energy is a summation of the interactions over all pairs of molecules. However, for many fluids such as water, methanol, and other highly polar molecules, pairwise additivity does not properly describe molecular interactions, especially in dense phases. This is why an approximate potential with one set of empirically adjusted or "effective" parameters cannot be made to reproduce a wide range of experimental data. Until recently, most simulations used the assumption of pairwise additivity.

Instead of using empirical potentials and the pairwise additivity assumption, rigorous *ab initio* molecular simulations can be performed. *Ab initio* methods, based on solving the Schrödinger equation, are the most fundamental way to describe interaction between molecules. In *ab initio* molecular simulations one solves the Schrödinger equation for the interaction energy among all the molecules at each stage of a simulation (move in MC or time step in molecular dynamics). Recently, *ab initio* molecular simulations for simple, small molecules (Car and Parrinello, 1995; Benoit *et al.*, 1996) have been performed, and this is one of the forefront areas of computational chemistry. Since all the molecules are considered simultaneously in the method, the pairwise additive assumption is not used. However, the computational cost is extremely high, and the method is impractical with the currently available computer resources for the reasonably complex systems of interest to chemical engineers.

An alternative is the sequential use of *ab initio* quantum chemistry and molecular simulation calculations. In this method quantum chemistry calculations are used to compute points on the potential energy surface of two interacting molecules, which are then fit to a potential form suitable for simulation purposes. To improve the prediction of macroscopic properties, nonpairwise additive forces can be included in the form of reaction field polarization or fluctuating charge models. Such simulations are less computationally intensive than fully *ab initio* molecular simulations, though the nonpairwise contributions add considerably to the computational load (Liu *et al.*, 1998). It is the sequential combination of quantum chemistry and molecular simulation that is discussed in what follows.

Popkie *et al.* (1973) performed the pioneering MC simulation of water with a potential derived from *ab initio* calculations for more than a hundred water pair configurations. The *ab initio* calculations were at the Hartree–Fock level with a minimal basis set that is now considered too simplistic. Their work was the first to provide a link between *ab initio* calculations on isolated molecules and macroscopic properties prediction. With rapid advances and ready availability of computational resources, the calculation of more accurate potential energy surfaces for a pair of molecules has become practical but is still not widely used in simulations. Detailed calculations of potential energy surfaces are still restricted to reasonably small molecules. Several molecules for which potential energy surfaces have been computed are water (Mas *et al.*, 1997; Niesar *et al.*, 1989; Matsuoka *et al.*, 1976), argon (Lotrich and Szalewicz, 1997a,b), carbon dioxide (Bukowski *et al.*, 1999a), acetonitrile (Cabaleiro-Lago and Ríos, 1997; Bukowski *et al.*, 1999b), methanol (Bukowski *et al.*, 1999b; Mooij *et al.*, 1999), and acetone (Hermida-Ramón and Ríos, 1998).

In principle, it is possible to calculate the potential energy surface between any two molecules (of the same or different species). However, it may not be practical because of the several hundreds of configurations that are required even for small molecules, and this number increases for large molecules with increased degrees of freedom, including molecular flexibility. Also, a reasonable level of theory and a large basis set, both of which are very demanding of computational resources and time, are required for accurate interaction energies.

There are two approaches often used for computing interaction energies: supermolecular and perturbation methods (for reviews see Chałasiński and Gutowski, 1988; Chałasiński and Szczęśniak, 1994). In the supermolecular method the interaction energy of a pair of molecules E_{AB}^{int} is the difference between the energy of the pair E_{AB} and the energy of the individual molecules E_A and E_B,

$$E_{AB}^{\text{int}} = E_{AB}\{AB\} - E_A\{AB\} - E_B\{AB\}, \tag{1}$$

where $\{AB\}$ refers to using the combined basis set of molecules A and B in the quantum chemical calculations. These single-point energies can be calculated with commonly available quantum chemistry packages [e.g.,

Gaussian (Gaussian, Inc.), Jaguar (Schrödinger, Inc.), Q-Chem (Q-Chem, Inc.), MolPro (Werner and Knowles), GAMESS (Schmidt *et al.*, 1993)]. In this equation, the energies on the right-hand side (RHS) are usually of the order of tens to hundreds of Hartrees, whereas their difference yields an interaction energy of the order of milli-Hartrees. Therefore, the proper selection of the level of theory and basis set is crucial in the supermolecular approach. A special problem is basis set superposition error (BSSE), a nonphysical error associated with the use of a finite basis set to represent the wave functions of the electrons. It has been shown that using large basis sets, in particular, diffuse functions, can attenuate the BSSE. Further discussion of BSSE is given by van Duijneveldt *et al.* (1994) and Kestner and Combariza (1999). The level of theory used in the interaction energy calculations must be chosen so as to model properly the interactions in the system. Typically, at a minimum, second-order Møller–Plesset theory (Levine, 1991; Szabo and Ostlund, 1996; Schleyer, 1998) is used, as Hartree–Fock theory at best may be useful only for systems with strong electrostatic interactions and when electron correlation effects are unimportant.

The second approach to calculate interaction energies is by perturbation methods, such as symmetry-adapted perturbation theory (SAPT). Detailed description of SAPT is beyond the scope of this review, and the interested reader is referred to the original papers by Jeziorski *et al.* (1994), Szalewicz and Jeziorski (1997), and Jeziorski and Szalewicz (1998). Perturbation methods have two important advantages over the supermolecular approach. The first is that all contributions to the total interaction energy, that is, induction, dispersion, exchange, and electrostatics forces, are separated and one can choose the degree of perturbation theory to which each is calculated. Also, knowing these different contributions to the interaction energy, one may more easily devise a suitable potential form. The second advantage is that the method does not suffer BSSE, thus allowing flexibility in the choice of a basis set to achieve good accuracy.

We next consider whether such two-body potentials are sufficient to produce macroscopic properties predictions by molecular simulation that are accurate enough for chemical engineering applications by considering two example systems. Recently, Bukowski *et al.* (1999b) calculated potential energy surfaces for several pairs of molecules (acetonitrile + acetonitrile, methanol + methanol, acetonitrile + carbon dioxide, methanol + carbon dioxide, dimethylnitramine + dimethylnitramine, dimethylnitramine + acetonitrile, dimethylnitramine + methanol, and dimethylnitramine + carbon dioxide). The interaction energies were calculated with SAPT using a reasonably large basis set [Dunning's (1989) cc-pVDZ, with the addition of d and p polarization functions] along with bond functions (these have been shown to improve the interaction energy description, especially dispersion forces). In the discussion to follow, we consider only the acetonitrile and methanol pairs. Bukowski *et al.* (1999b) studied many configurations, with energies ranging from -24.380 to $+2880$ kJ/mol for acetonitrile and from -20.790 to $+1150$ kJ/mol for methanol, providing a comprehensive scan over the potential energy surface for these two molecules. They then fitted the energies to a site–site potential of the form

$$E_{AB}^{int} = \sum_{a \in A} \sum_{b \in B} \left[\left(\frac{1}{r_{ab}} + A_{ab} \right) e^{\alpha_{ab} - \beta_{ab} r_{ab}} + f_1 \left(\delta_1^{ab} r_{ab} \right) \frac{e_a e_b}{r_{ab}} \right.$$

$$\left. - \sum_{n=6,8,\ldots} f_n \left(\delta_n^{ab} r_{ab} \right) \frac{C_n^{ab}}{r_{ab}^n} \right], \quad (2a)$$

$$f_n(x) = 1 - e^{-x} \sum_{k=0}^{n} \frac{x^k}{k!}, \quad (2b)$$

where r_{ab} is the distance between site a and site b of molecules A and B, respectively; e_a indicates the charge on site a; and α_{ab}, β_{ab}, C_n^{ab}, and δ_n^{ab} are parameters optimized to fit the calculated energies. The function $f_n(x)$ is the Tang–Toennies (1984) damping function, often used to represent the asymptotic behavior at small distances. This proposed interaction potential has 100 parameters for acetonitrile and 183 for methanol. Figure 1 shows how well Bukowski *et al.* (1999b) were able to describe the calculated interaction energies for acetonitrile and methanol with Eqs. (2). (Note that only energies up to 200 kJ/mol are shown in the figure.) These potentials can be considered to be the most accurate currently available. No other *ab initio* derived

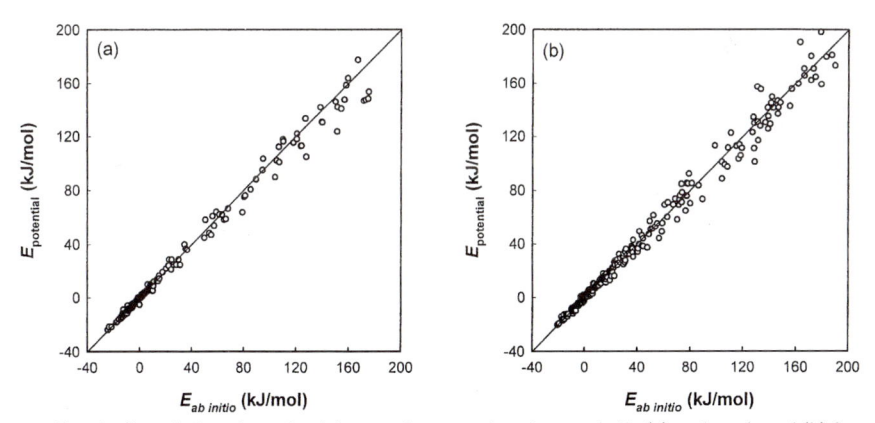

FIG. 1. Correlation plots of pair interaction energies of acetonitrile (a) and methanol (b) for Bukowski *et al.* (1999) potential [energies from Eq. (2) in the text]. *Ab initio* energies from Bukowski *et al.* (1999).

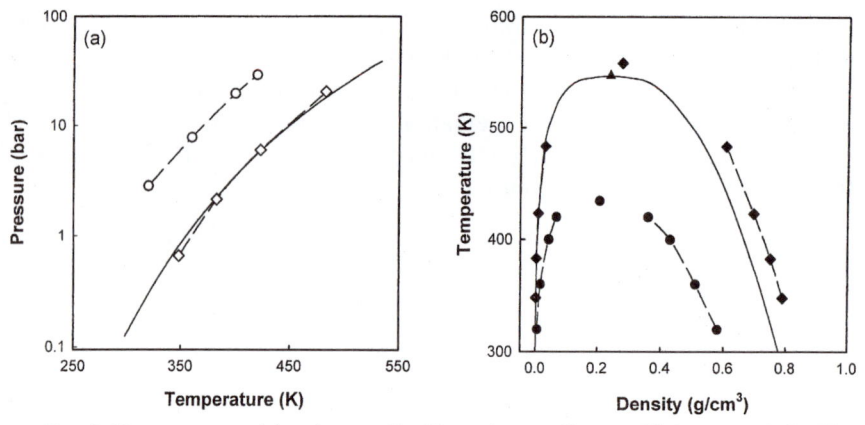

FIG. 2. Vapor pressure (a) and vapor–liquid coexistence diagram (b) for acetonitrile. The solid curve represents experimental data. Symbols are GEMC simulation results with the following potentials: diamonds, potential by Bukowski *et al.* (1999); circles, potential by Cabaleiro-Lago and Ríos (1997).

potential has been published for methanol, though Cabaleiro-Lago and Ríos (1997) presented another potential for acetonitrile obtained using the supermolecular method at the MP2/6-311+G* level, and fitted the energies with a 30-parameter site–site exponential-6 type plus Coulomb potential.

These potentials were used in our Gibbs ensemble Monte Carlo (GEMC) simulations (Panagiotopoulos, 1987) to obtain the predicted phase behavior. The simulations were performed in the NVT ensemble with 512 molecules and periodic boundary conditions and were corrected for long-range cutoff of the energy and pressure. The cutoff radius was set at 13 Å, which is less than half of the box length. Simulations were allowed to equilibrate, followed by production runs in which the particle distribution, densities, pressure, energies, and chemical potential were block averaged. The phase diagrams obtained and the experimental values are shown in Fig. 2 for acetonitrile and Fig. 3 for methanol. Critical properties were also calculated from the simulated equilibrium points using the renormalization group theory scaling law parameters and the law of rectilinear diameters,

$$\rho_l - \rho_v = A(T_c - T)^\beta$$
$$\frac{\rho_l + \rho_v}{2} = \rho_c + B(T_c - T), \tag{3}$$

where ρ is the density of the liquid (l) and vapor (v) phases, T_c the critical temperature, ρ_c the critical density, and β the critical exponent, and A and B are fitting constants. The estimated critical temperature and density are presented in Table I.

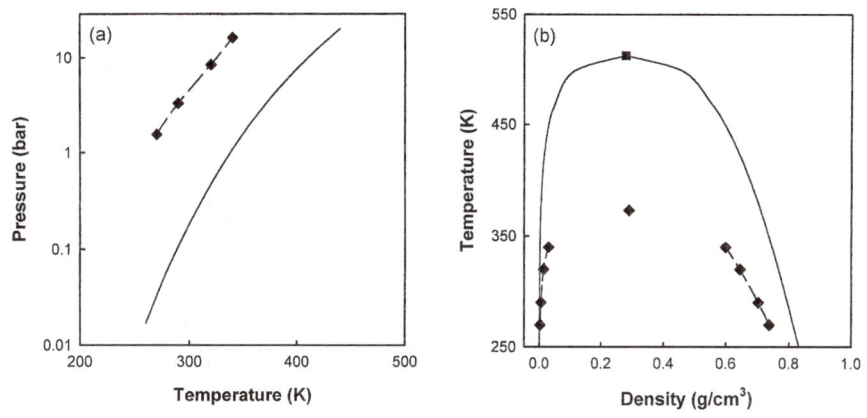

FIG. 3. Vapor pressure (a) and vapor–liquid coexistence diagram (b) for methanol. The solid curve represents experimental data, and symbols are GEMC simulation results with potential by Bukowski et al. (1999).

As shown in Fig. 2, the phase behavior for acetonitrile is reasonably well predicted using the potential of Eq. (2) with vapor pressures in agreement with experimental values, liquid densities slightly too high, and the critical point close to experimental data. This example shows that, in some cases, potentials calculated using accurate quantum chemical methods can lead to thermodynamic property predictions that can be of use for engineering applications. However, the acetonitrile predictions using the Cabaleiro-Lago and Ríos (1997) potential are not in good agreement with the experimental data, with predicted vapor pressures that are too high and liquid densities too low, suggesting that the potential is not sufficiently attractive. This is a result of the use of MP2 theory with a low-level basis set in their calculations. (The predicted interaction energies of the Cabaleiro-Lago and Ríos potential are systematically higher than those calculated by Bukowski et al.) This example illustrates the dependence of the predicted thermodynamic properties on the accuracy of the pair energies.

TABLE I
Estimated Critical Properties from GEMC Simulations of Vapor–Liquid Equilibria

		T_c (K)	ρ_c (g/cm^3)
Acetonitrile	Experimental	547.3	0.237
	Bukowski et al. (1999b)	558.5	0.276
	Cabaleiro-Lago and Ríos (1997)	434.3	0.206
Methanol	Experimental	512.7	0.278
	Bukowski et al. (1999b)	379.9	0.282

For methanol, however, the results are less satisfactory. The predicted properties using the Bukowski *et al.* (1999b) potential result in liquid densities that are generally too low and vapor pressures that are too high compared to experimental values. These results suggest that the potential for methanol is not sufficiently attractive. Since the pair potential was obtained from a high-level theory that has been shown to be accurate for acetonitrile, a plausible explanation is that nonpairwise interactions are important for methanol in a dense fluid. Methanol is a strong hydrogen-bonding fluid similar to water, and it is known that multibody effects are required to model liquid water properly.

One approach to incorporating multibody interactions is by using quantum chemical methods to compute three-body and higher-order interaction energies and then fit these with complicated potential functions that can be used in simulation. Three-body potentials have been proposed for water (Lotrich *et al.*, 1998) and some noble gases (Lotrich and Szalewicz, 1997a,b), and their use in simulation and in virial coefficient calculations has been shown to improve gas-phase property predictions. However, the quantum chemical calculations involved, the fitting of these multibody potentials, and their use in simulation are computationally prohibitive at present for molecules and mixtures of engineering interest.

A less computationally demanding alternative is the use of polarizable potentials, an approach that has been used mainly for strongly interacting systems such as water (Niesar *et al.*, 1989, 1990; Yoshii *et al.*, 1998; Dang and Chang, 1997; Sprik and Klein, 1988; Kuwajima and Warshel, 1990; Bernardo *et al.*, 1994; Stillinger and David, 1978; Svishchev and Hayward, 1999; Kiyohara *et al.*, 1998; Ahlström *et al.*, 1989; Millot *et al.*, 1998; Watanabe and Klein, 1989; Detrich *et al.*, 1984; Chialvo and Cummings, 1996, 1998). Polarizable potentials approximately account for multibody interactions by adding a contribution to the total interaction energy of the system as a result of the electric field created due to the induced dipoles. This nonadditive polarizable energy is a multibody effect since the induced dipole of one molecule generates an electric field that affects all other molecules in the system. For a complete discussion of polarization see Böttcher (1973). The polarizable energy of a system (U_{pol}) is

$$U_{\text{pol}} = -\frac{1}{2} \sum_i \mu_i^{ind} \cdot \boldsymbol{E}_i^o \quad (4)$$

where $\boldsymbol{E}_i^o$ is the electric field at site i generated by the fixed charges (e_j) in the system,

$$\boldsymbol{E}_i^o = \sum_{j \neq i} \frac{e_j \boldsymbol{r}_{ij}}{r_{ij}^3} \quad (5)$$

and if the electric field is not large, the induced dipole μ_i^{ind} at site i can be expressed as a linear response to the field,

$$\mu_i^{ind} = \tilde{\alpha}_i \cdot E_i \tag{6}$$

and

$$E_i = E_i^o + \sum_{j \neq i} \tilde{T}_{ij} \cdot \mu_i^{ind}, \tag{7}$$

where $\tilde{\alpha}$ is the polarizability tensor, E_i is the total electric field, $\tilde{T}_{ij}$ is the dipole tensor,

$$\tilde{T}_{ij} = \frac{1}{r_{ij}^3}\left(\frac{3r_{ij}r_{ij}}{r_{ij}^2} - 1\right) \tag{8}$$

$r_{ij} = r_i - r_j$ is the separation distance vector, and r_{ij} is the distance between site i and site j. These equations can also be written in tensorial form as linear equations that must be solved simultaneously. More commonly, Eqs. (6) and (7) are solved iteratively in a self-consistent manner for the induced dipoles, as follows.

(i) For a given configuration of the particles in the system, the permanent field is calculated with Eq. (5).
(ii) These calculated electric fields are used with an initial guess for the induced dipoles to calculate the total electric field [Eq. (7)].
(iii) The total field is then used to compute the new induced dipoles in Eq. (6).

Steps ii and iii are repeated until a specified convergence of the dipoles (usually 10^{-5} D) is achieved on two sequential iterations, and these induced dipoles are then used in Eq. (4) to obtain the polarization energy. The iterative procedure will usually converge within few cycles. The polarizability tensor, a static property of molecules, is often treated as a scalar quantity assuming isotropic polarizability, that is, the polarizability is taken to be the average of the principal diagonal of the polarizability tensor. This assumption is not unreasonable as long as anisotropic effects are not large. The polarizability of a molecule can, if necessary, also be obtained from a quantum chemical calculation for a single molecule.

The combination of a chosen pair potential and the polarization term provides a better interaction energy model, and one that incorporates multibody/nonadditive effects that generally lower the total energy of the system (that is, it makes the energy more attractive). However, adding the polarizable contribution comes at the expense of a substantial increase in computation time. Simulation times increase by a factor of 3 to 10 times using an interaction model with polarization over that when using only a pairwise additive potential. Most simulation studies using a polarizable nonadditivity term have been of liquid water, though acetonitrile (Hloucha and Deiters, 1997), trifluoromethane (Hloucha and Deiters, 1998), chloroform (Chang et al., 1997), carbon tetrachloride (Soetens et al., 1999; Veldhuizen and de Leeuw, 1996; Chang et al., 1995), and ethanol (González et al., 1999) have also been studied by simulation.

Hloucha and Deiters (1997) performed MC simulations of the effect of isotropic and anisotropic polarizability on acetonitrile modeled simply as fused hard spheres. Their findings suggest that anisotropic polarizabilities have no significant effect on the dielectric constant of that system, and the isotropic models resulted in a higher total energy than with anisotropic polarizability. Hloucha and Deiters (1998) also used simulation to study an anisotropic polarization model for liquid trifluoromethane described by a five-center Lennard–Jones model and found reasonable predictions for the thermodynamic, dieletric, and structural properties at temperatures between the melting and the triple point but large deviations at lower and higher temperatures.

Chang et al. (1995), using MD simulation, optimized potential parameters in an interaction model including polarization to predict thermodynamic and structural properties of the liquid and the liquid/vapor interface of carbon tetrachloride. Their simulation results reproduced the experimental data for pure carbon tetrachloride reasonably well and showed that the polarization energy was relatively low for the liquid/vapor interface, whereas for solvation of an ion by carbon tetrachloride, the total energy was due mainly to polarization. In a similar study, Chang et al. (1997) studied chloroform and, again by optimizing potential parameters in a polarizable model, obtained good agreement for thermodynamic and structural properties of chloroform. Carbon tetrachloride was also studied by Soetens et al. (1999) using a polarizable model with potential parameters determined from selected *ab initio* energy calculations for a pair of molecules. The results of their MD simulations of liquid carbon tetrachloride were found to be in good agreement with measured thermodynamic and dynamical properties.

Veldhuizen and de Leeuw (1996) used the OPLS parameters for methanol and both a nonpolarizable and a polarizable model for carbon tetrachloride for MD simulations over a wide range of compositions. The polarization contribution was found to be very important for the proper description of mixture properties, such as the heat of mixing. A recent study by González et al. (1999) of ethanol with MD simulations using the OPLS potential concluded that a nonpolarizable model for ethanol is sufficient to describe most static and dynamic properties of liquid ethanol. They also suggested that polarizabilities be introduced as atomic properties instead of the commonly approach of using a single molecular polarizability.

The most widely used models for water are based on effective potential formulations, such as SPC (Berendsen *et al.*, 1981), SPC/E (Berendsen *et al.*, 1987), TIP4P (Jorgensen *et al.*, 1983), and others for which parameters were optimized to experimental data. A polarization contribution has been added to these models to overcome some of their shortcomings. In some cases, polarization has been shown to improve some of the predicted properties of water (Sprik and Klein, 1988; Ahlström *et al.*, 1989; Watanabe and Klein, 1989; Bernardo *et al.*, 1994; Kiyohara *et al.*, 1998; Svishchev and Hayward, 1999). Others have added polarization to water by readjusting the potential parameters while still maintaining the framework of the SPC/E and TIP4P model (Chialvo and Cummings, 1996, 1998; Dang and Chang, 1997; Yoshii *et al.*, 1998). However, most of these approaches were of only limited success in improving properties predictions, as they usually were not able to describe simultaneously the thermodynamic, structural, and dielectric properties. A more rigorous approach has been to add polarization into pair potentials for water derived from *ab initio* interaction energies. In several studies, this led to greater success in reproducing the thermodynamic and/or structural properties (Stillinger and David, 1978; Detrich *et al.*, 1984; Niesar *et al.*, 1989, 1990; Kuwajima and Warshel, 1990; Millot *et al.*, 1998). Most of the studies were done using MD methods, and only a few by MC techniques (Detrich *et al.*, 1984; Kiyohara *et al.*, 1998).

As seen from the two examples here, pair potentials derived from *ab initio* interaction energies may not be sufficient for the prediction of dense fluid phase properties. This is particularly true for molecules with strong electrostatic interactions, where nonadditive forces are an important component of the total interaction energy. The methods for obtaining accurate fluid-phase properties from *ab initio* derived potentials are still under development, but as is evident in the case of acetonitrile, by knowing only the arrangement of the constituent atoms, one can make reasonable prediction of its phase behavior. One can foresee that in the not too distant future, such computational chemistry methods will be used routinely by chemical engineers. The progression will probably be, first, separate calculations of potentials that will be used in simulation and, then, as computer resources improve, complete *ab initio* simulations.

III. Infinite Dilution Activity Coefficients and Partition Coefficients from Quantum Mechanical Continuum Solvation Models

Solvation models based on statistical mechanics, as presented by Ben-Naim (1987), provide a different route to thermodynamic properties predictions in that the focus is on the solvation of one molecule by others in the system. The solvation process is defined as transferring a solute molecule from a fixed position in an ideal gas to a fixed position in a solution; the corresponding free energy change is referred to as the solvation free energy, $\Delta G^{*\text{sol}}$. Thermodynamic properties such as the activity coefficient γ and vapor pressure P^{vap} can be determined from this free energy as follows:

$$\ln \gamma_{1/2} = \frac{\Delta G^{*\text{sol}}_{1/2} - \Delta G^{*\text{sol}}_{1/1}}{RT} + \ln \frac{c_1}{x_1 c_1^0}, \tag{9}$$

and

$$\ln P_1^{\text{vap}} = \frac{\Delta G^{*\text{sol}}_{1/1}}{RT} + \ln c_1^0 \, RT, \tag{10}$$

where c_i is the molar concentration of component i, x_i is the mole fraction of i, superscript 0 indicates a pure liquid, and subscript i/j denotes the property of molecule i in solution j. Other properties such as the infinite dilution activity coefficient, Henry's law constant, the solubility, and partition coefficients can be determined from these properties.

The solvation free energy can be computed using molecular simulation methods. Molecular simulations for solvation free energy, sometimes called computational alchemy (Straatsma and McCammon, 1992; Levy and Gallicchio, 1998), require a multistage growth of the solute from a point or transformation from a solvent molecule, and at each stage a simulation must be performed. These methods provide the enthalpy, entropy, and free energy of solvation, the intramolecular structure, and, in MD simulations, also the dynamic properties of the system. However, most simulations neglect solute polarization, which is known to be important, as its inclusion is computationally intensive. Moreover, the use of empirical force fields in simulation and the limited system size reduce the reliability of the predicted solvation free energy, especially in the limit of infinite dilution.

An alternative approach to calculate the solvation free energy is to decompose the solvation process into two steps. First, the charges on the solute are turned off and the remaining hard particle is inserted into the solvent; this is equivalent to creating a cavity in the solvent to accommodate the solute. Then, after the solute is embedded in the solvent, the charges are turned on and the electronic configuration of the solute is restored. This decomposition separates the repulsive (cavity formation) and attractive (charging) interactions between the solute and the solvent molecules, allowing the calculation of each separately. More importantly, in contrast to explicit solvent models, the free energy change in the charging step can be computed efficiently by approximating the solvent as a homogeneous dielectric continuum. Using continuum solvation methods, sometimes called implicit solvent models, solvation effects such as the shift of electronic and vibrational spectra, structure change, and reaction equilibria of the solute can be calculated using a high level of quantum mechanics. The computational load is similar to that for an isolated molecule calculation (Cramer and Truhlar, 1999). However, the repulsive and attractive contributions are of the opposite sign, large and usually of the same order of magnitude, resulting in a low net free energy. Therefore, each stage of the calculation must be done carefully, so that an accurate solvation free energy is obtained. Here we demonstrate the use of this calculational method to obtain activity coefficients and partition coefficients at infinite dilution.

Infinite dilution activity coefficients, denoted γ^∞, are of great use in chemical engineering practice. For example, the two infinite dilution activity coefficients for a binary mixture can be used to fit the two adjustable parameters in an activity coefficient model and then make predictions of the vapor–liquid equilibria over the entire concentration range. Also, the reciprocal of γ^∞ provides a good estimate of the mole fraction solubility of an essentially immiscible solute in a liquid solvent (Sandler, 1996). At low and moderate pressures the product of γ^∞ and the vapor pressure of a pure solute gives Henry's law constant, which is a key parameter in industrial scrubbing processes. Moreover, from the knowledge of γ^∞ of chemical pollutant in water, one can estimate how it will partition in the environment (Sandler, 1996).

Starting from Eq. (9), the infinite dilution activity coefficient of component 1 in component 2 is

$$\ln \gamma_{1/2}^\infty = \frac{\Delta G_{1/2}^{*\,\text{chg}} - \Delta G_{1/1}^{*\,\text{chg}}}{RT} + \frac{\Delta G_{1/2}^{*\,\text{cav}} - \Delta G_{1/1}^{*\,\text{cav}}}{RT} + \ln \frac{c_2^0}{c_1^0}, \qquad (11)$$

where $\Delta G_{i/j}^{*\,\text{chg}}$ and $\Delta G_{i/j}^{*\,\text{cav}}$ are the charging and cavity formation free energies of solute i in solvent j, respectively. The traditional way of determining the cavity formation free energy is either from the scaled particle theory (SPT) of Pierotti (1976) or its modification (Claverie, 1978), both of which approximate the solvent molecules as being spherical, or from a simple linear correlation based on the surface area of the solute. These methods are not adequate for an accurate prediction of γ^∞. For example, the study by Floris et al. (1997) shows that the free energy of forming a pentane-shaped cavity in water from SPT is about 2.8 kcal/mol higher than that from MC simulation, resulting in an overestimation of γ^∞ by more than 11,000%. This error can be eliminated by reducing the effective water radius by approximately 5% or by reducing the cavity radii for methyl and methylene groups by about 10%, which indicates the high accuracy needed for these artificially defined quantities. Another approach to obtain $\Delta G^{*\,\text{cav}}$ is by computing the excess combinatorial entropy. Sayegh and Vera (1980) compared various methods for excess entropy calculations and found that the Starverman–Guggenheim combinatorial term (Staverman,

1950; Guggenheim, 1952), which was derived based on the lattice theory, is both simple and accurate compared to more sophisticated methods. Therefore, we use the Starverman–Guggenheim combinatorial term for the cavity formation free energy, and Eq. (11) becomes

$$\ln \gamma_{1/2}^\infty = \ln \gamma_{1/2}^\infty(\text{SG}) + \frac{\Delta G_{1/2}^{*\text{chg}} - \Delta G_{1/1}^{*\text{chg}}}{RT} \quad (12)$$

with

$$\ln \gamma_{1/2}^\infty(\text{SG}) = \ln \frac{r_1}{r_2} + \frac{z}{2} q_1 \ln \frac{q_1 r_2}{q_2 r_1} + \frac{z}{2}\left(\frac{r_1}{r_2} q_2 - q_1\right) + \left(1 - \frac{r_1}{r_2}\right), \quad (13)$$

where r_i and q_i are the volume and surface area parameters for species i, and z is the coordination number, taken to be 10.

The charging free energy is determined from continuum solvation models, which have been a rapidly developing branch of computational quantum chemistry (Tomasi and Persico, 1994; Rivail and Rinaldi, 1995; Cramer and Truhlar, 1999). These methods, as mentioned earlier, treat the solvent to be a homogeneous continuum and assume a linear response of the solvent molecules to the electric field established by the solute molecule. The electrostatic interaction between the solute and the solvent is governed by the Poisson equation. Various assumptions can be made to simplify this equation or the boundary conditions used, which lead to different models for the electrostatic component of the charging free energy (Dillet et al., 1993; Klamt and Schüürmann, 1993; Marten et al., 1996; Cossi et al., 1996; Qiu et al., 1997; Hawkins et al., 1998). Recent work by Amovilli and Mennucci (1997) includes nonelectrostatic interactions, i.e., dispersion and repulsion, in the polarizable continuum model (PCM) of Tomasi (Tomasi and Persico, 1994), and no empirical or fitting parameters are needed. We have used the latter approach for calculating the charging free energy.

We have developed the following procedure for determining the parameters in Eq. (12) from the quantum chemistry package GAMESS (Schmidt et al., 1993).

1. The structure of the solute in vacuum was optimized (energy minimized) using the Hartree–Fock (HF) method with the DZPsp(df) basis set (Lin and Sandler, 1999a). The same geometry was used in subsequent solvation calculations without further optimization.
2. A PCM calculation was performed using the HF method at the DZPsp(df) level for the electrostatic and induction contributions, and the method of Amovilli and Mennucci (1997) for the dispersion and repulsion contributions. Using PCM, the solute molecule is embedded in the homogeneous dielectric solvent, with its shape described by fused spheres centered on the atomic nuclei. The radii of these spheres are

the van der Waals radii of the atoms multiplied by a group scale factor α that accounts for the inhomogeneity in the first solvation layer. (The values for α we have determined so far are listed in Table II.) This calculation requires the dielectric constant, ionization potential, refractive index, and density of the solvent.

3. The volume and surface area of a molecule for use in Eq. (13) were calculated with the optimized geometry from step 1 and the van der Waals radius for each atom. For convenience, these values were then normalized using the van der Waals volume and surface area of a standard segment,

$$r = \frac{\text{van der Waals volume}}{15.17 \text{ cm}^3/\text{mol}} \quad \text{and} \quad q = \frac{\text{van der Waals surface area}}{2.5 \times 10^9 \text{ cm}^2/\text{mol}}.$$

[For water, we used $r_W = 0.92$ and $q_W = 1.40$ as in the UNIQUAC model (Abrams and Prausnitz, 1975).]

The use of scale factor α in the charging free energy calculations deserves some discussion. In most continuum models, such as PCM, a scale factor is used to account for the fact that the properties of the first solvation are different from the solvent properties in the bulk. This parameter also compensates for the failure of a simple linear response at the solute–solvent boundary (Klamt et al., 1998). In the use of PCM by others the same scale factor has been assigned to all atoms in a molecule. However, this leads to inaccurate predictions of γ^∞ for most substances. The scale factor α is used to characterize the dielectric behavior of the solvent in the vicinity of the solute, and one does not expect, for example, the distribution, orientation, and polarization of the water molecules around a methylene group in ethanol to be the same as those around the hydroxyl group. Therefore, we introduced a unique scale factor for each functional group and solvent, and a group contribution calculation for $\Delta G_{i/j}^{*\text{chg}}$, and developed the group contribution solvation (GCS) model (Lin and Sandler, 1999b). A comparison of the prediction of 29 values of γ^∞ from the GCS, UNIFAC (Fredenslund et al., 1975), and modified UNIFAC (Gmehling et al., 1993) models is given in Table III. The absolute average percentage deviation (AA%D) for 25 experimental data points is 7% using the GCS model, which is a significantly smaller error than that from the UNIFAC (39%) and modified UNIFAC (25%).

Since the mole fraction-based partition coefficient of a dilute species between two liquid phases is the inverse of the ratio of its infinite dilution activity coefficients in each phase, the GCS model can also be used for such predictions. Among the many possible partition coefficients, the octanol–water partition coefficient, K_{OW}, is of special interest. For example, K_{OW} is a key parameter in determining the fate of a persistent organic pollutant

TABLE II
Optimized Values of the Group Scale Factor α for Functional Groups in Different Solvents[a]

Functional groups	Solvent										
	CH$_3$OH	C$_2$H$_5$OH	C$_4$H$_9$OH	C$_8$H$_{17}$OH	CH$_3$CH	C$_5$H$_{12}$	C$_6$H$_{14}$	C$_7$H$_{16}$	H$_2$O	DMSO[b]
CH$_n$ ($n=0,1,2,3$)	1.12	1.13	1.15	1.14	1.24	1.09	1.10	1.103	1.42	1.196
Cyclic CH$_2$				1.14	1.21		1.10		1.35	
ACH$_n$ ($n=0,1$)[c]				1.22	1.24		1.10		1.33	
H$_2$O					1.19		1.10		1.11	
OH	1.07	1.106		1.15	1.15		1.10	1.10	0.88	
CN				1.38	1.23		1.10		1.10	
NO$_2$				1.33			1.10		1.12	
C=O				1.32			1.10		0.88	
OC=O				1.31			1.10		0.85	
Cl(C$_n$H$_{2n+1}$Cl)				1.28	1.18		1.10		1.17	
Cl(C$_n$H$_n$CHCl$_2$)				1.24	1.18		1.10		1.20	
Cl(C$_n$H$_n$CCl$_3$)				1.20	1.18		1.10		1.23	
Cl(CCl$_4$)				1.16	1.18		1.10		1.26	

[a]Examples. C$_2$H$_5$CN in water: $\alpha = 1.42$ for each atom in group CH$_3$ and CH$_2$; $\alpha = 1.10$ for each atom in the CN group. CH$_3$CHCl$_2$ in water: $\alpha = 1.42$ for all C and H atoms; $\alpha = 1.20$ for the two Cl atoms.
[b]Dimethyl sulfoxide.
[c]ACH$_n$ represents the CH$_n$ group in an aromatic ring.

TABLE III
Comparison of Predicted Infinite Dilution Activity Coefficients from the GCS, UNIFAC, and Modified UNIFAC at 298.15 K

Solute/solvent	GCS $\gamma_{1/2}^\infty$	%D[a]	UNIFAC $\gamma_{1/2}^\infty$	%D	Modified UNIFAC $\gamma_{1/2}^\infty$	%D	Experiment $\gamma_{1/2}^\infty$
CH_3CN/H_2O	1.7×10^1	6	1.32×10^1	19	1.19×10^1	7	1.11×10^1
CH_3OH/H_2O	1.31×10^0	−10	2.24×10^0	54	1.97×10^0	35	1.46×10^0
C_2H_5OH/H_2O	3.52×10^0	−8	7.62×10^0	99	4.79×10^0	25	3.83×10^0
C_4H_9OH/H_2O	4.88×10^1	−9	5.40×10^1	1	4.19×10^1	−21	5.33×10^1
$C_8H_{17}OH/H_2O$	1.01×10^4	−13	3.16×10^3	−73	4.08×10^3	−65	1.16×10^4
C_5H_{12}/H_2O	8.54×10^4	−15	3.21×10^3	−97	1.92×10^3	−98	1.00×10^5
C_6H_{14}/H_2O	3.97×10^5	−1	1.06×10^4	−97	6.62×10^3	−98	4.00×10^5
H_2O/CH_3CN	1.23×10^1	9	8.40×10^0	−26	8.46×10^0	−25	1.13×10^1
$C_8H_{17}OH/CH_2CN$	1.93×10^0		5.88×10^0		8.40×10^0		
C_5H_{12}/CH_3CN	1.84×10^1	11	1.72×10^1	4	2.05×10^1	24	1.66×10^1
C_6H_{14}/CH_3CN	2.50×10^1	−2	2.41×10^1	−6	2.60×10^1	2	2.55×10^1
C_5H_{12}/CH_3OH	2.07×10^1	6	1.44×10^1	−26	1.77×10^1	−9	1.95×10^1
C_5H_{12}/C_2H_5OH	8.83×10^0	16	6.06×10^0	−21	8.55×10^0	12	7.64×10^0
C_6H_{14}/C_2H_5OH	1.01×10^1		7.75×10^0		1.03×10^1		
C_5H_{12}/C_4H_9OH	4.32×10^0	2	3.16×10^0	−26	4.24×10^0	0	4.25×10^0
$CH_3CN/C_8H_{17}OH$	8.38×10^0	12	2.79×10^0	−63	5.97×10^0	−20	7.48×10^0
$C_5H_{12}/C_8H_{17}OH$	2.15×10^0	−2	1.65×10^0	−25	2.32×10^0	6	2.19×10^0
$C_6H_{14}/C_8H_{17}OH$	3.01×10^0	7	1.95×10^0	−31	2.54×10^0	−10	2.81×10^0
C_2H_5OH/C_5H_{12}	9.71×10^1		2.89×10^1		6.00×10^1		
H_2O/C_6H_{14}	1.63×10^3	−1	1.40×10^3	−15	1.36×10^2	−92	1.65×10^3
CH_3CN/C_6H_{14}	2.99×10^1	8	2.76×10^1	0	3.43×10^1	24	2.76×10^1
C_2H_5OH/C_6H_{14}	5.98×10^1	1	2.77×10^1	−53	5.32×10^1	−10	5.94×10^1
C_4H_9OH/C_6H_{14}	3.55×10^1	−2	1.84×10^1	−49	3.96×10^1	10	3.61×10^1
$C_8H_{17}OH/C_6H_{14}$	1.37×10^1		9.40×10^0		2.65×10^1		
C_5H_{12}/C_6H_{14}	9.74×10^{-1}	4	9.89×10^{-1}	5	9.99×10^{-1}	6	9.40×10^{-1}
CH_3OH/C_7H_{16}	9.82×10^1	28	1.92×10^1	−75	7.46×10^1	−3	7.70×10^1
C_2H_5OH/C_7H_{16}	4.90×10^1	−1	2.64×10^1	−46	4.85×10^1	−2	4.92×10^1
C_5H_{12}/C_7H_{16}	9.15×10^{-1}	−1	9.63×10^{-1}	5	9.95×10^{-1}	8	9.20×10^{-1}
$C_5H_{12}/DMSO$	6.22×10^1	1	2.61×10^1	−57	5.35×10^1	−13	6.13×10^1
AA%D[b]		7		39		25	

[a] Percentage deviation $= [(\gamma_{predicted}^\infty - \gamma_{experiment}^\infty)/\gamma_{experiment}^\infty] \times 100$.
[b] Absolute average percentage deviation.

(POP) in the environment, that is, how the chemical in water will be taken up by animal and vegetable life and by soils and sediments. This coefficient is also used as an indicator of chemical hydrophobicity in most quantitative structure–activity relationship (QSAR) models for designing new drugs (Franke, 1984). Experimental values of the octanol–water partition coefficient for the hydrophobic chemicals common in the chemical industry vary over eight orders of magnitude, and because of the very low concentrations of hydrophobic chemicals in water, these are very difficult to measure accurately.

An empirical correlation (Lin and Sandler, 1999b) relates the octanol–water partition coefficient of solute i (which is the partitioning of the solute between water and water-saturated octanol) and the infinite dilution activity coefficients of the solute in water and in pure (not water saturated) octanol,

$$\log K_{OW,i} = -0.68 + 0.91 \cdot \log \frac{\gamma_{i/W}^\infty}{\gamma_{i/O}^\infty}. \qquad (14)$$

Using the GCS model for the infinite dilution activity coefficients, we have

$$\log K_{OW,i} = -0.68 + 0.395 \left[\ln \frac{\gamma_{i/W}^\infty(SG)}{\gamma_{i/O}^\infty(SG)} + \frac{\Delta G_{i/W}^{*\,chg} - \Delta G_{i/O}^{*\,chg}}{RT} \right]. \qquad (15)$$

Predictions of the octanol–water partition coefficients for 28 linear and 12 nonlinear solutes from the above equation are given by Lin and Sandler (1999b). The predictions are in good agreement with experimental data, with the overall root mean square (RMS) deviation on $\log K_{OW}$ being only 0.15 (or an average deviation of 41% in K_{OW}).

An insight obtained from the PCM calculations was that for monofunctional solutes (that is, molecules with only one nonalkyl group), all parameters in the GCS model (r, q, and ΔG^{*chg}) could be determined in a group contribution manner. Based on this observation, Lin and Sandler (1999b) generalized the GCS model to multiple components and derived the following exact expression for the octanol–water partition coefficient, the GCSKOW model (Lin and Sandler, 1999b):

$$\log K_{OW,i} = \frac{1}{2.303} \left[\ln \frac{c_{OR}}{c_W} + \ln \frac{\gamma_{i/W}^\infty(SG)}{\gamma_{i/OR}^\infty(SG)} + \frac{\Delta G_{i/W}^{*\,chg} - \Delta G_{i/OR}^{*\,chg}}{RT} \right], \qquad (16)$$

where the subscript OR represents the octanol-rich (water-saturated octanol) phase. Under ambient conditions, this equation simplifies to

$$\log K_{OW,i} = -0.126 + 1.031 r_i - 1.208 q_i + \frac{\Delta G_{i/W}^{*\,chg} - \Delta G_{i/OR}^{*\,chg}}{1.364}. \qquad (17)$$

In the GCSKOW model, the molecular structure parameters, r_i and q_i, and the free energy parameter, $\Delta G_{i/W}^{*\,chg} - \Delta G_{i/OR}^{*\,chg}$, are determined using the group contribution method, that is, $r_i = \sum_{k=1}^{N_i} R_k$, $q_i = \sum_{k=1}^{N_i} Q_k$, and $\Delta G_{i/W}^{*\,chg} - \Delta G_{i/OR}^{*\,chg} = \sum_{k=1}^{N_i} \Delta \Delta G_{k/W\text{-}OR}^{*\,chg}$, where the summation is over the N_i functional groups contained in species i. Also, R_k, Q_k, and $\Delta \Delta G_{k/W\text{-}OR}^{*\,chg}$ are the volume, surface area, and charging free energy contributions of functional group k (these parameters are tabulated by Lin and Sandler, 1999b). This model is very accurate for monofunctional molecules. The RMS deviation on $\log K_{OW}$ for 226 monofunctional compounds from the GCSKOW model is 0.14 (38% in K_{OW}), compared to 0.18 (51%) from the ClogP model (Hansch and Leo, 1995), 0.21 (62%) from the KOW–UNIFAC model (Wienke and Gmehling, 1998), and 0.23 (71%) from LSER (Kamlet et al., 1988).

An application of continuum solvation calculations that has not been extensively studied is the effect of temperature. A straightforward way to determine the solvation free energy at different temperatures is to use the known temperature dependence of the solvent properties (dielectric constant, ionization potential, refractive index, and density of the solvent) and do an *ab initio* solvation calculation at each temperature. Elcock and McCammon (1997) studied the solvation of amino acids in water from 5 to 100°C and found that the scale factor α should increase with temperature to describe correctly the temperature dependence of the solvation free energy. Tawa and Pratt (1995) examined the equilibrium ionization of liquid water and drew similar conclusions. An alternative way to study temperature effect is through the enthalpy of solvation. The temperature dependence of γ^∞ is related to the partial molar excess enthalpy at infinite dilution, $\bar{H}^{ex,\infty}$

$$\left(\frac{\partial \ln \gamma_{1/2}^\infty}{\partial T}\right)_P = -\frac{\bar{H}_{1/2}^{ex,\infty}}{RT^2}. \tag{18}$$

For a limited temperature range, $\bar{H}^{ex,\infty}$ can be taken to be constant so that γ^∞ at different temperatures can be calculated from its value at 25°C by integrating Eq. (18). It has been shown (Ben-Naim, 1987) that the excess enthalpy can be determined from the energy of solvation ΔU^{*sol},

$$\frac{\bar{H}_{1/2}^{ex,\infty}}{RT^2} = \frac{(\Delta U_{1/2}^{*sol} - \Delta U_{1/1}^{*sol})}{RT^2} + \frac{P(\bar{V}_{1/2} - \bar{V}_{1/1})}{RT^2} - \frac{P(\kappa_{T,2} - \kappa_{T,1})}{T}$$
$$- (\alpha_{T,1} - \alpha_{T,2}), \tag{19}$$

where κ_T is the isothermal compressibility, and α_T is the coefficient of thermal expansion. For most compounds under ambient conditions, the first term on the RHS of Eq. (19) (contributions from energy of solvation) is about two orders of magnitude larger than the other terms. Therefore we can approximate $\bar{H}^{ex,\infty}$ from ΔU^{*sol}.

Table IV shows the prediction of the infinite dilution activity coefficient of ethanol in heptane at various temperatures using both methods discussed above, together with predictions from the UNIFAC and modified UNIFAC

TABLE IV
Infinite Dilution Activity Coefficient of Ethanol in Heptane at Different Temperatures

	GCS[a]		GCS($\bar{H}^{ex,\infty}$)[b]		UNIFAC		Modified UNIFAC		Experiment
T (K)	$\gamma_{1/2}^\infty$	%D	$\gamma_{1/2}^\infty$	%D	$\gamma_{1/2}^\infty$	%D	$\gamma_{1/2}^\infty$	%D	$\gamma_{1/2}^\infty$
293.15	52.6	3	59.6	17	28.2	−45	55.7	9	51.0
298.15	49.2	0	49.2	0	26.4	−46	48.5	−2	49.2
314.45	39.9	10	27.3	−25	21.6	−41	31.8	−13	36.3
323.15	36.0	108	20.4	18	19.5	13	25.7	49	17.3
333.15	32.2	101	14.9	−7	17.5	9	20.5	28	16.0
343.15	29.0	97	11.1	−24	15.8	8	16.4	12	14.7
353.15	26.2	97	8.4	−37	14.4	8	13.3	0	13.3
363.15	23.9	96	6.5	−47	13.2	8	10.9	−11	12.2
373.15	21.8	95	5.0	−55	12.1	8	9.0	−20	11.2
AA%D		64		22		22		15	

[a]Infinite dilution activity coefficient determined by varying solvent properties at each temperature.
[b]Infinite dilution activity coefficient determined from Eq. (18) with $\bar{H}^{ex,\infty}$ calculated at 298.15 K.

models. The modified UNIFAC model with its parameters fitted to experimental $\bar{H}^{ex}$ data gives the best predictions. Using the GCS calculation at each temperature underestimates the temperature dependency of γ^∞, whereas the use of Eq. (18) with $\bar{H}^{ex,\infty}$ calculated at 298.15 K tends to overestimate the temperature dependency. The UNIFAC model, while accurate at high temperatures, also underestimates the temperature effect on γ^∞. This is an indication of the limits of the current models.

The lack of knowledge of the scale factor and its temperature dependence remains a central problem for the broader application using PCM methods. In the GCS model, the scale factors are determined from fitting experimental γ^∞ for small monofunctional solutes. The use of these values in predicting infinite dilution activity coefficients for multifunctional compounds as listed in Table V shows that, even though being more accurate than the simple group contribution GCSKOW model, the continuum solvation calculations give large errors for compounds containing two strong nonalkyl groups in close proximity ($n = 1$, where n is the number of intervening methylene groups). This failure is because the interaction of one functional group with the solvent molecule is altered by the presence of another nonalkyl functional group on the same molecule. One way to account for this is to change the values of the scale factors α to compensate for the inhomogeniety and nonlinear response of the first solvation shell. Efforts have been made to determine the scale factor, or equivalent atomic

TABLE V
PREDICTION OF THE $X(CH_2)_nY$ TYPE OF SOLUTES FROM THE GCS MODEL USING EQ. (15) AND THE GCSKOW MODEL

			Equation (15)		GCSKOW		Experiment
X	Y	n	$\log K_{OW}^{calc}$	Dev.	$\log K_{OW}^{calc}$	Dev.	$\log K_{OW}^{expt}$
–CN	–CN	1	–1.90	–1.40	–2.04	–1.54	–0.50
		2	–1.13	–0.14	–1.51	–0.52	–0.99
		3	–0.87	–0.15	–0.98	–0.26	–0.72
–Cl	–Cl	1	0.82	–0.43	0.47	–0.78	1.25
		2	1.31	–0.17	1.00	–0.47	1.48
		3	1.89	–0.11	1.53	–0.47	2.00
–CN	–Cl	1	–0.40	–0.85	–0.79	–1.24	0.45
		2	0.24	0.06	–0.26	–0.44	0.18
		3	0.58	0.02	0.27	–0.29	0.56

radii, based on detailed molecular theories. Smith (1999) derived a set of atomic radii for amino acids based on the solute–solvent interaction potential. Nina *et al.* (1997) and Babu and Lim (1999) used MD simulation to find the relationship between the solvent radial distribution and the atomic radii for continuum solvation calculations. Goncalves and Livotto (1999) allowed the radii to change depending on the charges of the atoms, and Zhan and Chipman (1998) studied the use of solute electronic isodensity surface as the solvation cavity. The dielectric saturation and electrostriction resulting in a nonlinear response of the solvent molecules have also been addressed recently (Aqvist and Hansson, 1996; Bader *et al.*, 1997; Hyun and Ichiye, 1998). We discuss another solution to this general problem of proximity effects in the next section.

IV. Use of Computational Quantum Mechanics to Improve Thermodynamic Property Predictions from Group Contribution Methods

Group contribution methods, such as the one discussed above, have been reasonably successful for estimating many physical and thermodynamic properties of pure substances and mixtures, especially when each molecule contains no more than one nonalkyl functional group. These methods dissect a molecule into functional groups that are assumed to be independent of each other. That is, a functional group is assumed to behave the same in its interactions with other functional groups independent of the molecule of which it is a part. The properties of the system are then obtained by summing the contributions from all the groups. This deconstruction of molecules greatly reduces the number of parameters needed to describe the properties of systems containing these functional groups. Once the group parameters for the property of interest have been determined from available experimental data, they can be used to predict the properties of new, more complex systems. Such methods, when accurate, not only provide a simple and systematic method for approximating the properties of new systems but, in reverse engineering, can also be used to design new compounds with desired properties.

However, group contribution methods have three important shortcomings. First, the definition of groups is empirical and sometimes not intuitive, with different methods using different groups to represent the same molecule. This raises the question of finding a "best" set of functional groups to describe the system properties. The second problem is that these methods are incapable of distinguishing between isomers. For example, the predicted values of K_{OW} for the isomers 2,2-dimethyl propanol and 2-methyl-2-butanol are identical from these methods, while the measured values differ by a factor of 2. We refer to this type of deficiency as the structure effect. Finally, and most important, is that group contribution methods fail when a molecule contains two or more strong functional groups in close proximity. This is because the interaction of a strong functional group with the others, which is a result of electrostatic forces, is affected by the presence of other neighboring strong functional groups; thus the underlying assumption that each group is independent of others is no longer valid. This failure of group contribution methods is referred to as the (intramolecular) proximity effect.

Most group contribution models correct for the structure and proximity effects by adjusting values of the group parameters depending on the nearby groups, by constructing new groups, or by a combination of these. For example, Kehiaian (1983), in his DISQUAC model, accounted for proximity effects by empirically varying the values of interaction parameters for a group depending on its neighboring groups (Kehiaian and Marongiu, 1985). Wu and Sandler (1991a,b) used quantum mechanical calculations to determine the charges on the atoms of a molecule and suggested that a better definition for groups was that each be approximately electroneutral. This method resolved the proximity effects by forming larger groups that contained the interacting functional groups. Gani and co-workers (Constantinou and Gani, 1994; Abildskov et al., 1996, 1999) proposed the inclusion of second-order groups in addition to the common functional groups, which they referred to as first-order groups. Second-order groups served as an empirical correction for structure and proximity effects, and interestingly, many of the second-order groups were found to be identical to those defined by Wu and Sandler (1991a).

Although providing improved accuracy, Kehiaian's approach introduces a large number of empirical correction factors that can be obtained only from fitting to very precise experimental data [in this method there are potentially $(N(N+1)/2) \times m$ additional parameters for a set of N groups when up to the mth nearest neighbors are considered]. Consequently, DISQUAC parameters are available for only a very limited number of functional groups. The Wu and Sandler approach provides a basis for choosing groups, however, this method requires the creation of new groups whenever a new proximity effect is encountered.

The failure of simple group contribution methods results from the distortion of the electron cloud of a functional group when a strongly electronegative or electropositive group is located on the same molecule. *Ab initio* molecular orbital calculations can quantify such distortions of the electron distribution. For example, the standard Mulliken (1955) population analysis can be used to determine the charge on each atom in a functional group and the multipole moments for each group in a molecule. As the charge distribution determines the electrostatic interactions between the functional groups, and varies depending on the molecule, this information can be used to account for structure and proximity effects.

As an example of how this may be used, we return to the group contribution model for the octanol–water partition coefficient discussed above. As already shown, this model was quite good for monofunctional (i.e., only one nonalkyl group) solutes; when applied to multiple functional solutes, the large deviations shown in Fig. 4a were found. The failure of the GCSKOW model is the result of strong proximity effects in multifunctional compounds.

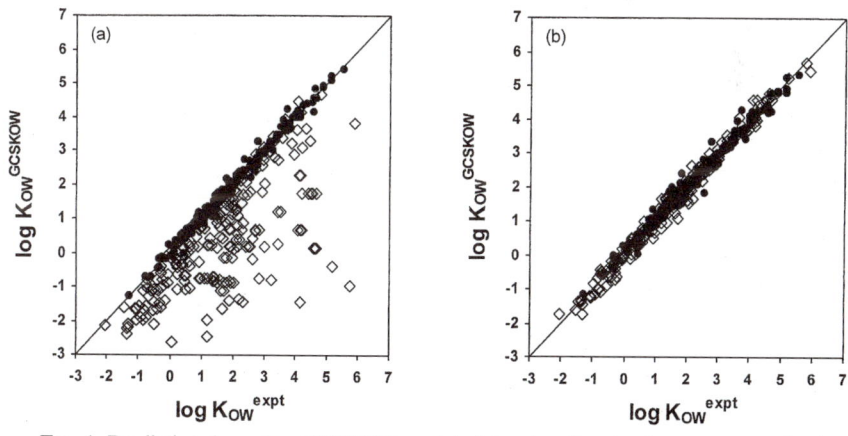

FIG. 4. Prediction from the GCSKOW model: (a) using the simple group contribution method; (b) using the multipole correction method. In these figures, the filled circles are monofunctional solutes and the open diamonds are multifunctional solutes.

However, we can account for such proximity effects by correcting the electrostatic contribution to the charging free energy.

The charging free energy, originating from the attractive interaction between solute i and solvent j, has three contributions: electrostatic, dispersion, and repulsion (Tomasi and Persico, 1994),

$$\Delta G_{i/j}^{*\,\mathrm{chg}} = \Delta G_{i/j}^{*\,\mathrm{es}} + \Delta G_{i/j}^{*\,\mathrm{disp}} + \Delta G_{i/j}^{*\,\mathrm{rep}}. \tag{20}$$

The nonelectrostatic contributions are usually assumed to be dependent on the molecular surface area (Tannor et al., 1994; Qiu et al., 1997; Klamt et al., 1998; Cramer and Truhlar, 1999) and do not vary with different solute electronic configurations. The electrostatic contribution, however, is very sensitive to the charge distribution in a molecule. Kirkwood (1934) derived a general equation for the electrostatic interaction between a distribution of discrete point charges in a spherical cavity and in an isotropic dielectric medium. The result was expressed in terms of the multipole expansion at the center of the cavity (see also Rivail and Rinaldi, 1976). Truncating this expression at the second-order term and assuming that the charge distribution of a functional group of the solute is independent of the solvent, the electrostatic contribution of functional group k in solvent j is

$$\Delta G_{k/j}^{*\,\mathrm{es}} = -\frac{\varepsilon_j - 1}{2\varepsilon_j} \frac{e_k^2}{a_{k/j}} - \frac{\varepsilon_j - 1}{1 + 2\varepsilon_j} \frac{\mu_k^2}{a_{k/j}^3}, \tag{21}$$

where e_k is the net charge of group k, μ_k is the dipole moment evaluated at the center of the cavity, ε_j is the dielectric constant of solvent j, and $a_{k/j}$ is the effective cavity radius of group k in solvent j. Therefore the charging free energy contribution of functional group k becomes

$$\Delta\Delta G_{k/\mathrm{W\text{-}OR}}^{*\,\mathrm{chg}} = \Delta\Delta G_{k/\mathrm{W\text{-}OR}}^{*\,\mathrm{non-es}} + C_k^e e_k^2 + C_k^\mu \mu_k^2, \tag{22}$$

where $\Delta\Delta G_{k/\mathrm{W\text{-}OR}}^{*\,\mathrm{non-es}} = \Delta\Delta G_{k/\mathrm{W\text{-}OR}}^{*\,\mathrm{disp}} + \Delta\Delta G_{k/\mathrm{W\text{-}OR}}^{*\,\mathrm{rep}}$ and

$$C_k^e = -\frac{\varepsilon_\mathrm{W} - 1}{2\varepsilon_\mathrm{W} a_{k/\mathrm{W}}} + \frac{\varepsilon_\mathrm{OR} - 1}{2\varepsilon_\mathrm{OR} a_{k/\mathrm{OR}}}$$

$$C_k^\mu = -\frac{\varepsilon_\mathrm{W} - 1}{(1 + 2\varepsilon_\mathrm{W}) a_{k/\mathrm{W}}^3} + \frac{\varepsilon_\mathrm{OR} - 1}{(1 + 2\varepsilon_\mathrm{OR}) a_{k/\mathrm{OR}}^3}. \tag{23}$$

The use of the GCSKOW model with multipole corrections requires the charge, dipole moment, and coefficients C_k^e and C_k^μ for each group in the molecule. The volume, surface area, and free energy parameter $\Delta\Delta G_{k/\mathrm{W\text{-}OR}}^{*\,\mathrm{non-es}}$ are obtained from the summation of group contributions. The *Gaussian* program (Frisch et al., 1995) was used to compute the group charges and dipoles of each isolated solute molecule. Geometry optimization in vacuum was performed using the HF method with a 6-31G(d,p) basis set, followed by a Mulliken population analysis to obtain the charge on each atom. The net group charge e_k and dipole $\bar{\mu}_k$ were then calculated from $e_k = \sum_{m=1}^{N_k} e_m$ and $\bar{\mu}_k = \sum_{m=1}^{N_k} e_m \vec{r}_m$, where the summation is over the N_k atoms contained in group k; e_m is the Mulliken charge of atom m; $\vec{r}_m$ is the position vector of atom m originating from the center of gravity of group k. The group R_k and Q_k were computed from the van der Waals radius of each atom. Finally, the free energy parameter $\Delta\Delta G^{*\,\text{non-es}}_{k/\text{W-OR}}$ and multipole coefficients C_k^e and C_k^μ for each group were obtained by minimization of the RMS deviation of $\log K_{\text{OW}}$, and the results are given by Lin and Sandler (2000).

Figure 4b shows the predictions from the GCSKOW model with multipole corrections for a total of 450 compounds, including 246 multifunctional molecules. Compared with Fig. 4a, it is seen that the quantum mechanically based multipole correction method substantially reduces the error in prediction for multifunctional compounds. The effectiveness of such corrections can be seen by comparing some example predictions using both methods presented in Table VI. With multipole corrections, not only are the

TABLE VI
EXAMPLES OF PREDICTIONS FROM THE GCSKOW MODEL WITH AND WITHOUT MULTIPOLE CORRECTIONS

	GCSKOW		
Solute	Without multipole corrections	With multipole corrections	Experiment
Monofunctional compounds			
Pentanol	1.44	1.47	1.56
Hexanol	1.98	2.01	2.03
Heptanol	2.52	2.55	2.62
Monofunctional isomers (structure effects)			
2,2-Dimethylpropanol	1.03	1.25	1.36
2-Methyl-2-butanol	1.03	0.94	0.89
Multifunctional compounds (proximity effects)			
CN(CH$_2$)$_0$CN	−2.66	0.07	0.07
CN(CH$_2$)$_1$CN	−2.12	−0.51	−0.50
CN(CH$_2$)$_2$CN	−1.57	−0.85	−0.99
CN(CH$_2$)$_3$CN	−1.03	−0.53	−0.72
CN(CH$_2$)$_4$CN	−0.49	−0.21	−0.32
Pharmaceutical compounds			
Nicotine	1.45	0.98	1.17
Theophylline	−10.02	−0.02	−0.02
Caffeine	−9.54	−0.07	−0.07
Piracetam	−4.77	−1.60	−1.54
Mexiletine	0.45	2.06	2.15

predictions for monofunctional compounds improved, but also the model now correctly accounts for structure effects in isomers and proximity effects in multifunctional solutes.

The influence of proximity effects, which are the major cause of the failure of most simple group contribution models, can be illustrated using the $CN(CH_2)_nCN$ series. The octanol–water partition coefficients for this series deviate from linear incremental behavior with the number of CH_2 groups and exhibit an unexpected minimum as a function of the number of the methylene groups separating the two CN groups. This behavior can be understood by studying the variation of the free energy parameter ($\sum_{k=1}^{N_i} \Delta G^{*\,\text{non-es}}_{k/\text{W-OR}}$), the charge correction ($\sum_{k=1}^{N_i} C_k^e e_k^2$), and the dipole correction ($\sum_{k=1}^{N_i} C_k^\mu \mu_k^2$) contributions shown in Fig. 5. The contribution from the free energy term increases with the number of methylene groups contained in the molecule, resulting in a linear increase in $\log K_{\text{OW}}$. However, the variation of the other correction terms with n is not linear since the charge and dipole moment of the CN and CH_2 groups vary with the number of methylene groups as shown in Fig. 6, resulting in a nonlinear behavior of $\log K_{\text{OW}}$. Consequently, the proximity effect appears as a nonlinear contribution to K_{OW} resulting from the electrostatic interactions between groups.

A more severe test of group contribution methods is in their application to large compounds that, because of their molecular complexity, make

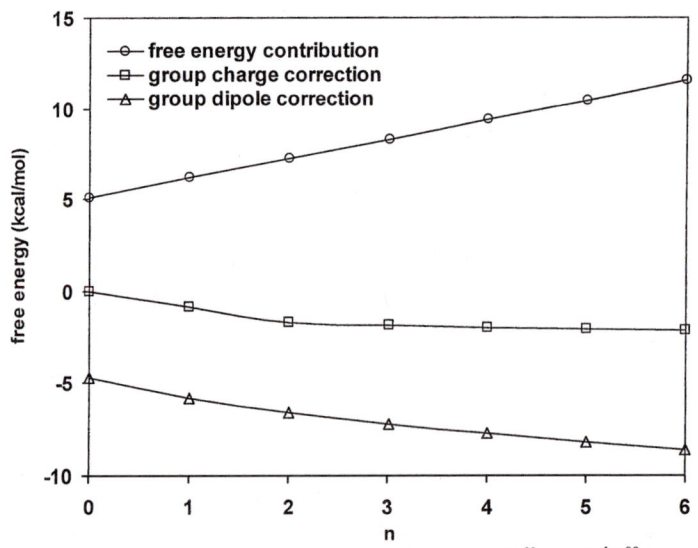

FIG. 5. The variation of the free energy contribution ($\sum_{k=1}^{N_i} \Delta G^{*\,\text{chg},00}_{k/\text{W-OR}}$), the group charge contribution ($\sum_{k=1}^{N_i} C_k^e e_k^2$), and the group dipole contribution ($\sum_{k=1}^{N_i} C_k^\mu \mu_k^2$) to K_{OW} in the $CN(CH_2)_nCN$ series.

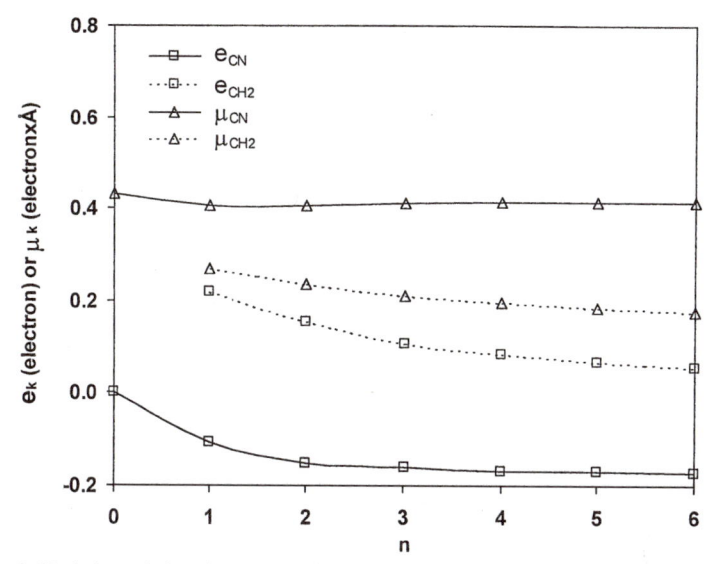

FIG. 6. Variation of the charge and dipole moment on the CN and CH$_2$ groups in the CH(CH$_2$)$_n$CN series.

accurate predictions using either traditional group contribution methods or quantum chemical calculations difficult. Table VI shows the good predictions of logK_{OW} for five small to medium-sized pharmaceutical-like compounds using the GCSKOW model. Also, we have compared the predictions of this method with other group contribution methods for 450 compounds. The GCSKOW model, with a RMS deviation of 0.18 in logK_{OW} or 52% deviation in K_{OW}, is superior to the KOW–UNIFAC model (0.28 or 92%) and is comparable to the more empirical, multiparameter ClogP model (0.18 or 52%).

V. Use of *ab Initio* Energy Calculations for Phase Equilibrium Predictions

Activity coefficients, which play a central role in chemical thermodynamics, are usually obtained from the analysis of phase equilibrium measurements. However, with shifts in the chemical industry and the use of combinatorial chemistry, new chemicals are being introduced for which the needed phase equilibrium data may not be available. Therefore, predictive methods for estimating activity coefficients and phase behavior are needed. Group contribution methods, such as the ASOG [analytical solution of groups

(Kojima and Togichi, 1979)] model and UNIFAC [UNIQUAC functional-group activity coefficient (Fredenslund et al., 1977)] model discussed earlier, have been developed. However, these models are limited to the range of classes of compounds and conditions of the regressed experimental data used in their development.

In Section II we discussed how *ab initio* quantum chemical calculations to obtain interaction potential information could be used with computer simulation to predict thermodynamic properties. However, such calculations are very time-consuming and, as already mentioned, not always of a high accuracy. Monte Carlo or MD simulations were required in those calculations because the optimum (minimum energy) configuration of an isolated pair of molecules is generally not representative of the configurations of molecules in the dense fluid state. Usually a large region of phase space must be sampled to get accurate thermodynamic properties, which is the reason that statistical mechanical simulation methods are used. However, if the interactions are very strong, for example hydrogen-bonding interactions, then a small number of preferred interaction states among a collection of a few molecules may be representative of the dense fluid state.

The discussion that follows describes a procedure for using *ab initio* quantum chemical calculations for small clusters of strongly interacting molecules to make *a priori* predictions of activity coefficients without any adjustable parameters. The central idea is the use of quantum mechanics to determine the parameters in activity coefficient models (Sum and Sandler, 1999a,b). This is done by using *ab initio* methods to determine a minimum energy configuration of a cluster of molecules and then to compute the average interaction energies between sets of like and unlike pairs of molecules within that cluster. These energies are then used as the average interaction energy parameters that appear in activity coefficient models such as the UNIQUAC (Abrams and Prausnitz, 1975; Maurer and Prausnitz, 1978) and Wilson (1964) models. A similar approach has also been used by Jónsdóttir *et al.* (1994, 1998, 1999; Jónsdóttir and Klein, 1997; Jónsdóttir and Rasmussen, 1996, 1999), however, they used molecular mechanics and empirical force fields to determine these average interaction energies.

In usual applications, the parameters of the UNIQUAC and Wilson activity coefficient models are fitted to experimental phase equilibria data. However, in the development of these models, the adjustable parameters correspond to the difference of interaction energies between the like and the unlike species,

$$\Delta u_{ij} = E_{ij}^{\text{int}} - E_{jj}^{\text{int}} \tag{24a}$$

$$\Delta \lambda_{ij} = E_{ij}^{\text{int}} - E_{ii}^{\text{int}} \tag{24b}$$

in the UNIQUAC and Wilson models, respectively, where E_{ij}^{int} is the interaction energy between molecule i and molecule j. If these energies are indeed the interaction energies between the molecules, they should be obtainable from first principles calculations, for example, from the combination of quantum chemistry and simulation for weakly interacting systems, or, as we consider here, from only quantum chemistry calculations for strongly interacting systems. Also, the theoretical foundation of activity coefficient models such as UNIQUAC and Wilson have been questioned (Gierycz and Nakanishi, 1984; Gierycz *et al.*, 1984; Haile, 1986, Hamad, 1998; Heyes, 1991; K.-H. Lee *et al.*, 1986; L. L. Lee *et al.*, 1983, 1986; Netemeyer and Glandt, 1988). Several simulation studies have been performed with idealized models (e.g., hard spheres, square-well fluids) to study the local composition concept that underlies these models. However, the validity of these models remained unresolved, and although new thermodynamic models have been proposed over the 30 years since the introduction of the UNIQUAC and Wilson models, these older models are still the most widely used.

An outline of the quantum chemical procedure used to determine the interaction energies needed in Eq. (24) is as follows [see Sum and Sandler (1999a,b) for details of the calculations].

1. A cluster composed of eight molecules (four of each kind) was constructed and its optimum geometry found by minimizing the cluster energy using the semiempirical PM3 method (Stewart, 1989a,b).
2. A further cluster geometry optimization (energy minimization) was performed using the Hartree–Fock (HF) method with the 6–31G(d,p) basis set [HF/6-31G(d,p)].
3. Directly interacting molecular pairs (by hydrogen bond or in close proximity) were selected from this optimized cluster, and their separation distances and relative orientations recorded.
4. The interaction energy of each molecular pair was then computed using the supermolecular method [Eq. (1)] at the HF/6-311++G(3d,2p) level at the separation and orientation obtained in the previous step.
5. Finally, the energies of the sets of the same molecular pairs were linearly averaged to obtain the energy parameters to be used in the activity coefficient models.

It was found that for clusters smaller than eight molecules, the above procedure would have to be repeated for several initial configurations, as the results were dependent on the initial arrangement of the molecules. With eight molecules, the results obtained were less dependent on the initial configuration of the cluster, and this system size was found to be a reasonable compromise between computational cost and a proper representation of the phase space of a dense, liquid-like fluid.

The energy minimization step is the most time-consuming part of the calculations. For systems with large molecules, the optimization required several days to over a week on a multiprocessor supercomputer even at the HF level. This level of theory is known to result in reasonably accurate geometries of molecules and, with density functional theory (DFT) (Parr and Yang, 1989; Schleyer, 1998; and references therein), is the least expensive computational method. Density functional theory was not used here because the available functionals were not optimized for hydrogen bonding of importance in the systems of interest here. Also, the use of more sophisticated methods [e.g., Møller–Plesset perturbation theory (Levine, 1991; Szabo and Ostlund, 1996; Schleyer, 1998)] did not justify the very large additional computational cost for the slight improvement that would be obtained in the final optimized geometry. The cluster optimized at the HF/6-31G(d,p) level was taken to represent a sample of the phase space of a dense fluid. (However, we cannot assure that the final configuration obtained is the absolute minimum energy configuration since, for systems as complex as the ones under study, many local minima may exist.)

Next, the energies were calculated using the HF method with the 6-311++G(3d,2p) basis set. Hartree–Fock theory does not account for

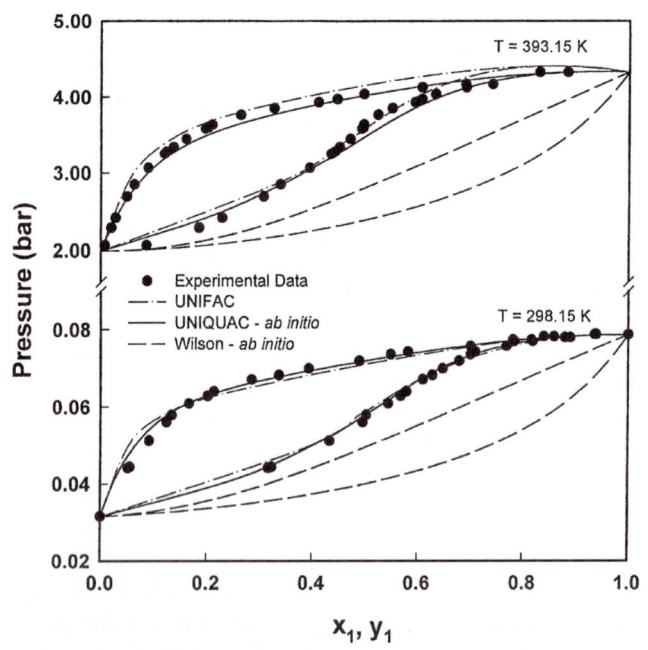

FIG. 7. Vapor–liquid equilibrium diagram for ethanol (1) + water (2). UNIQUAC and Wilson parameters from *ab initio* calculations. Experimental data from Gmehling *et al.* (1977 onward).

electron correlation, only electrostatic interactions. However, because the final quantity of interest is the difference of interaction energies, it was assumed that the electron correlation would largely cancel, so that it need not explicitly be considered. Methods accounting for electron correlation include DFT, which could not be used as discussed above, and Møller–Plesset perturbation theory, which was not used because of the computational cost for the large systems considered. The large basis set was used to minimize the BSSE (other, smaller basis sets give similar interaction energies, but the BSSE is a large fraction of these energies). All the *ab initio* computations were performed using the *Gaussian* program (Frisch *et al.*, 1995) on multi-processor computers (Cray J-90 and SGI Origin2000).

Once the interaction energies were obtained, they were used to calculate the parameters in the UNIQUAC and Wilson models given by Eq. (24). To test the validity of the method, low-pressure vapor–liquid equilibrium (VLE) predictions were made for several binary aqueous systems. The calculations were done using the usual γ–ϕ method assuming an ideal vapor phase (Sandler, 1999). Figures 7 and 8 show the low-pressure VLE diagrams for the binary aqueous mixtures of ethanol and acetone [see Sum and Sandler (1999a,b) for results for additional systems and values of the

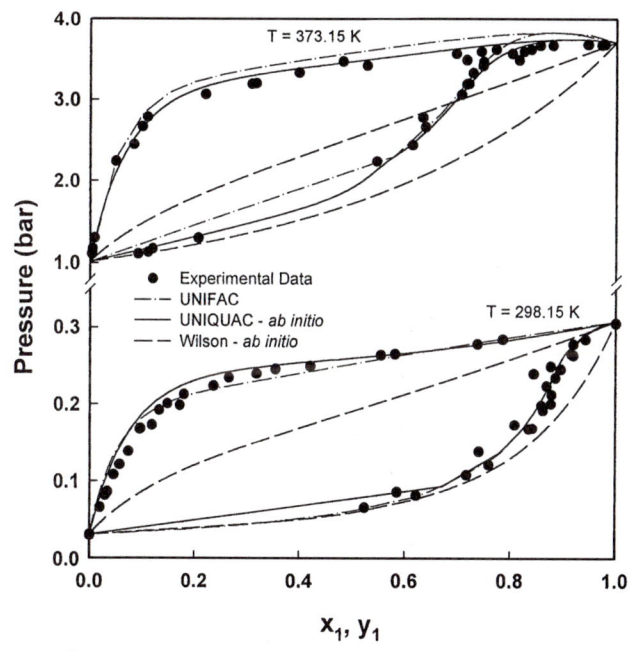

FIG. 8. Vapor–liquid equilibrium diagram for acetone (1) + water (2). UNIQUAC and Wilson parameters from *ab initio* calculations. Experimental data from Gmehling *et al.* (1977 onward).

calculated parameters]. Only binary aqueous systems were studied, as water is a small molecule (has few electrons) and its interactions with other molecules is mainly through hydrogen bonding. For the interaction energies calculated by the HF method to be meaningful, we could consider only hydrogen-bonding systems in which electron correlation resulting in van der Waals or dispersion energies could be neglected.

As shown in Figs. 7 and 8, the low-pressure predictions based on the UNIQUAC model using the calculated interaction parameters are very good. However, predictions with the Wilson model, using the same interaction energies, did not give satisfactory results for any of the systems studied. In the development of both the UNIQUAC and the Wilson models, the adjustable parameters are interpreted as interaction energies. From these and other results (Sum and Sandler, 1999a,b) it appears the UNIQUAC model has a reasonable theoretical basis since the energies that appear in this model are the same as those calculated from quantum mechanics. In contrast, the parameters in the Wilson model must be treated as completely adjustable to obtain good agreement with experimental data.

Hydrogen bonding results in less configurational freedom among the molecules, and this is probably the reason the method used here leads to reasonable predictions for small-molecule aqueous systems. It was found that as the nonaqueous molecule increased in size and flexibility, it was more difficult to find the minimum energy cluster, due mainly to the large number of geometric degrees of freedom. Therefore, it is important to sample phase space using many possible interactions between the molecules to obtain the correct interaction energies, especially for systems with large non-hydrogen-bonding molecules. In such cases, the combination of quantum chemistry and molecular simulation discussed in Section II is the appropriate way to proceed, rather than considering only the minimum energy configurations as was done here.

It is useful to note the main approximations made in the calculations just described. First, electron correlation was neglected by using HF theory. Second, a cluster size of eight molecules optimized at HF/6-31G(d,p) at 0 K and in vacuum has been assumed to be a good representation of the system. This assumption implies that the configurations and average interaction energies of a hydrogen-bonding fluid at the temperatures of interest can be approximated from the equilibrium configuration of a sample cluster at 0 K. Third, the quantum mechanically calculated interaction energies obtained from the minimum energy configurations were assumed to be equal to the energy parameters in the UNIQUAC and Wilson models.

The methodology presented here was also tested for several other nonaqueous systems, but with only limited and inconsistent success. For weakly interacting systems, calculation of interaction energies at the HF method level does not account for the van der Waals and dispersion energies. Also, for such systems the small cluster used is not sufficient to sample properly the configurational phase space available to the molecules; complete simulations need to be done.

VI. Conclusions

Quantum chemistry computational packages are now readily available for most computer platforms. Considerable care, based on knowledge, is needed to choose the appropriate level of theory, basis sets, and methodology to give acceptable results for the calculated energies, molecular geometry, etc., which are the typical ways in which such programs are used. In this review, we have considered a quite different question: How can such quantum chemistry programs be applied to some areas of chemical engineering? We have considered several here, based largely on our own work. The first was a straightforward, but computationally intensive procedure of using quantum chemistry to obtain accurate, detailed intermolecular potential functions that are then used in computer simulation. The second application, less computer intensive than the first, was the use of quantum polarization continuum solvation models to predict infinite dilution activity coefficients and octanol–water partition coefficients of solutes. By using PCM calculations, large solute molecules could be studied. The result was an easy-to-use group contribution model for octanol–water partition coefficients for solutes with only a single nonalkyl functional group.

However, that group contribution method, as all others of its genre, does not result in accurate predictions when there is more than one nonalkyl functional group in a molecule. This is a result of the change in the electron distribution of a functional group depending on other functional groups in the same molecule. In this case, we have shown how a hybrid model that combines a group contribution calculation with a quantum chemical calculation for an isolated molecule can be used to correct for the failure of using only a group contribution method for the calculation of octanol–water partition coefficients of complex molecules. Research is now under way to extend this idea to other classes of group contribution methods. Finally, we have shown how the average energies of interaction computed for strongly interacting, hydrogen-bonding systems can be used directly to determine the parameters in activity coefficient models and then to predict the phase behavior.

While this review is not comprehensive in its scope, it is our hope that the message that comes through is that quantum chemistry calculations do have an increasing role to play in applied chemical engineering.

ACKNOWLEDGMENTS

The authors thank the National Science Foundation (CTS-9521406) and the Department of Energy (DE-FG02-85ER13436) for financial support of this research.

REFERENCES

Abildskov, J., Constantinou, L., and Gani, R., *Fluid Phase Equil.* **118,** 1–12 (1996).
Abildskov, J., Gani, R., Rasmussen, P., and O'Connell, J. P., *Fluid Phase Equil.* **160,** 349–356 (1999).
Abrams, D. S., and Prausnitz, J. M., *AIChE J.* **21,** 116–128 (1975).
Ahlström, P., Wallqvist, A., Engström, S., and Jönsson, B., *Mol. Phys.* **68,** 563–581 (1989).
Amovilli, C., and Mennucci, B., *J. Phys. Chem. B* **101,** 1051–1057 (1997).
Aqvist, J., and Hansson, T., *J. Phys. Chem.* **100,** 9512–9521 (1996).
Babu, C. S., and Lim, C., *J. Phys. Chem. B* **103,** 7958–7968 (1999).
Bader, J. S., Cortis, C. M., and Berne, B. J., *J. Chem. Phys.* **106,** 2372–2387 (1997).
Ben-Naim, A., "Solvation Thermodynamics," Plenum Press, New York, 1987.
Benoit, M., Bernasconi, M., Focher, P., and Parrinello, M., *Phys. Rev. Lett.* **76,** 2934–2936 (1996).
Berendsen, H. J. C., Postma, J. P. M., van Gunsteren, W. F., and Hermans, J., *in* "Intermolecular Forces: Proceedings of the Fourteenth Jerusalem Symposium on Quantum Chemistry and Biochemistry" (B. Pullman, Ed.), Reidel, Dordrecht, 1981, pp. 331–342.
Berendsen, H. J. C., Grigera, J. R., and Straatsma, T., *J. Phys. Chem.* **91,** 6269–6271 (1987).
Bernardo, D. N., Ding, Y., Krogh-Jespersen, K., and Levy, R. M., *J. Phys. Chem.* **98,** 4180–4187 (1994).
Böttcher, C. J. F., "Theory of Electric Polarization," 2nd ed., Elsevier, Amsterdam, 1973.
Brooks, B. R., Bruccoleri, R. E., Olafson, B. D., States, D. J., Swaminathan, S., and Karplus, M., *J. Comp. Chem.* **4,** 187–217 (1983).
Bukowski, R., Sadlej, J., Jeziorski, B., Jankowski, P., and Szalewicz, K., *Chem. Phys.* **110,** 3785–3803 (1999a).
Bukowski, R., Szalewicz, K., and Chabalowski, C. F., *J. Phys. Chem. A* **103,** 7322–7340 (1999b).
Cabaleiro-Lago, E. M., and Ríos, M. A., *J. Phys. Chem. A* **101,** 8327–8334 (1997).
Car, R., and Parrinello, M., *Phys. Rev. Lett.* **55,** 2471–2474 (1985).
Chałasiński, G., and Gutowski, M., *Chem. Rev.* **88,** 943–962 (1988).
Chałasiński, G., and Szczęśniak, M. M., *Chem. Rev.* **94,** 1723–1765 (1994).
Chang, T.-M., Peterson, K. A., and Dang, L. X., *J. Chem. Phys.* **103,** 7502–7513 (1995).
Chang, T.-M., Dang, L. X., and Peterson, K. A., *J. Phys. Chem. B* **101,** 3413–3419 (1997).
Chialvo, A. A., and Cummings, P. T., *J. Chem. Phys.* **105,** 8274–8281 (1996).
Chialvo, A. A., and Cummings, P. T., *Fluid Phase Equil.* **150–151,** 73–81 (1998).
Claverie, P., *in* "Intermolecular Interactions: from Diatomics to Biomolecules" (Pullman, B., Ed.), Wiley, Chichester, 1978.
Constantinou, L., and Gani, R., *AIChE J.* **40,** 1697–1710 (1994).
Cornell, W. D., Cieplak, P., Bayly, C. I., Gould, I. R., Merz, K. M., Ferguson, D. M., Spellmeyer, D. C., Fox, T., Caldwell, J. W., and Kollman, P. A., *J. Am. Chem. Soc.* **117,** 5179–5197 (1995).
Cossi, M., Barone, V., Cammi, R., and Tomasi, J., *Chem. Phys. Lett.* **255,** 327–335 (1996).
Cramer, C. J., and Truhlar, D. G., *Chem. Rev.* **99,** 2161 (1999).
Dang, L. X., and Chang, T.-M., *J. Chem. Phys.* **106,** 8149–8159 (1997).

Detrich, J., Corongiu, G., and Clementi, E., *Chem. Phys. Lett.* **112,** 426–430 (1984).
Dillet, V., Rinaldi, D., Àngyàn, J. G., and Rivail, J.-L., *Chem. Phys. Lett.* **202,** 18–22 (1993).
Dunning, T. H. Jr., *J. Chem. Phys.* **90,** 1007–1023 (1989).
Elcock, A. H., and McCammon, J. A., *J. Phys. Chem. B* **101,** 9624–9634 (1997).
Floris, F., Selmi, M., Tani, A., and Tomasi, J., *Chem. Phys.* **107,** 6353–6365 (1997).
Franke, R., "Theoretical Drug Design Methods," Akademie Verlag, Berlin, 1984.
Fredenslund, A., Jones, R. L., and Prausnitz, J. M., *AIChE J.* **21,** 1086–1099 (1975).
Fredenslund, A., Gmeling, J., and Rasmussen, P., "Vapor-Liquid Equilibrium Using UNIFAC," Elsevier, Amsterdam, 1977.
Frisch, M. J., Trucks, G. W., Schlegel, H. B., Gill, P. M. W., Johnson, B. G., Robb, M. A., Cheeseman, J. R., Keith, T. A., Petersson, G. A., Montgomery, J. A., Raghavachari, K., Al-Laham, M. A., Zakrzewski, V. G., Ortiz, J. V., Foresman, J. B., Peng, C. Y., Ayala, P. A., Wong, M. W., Andres, J. L., Replogle, E. S., Gomperts, R., Martin, R. L., Fox, D. J., Binkley, J. S., Defrees, D. J., Baker, J., Stewart, J. P., Head-Gordon, M., Gonzalez, C., and Pople, J. A., "Gaussian 94," Gaussian, Inc., Pittsburgh, PA, 1995.
Gaussian, Inc., http://www.gaussian.com/.
Gierycz, P., and Nakanishi, K., *Fluid Phase Equil.* **16,** 255–273 (1984).
Gierycz, P., Tanaka, H., and Nakanishi, K., *Fluid Phase Equil.* **16,** 241–253 (1984).
Gmehling, J., Li, J., and Schiller, M., *Ind. Eng. Chem. Res.* **32,** 178–193 (1993).
Gmehling, J., Onken, U., and Arlt, W., "Vapor-Liquid Equilibrium Data Collection," DECHEMA, Frankfurt, 1977 and onward.
Goncalves, P. F. B., and Livotto, P. R., *Chem. Phys. Lett.* **304,** 438–444 (1999).
González, M. A., Enciso, F. J., and Bée, M., *J. Chem. Phys.* **110,** 8045–8059 (1999).
Guggenheim, E. A., "Mixtures," Clarendon Press, Oxford, 1952, p. 196.
Haile, J. M., *Fluid Phase Equil.* **26,** 103–127 (1986).
Hamad, E. Z., *Fluid Phase Equil.* **142,** 163–184 (1998).
Hansch, C., and Leo, A., "Exploring QSAR: Fundamentals and Applications in Chemistry and Biology," American Chemistry Society, Washington, DC, 1995.
Hawkins, G. D., Cramer, C. J., and Truhlar, D. G., *J. Phys. Chem. B* **102,** 3257–3271 (1998).
Hermida-Ramón, J. M., and Ríos, M. A., *J. Phys. Chem. A* **102,** 2594–2602 (1998).
Heyes, D. M., *J. Chem. Soc. Faraday Trans.* **87,** 3373–3377 (1991).
Hloucha, M., and Deiters, U. K., *Mol. Phys.* **90,** 593–597 (1997).
Hloucha, M., and Deiters, U. K., *Fluid Phase Equil.* **149,** 41–56 (1998).
Hyun, J. K., and Ichiye, T., *J. Chem. Phys.* **109,** 1074–1083 (1998).
Jeziorski, B., and Szalewicz, K., in "Encyclopedia of Computational Chemistry" (P. von R. Schleyer, Ed.), Wiley, Chichester, 1998.
Jeziorski, B., Moszyński, R., and Szalewicz, K., *Chem. Rev.* **94,** 1887–1930 (1994).
Jónsdóttir, S. Ó., and Klein, R. A., *Fluid Phase Equil.* **132,** 117–137 (1997).
Jónsdóttir, S. Ó., and Rasmussen, K., *Fluid Phase Equil.* **115,** 59–72 (1996).
Jónsdóttir, S. Ó., and Rasmussen, P., *Fluid Phase Equil.* **160,** 411–418 (1999).
Jónsdóttir, S. Ó., Rasmussen, K., and Fredenslund, A., *Fluid Phase Equil.* **100,** 121–138 (1994).
Jónsdóttir, S. Ó., Rasmussen, K., Rasmussen, P., and Welsh, W. J., *Comput. Theor. Polym. Sci.* **8,** 75–81 (1998).
Jónsdóttir, S. Ó., Welsh, W. J., Rasmussen, K., and Klein, R. A., *New J. Chem.* **23,** 153–163 (1999).
Jorgensen, W. L., and Tirado-Rives, J., *J. Am. Chem. Soc.* **110,** 1657–1666 (1988).
Jorgensen, W. L., Chadandrasekhar, J., Madura, J. D., Impey, E. W., and Klein, M. L., *J. Chem. Phys.* **79,** 926–935 (1983).
Kamlet, M. J., Doherty, R. M., Abraham, M. H., Marcus, Y., and Taft, R. W., *J. Phys. Chem.* **92,** 5244–5255 (1988).

Kehiaian, H. V., *Fluid Phase Equil.* **13**, 243–252 (1983).
Kehiaian, H. V., and Marongiu, B., *Fluid Phase Equil.* **21**, 197–209 (1985).
Kestner, N. R., and Combariza, J. E., *in* "Reviews in Computational Chemistry" (K. B. Lipkowitz and D. B. Boyd, Eds.), Wiley–VCH, New York, 1999, pp. 99–132.
Kirkwood, J. G., *J. Chem. Phys.* **2**, 351–361 (1934).
Kiyohara, K., Gubbins, K. E., and Panagiotopoulos, A. Z., *Mol. Phys.* **94**, 803–808 (1998).
Klamt, A., and Schüürmann, G., *J. Chem. Soc. Perkin Trans.* **2**, 799–805 (1993).
Klamt, A., Jonas, V., Bürger, T., and Lohrenz, J. C. W., *J. Phys. Chem. A* **102**, 5074–5085 (1998).
Kojima, K., and Tochigi, T., "Prediction of Vapor-Liquid Equilibrium by the ASOG Method," Elsevier, Amsterdam, 1979.
Kuwajima, S., and Warshel, A., *J. Phys. Chem.* **94**, 460–466 (1990).
Lee, K.-H., Sandler, S. I., and Monson, P. A., *Int. J. Thermophys.* **7**, 367–379 (1986).
Lee, L. L., Chung, T. H., and Starling, K. E., *Fluid Phase Equil.* **12**, 105–124 (1983).
Lee, L. L., Chung, F. T. H., and Landis, L. H., *Fluid Phase Equil.* **31**, 253–272 (1986).
Levine, I. N., "Quantum Chemistry," 4th ed., Prentice–Hall, Englewood Cliffs, NJ, 1991.
Levy, M., and Gallicchio, E., *Annu. Rev. Phys. Chem.* **49**, 531–567 (1998).
Lin, S.-T., and Sandler, S. I., *AIChE J.* **45**, 2606–2618 (1999a).
Lin, S.-T., and Sandler, S. I., *Ind. Eng. Chem. Res.* **38**, 4081–4091 (1999b).
Lin, S.-T., and Sandler, S. I., submitted for publication (2000).
Liu, Y.-P., Kim, K., Berne, B. J., Friesner, R. A., and Rick, S. W., *J. Chem. Phys.* **108**, 4739–4755 (1998).
Lotrich, V. F., and Szalewicz, K., *J. Chem. Phys.* **106**, 9668–9687 (1997a).
Lotrich, V. F., and Szalewicz, K., *J. Chem. Phys.* **106**, 9688–9702 (1997b).
Lotrich, V. F., Szalewicz, K., and Jeziorski, B., *Polish J. Chem.* **72**, 1826–1848 (1998).
Lowe, J. P., "Quantum Chemistry," 2nd ed., Academic Press, San Diego, 1993.
Marten, B., Kim, K., Cortis, C., Friesner, R. A., Murphy, R. B., Ringnalda, M. N., Sitkoff, D., and Honig, B., *J. Phys. Chem.* **100**, 11775–11788 (1996).
Mas, E. M., Szalewicz, K., Bukowski, R., and Jeriorski, B., *J. Chem. Phys.* **107**, 4207–4218 (1997).
Matsuoka, O., Clementi, E., and Yoshimine, M., *J. Chem. Phys.* **64**, 1351–1361 (1976).
Maurer, G., and Prausnitz, J. M., *Fluid Phase Equil.* **2**, 91–99 (1978).
Metropolis, N., Rosenbluth, A. W., Rosenbluth, M. N., Teller, A. H., and Teller, E., *J. Chem. Phys.* **21**, 1087–1092 (1953).
Millot, C., Soetens, J.-C., Costa, M. T. C. M., Hodges, M. P., and Stone, A. J., *J. Phys. Chem. A* **102**, 754–770 (1998).
Mooji, W. T. M., van Duijneveldt, F. B., van Duijneveldt-van de Ridt, J. G. C. M., and van Eijck, B. P., *J. Phys. Chem. A* **103**, 9872–9882 (1999).
Mulliken, R. S., *Chem. Phys.* **12**, 1833 (1955).
Netemeyer, S. C., and Glandt, E. D., *Ind. Eng. Chem. Res.* **27**, 1516–1524 (1988).
Niesar, U., Corongiu, G., Huang, M. J., Dupius, M., and Clementi, E., *Int. J. Quantum Chem. Quantum Chem. Symp.* **23**, 421–443 (1989).
Niesar, U., Corongiu, G., Clementi, E., Kneller, G. R., and Bhattacharya, D. K., *J. Phys. Chem.* **94**, 7949–7956 (1990).
Nina, M., Beglov, D., and Roux, B., *J. Phys. Chem. B* **101**, 5239–5248 (1997).
Panagiotopoulos, A. Z., *Mol. Phys.* **61**, 813–826 (1987).
Parr, R. G., and Yang, W., "Density-Functional Theory of Atoms and Molecules," Oxford University Press, Oxford, 1989.
Pierotti, R. A., *Chem. Rev.* **76**, 717–726 (1976).
Popkie, H., Kistenmacher, H., and Clementi, E., *J. Chem. Phys.* **59**, 1325–1336 (1973).
Q-Chem, Inc., http://www.q-chem.com/.
Qiu, D., Shenkin, P. S., Hollinger, F. P., and Still, W. C., *Phys. Chem. A* **101**, 3005–3014 (1997).

Rivail, J. L., and Rinaldi, D. A., *Chem. Phys.* **18,** 233–242 (1976).
Rivail, J.-L., and Rinaldi, D., *in* "Computational Chemistry: Reviews of Current Trends" (Leszczynski, J., Ed.), World Scientific, New York, 1995, pp. 139–174.
Rosenbluth, M. N., and Rosenbluth, A. W., *J. Chem. Phys.* **22,** 881–884 (1954).
Sandler, S. I., *Fluid Phase Equil.* **116,** 343–353 (1996).
Sandler, S. I, "Chemical and Engineering Thermodynamics," 3rd ed., Wiley, New York, 1999.
Sayegh, S. G., and Vera, J. H., *Chem. Eng. J.* **19,** 1–10 (1980).
Schleyer, P. von R., (Ed.), "Encyclopedia of Computational Chemistry," Wiley, Chichester, 1998.
Schmidt, M. W., Baldridge, K. K., Boatz, J. A., Elbert, S. T., Gordon, M. S., Jensen, J. H., Koseki, S., Matsunaga, N., Nguyen, K. A., Su, S., Windus, T. L., Dupuis, M., and Montgomery, J. A., *J. Comput. Chem.* **14,** 1347–1363 (1993).
Schrödinger, Inc., http://www.schrodinger.com/.
Smith, B. J., *J. Comput. Chem.* **20,** 428–442 (1999).
Soetens, J.-C., Jansen, G., and Millot, C., *Mol. Phys.* **96,** 1003–1012 (1999).
Sprik, M., and Klein, M. L., *J. Chem. Phys.* **89,** 7556–7560 (1988).
Staverman, A. J., *Rec. Trav. Chim. Pays-Bas.* **69,** 163 (1950).
Stewart, J. J. P., *J. Comput. Chem.* **10,** 209–220 (1989a).
Stewart, J. J. P., *J. Comput. Chem.* **10,** 221–264 (1989b).
Stillinger, F. H., and David, C. W., *J. Chem. Phys.* **69,** 1473–1484 (1978).
Straatsma, T. P., and McCammon, J. A., *Annu. Rev. Phys. Chem.* **43,** 407–435 (1992).
Sum, A. K., and Sandler, S. I., *Fluid Phase Equil.* **160,** 375–380 (1999a).
Sum, A. K., and Sandler, S. I., *Ind. Eng. Chem. Res.* **38,** 2849–2855 (1999b).
Svishchev, I. M., and Hayward, T. M., *J. Chem. Phys.* **111,** 9034–9038 (1999).
Szabo, A., and Ostlund, N. S., "Modern Quantum Chemistry—Introduction to Advanced Electronic Structure Theory," McGraw–Hill, New York, 1989.
Szalewicz, K., and Jeziorski, B., *in* "Molecular Interactions—From van der Waals to Strongly Bound Complexes" (S. Scheiner, Ed.), Wiley, New York, 1997, pp. 3–43.
Tang, K. T., and Toennis, J. P., *J. Chem. Phys.* **80,** 3726–3741 (1984).
Tannor, D. J., Marten, B., Murphy, R., Friesner, R. A., Sitkoff, D., Nicholls, A., Ringnalda, M., Goddard, W. A. III, and Honig, B., *J. Am. Chem. Soc.* **116,** 11875–11882 (1994).
Tawa, G. J., and Pratt, L. R., *J. Am. Chem. Soc.* **117**(5), 1625–1628 (1995).
Tomasi, J., and Persico, M., *Chem. Rev.* **94,** 2027–2094 (1994).
van Duijneveldt, F. B., van Duijneveldt-van de Rijdt, J. G. C. M., and van Lenthe, J. H., *Chem. Rev.* **94,** 1873–1885 (1994).
Veldhuizen, R., and de Leeuw, S. W., *J. Chem. Phys.* **105,** 2828–2836 (1996).
Watanabe, K., and Klein, M. L., *Chem. Phys.* **131,** 157–167 (1989).
Werner, H.-J., and Knowles, P. J., MOLPRO, http://www.tc.bham.ac.uk/molpro/.
Wienke, G., and Gmehling, J., *Toxicol. Environ. Chem.* **65,** 57–86 (1998).
Wilson, G. M., *J. Am. Chem. Soc.* **86,** 168–176 (1964).
Wood, W. W., and Parker, F. R., *J. Chem. Phys.* **27,** 720–733 (1957).
Wu, H. S., and Sandler, S. I., *Ind. Eng. Chem. Res.* **30,** 881–889 (1991a).
Wu, H. S., and Sandler, S. I., *Ind. Eng. Chem. Res.* **30,** 889–897 (1991b).
Yoshii, N., Yoshie, H., Miura, S., and Okazaki, S., *J. Chem. Phys.* **109,** 4873–4884 (1998).
Zhan, C.-G., and Chipman, D. M., *J. Chem. Phys.* **109,** 10543–10558 (1998).

CAR–PARRINELLO METHODS IN CHEMICAL ENGINEERING: THEIR SCOPE AND POTENTIAL

Bernhardt L. Trout

Department of Chemical Engineering, Massachusetts Institute of Technology, Cambridge, Massachusetts 02139

I. Introduction	353
II. Objectives and Description of This Article	355
III. Objectives of Car–Parrinello Methods and Classes of Problems to Which They Are Best Applicable	356
IV. Methodology	357
A. Classical Molecular Dynamics	357
B. Density-Functional Theory	358
C. Choice of Model and Solution of the Equations Using Plane-Wave Basis Sets and the Pseudopotential Method	362
D. Car–Parrinello Molecular Dynamics	368
V. Applications	370
A. Gas-Phase Processes	372
B. Processes in Bulk Materials	376
C. Properties of Liquids, Solvation, and Reactions in Liquids	378
D. Heterogeneous Reactions and Processes on Surfaces	382
E. Phase Transitions	386
F. Processes in Biological Systems	389
VI. Advances in Methodology	392
VII. Concluding Remarks	393
Appendix A: Further Reading	393
Appendix B: Codes With Capabilities to Perform Car–Parrinello Molecular Dynamics	394
References	394

I. Introduction

It is now 15 years after the publication of the famous paper by Roberto Car and Michele Parrinello (1985) on the integration of the quantum mechanical method, density-functional theory, with classical molecular dynamics. By

Moore's law, computational power has increased by a factor of 1000 since then. In addition, with the development of many computer codes that employ the Car–Parrinello method, combined with the enhanced accuracy of density functionals that have been devised since 1985, there has been widespread use of the Car–Parrinello method. Between 1997 and the beginning of 2000, over 500 papers that incorporate results found using Car–Parrinello simulations have been published in major journals. Most of these papers have been published by groups considered primarily to be in the field of physics. On the other hand, the topics of these papers span the spectrum of disciplines as traditionally classified: physics, industrial chemistry, catalysis, materials engineering, microelectronic materials, polymer science, biology, and geology.

The time is ripe for the broader chemical engineering community to be exposed to Car–Parrinello methods and their potential for chemical engineering. Computational methods, even very advanced ones, will continue to become more and more routine for nonspecialists. Having these computational methods in the chemical engineer's toolbox will necessarily lead to enhanced ability to make significant advances, both experimental and theoretical.

Moreover, chemical engineers think differently than scientists or other engineers, and this different way of thinking will lead to advances in Car–Parrinello methods that can be used in other fields. As chemical engineers, we always deal with inherently complex systems and have always focused on treating multiple scales, starting at the molecular level. Multiscale modeling, starting at the molecular scale and going to the macroscopic scale, is one of the current Grand Challenge Problems, spanning fields of science and engineering. Car-Parrinello methods present extremely powerful ways of treating the molecular scale and of scaling up to the macroscopic scale. Chemical engineers will help to solve the Grand Challenge Problem of multiscale modeling, and having methods such as Car–Parrinello methods in their toolbox is a prerequisite to solving them. Finally, many chemical engineers have asked me about recommended reviews on this subject. While many excellent reviews exist (see Appendix A), they do not address the field directly in a way that makes their potential readily evident to chemical engineers who are not already experts in these methods.

In this review, I have interpreted the term "Car–Parrinello methods" in the broad sense to mean those which combine first-principles quantum mechanical methods with molecular dynamics methods. I use this term synonymously with *"ab initio* molecular dynamics," "first-principles molecular dynamics," and *"ab initio* simulations." Thus, ways of solving the many-body electronic problem, such as Hartree–Fock and correlation methods, are included, in addition to the projector-augmented plane-wave method. In the original Car–Parrinello method, molecular motion is treated classically via molecular dynamics and the wavefunction is propagated via fictitious electronic dynamics. I include methods based on this idea, in addition to other ways of keeping the wavefunction on the Born–Oppenheimer surface, such as optimization of the wavefunction at each molecular dynamics step. The latter method is often called "Born–Oppenheimer molecular dynamics."

I have designated "Car-Parrinello methods" in the plural for two main reasons. One is that, in deference to the impact of Car and Parrinello's classic 1985 paper, I am using this term to designate all ways of simulating the motion of atoms via combining a classical molecular dynamics treatment of the nuclei with a first-principles treatment of the electrons. The second, and perhaps most important, reason is that there are many molecular dynamics methods that have been and are being developed specifically for the simulation of atoms and molecules using forces calculated from derivatives of the quantum mechanical energy of these systems. It is the development of these methods that Car and Parrinello initiated.

II. Objectives and Description of This Article

The objective of this article is to expose the chemical engineering community to Car–Parrinello methods, what they have accomplished, and what their potential is for chemical engineering. Consistent with this objective, in Section IV, I give an overview of the most widely used quantum mechanical method for solving the many-body electronic problem, density-functional theory, but describe other methods only cursorily. I also describe the practical solution of the equations of density-functional theory for molecular and extended systems via the plane-wave pseudopotential method, mentioning other methods only cursorily. Finally, I end this section with a description of the Car–Parrinello method itself.

In Section V, the longest one, I present a summary of the literature that presents methods and applications that tend to have the greatest relevance to chemical engineering, with an emphasis on work since 1997. I group the applications into several categories: gas-phase processes, processes in bulk materials, properties of liquids and processes in liquids, heterogeneous reactions and processes on surfaces, phase transitions, and processes in biological systems. In each category, I choose one or two case studies to pursue in detail and mention a few other relevant studies. Of course, there is an element of arbitrariness in choosing to focus on a few studies out of thousands, and certainly my exposure to the Parrinello group, both past and present, has made me more familiar with their work and more likely to choose applications from them. My hope is that chemical engineers will see how the methods used in these applications can have relevance to their own areas of interest and will thus benefit from this review.

III. Objectives of Car–Parrinello Methods and Classes of Problems to Which They Are Best Applicable

The reader can envision that the accurate calculation of the energy of a many-electron system is a computationally intensive endeavor, particularly when it must be performed thousands or tens of thousands of times, throughout a molecular dynamics trajectory. This section addresses the reasons one would wish to use these methods, in contrast to both experimental methods and molecular dynamics methods based on classical potentials. In its broadest sense, the reason for using Car–Parrinello methods is to determine properties of systems with as few fitted parameters or *a priori* assumptions as possible.

Car–Parrinello methods contrasted with experimental approaches: They can be used to study details of chemical systems that are difficult to address directly using experimental methods. One example is the nature and rate of elementary steps of catalytic processes. These are almost always very difficult to isolate experimentally.

Car–Parrinello methods contrasted with classical molecular dynamic methods: They can be used to obtain information about systems computationally, in which it is expected that conditions of the system change such that the range of validity of fitted potentials is exceeded. This is almost always the case when chemical bonds are broken and/or formed. It is also true for phase transitions or extreme states of high/low pressures and/or temperatures where not many experimental data exist. The accuracy of Car–Parrinello methods is not a strong function of the conditions to which they are applied (although they will be a function of the system to which they are applied).

Car–Parrinello methods contrasted with static (0 K temperature) computational quantum mechanical methods: They can treat entropy accurately without the need to use models such as the harmonic approximation for degrees of freedom of atomic motions. They can be used to sample potential energy surfaces on picosecond time scales, which is essential for treating liquids and aqueous systems. They can be used to sample reaction pathways or other chemical processes with a minimum of *a priori* assumptions. In addition, they can be used to find global minima [in conjunction with methods of optimization such as simulated annealing (Kirkpatrick *et al.*, 1983)] and to step out of local minima.

In summary, they are used both to understand trends in chemical systems as functions of composition, morphology, and thermodynamic conditions and to obtain information on the properties of specific chemical systems.

IV. Methodology

Integral to Car–Parrinello methods is the use of computational quantum mechanics to determine the state of a number of electrons in the presence of any configuration of atomic nuclei. Determining the electronic state of the system quantum mechanically can be contrasted with using empirically derived potentials, such as Lennard–Jones or Morse potentials, used in classical methods. Once the electronic state has been computed, all properties of the system can be found. For molecular dynamics simulations, the most important properties are the absolute energy of the system and the forces on the individual atomic nuclei. Once these forces are computed, the nuclei can be propagated using classical equations of motion.

By far the major computational quantum mechanical method used to compute the electronic state in Car–Parrinello simulations is density-functional theory (DFT) (Hohenberg and Kohn, 1964; Kohn and Sham, 1965; Parr and Yang, 1989). It is the method used originally by Roberto Car and Michele Parrinello in 1985, and it provides the highest level of accuracy for the computational cost. For these reasons, in this section the only computational quantum mechanical method discussed is DFT. Section A consists of a brief review of classical molecular dynamics methods. Following this is a description of DFT in general (Section B) and then a description of practical DFT computations of chemical systems using the plane-wave pseudopotential method (Section C). The section ends with a description of the Car–Parrinello method and some basic issues involved in its use (Section D).

A. CLASSICAL MOLECULAR DYNAMICS

In classical molecular dynamics simulations, atoms are generally considered to be points which interact with other atoms by some predefined potential form. The forms of the potential can be, for example, Lennard–Jones potentials or Coulomb potentials. The atoms are given velocities in random directions with magnitudes selected from a Maxwell–Boltzman distribution, and then they are allowed to propagate via Newton's equations of motion according to a finite-difference approximation. See the following references for much more detailed discussions: Allen and Tildesley (1987) and Frenkel and Smit (1996). To determine the magnitude and direction of motion of the nuclei at each time step, the force is computed on each nucleus by taking the negative of the gradient of the potential arising from the interaction of the atom with all other atoms in the simulation.

B. Density-Functional Theory

To avoid using a predefined form for the interaction potential in molecular dynamics simulations, the quantum mechanical state of the many-electron system can be determined for a given nuclear configuration. From this quantum mechanical state, all properties of the system can be determined, in particular, the total electronic energy and the force on each of the nuclei. The quantum mechanically derived forces can then be used in place of the classically derived forces to propagate the atomic nuclei. This section describes the most widely used quantum mechanical method for computing these forces used in Car–Parrinello simulations.

Atomic nuclei are much heavier than electrons and can, in general, be treated accurately using a classical approach. Electrons, of course, must be treated quantum mechanically, and they are considered to move via the equations of quantum mechanics within the fixed external potential of the positively charged nuclei. Because of the relative speed of the motion of the electrons compared to that of the nuclei, their motion is, to an excellent approximation, separate from that of the nuclei in what is called the Born–Oppenheimer approximation. Moreover, excited electronic states are usually irrelevant at temperatures of interest to chemical engineers (<10,000 K), so only their ground state (minimum energy state) needs to be considered. (I do not consider here the interaction of radiation with matter, the treatment of which is not readily possible at this time using Car–Parrinello methods.)

One would then wish to solve the Schrödinger equation for the wavefunction of the electrons in the potential of the nuclei. This wavefunction would define the state of the system and can be used to obtain all observable properties of the system. The problem is that the many-body Schrödinger equation cannot be solved accurately, even using the most powerful of computers, except for the simplest of systems, consisting of only a few electrons.

Computational quantum mechanical methods, such as the Hartree–Fock method (Hehre *et al.*, 1986; Szabo and Ostlund, 1989; Levine, 2000), were developed to convert the many-body Schrödinger equation into a single-electron equation, which can then be solved tractably with modern computational power. The single-electron equation is an approach by which the state (or wavefunction) of each electron is computed within the field of all of the other electrons. The major limitation of this approach is that electronic motion is correlated, and this electronic correlation cannot be treated exactly, except for very small systems. A tremendous amount of research has gone into treating electronic correlation (Szabo and Ostlund, 1989), and the general result is that one can consistently treat this correlation more and more accurately with a corresponding increase in computational cost.

An alternative to computing the properties of many-body electronic systems via the wavefunction is computing these properties via the electronic density. This approach was invented by Hohenberg and Kohn, who showed that the total energy of the many-body electronic system can be expressed as a functional of the density in the following way:

$$E[\rho(\mathbf{r})] = T[\rho(\mathbf{r})] + E_{e-e}[\rho(\mathbf{r})] + E_{e-n}[\rho(\mathbf{r})], \quad (1)$$

where $E[\rho(\mathbf{r})]$ is the total electronic energy, $T[\rho(\mathbf{r})]$ is the electronic kinetic energy, $E_{e-e}[\rho(\mathbf{r})]$ is the contribution to the total energy from electron–electron interactions, and $E_{e-n}[\rho(\mathbf{r})]$ is the contribution to the total energy from electron–nuclear interactions. (By convention, I have not written explicitly the nuclear–nuclear interaction, which is generally added directly to $E[\rho(\mathbf{r})]$ after the computation of $\rho(\mathbf{r})$ and $E[\rho(\mathbf{r})]$.)

The form of $E_{e-n}[\rho(\mathbf{r})]$ is generally expressed as $\int v(\mathbf{r})\rho(\mathbf{r})\,d\mathbf{r}$, where $v(\mathbf{r})$ is the nuclear potential (*vide infra*), generally taken to be a Coulombic interaction. Finding accurate functional forms for $T[\rho(\mathbf{r})]$ and for $E_{e-e}[\rho(\mathbf{r})]$ is a continuing challenge. In fact, a *major difference* between approaches based on the wavefunction and those based on the electronic density is that, in the latter case, accuracy cannot be systematically increased as it can in the former case.

To show the validity of using Eq. (1) to compute the total electronic energy of a many-electron system, Hohenberg and Kohn, in their famous 1964 paper, presented two proofs that provided the foundation for DFT. In the first, they proved that an external potential (such as classical nuclei distributed in space) is a unique functional of the electron density (apart from a trivial additive constant). For most practical purposes, the converse is true, and the electron density of N electrons in an external potential is considered to result uniquely from that potential. Parr and Yang (1989) give an in-depth discussion of these issues, in addition to providing the staple text on DFT. We also remind the reader that a functional maps a set of functions to a set of numbers, in contrast to a function, which maps one set of numbers to another set of numbers.

In the second proof, Hohenberg and Kohn showed that there exists a universal functional of the electronic density, $F[\rho(\mathbf{r})]$, independent of the external potential, $v(\mathbf{r})$, such that the energy of the system,

$$E \equiv \int v(\mathbf{r})\rho(\mathbf{r})\,d\mathbf{r} + F[\rho(\mathbf{r})], \qquad (2)$$

has its minimum at the correct ground-state energy. For the systems of interest here,

$$v(\mathbf{r}) = \sum_{I}^{N_I} \frac{1}{|\mathbf{R}_I - \mathbf{r}|}, \qquad (3)$$

where N_I is the number of atomic nuclei (ions) and $\mathbf{R}_I$ is the position of each nucleus.

The consequences of what Hohenberg and Kohn showed are that if $F[\rho(\mathbf{r})]$ can be found, the unique ground-state electronic density can be found for any configuration of nuclei [i.e., external potential $v(\mathbf{r})$], and this ground-state electronic density can then be used in place of the wavefunction to define the electronic state of the system and, thus, all observable properties of the system.

$F[\rho(\mathbf{r})]$ can be divided into a contribution from the kinetic energy of the electrons, $T[\rho(\mathbf{r})]$, and the energy of interaction of all of the electrons, $E_{e-e}[\rho(\mathbf{r})]$:

$$F[\rho(\mathbf{r})] = T[\rho(\mathbf{r})] + E_{e-e}[\rho(\mathbf{r})]. \qquad (4)$$

$E_{e-e}[\rho(\mathbf{r})]$ can be further divided into three terms:

$$E_{e-e}[\rho(\mathbf{r})] = J[\rho(\mathbf{r})] + E_x[\rho(\mathbf{r})] + E_c[\rho(\mathbf{r})], \qquad (5)$$

where $J[\rho(\mathbf{r})]$ is the mean-field electronic interaction term, traditionally designated with the symbol J, from Hartree–Fock theory,

$$J[\rho(\mathbf{r})] = \iint \frac{\rho(\mathbf{r})\rho(\mathbf{r}')}{|\mathbf{r}-\mathbf{r}'|}\,d\mathbf{r}\,d\mathbf{r}'. \qquad (6)$$

$E_x[\rho(\mathbf{r})]$ is the correction due to Pauli exchange (correlation between electrons of the same spin), and $E_c[\rho(\mathbf{r})]$ is the correction due to all other electronic correlations (primarily between electrons of differing spins). We note that, here and below, all quantities are expressed as atomic units (Hehre et al., 1986), so that no conversion constants need to be included in the equations. Also, as stated above, the nuclear–nuclear repulsion can be added trivially and is considered implicit in the equations presented.

There are two major issues involved in applying the formalism described above. (i) There exists no good functional for the electronic kinetic energy, $T[\rho(\mathbf{r})]$, and (ii) there exists no exact expression for $E_x[\rho(\mathbf{r})]$ or for $E_c[\rho(\mathbf{r})]$. Issue ii has been addressed by the continuing development of more and more accurate functionals for $E_x[\rho(\mathbf{r})]$ and for $E_c[\rho(\mathbf{r})]$. A bit more detail on these functionals is described below. Issue i is drastic and was overcome by Kohn and Sham (1965) in a way that allows the computation of almost the exact electronic kinetic energy, but with a dramatic increase in the computational cost of a density functional theory calculation.

Kohn and Sham's approach is to replace the kinetic energy functional with the exact kinetic energy functional for a noninteracting reference state and then to incorporate into $E_c[\rho(\mathbf{r})]$ the correction in the kinetic energy between the noninteracting reference state and the interacting state. The kinetic energy of the noninteraction reference state is expressed as

$$T_s[\rho(\mathbf{r})] = \sum_i^N \int \psi_i^*(\mathbf{r}) \left(-\frac{1}{2}\nabla^2\right) \psi_i(\mathbf{r})\,d\mathbf{r}, \tag{7}$$

where the $\psi_i(\mathbf{r})$'s are the one-electron (orthonormal) Kohn–Sham orbital wavefunctions. Recall that

$$\rho(\mathbf{r}) = \sum_i^N |\psi_i(\mathbf{r})|^2. \tag{8}$$

(We note that, here and below, we have developed the formalism for spin-paired systems, e.g., systems for which the net spin is zero. It is straightforward to generalize this formalism to treat systems with nonzero spin.)

Now the total energy of the many-electron system can be expressed as

$$E[\rho(\mathbf{r})] = T_s[\rho(\mathbf{r})] + \int v(\mathbf{r})\rho(\mathbf{r})\,d\mathbf{r} + J[\rho(\mathbf{r})] + E_{xc}[\rho(\mathbf{r})], \tag{9}$$

where

$$E_{xc}[\rho(\mathbf{r})] = T[\rho(\mathbf{r})] - T_s[\rho(\mathbf{r})] + E_x[\rho(\mathbf{r})] + E_c[\rho(\mathbf{r})]. \tag{10}$$

The ground state $\rho(\mathbf{r})$ which is the solution to Eq. (9) is then that which minimizes $E[\rho(\mathbf{r})]$, subject to the constraint that the one-electron orbital wavefunctions, ψ_i's, which constitute $\rho(\mathbf{r})$ as expressed in Eq. (8), are orthonormal:

$$\int \psi_i^*(\mathbf{r})\psi_j(\mathbf{r})\,d\mathbf{r} = \delta_{ij}, \tag{11}$$

where δ_{ij} is the Krönicker δ function.

Combining Eqs. (7) and (8) with Eq. (9) yields an expression for the total energy as a functional of the Kohn–Sham orbitals. This expression can be solved subject to the constraints in Eq. (11) using the method of Lagrange's undetermined multipliers. We can write

$$\mathcal{L} = E[\rho(\mathbf{r})] - \sum_i^N \sum_j^N \varepsilon_{ij} \int \psi_i^*(\mathbf{r})\psi_j(\mathbf{r})\,d\mathbf{r}, \tag{12}$$

where the ε_{ij}'s are the Lagrange multipliers.

$\mathcal{L}$ is then functionally minimized with respect to each ψ_i, yielding N coupled Euler–Lagrange equations of the form

$$\left[-\frac{1}{2}\nabla^2 + v_{eff}(\mathbf{r})\right]\psi_i(\mathbf{r}) = \varepsilon_i \psi_i(\mathbf{r}), \qquad i = 1, 2, 3, \ldots, N, \tag{13}$$

where

$$v_{eff}(\mathbf{r}) = v(\mathbf{r}) + \int \frac{\rho(\mathbf{r}')}{|\mathbf{r} - \mathbf{r}'|} d\mathbf{r}' + v_{xc}(\mathbf{r}), \tag{14}$$

and

$$v_{xc}(\mathbf{r}) = \frac{\delta E_{xc}[\rho(\mathbf{r})]}{\delta \rho(\mathbf{r})} \tag{15}$$

the functional derivative of $E_{xc}[\rho(\mathbf{r})]$.

All that remains is to choose a functional form for $E_{xc}[\rho(\mathbf{r})]$. There are quite a number that have proposed and tested, and developing more and more accurate exchange-correlation functionals is an area of ongoing research. The functionals are usually labeled by the agglomeration of the initials of the surnames of the inventors. Typical examples are BLYP (Becke, 1988; Lee et al., 1988), PW (Perdew and Wang, 1992), PBE (Perdew et al., 1996), and HCTH/120 (Boese et al., 2000). These are all universal functionals with fitted parameters that are fixed and used for any system. There is little known on how to choose a given functional for a particular problem to maximize accuracy, although extensive studies on the heats of formation of gas-phase molecules have been performed by Curtiss et al. (1997). Often several functionals are tested for a particular system, and the one that best matches experimental results is chosen. Othertimes, a particular functional is chosen and then its accuracy for calculating the properties of a particular system is determined by comparison to related experimental results. As a rule of thumb, bond energies computed using these functionals have an error of ~5 kcal/mol, and bond lengths and angles have an error of ~1% compared to experiment. It is strongly advised, however, to compare whatever experimental data are available to the computed result for the particular system studied.

C. Choice of Model and Solution of the Equations Using Plane-Wave Basis Sets and the Pseudopotential Method

Once the exchange-correlation functional is chosen, the N one-electron coupled, nonlinear Eqs. (13) can be solved self-consistently for the ψ_i's. To do this, a suitable expansion is chosen for the ψ_i's. This expansion is called a "basis set." (Note that the term "basis set" used in this context differs from that which mathematicians would term a basis set.)

The form of the basis set expansion is chosen to be convenient to model the system and to perform the integrals needed to solve the coupled Eqs. (13). Examples of convenient forms are Gaussians, numerical grids, and plane waves. Plane waves have often (but not exclusively) been chosen as the basis set expansion used in Car–Parrinello simulations for two reasons: (i) It is generally desired to simulate extended systems, such as bulk materials, surfaces, and liquids, and plane waves provide a convenient way to model these systems using periodic boundary conditions; (ii) forces on atomic nuclei can be calculated very efficiently if the electrons are described by plane waves by making use of the Hellman–Feynman theorem (*vide infra*).

Typically the model chosen is a collection of nuclei with corresponding electrons in a periodically repeated cell, called a "supercell." This supercell can have any space filling symmetry. If the system is a crystal, having translational symmetry, the supercell may have one or more unit cells. If the system is an amorphous solid or a liquid, a large enough supercell must be chosen so that correlations introduced by the periodic symmetry do not affect greatly the properties to be determined in the calculation. Finally, if the system is a surface, generally a "slab model," several atomic layers thick, is chosen and enough vacuum space is left so that spurious periodic interactions to not occur. Generally 5–6 Å of vacuum is enough. Examples of supercells with repeated images for each of these three cases are presented in Figs. 1–3.

Plane waves have the following form:

$$f_i = \frac{1}{\sqrt{\Omega}} e^{i\mathbf{g}\cdot\mathbf{r}}, \qquad (16)$$

where Ω is the volume of the cell, and $\mathbf{g}$ is a reciprocal lattice vector of the cell. The ψ_i's in Eqs. (11)–(13) cannot, however, simply be expressed as a summation of f_i's over reciprocal lattice vectors, because application of the Bloch theorem to electronic orbitals in periodic potentials leads to the introduction of wave vectors, $\mathbf{k}$, lying in the first Brillouin zone (see, e.g., Ashcroft and Mermin, 1976).

Thus, using periodic boundary conditions, the ψ_i's, in Eqs. (11)–(13) are expanded at each wave vector $\mathbf{k}$ (also stated "at each k-point") in terms of plane waves as follows:

$$\psi_i^{\mathbf{k}} = \sum_{\mathbf{g}} c_i^{\mathbf{k}}(\mathbf{g}) e^{i(\mathbf{g}+\mathbf{k})\cdot\mathbf{r}}, \qquad (17)$$

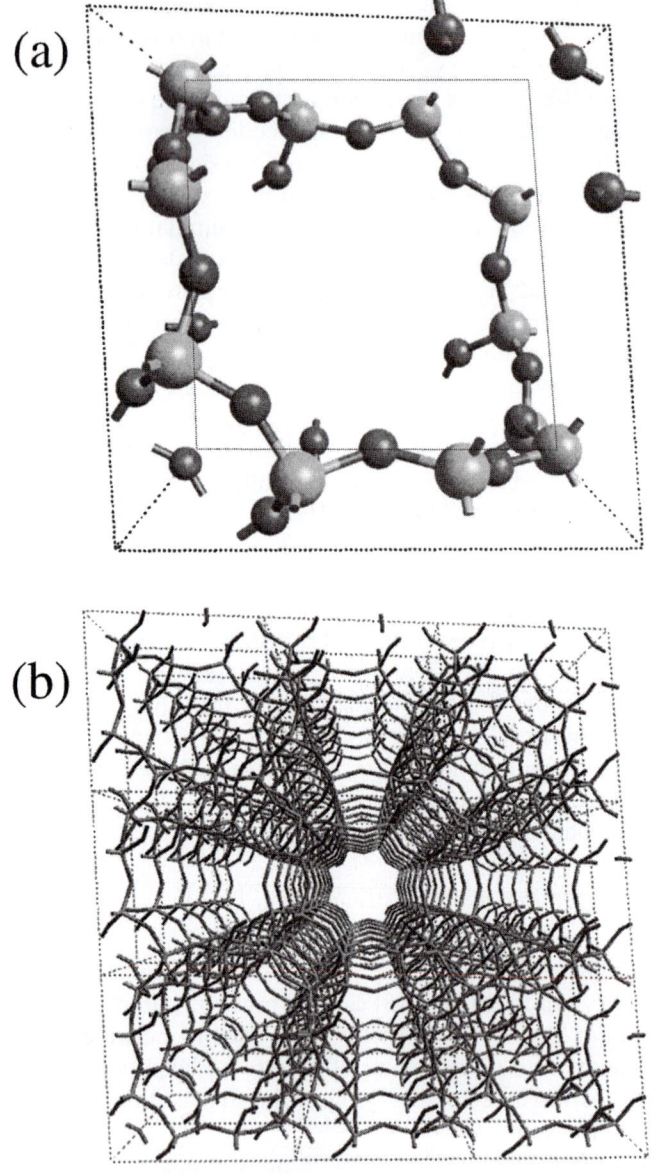

FIG. 1. (a) A trigonal unit cell of the zeolite chabazite, consisting of 36 atoms. This unit cell is also chosen as the supercell. (b) A perspective view down a channel of chabazite; 54 unit cells are displayed.

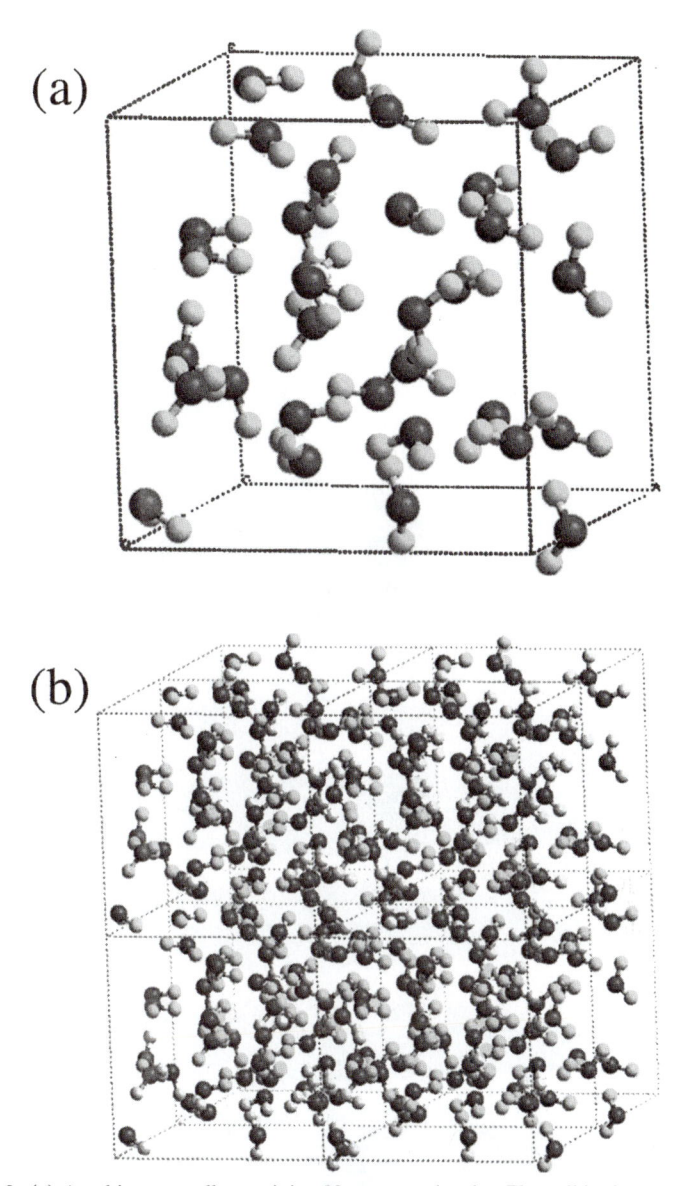

FIG. 2. (a) A cubic supercell containing 32 water molecules. The cell is chosen so that the density is 1.00 g/cm^3. (b) Eight supercells demonstrating the disorder of the system.

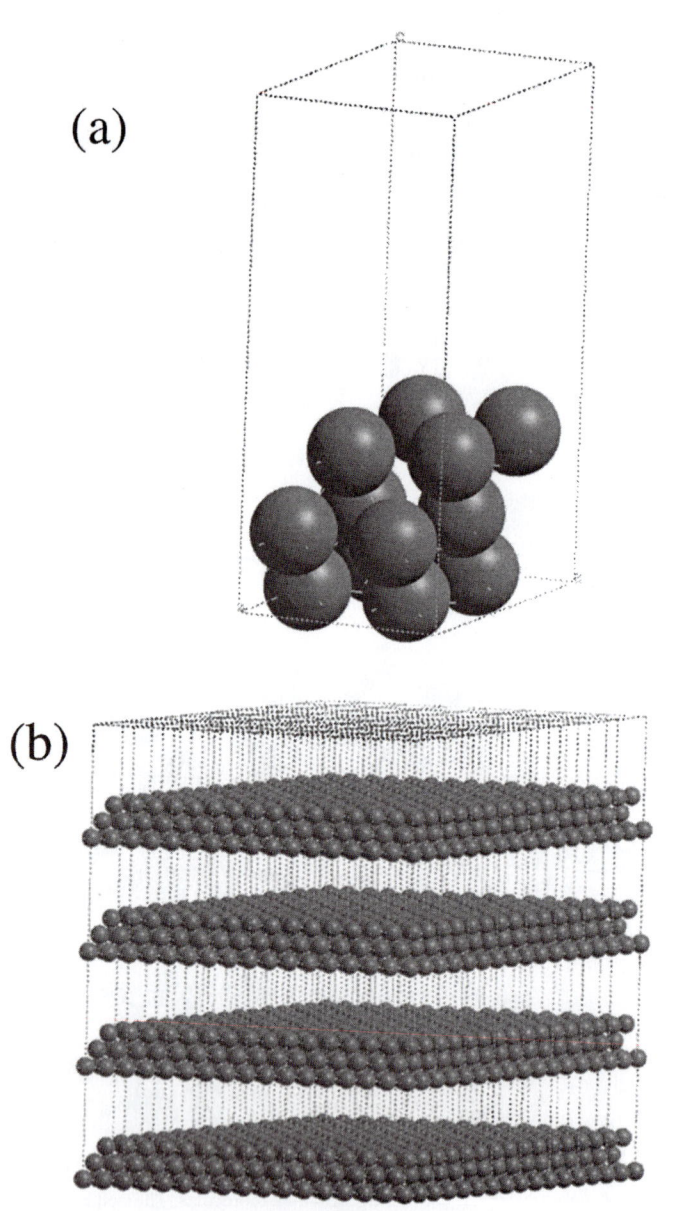

FIG. 3. (a) A slab model of the platinum (111) surface, consisting of 12 atoms. The symmetry of the supercell is hexagonal. (b) Two hundred fifty-six supercells.

where the $c_i^k(\mathbf{g})$'s are coefficients to be determined in the calculation. Now the ψ_i's in Eqs. (6)–(12) must be replaced by $\psi_i^{(\mathbf{k})}$'s and the electron density [Eq. (7)] becomes

$$\rho(\mathbf{r}) = \sum_{\mathbf{k}} w_{\mathbf{k}} \sum_i y_i^{\mathbf{k}} |\psi_i^{\mathbf{k}}|^2, \tag{18}$$

where $w_{\mathbf{k}}$ is the weight of each k-point, and $y_i^{\mathbf{k}}$ is the occupation number of each orbital i at each k-point. For these expansions to be exact, the summations over $\mathbf{k}$ and $\mathbf{g}$ should be infinite. In practice, the number of k-points in the summation can often be small, since sets of special k-points can be chosen for a given lattice symmetry. The summation over $\mathbf{g}$ is then performed to include all of those reciprocal lattice vectors with energy $E = \frac{1}{2}(\mathbf{k}+\mathbf{g})^2$ less than a cutoff energy, E_{cut}. E_{cut} is generally expressed in units of Rydbergs and is generally roughly between 10 and 100 Ry. For a 1-nm^3 cell this corresponds to 10,000 to 100,000 plane waves per k-point for each Kohn–Sham orbital.

Plane waves, while natural for describing electronic orbitals in periodic potentials, do not describe the orbitals well near the nuclei, where electronic gradients can be quite large. In fact, to describe properly these orbitals near the nuclei, expansions much larger than those mentioned above would have to be used, unless a special technique were used to treat the core electronic region. This technique is called the "pseudopotential method." In this method, the core electrons are replaced with a fixed potential (which is generally nonlocal), and only the valence electrons are treated explicitly. Such an approach is validated by the fact that, in general, the energy levels of core electrons do not change under different chemical environments as evidenced by numerous XPS data. When plane waves are the chosen basis functions and pseudopotentials are used, this method is called the "plane-wave pseudopotential method."

Using pseudopotentials has several major beneficial consequences: (i) Only the valence electrons must be treated explicitly, thus the number of equations to be solved [Eqs. (13)] can be reduced drastically; (ii) the pseudo-orbitals are very smooth near the atomic core, and thus E_{cut} can be reduced drastically; and (iii) important relativistic effects of the core electrons of heavy elements such as the 5d elements can be included in nonrelativistic calculations. The major downsides are that the potential $v(\mathbf{r})$ in Eq. (3) must be replaced with a more complicated and computationally expensive nonlocal pseudopotential and, more importantly, that the transferability of the pseudopotential, i.e., its accuracy in different bonding environments, may not be perfect. Developing highly transferable pseudopotentials that can be used at as low an E_{cut} as possible is a major current topic of research.

Once the k-points and the basis set expansion for the $\psi_i^\mathbf{k}$'s are chosen, the coupled Eqs. (13) can be solved self-consistently for the coefficients, $c_i^\mathbf{k}(\mathbf{g})$. The total electronic energy of the system can then be calculated via

Eq. (9), and the forces on the nuclei can also be determined. More details on pseudopotential methods, on methods of solving the Kohn–Sham equations, and on the algorithms used can be found in the references described in Appendix A.

As noted above, when the orbitals are expanded in terms of plane waves, forces can be calculated accurately and efficiently using the Hellman–Feynman theorem. This theorem states that the sum of the last two terms of the derivative of the total energy with respect to a nuclear coordinate λ, expressed in Dirac notation as follows,

$$\frac{d}{d\lambda}E(\lambda) = \langle \psi(\lambda)| \frac{d}{d\lambda}H(\lambda)|\psi(\lambda)\rangle + \left[\frac{d}{d\lambda}\langle \psi(\lambda)|\right] H(\lambda)|\psi(\lambda)\rangle$$
$$+ \langle \psi(\lambda)|H(\lambda)\left[\frac{d}{d\lambda}|\psi(\lambda)\rangle\right] \quad (19)$$

are equal to zero. Thus,

$$\frac{d}{d\lambda}E(\lambda) = \langle \psi(\lambda)| \frac{d}{d\lambda}H(\lambda)|\psi(\lambda)\rangle. \quad (20)$$

This theorem is, of course, valid only when ψ is described exactly. Because plane waves describe all of space evenly and are not centered on atomic nuclei, Eq. (20) can be used to calculate forces accurately for approximate plane-wave expansions of ψ. For expansions of ψ in terms of atomic centered basis functions such as Gaussians and atomic centered grids, Eq. (19) must be used. This is because space is not covered uniformly by atomic centered basis functions, and thus, errors in ψ are nonuniform, and the last two terms in Eq. (19) do not add to zero. The added expense of calculating forces via Eq. (19) often prohibits the use of atomic centered basis functions in molecular dynamics simulations.

D. Car–Parrinello Molecular Dynamics

The forces on each atomic nucleus, calculated quantum mechanically for a given nuclear configuration, can then be used in one of the finite-difference methods for propagating the atomic nuclei described in Section A above. One might envision solving for the orbitals and forces for a given nuclear configuration as described in Section B, advancing the nuclear positions in time, computing the orbitals and forces for the new nuclear configuration, etc. This method, called "Born–Oppenheimer molecular dynamics" is currently used by many researchers, but it is different from the original method developed by Car and Parrinello, which is also the primary method used today.

The Car–Parrinello method presents a trick to propagate the orbitals along with the nuclei without reoptimizing them at each molecular dynamics step. It does so by introducing a fictitious orbital mass and propagating the orbitals *with* the nuclei via appropriate equations of motion. The Car–Parrinello Lagrangian is

$$L = \sum_i \frac{1}{2} \int d\mathbf{r} \mu |\dot{\psi}_i(\mathbf{r})|^2 + \sum_I \frac{1}{2} M_I \dot{R}_I^2 - E[\{\psi_i(\mathbf{r})\}, R_I]$$
$$+ \sum_{ij} \Lambda_{ij} \left(\int d\mathbf{r} \psi_i^*(\mathbf{r}) \psi_j(\mathbf{r}) - \delta_{ij} \right). \tag{21}$$

(Note that for simplicity, I have not written explicitly the k-point indices or the orbital occupancies.) In Eq. (21), the second term is the kinetic energy of the nuclei and the third term is the potential energy of the nuclei, which is also the electronic energy [Eq. (9) if calculated via DFT]. The first term is the fictitious kinetic energy of the orbitals, where μ is the fictitious mass of the orbitals. The fourth term is a set of constraints which keep the orbitals orthonormal, where Λ_{ij} are the undetermined multipliers. Note that additional constraints can be added into this Lagrangian (*vide infra*).

From this Lagrangian, the following equations of motion are generated:

$$\mu \ddot{\psi}_i(\mathbf{r}, t) = -\frac{\delta E[\{\psi_i(\mathbf{r}, t)\}, R_I]}{\delta \psi_i^*(\mathbf{r}, t)} + \sum_j \Lambda_{ij} \psi_j(\mathbf{r}, t) \tag{22a}$$

$$M_I \ddot{R}_I = -\nabla_{R_I} E[\{\psi_i(\mathbf{r}, t)\}, R_I]. \tag{22b}$$

Thus, the orbitals are propagated with the nuclei, although while the nuclear motion is physical, the motion of the orbitals is only a means to adjust the orbitals so that they remain on the ground-state (Born–Oppenheimer) surface. The original idea of Car and Parrinello was to use these equations of motion to optimize simultaneously the orbitals and the nuclear configuration, using a method of dynamic simulated annealing (Kirkpatrick *et al.*, 1983). As it turns out, Car–Parrinello calculations are rarely performed to do exactly this, although they are often used to find structural properties of materials by allowing the nuclei to jump over small barriers. Their primary power, however, comes in their ability to investigate the short-time dynamics of systems and to sample reaction pathways within various ensembles.

We might ask why this method works and, even more importantly from a practical standpoint, when it will not work. The motion of the orbitals for small deviations from the ground state can be described by a superposition of oscillations with frequencies:

$$\omega_{ik} = \left(\frac{2(\varepsilon_k - \varepsilon_i)}{\mu}\right)^{\frac{1}{2}}, \qquad (23)$$

where ε_k is the eigenvalue of an unoccupied state and ε_i is the eigenvalue of an occupied state. Thus, the lowest orbital frequency is $\omega_{min} = (2E_g/\mu)^{\frac{1}{2}}$, where E_g is the energy gap (HOMO–LUMO gap for molecules). Choosing a typical value of μ of 400 a.u. and for typical values of E_g of ~2 eV, ω_{min} is of the order of 1000 THz, versus the highest frequency of the nuclear motion of about 100–200 THz (Galli and Parrinello, 1991). This separation in the rate of orbital and nuclear motion allows the orbitals to follow the motion of the nuclei, while remaining on their ground-state (Born–Oppenheimer) surface. The values of μ and M_I are major factors in determining the minimum time step necessary for accurate integration in the molecular dynamics simulation. A typical value of the time step in a Car–Parrinello simulation for the situation described above is ~0.1 fs.

As E_g gets smaller, μ must be set smaller, and consequently, the molecular dynamics time step must be set smaller. If the time step is too large, there will be an exchange of energy between the nuclear and the orbital motion, and the orbitals will not remain on the ground-state (Born–Oppenheimer) surface. When this occurs, the effect on the simulation is disastrous. Obviously, the smaller the molecular dynamics time step is, the less efficient the simulation is. This will be a problem for metals in particular, where E_g will be very small (0 for an infinite system). There are two ways of addressing this problem. One is by attaching a thermostat to control the motion of the orbitals (Galli and Parrinello, 1991), and the other is by performing Born–Oppenheimer molecular dynamics as described at the beginning of Section D.

I close this section by noting that the choice of performing Car–Parrinello molecular dynamics, using Eqs. (22a) and (22b) versus Born–Oppenheimer molecular dynamics should be made as a trade-off between efficiency and accuracy. There have been only a few studies on this choice, an excellent description of which is given by Marx (2000).

V. Applications

The methodology described above provides the ability to perform molecular dynamics simulations without choosing forms and parameters for interaction potentials. Because Car–Parrinello simulations are molecular dynamics simulations, all methodologies developed for classical molecular dynamics simulations can be used in Car–Parrinello methods. Thus, (N,V,T) simulations can be performed in addition to (N,P,T) simulations, etc. When the interaction among atoms is determined quantum mechanically, the breaking and formation of bonds can be treated accurately, and charge distributions, polarizability, and charge transfer, in addition to other properties, can be computed.

The downside to the power of Car–Parrinello calculations is that they are computationally costly. Typically, the largest systems that can be treated are of the order of 100 atoms, and the time scale of the simulations is of the order of picoseconds. The purposes of Car–Parrinello simulations generally fall under three main categories: (i) simulations as a means of optimization to determine structural properties, (ii) direct simulations of processes occurring over short time scales, and (iii) simulations to sample equilibrium properties.

The majority of Car–Parrinello simulations performed so far fall under category i. Car–Parrinello simulations allow nuclei to hop over small energetic barriers, whereas typical geometry optimizations (with dynamics) yield only the closest local minimum to the starting point. Examples falling under this category include the calculation of elastic constants, interfacial structures of electronic materials and grain boundaries, molecular crystals, minerals, and structures of various defects. Two notable examples are given by Haase *et al.* (1997) and by Janotti *et al.* (1997) and Papoulias *et al.* (1997).

Haase *et al.* (1997) studied methanol adsorption at active sites of an acidic chabazite catalyst (see Fig. 1). They addressed a long-standing controversy of whether methanol is chemisorbed at the acid site (the acidic proton is transferred to the methanol) or physisorbed (the acidic proton is not transferred to the methanol). Previous quantum mechanical studies were inconclusive, primarily because the highly corrugated potential energy surface had many local minima in the region of interest, and this surface was very difficult to sample comprehensively. By allowing the entire system to move dynamically at 400 K, Haase *et al.* (1997) were able to sample a range of local minima and showed that the lowest energy of adsorption of methanol was the physisorbed state.

Both Janotti *et al.* (1997) and Papoulias *et al.* (1997) studied the formation of defects in GaAs. Using Car–Parrinello methods to optimize simultaneously the orbitals and the geometry of their system, Janotti *et al.* (1997) predicted the stability of various defect pairs over others, in addition to the properties of those defect pairs. Papoulias *et al.* (1997) studied As interstitial pairs in As-rich GaAs. Previous static studies had shown that there are many different defect pairs possible with stable minima. Papoulias *et al.* (1997) showed that the lowest energy defect consisted of As atoms forming a pair of split interstitials centered on nearest-neighbor sites. They also investigated the electronic properties of these As pairs and concluded that they are electrically inactive, thus explaining experimental observations of a large number of inactive As in GaAs grown using certain methods.

Studies under categories ii and iii provide more poignant examples of the power of Car–Parrinello methods. Because of the magnitude of the literature on applications of Car–Parrinello simulations, I have chosen to focus on a few case studies to illustrate the potential of simulations under these categories for problems of interest to chemical engineers. The areas that I have chosen are (A) gas-phase processes; (B) processes in bulk materials; (C) properties of liquids, solvation, and reactions in liquids; (D) heterogeneous reactions and processes on surfaces; (E) phase transitions; and (F) processes in biological systems.

A. Gas-Phase Processes

Gas-phase systems, if they are small enough, are convenient to study, because of the relatively small computational time. To study such systems using the plane-wave pseudopotential method, large supercells must be chosen to avoid spurious interactions from periodic potentials. Of course, if localized basis functions are chosen, such as Gaussians, there is no periodicity.

Two example studies used Car–Parrinello methods to study gas-phase reactions directly. These fall under category ii described above. Yamataka *et al.* (1999) simulated the reaction of the formaldehyde radical anion and methyl chloride. Frank *et al.* (1998) simulated the reaction of OH radicals with ketones.

Since the reaction of the formaldehyde radical anion with methyl chloride involves overcoming a barrier over time scales much larger than a picosecond, Yamataka *et al.* (1999) could not directly simulate the entire reaction process. Instead, they started at a transition-state structure, known from static quantum mechanical calculations; chose random initial velocities; and performed nine simulations at 298 K. Three of these simulations resulted in the formation of the reactants, three resulted in the formation of products via an electron-transfer reaction, and three resulted in the formation of products via a carbon-substituted S_N2 reaction. These three sets of resulting structures are shown in Fig. 4.

The trajectories of the electron-transfer process consisted of three phases as shown in Fig. 5. During the first phase, up to about 30 fs, there was not much change. During the second phase, which occurred over the next 200 fs, the chlorine atom starts separating from its adjacent carbon atom, but after an initial drop, the total energy remains constant. After this period, during phase 3, the carbon–carbon distance increases, and both the charge and the

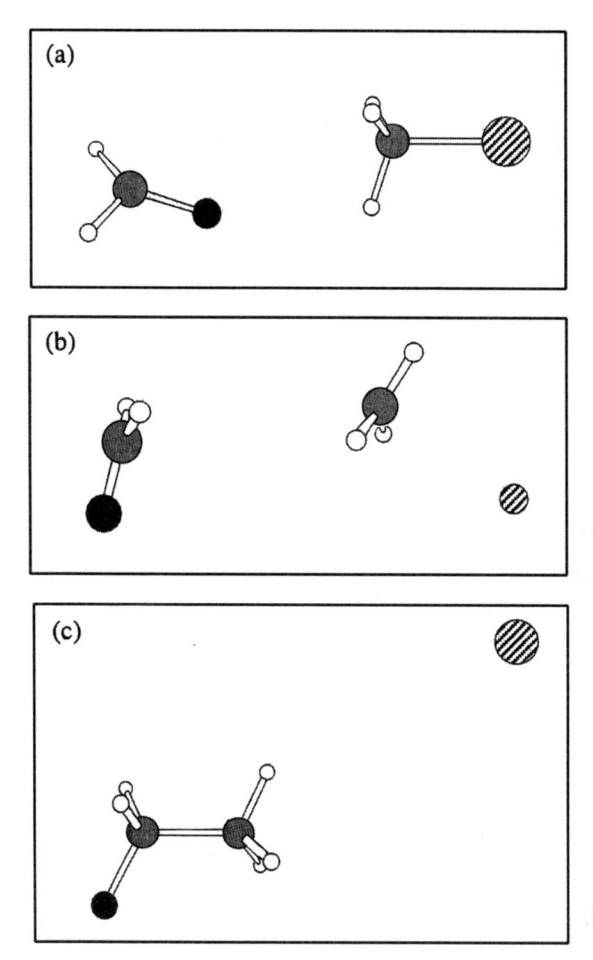

FIG. 4. Snapshots of structures obtained (a) after 500 fs in a trajectory leading to the reactant state, (b) after 300 fs in a trajectory leading to the electron-transfer product, and (c) after 300 fs in a trajectory leading to the S$_N$2 products. Reprinted with permission from Yamataka *et al.* (1999).

spin density on each species stabilize. The simulations by Yamataka *et al.* (1999) support the notion that products could be formed either via an S$_N$2 process or via an electron-transfer process at 298 K. They also produced a detailed analysis of the nature of the reaction process.

Ketones are organic pollutants which react with OH in the atmosphere. As an alternative to performing expensive and time-consuming experimental studies on these reactions, Frank *et al.* (1998) chose to evaluate the reactivity

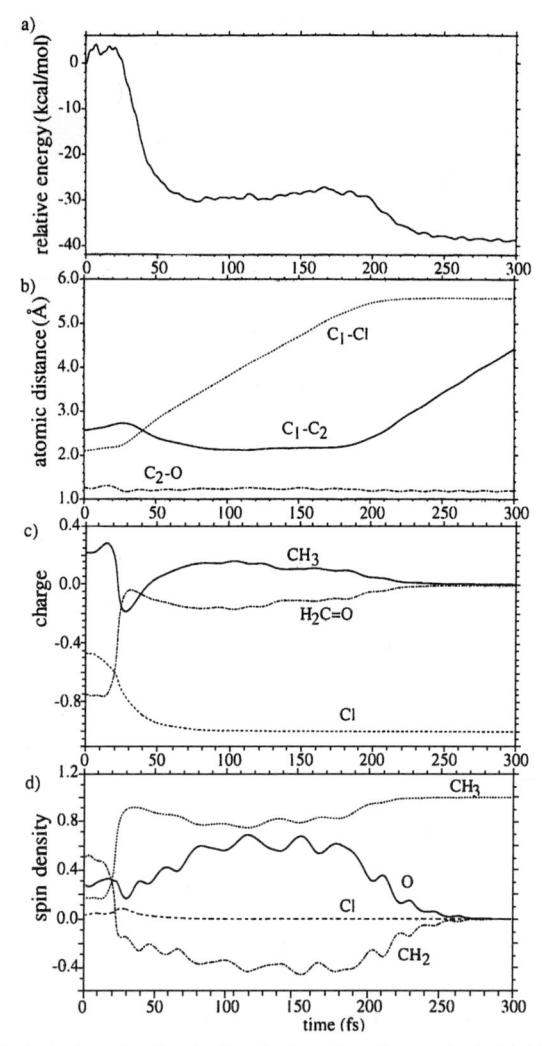

FIG. 5. Sample trajectory leading to the electron-transfer product. (a) Electronic energy, (b) atomic distances, (c) Mulliken group charges, and (d) spin density. Reprinted with permission from Yamataka *et al.* (1999).

of various ketones with OH using Car–Parrinello simulations. As in the previous example, reaction barriers prevent simulating an entire reaction process at ambient temperatures. Thus, to evaluate relative reactivities of various ketones, Frank *et al.* (1998) chose initial reactant configurations and gave the species relatively high kinetic energies with velocities sampled from Maxwell–Boltzman distributions. In this way, they were able to delineate two

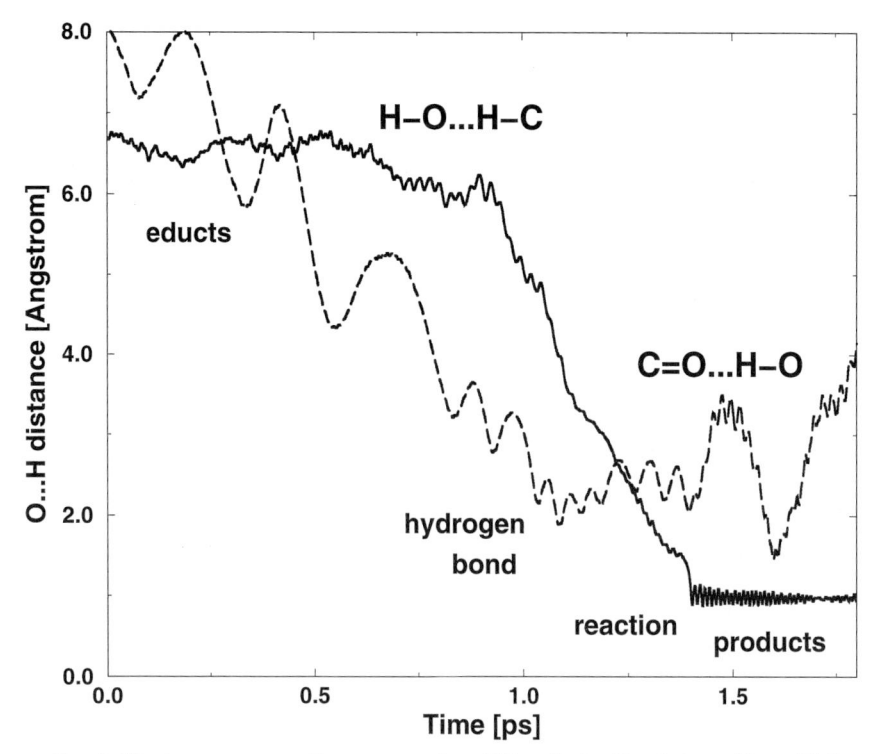

FIG. 6. The low-energy reaction process of an OH radical with a ketone. The solid line shows, as a function of time, the distance between the oxygen atom in the OH radical and the hydrogen atom in the ketone molecule. The dashed line shows, as a function of time, the distance between the oxygen atom in the ketone molecule and the hydrogen atom in the OH radical. Reprinted with permission from Frank *et al.* (1998).

reaction mechanisms for the process, depending on the kinetic energy. This implies that the product distribution will be a strong function of the kinetic energy of the reactants.

The mechanism at high kinetic energies was a simple abstraction process. At lower kinetic energies, a more complex mechanism ensues, as illustrated in Figs. 6 and 7. Initially, a hydrogen bridge is formed. Next, the oxygen atom in the OH approaches one of the hydrogen atoms in the ketone, leading to a ring structure. Finally, the hydrogen bridge is broken, leading to a free water molecule and a ketyl radical. It should be noted that the ring structure (Fig. 7, bottom left) does not correspond to a local minimum and would not have been observed by static quantum calculations.

Other gas-phase studies include the thermal dissociation of acetic acid (Liu *et al.*, 1999b), and the study of the reaction $Cl^- + CH_3Br \rightarrow CH_3Cl + Br^-$

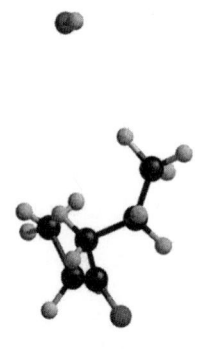

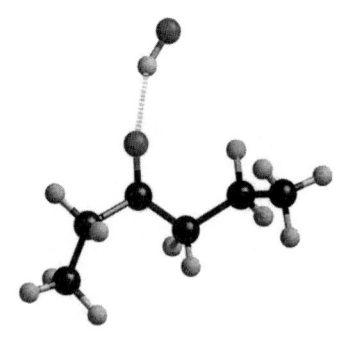

0 fs: educts **1149 fs: hydrogen bond**

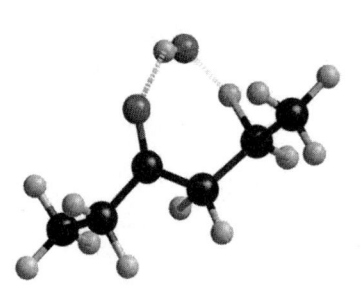

1391 fs: reaction **1451 fs: products**

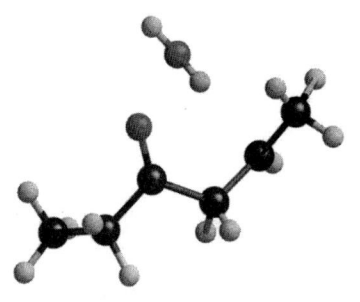

FIG. 7. Snapshots of the low-energy reaction process of an OH radical with a ketone. Reprinted with permission from Frank *et al.* (1998).

(Raugei *et al.*, 1999). There have also been quite a number of studies on gas-phase clusters of varying types, in addition to several other gas-phase reactions, including isomerization reactions.

B. Processes in Bulk Materials

Aside from using Car–Parrinello methods to evaluate structural properties and relaxation processes in bulk materials, they have been used under category ii to evaluate directly diffusivities of ions in materials. Wengert *et al.* (1996) studied the diffusivities of Si, Mg, and Li in the superionic conductor, $Li_{2-2x}Mg_{1+x}Si$ ($x \sim 0.06$) at 600, 900, and 1400 K. Some of their results from their trajectories are presented in Fig. 8.

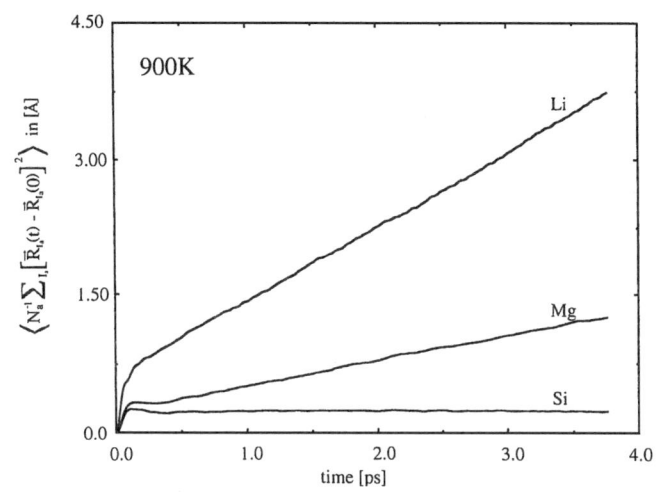

FIG. 8. Mean square displacements of Li, Mg, and Si at 900 K as a function of time. Reprinted with permission from Wengert et al. (1996).

Using the Einstein relation,

$$D = \lim_{t \to \infty} \frac{\langle R^2(t) \rangle}{6t}, \qquad (24)$$

the diffusivities of Li, Mg, and Si were determined to be 1.4×10^{-5}, 0.5×10^{-5}, and 0.0×10^{-5} cm^2/s, respectively. It can also be noted in the plots that the linear, diffusive regime was reached at short times (<1 ps). For diffusion that occurs via hopping of large barriers, the time to reach the diffusive regime might be much larger. In this example, the root mean square displacements are small, a few angstroms, over the time scale of the simulations, and one should always be cautious about extrapolating long-time behavior from results over short times. Nevertheless, these values seem reasonable, and Fig. 8 clearly shows that at times above ∼0.2 ps, the atoms are no longer moving in an inertial regime.

Molteni et al. (1996a) (see also Molteni et al., 1996b) studied the sliding of grain boundaries in germanium using *ab initio* simulations. They simulated the sliding process quasi-statically by shifting the relative positions of two grains and letting the geometry relax after each shift. By this means, they were able to calculate quantitatively the energy as a function of the shear displacement. Figure 9 shows the energy per unit area as a function of the displacement of the grain boundary. They found that the sliding takes place via a stick-slip mechanism, mediated by a rebonding process.

Other studies include the determination of the mechanism of growth of carbon nanotubes (Charlier et al., 1997; Bernholc et al., 1998) and the

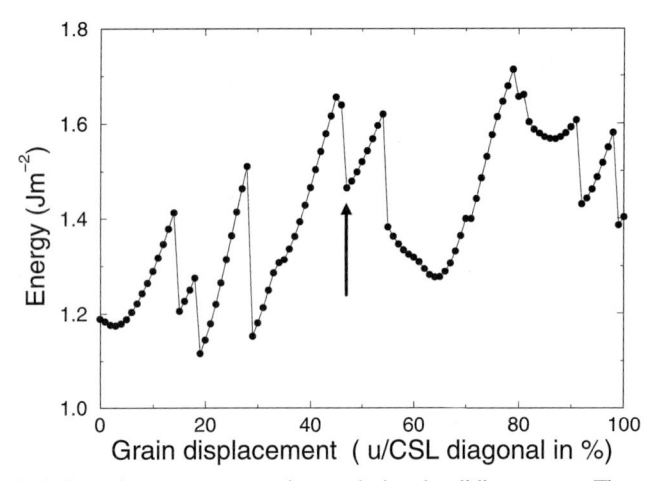

FIG. 9. Grain boundary energy per unit area during the sliding process. The arrow shows where the disorder starts to migrate away from the boundary interface. Reprinted with permission from Molteni et al. (1996).

diffusion and motion of various defects in bulk materials (Hamann, 1998; Valladares et al., 1998) including the process of radiation-induced formation of defects (Estreicher et al., 1999). A study by Debernardi et al. (1997) used Car–Parrinello simulations to reproduce the essential features of the IR adsorption spectrum of amorphous silicon.

C. Properties of Liquids, Solvation, and Reactions in Liquids

One of the major objectives of first-principles calculations has been to be able to study liquids. This was not possible before the advent of Car–Parrinello methods, since the disorder of the liquid could not be averaged properly. In recent years, quite a number of studies of reaction processes and structural properties (categories ii and iii) have been performed, primarily on water and on molten metals.

Silvestrelli and Parrinello (1999) studied the structural and bonding properties of liquid water using supercells similar to the one shown in Fig. 2. Note that for computational efficiency, they actually studied D_2O. Their work [building on the previous work of Sprik et al. (1996)] demonstrates the ability to compute pair correlation functions, as shown in Fig. 10. Comparison with experimental results is extremely good. It should also be noted that there are discrepancies between the two experimental results for the oxygen–oxygen pair correlation curves. (There has been considerable controversy in the literature regarding these curves.)

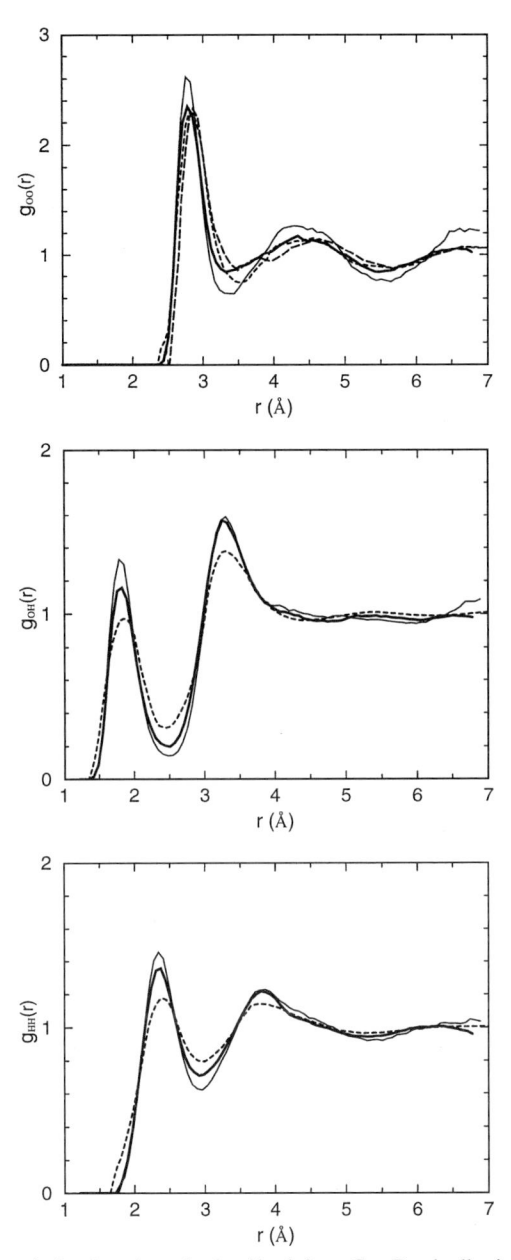

FIG. 10. Pair correlation functions obtained both from Car–Parrinello simulations and from experiments. The thick solid lines are simulation results obtained using a supercell with 64 water molecules. The thin solid lines are for the 32-water molecule simulation. The short-dashed line is from experimental neutron scattering results (Soper et al., 1997), and the long-dashed line is from an X-ray study. Reprinted with permission from Silvestrelli and Parrinello (1999).

In the same study, the diffusivity of water under ambient conditions was determined for the 64-water molecule system to be $2.8 \pm 0.5 \times 10^{-5}$ cm^2/s, very close to the experimental value of 2.4×10^{-5} cm^2/s. Also, the investigators were able to determine the distribution of dipole moments of the water molecules via generation of functions that describe the charge distribution on individual molecules. They found a broad distribution with an average of 3.0 D. This result, aside from being of fundamental scientific significance, has tremendous consequences for modeling of aqueous systems using classical potentials, since in these models the dipole moment of water must be included in the parameterization.

Silvestrelli et al. (1997) used Car–Parrinello molecular dynamics to obtain the IR spectrum of D_2O. They did so by determining the time autocorrelation function for the dipole moment of their cell and then relating this function to the absorption coefficient as a function of frequency. Their computed spectrum and comparison to experiment are displayed in Fig. 11, taking into account corrections introduced by the authors. The essential features of the experimental spectrum, particularly the low-frequency peak, are reproduced well. The authors were then able to assign specific modes of the spectrum.

This method of computing spectra should be contrasted with the more typical method of calculating harmonic vibrational modes from quantum calculations performed at 0 K. The latter method cannot be applied to liquid systems for which a single, static configuration will not properly describe the system. In addition, the method of Silvestrelli et al. (1997) incorporates anharmonic contributions and allows the computation of the IR spectrum directly from the motions of the dipole moment, instead of from nuclear vibrations. While computing IR spectra from first principles in this way is not yet entirely straightforward, Silvestrelli et al. (1997) have shown that accurate computations are possible, in addition to demonstrating the power of Car–Parrinello simulations to deconvolute the peaks and to assign them to specific modes.

Other work has been performed in liquid systems, including studying homogeneous catalytic processes, the transport of hydronium and hydroxyl ions in water, reactions in water, the solvation of ions in water, the structure of water/solid interfaces, and the viscosities of liquids.

Woo et al. (1997b) used Car–Parrinello molecular dynamics to calculate the free energy barriers for chain termination and chain branching processes during the polymerization of olefins on homogeneous constrained geometry catalysts. Later, the same authors used hybrid methods to study various other homogeneous polymerization catalysts (Woo et al., 1997a). Aagaard et al. (1998) studied the reaction mechanism of a ruthenium-based metathesis catalyst, lending support to a mechanism proposed in the literature for the process.

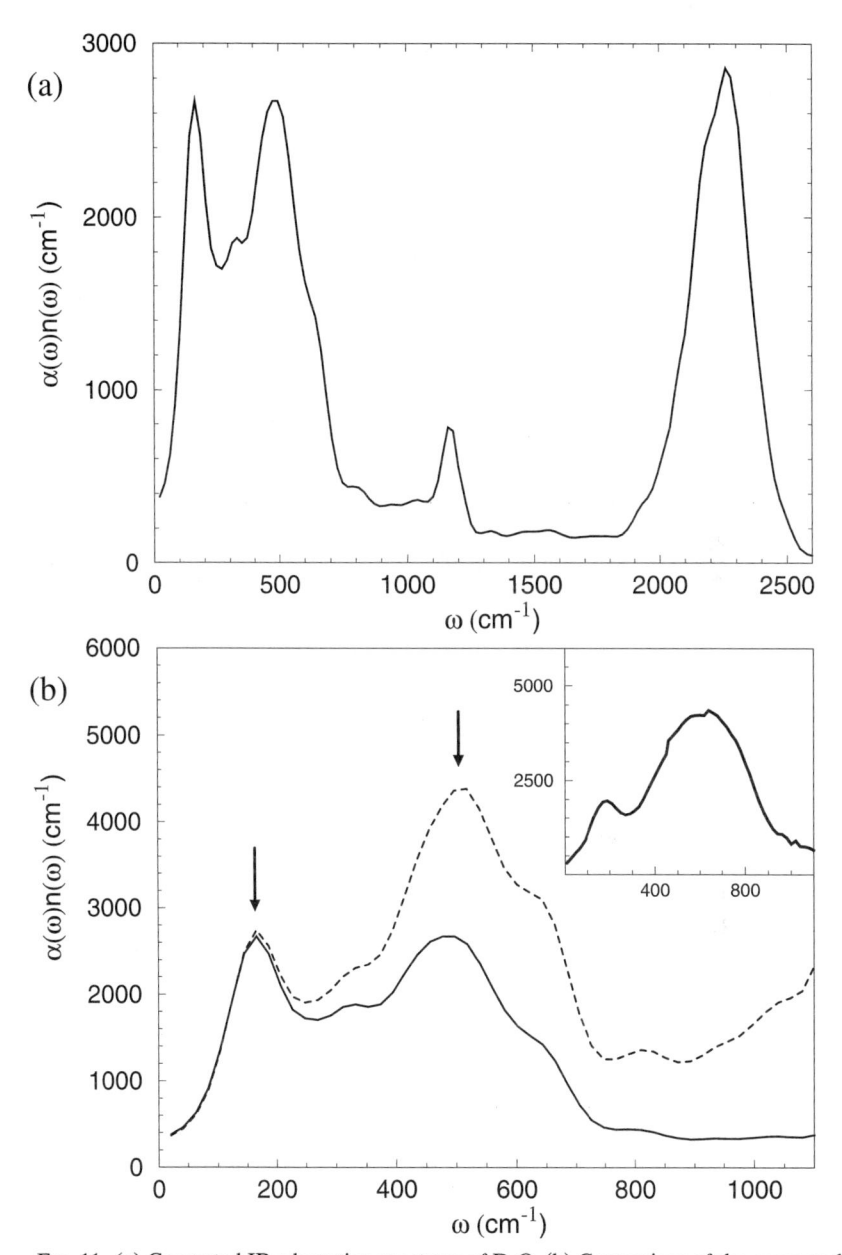

FIG. 11. (a) Computed IR adsorption spectrum of D_2O. (b) Comparison of the computed IR adsorption spectrum with experiment. The solid line is the computed spectrum, the dashed line is the computed spectrum with Egelstaff quantum corrections, and the inset shows an experimental IR spectrum of H_2O. Reprinted with permission from Silvestrelli et al. (1997).

Tuckerman et al. (1995) used a 32-water molecule system to investigate the structure and dynamics of proton transfer in aqueous systems at ambient conditions. For hydronium ions, they found dynamic ion complexes, which fluctuated between an $(H_5O_2)^+$ structure and an $(H_9O_5)^+$ structure. For hydroxyl ions, they found that the predominant structure is a fourfold coordinated and planar $(H_9O_5)^-$ complex. They found, however, that proton transfer occurs only via an $(H_7O_4)^-$ complex. They also estimated activation barriers to proton hopping, which are consistent with experimentally observed barriers. The fourfold planar coordination of hydroxyl came as a surprise, since this structure causes the hydrogen bonding network of water to be distorted. Trout and Parrinello (1999) explained this via electron localization arguments.

Liu et al. (1999a) studied the hydrolysis of Cl_2 in water, while Meijer and Sprik (1998a) investigated the reaction mechanism of water with formaldehyde in sulfuric acid. Laasonen and Klein (1997) studied the hydrolysis of HCl upon addition to water, and Meijer and Sprik (1998b) studied the addition of H_2O to SO_3 in solution. Finally, Trout and Parrinello (1998) evaluated the mechanism and free energy profile of the dissociation of H_2O in water.

The solvation of several ions in water has been investigated, including K^+ (Ramaniah et al., 1999), Cu^{2+} (Berces et al., 1999), and Be^{2+} (Marx et al., 1997). The structure of a water/silicon interface was studied (Ursenbach et al., 1997), in addition to a water/copper interface (Halley et al., 1998) and a water/palladium interface (Klesing et al., 1998). Finally, two studies have used Car–Parrinello simulations in conjunction with the Green–Kubo relations to calculate viscosities in liquid metals (Alfe and Gillan, 1998; Stadler et al., 1999).

D. HETEROGENEOUS REACTIONS AND PROCESSES ON SURFACES

There has been a variety of studies using Car–Parrinello simulations to determine the structure and energetics of adsorbates on semiconductor and insulating surfaces. Studies on metal surfaces are much rarer, and as far as I know, first-principles molecular dynamics simulations have not yet been used to study reactive processes on metals. The reason is primarily one of computational expense, because metals require the inclusion of a large number of k-points. There is, of course, a substantial body of work which uses static quantum mechanical calculations to study reactions on metal surfaces.

In this section, I have chosen to focus on two case studies of heterogeneous catalytic reaction processes. These fall under category iii, as defined above. The first one involves polymerization reactions on

Ziegler–Natta catalysts. The second one involves hydrocarbon coupling in the solid acid zeolite chabazite.

Boero et al. (1998) used Car–Parrinello molecular dynamics to study the polymerization of ethylene at titanium sites in $MgCl_2$-supported Ziegler–Natta catalysts. Their objectives were to evaluate the reaction mechanism, in addition to determining the free energy profile of the polymerization process. Obviously, the characteristic time scale of this process is much greater than the picosecond time scale directly accessible by the simulation. Thus, it is not possible to observe the polymerization process via a straightforward Car–Parrinello simulation.

To circumvent this problem of time scales, Boero et al. (1998) used a method of sampling the derivative of the free energy with respect to a reaction coordinate along a reaction pathway. They then integrated this curve to obtain the free energy profile for the reaction. This procedure is called the method of constraints within the "blue moon" ensemble (Carter et al., 1989). The authors chose the initial step in the polymerization process to be the addition of an ethylene molecule to a methyl group bound to the Ti site in the catalyst. Thus, a natural reaction coordinate to choose is the distance between a carbon atom from the ethylene molecule and the carbon atom in the methyl group. With this simple distance constraint, the derivative of the free energy is

$$\frac{dF}{dr_{C-C}} = -\langle f \rangle_{r_{C-C}}, \quad (25)$$

where F is the free energy, r_{C-C} is the value of the constraint, and $\langle f \rangle_{r_{C-C}}$ is the ensemble averaged force on the constraint, which is equivalent to the ensemble averaged Lagrange multiplier associated with the constraint in the equations of motion. Note that almost any constraint can be chosen as the reaction coordinate, but the resulting equation may be much more complicated than Eq. (24) (Sprik and Ciccotti, 1998).

Initially, Boero et al. (1998) performed an optimization of their entire system on a sixfold coordinated Ti site called the Corradini site. They then varied the values of their reaction coordinate, decreasing it by increments of 0.2 to 0.1 Å. At each value of the reaction coordinate, they performed Car–Parrinello simulations to obtain $\langle f \rangle_{r_{C-C}}$. The temperature of the simulations was 323 K.

The resulting free energy profile is displayed in Fig. 12. Also included in the figure is the total energy calculated at each point. The energy barrier of 14.8 kcal/mol can be compared to experimental estimates of between 6 and 12 kcal/mol. The computed energy of reaction of −6 kcal/mol, however, was much less negative than the experimental estimates of −22 kcal/mol. This discrepancy led the authors to test other sites for their possible catalytic

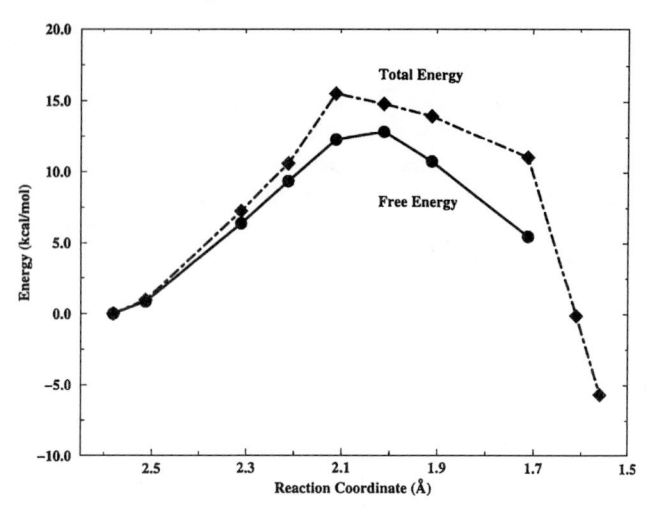

FIG. 12. Free energy and total energy profiles for the insertion of ethylene at the sixfold Corradini site. Reprinted with permission from Boero et al. (1998).

activity, including a fivefold coordinated Ti site. The results of energy calculations for this site compared with the sixfold coordinated site are displayed in Fig. 13. The barrier height for the reaction over the fivefold coordinated site is significantly lower than that of the sixfold coordinated site. In addition, the computed energy of reaction matches the measured one. Thus, the authors

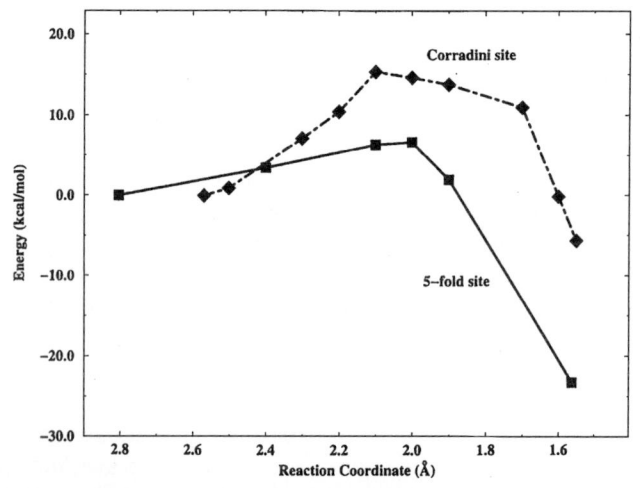

FIG. 13. Comparison of the total energy profiles for the sixfold Corradini site and the fivefold site. Reprinted with permission from Boero et al. (1998).

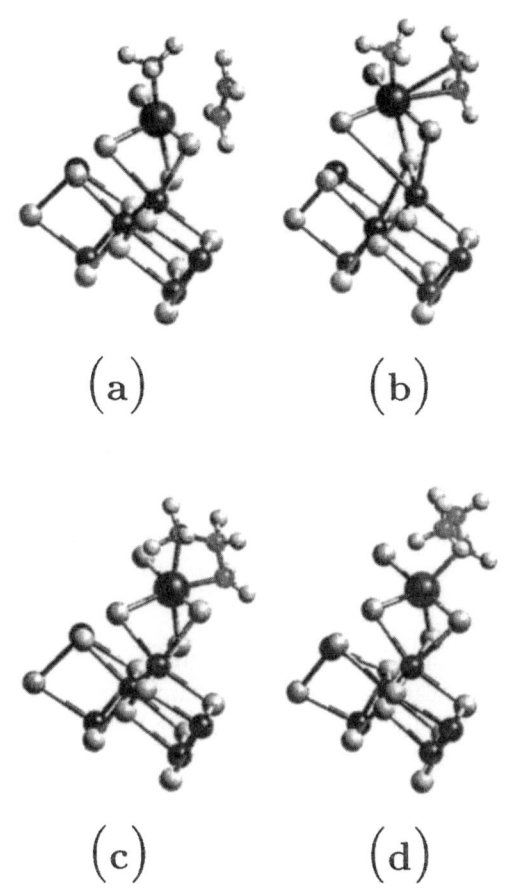

FIG. 14. Snapshots of important steps of the ethylene insertion process over the fivefold Ti site: (a) the approach of the ethylene, (b) formation of the π-complex, (c) the transition state, and (d) the formed chain. Reprinted with permission from Boero et al. (1998).

concluded that the fivefold coordinated site was the active one. Snapshots of the polymerization over the fivefold coordinated site are displayed in Fig. 14.

In our laboratory, we have studied the coupling of two methanol molecules at the acid site of chabazite (see Fig. 1) (Giurumescu and Trout, 2001). This is hypothesized to be an important elementary step in the formation of the first carbon–carbon bond in methanol-to-olefins processes. Because this step has a significant activation barrier, we have chosen to use the method of constraints, with the constraint being the carbon–carbon distance. Simulations were performed at 673 K for 1.5 ps at each point. Our free energy profile is shown in Fig. 15.

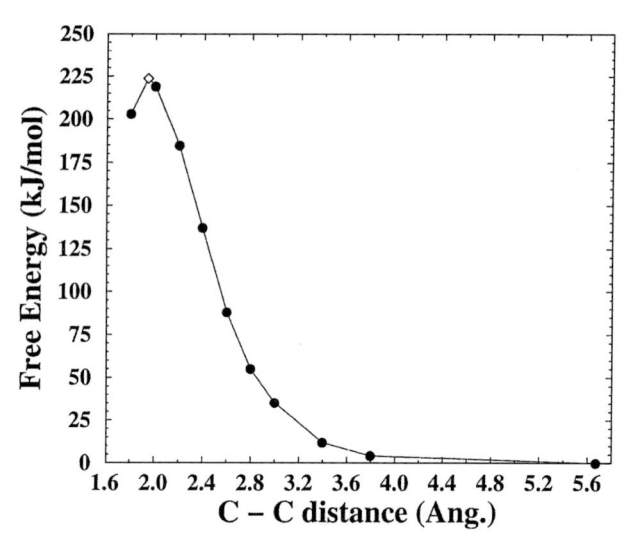

FIG. 15. Free energy profile of carbon–carbon bond formation between two methanol molecules in chabazite. The transition state, found via interpolation, is marked with an open circle.

Aside from quantifying the energetics and the free energy profile for this elementary acid-catalyzed process, we were able to show (1) that the zeolite stabilizes charged complexes, allowing them to move through the channels at 673 K, and (2) a plausible pathway for the reaction process. An important step along the reaction pathway is the formation of a methane-like species and a protonated formaldehyde, both associated with a water molecule. This complex is shown in Fig. 16, slightly before the transition state. As the final stage of the reaction proceeds, the methane-like species transfers a proton to the water molecule in a concerted process through the protonated formaldehyde. The resulting hydronium species moves away as the two remaining species form ethanol. At the very end, the hydronium transfers a proton to the ethanol molecule as it adsorbs at the acidic site, adjacent to the Al atom.

E. Phase Transitions

Car–Parrinello simulations present powerful ways to study phase diagrams and to find new phases of materials, particularly at high pressures. Typically, (N,T,P) simulations are performed using deformable supercells. Methodology can be found in the following references: Focher *et al.* (1994),

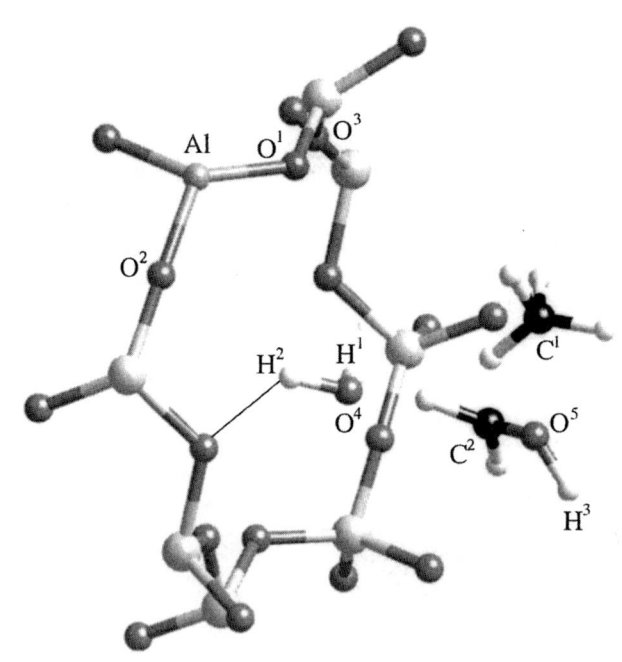

FIG. 16. Configuration close to the transition state, just before concerted proton transfer to the water molecule.

Bernasconi et al. (1995), and Bernasconi et al. (1996). An example of a relevant phase transition is the polymerization of acetylene under pressure (Bernasconi et al., 1997).

In this study, the authors used a 16-molecule cell, equilibrating it at 298 K and 3 GPa. They then increased the pressure at a rate of 25 GPa/ps, until reaching 9 GPa. At this pressure, they increased the temperature to 400 K and then stepped the pressure to 25 GPa. They observed the polymerization process as shown by the snapshots in Fig. 17. Polymerization did not start until the pressure was increased to 25 GPa. At that pressure, the molecules rapidly formed dimers as shown in Fig. 17b. The final product, shown in Fig. 17c, was a mixture of chains of *cis*- and *trans*-polyacetylene. The authors also analyzed the electronic and energetic characteristics of the system and predicted that the injection of triplet excitons would greatly enhance the rate of polymerization.

Two important studies have predicted the melting point of solid materials from first-principles calculations. Sugino and Car (1995) used Car–Parrinello molecular dynamics to estimate the melting point of silicon from first

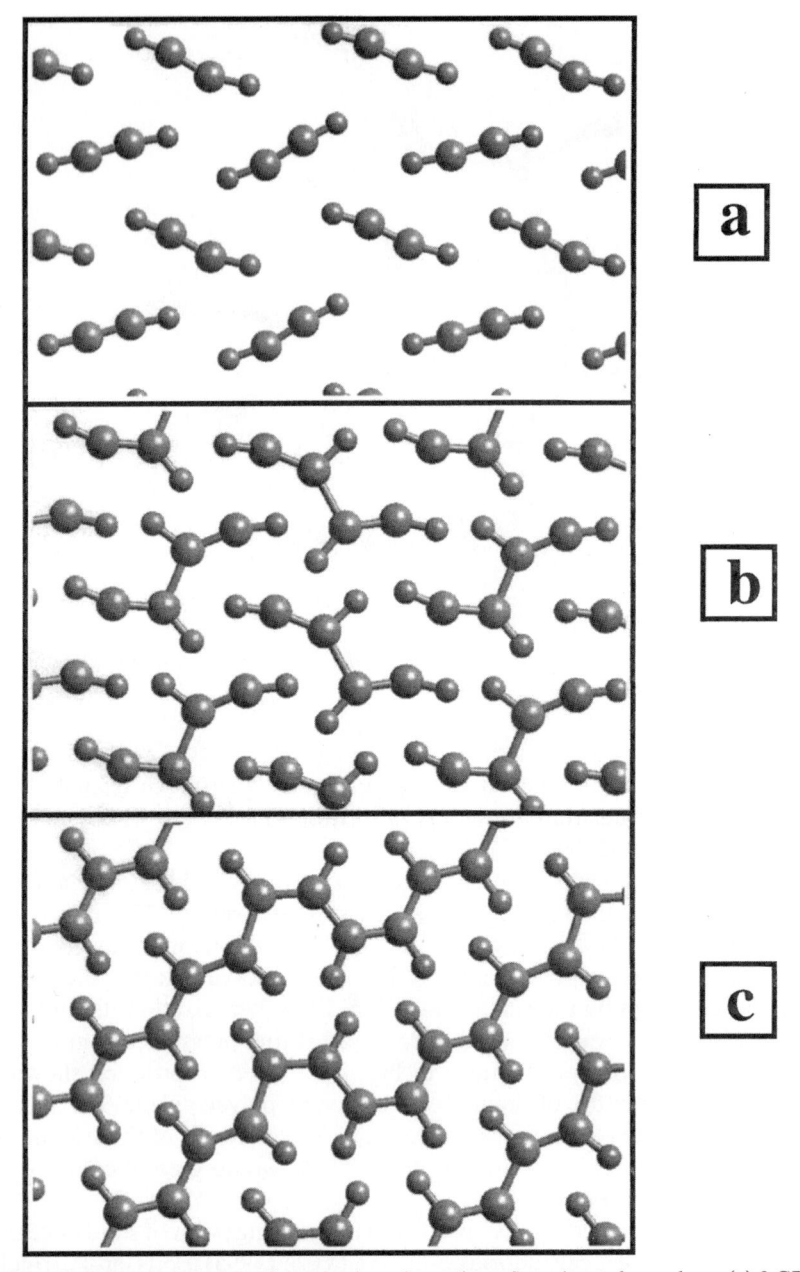

FIG. 17. Pressure-induced polymerization of acetylene. Snapshots of one plane: (a) 9 GPa and 400 K, (b) intermediate configuration at 25 GPa and 400 K, and (c) final configuration at 25 GPa and 400 K. Reprinted with permission from Bernasconi et al. (1997).

principles. Their computed value is 1350 K, compared with an experimental value of 1685 K. de Wijs *et al.* (1998) calculated the melting point of Al to be 890 K, compared to an experimental value of 933 K. Because of hysteresis, the computational time necessary to determine the melting points from direct simulations of the melting process is prohibitive. Thus, both groups of authors used thermodynamic integration to compute the free energies of both the liquid and the solid phases as functions of temperatures. The temperature at which these two curves intersect is the melting point.

Even using this approach, the time necessary to compute the free energy at each temperature from Car–Parrinello simulations would be prohibitive. Thus, both groups of authors used two sets of integration calculations. They used classical pair-potential calculations within the scheme of thermodynamic integration to determine the solid and liquid free energy curves, and at each temperature, they used thermodynamic integration to determine the free energy difference between the quantum mechanical system and the classical system.

Another triumph of Car–Parrinello simulations applied to phase transitions was the prediction of a new high-pressure phase of ice, ice XI (Benoit *et al.*, 1996). Other studies include the prediction of a new phase of amorphous silica (Wentzcovitch *et al.*, 1998), the investigation of phase IV of H_2S (Rousseau *et al.*, 1999) and of various phases of HBr (Ikeda *et al.*, 1999), the study of the polymerization of CO under pressure (Bernard *et al.*, 1998), and the study of the phase diagram of carbon at high temperatures and pressures (Grumbach and Martin, 1996). Finally, several studies have isolated metal–insulator transitions in bulk materials (Silvestrelli *et al.*, 1996), and one has investigated the ferroelastic transition in SiO_2 stishovite (Lee and Gonze, 1997).

F. Processes in Biological Systems

On the frontier of Car–Parrinello simulations is the application to biological systems. These systems are large and often require the incorporation of solvation structures, and energetics of solvation are generally important. Thus, computations of entire biomolecules would be too expensive. Nevertheless, several recent studies have isolated essential features of biological processes by studying carefully chosen models consisting of a tractable number of atoms.

A comprehensive review of the state of applying Car–Parrinello simulations to biological simulations has recently appeared (Carloni and Rothlisberger, 2001). For the purpose of this section, I illustrate the kind of models used and the potential of Car–Parrinello methods applied to biological

390 BERNHARDT L. TROUT

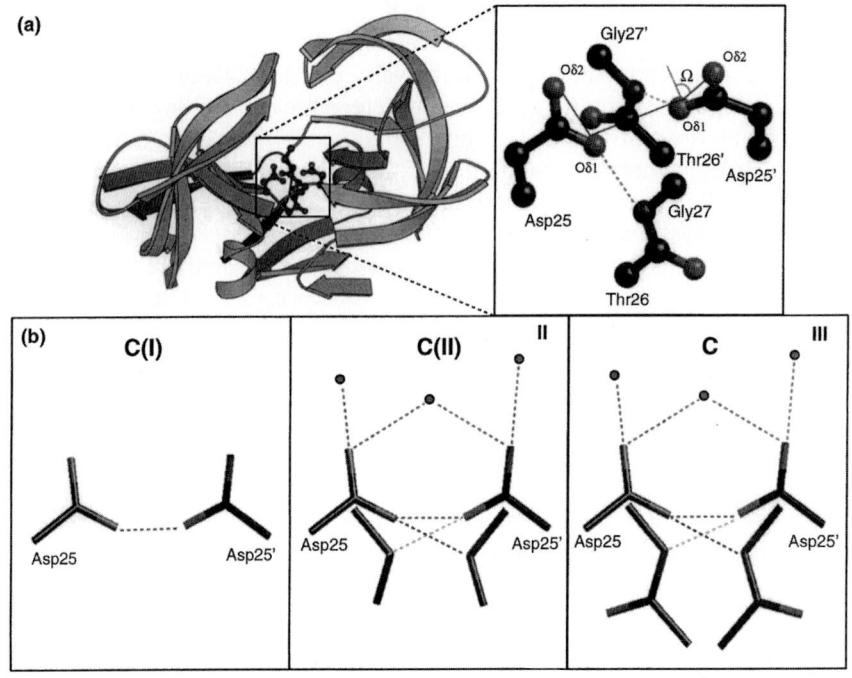

FIG. 18. (a) The structure of HIV-1 PR and its cleavage site; (b) models used in the Car–Parrinello simulations. Reprinted with permission from Carloni and Rothlisberger (2001).

systems, by describing an example of understanding processes in enzymes as a step in developing pharmaceuticals. In particular, the example is chosen from a recent study on HIV-1 protease (Piana and Carloni, 2000).

The structure of this protease is shown in Fig. 18a, including what is thought to be a crucial region for enzymatic function and for the binding of both substrate and inhibitors. Figure 18b shows several models chosen by Piana and Carloni (2000) to study via Car–Parrinello simulations. Figure 19 shows the dynamics of proton motion in the aspartyl dyad for the various models in Fig. 18b. As shown in Fig. 19b, for the C(I) model, the proton can hop between the two oxygen atoms, keeping them close to each other, but the authors conclude that the repulsion of the other oxygen atoms on the carboxylates renders the system unstable.

As shown in Fig. 19c, the choice of model C(II), with water molecules, to model the inclusion of the hydrogen bonding interactions with the neighboring glycines leads to an unstable system. Including the peptide link in model C stabilizes the system, as shown in Fig. 19d, and leads to the conclusion that the strong dipoles of the Thr26(26′)–Gly17(17′) units interact with the

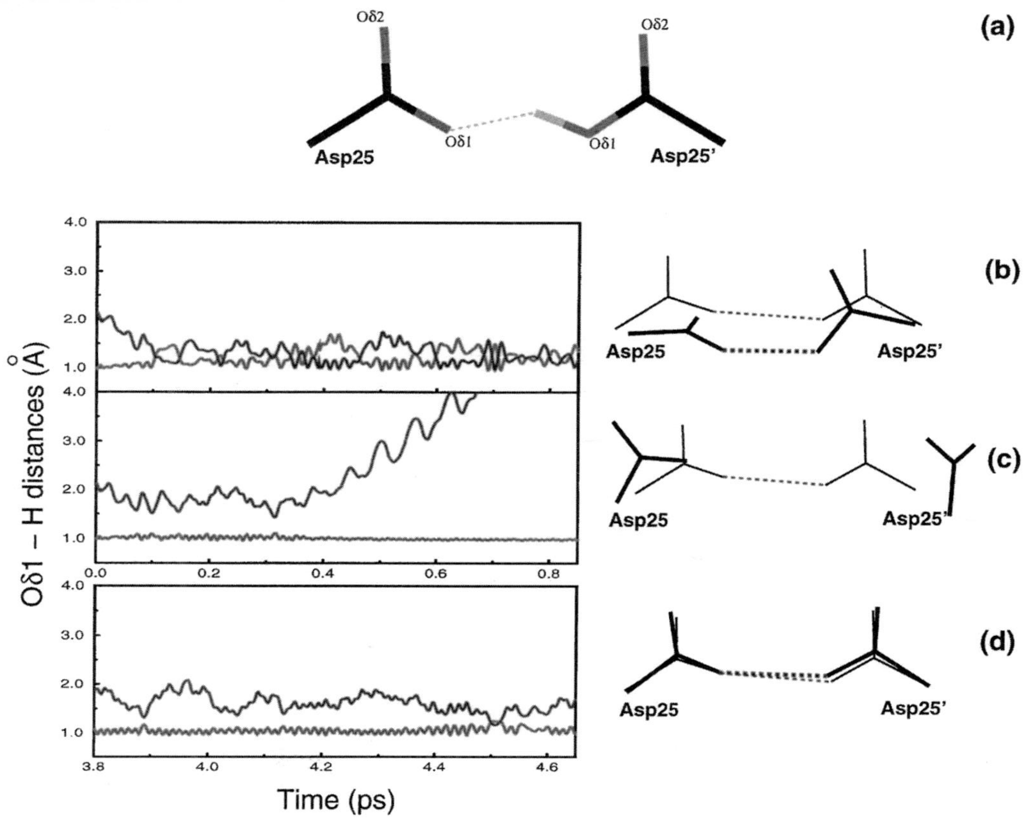

FIG. 19. (a) Location of the proton between the Asp dyad; (b, c, d) position of the proton as a function of time for models C(I), C(II), and C, respectively. Note that in d, only the last 0.9 ps is shown. Reprinted with permission from Carloni and Rothlisberger (2001).

negative charge of the aspartyl dyad, leading to stabilization. Thus, a careful choice of models and simulations by Piana and Carloni (2000) has led to fundamental insight into the stability of an important site of a medically relevant protease.

VI. Advances in Methodology

Advances in methodology fall into two categories: (1) new algorithms and methods for enhanced computational accuracy and/or performance and (2) ways of incorporating new physics. In terms of category 1, an important direction is the incorporation of classical mechanical potentials within the Car–Parrinello framework so that larger models can be treated, albeit without performing fully quantum mechanical simulations (Woo et al., 1997a, 1999; Rothlisberger, 1998). Thus, for example, many water molecules in a solvated system can be treated classically without much loss in accuracy but with a tremendous gain in speed. Another direction is to use hybrid basis functions, such as a combination of localized functions (e.g., Gaussians) with delocalized functions (e.g., plane waves) (Lippert et al., 1997, 1999). In this way, the advantages of plane waves can be combined with functions that more realistically treat the core regions, where electronic density gradients are large. Another recently proposed scheme is adaptive control of the fictitious orbital mass, μ in Eqs. (21) and (22), to minimize error and maximize efficiency of each time step of a Car–Parrinello simulation (Bornemann and Schutte, 1999).

Two examples of advanced methodologies which incorporate new physics are the fully quantum mechanical treatment of the nuclei, in addition to the electrons, and the treatment of dynamics on excited state surfaces. A quantum mechanical treatment of the nuclei can be accomplished via the incorporation of path integral methods into the Car–Parrinello scheme (Marx and Parrinello, 1996; Tuckerman et al., 1996). Of course, these methods are useful mainly for treating small systems containing hydrogen atoms. A few applications include the investigation of the $H_5O_2^+$ molecule (Tuckerman et al., 1997), the CH_5^+ molecule (Marx and Parrinello, 1999), and solid high-pressure phases of ice (Benoit et al., 1998a, 1999).

A disadvantage of using Car–Parrinello path integral methods is that the molecular dynamics is used only to compute averaged properties, the simulation dynamics having no direct physical meaning. A recently developed, albeit approximate method for generating fully quantum mechanical dynamics is the *ab initio* centroid molecular dynamics method (Marx et al., 1999; Pavese et al., 1999). The application of Car–Parrinello methods to treat excited state surfaces is extremely recent (Bittner and Kosov, 1999) but promises to become continually more prevalent.

VII. Concluding Remarks

We have seen the potential and possibilities of Car–Parrinello methods through examples in numerous fields including catalysis, gas- and liquid-phase chemistry, materials, microelectronics, phase behavior, and biology. Via exposure to these examples, which illustrate the state of the art of computational quantum mechanical methods, we can easily see how relevant Car–Parrinello methods are to chemical engineering, which continually demands more detailed understanding of complex molecular processes. I close by reiterating that chemical engineers have always been molecular engineers dealing with complex systems, and we have always been concerned with multiple scales. Not only is the time ripe for chemical engineers to embrace the use of Car–Parrinello methods, but also we are in the ideal position to make new contributions to the development of these methods for the broader engineering and scientific communities.

APPENDIX A: FURTHER READING

A comprehensive overview of quantum mechanics is given by Cohen-Tannoudji et al. (1977), and another good book is by Levine (2000). A staple text on solid-state physics is by Ashcroft and Mermin (1976). A thorough introduction to density-functional theory is given by Parr and Yang (1989). Two good books to learn more about molecular dynamics simulations are by Allen and Tildesley (1987) and Frenkel and Smit (1996). To learn more about pseudopotential methods, two sources with which to begin are by Pickett (1989) and Bachelet et al. (1982).

Several reviews on Car–Parrinello molecular dynamics have appeared recently. A comprehensive, book-sized review, focusing on the code developed by Parrinello and collaborators is by Marx and Hutter (2000). Two other good reviews are by Galli and Parrinello (1991) and Galli and Pasquarello (1993). A terse review, focusing on methodology is by Sandré and Pasturel (1997). Other reviews which combine theory with applications are by Parrinello (1997), Gillan (1997), and Radeke and Carter (1997). A review of Car–Parrinello applied to biological systems by Carloni and Rothlisberger (2001) has just appeared.

APPENDIX B: CODES WITH CAPABILITIES TO PERFORM CAR–PARRINELLO MOLECULAR DYNAMICS

Codes that have Car–Parrinello molecular dynamics implemented are CPMD (Hutter *et al.*, 1995–1999), CASTEP (Molecular Simulations, Inc.), VASP (Kresse and Furthmuller, 1996), HONDO 96 (Dupuis *et al.*), CP-PAW (Blöchl, 1994), fhi98md (Bockstedte *et al.*, 1997), and NWChem (developed and distributed by Pacific Northwest National Laboratory). There are likely many others that have been developed in research laboratories around the world.

ACKNOWLEDGMENTS

First and foremost, I acknowledge Michele Parrinello, without whom this article would not have been conceived. I also thank those who have helped to teach me both the theory and the practice of Car–Parrinello methods: M. Bernasconi, M. Boero, J. Hutter, D. Marx, C. Moleni, R. Rousseau, C. Rovira, P. Silvestrelli, and M. Tuckerman. Finally, praise for help with the literature search goes to P. Wen and J. Thompson.

REFERENCES

Aagaard, O. M., Meier, R. J., and Buda, F., *J. Am. Chem. Soc.* **120,** 7174–7182 (1998).
Alfe, D., and Gillan, M. J., *Phys. Rev. Lett.* **81,** 5161–5164 (1998).
Allen, M. P., and Tildesley, D. J., *Computer Simulation of Liquids,* Oxford University Press, Oxford, 1987.
Ashcroft, N. W., and Mermin, N. D., *Solid State Physics,* Saunders College, Philadelphia, 1976.
Bachelet, G. B., Hamman, D. R., and Schluter, M., *Phys. Rev. E* **26,** 4199–4228 (1982).
Becke, A. D., *Phys. Rev. A* **38,** 3098 (1988).
Benoit, M., Bernasconi, M., and Parrinello, M., *Phys. Rev. Lett.* **76,** 2934–2936 (1996).
Benoit, M., Marx, D., and Parrinello, M., *Comput. Mater. Sci.* **10,** 88–93 (1998a).
Benoit, M., Marx, D., and Parrinello, M., *Nature* **392,** 258–261 (1998b).
Benoit, M., Marx, D., and Parrinello, M., *Solid State Ion.* **125,** 23–29 (1999).
Berces, A., Nukada, T., Margl, P., and Ziegler, T., *J. Phys. Chem. A* **103,** 9693–9701 (1999).
Bernard, S., Chiarotti, G. L., Scandolo, S., and Tosatti, E., *Phys. Rev. Lett.* **81,** 2092–2095 (1998).
Bernasconi, M., Chiarotti, G. L., Focher, P., Scandolo, S., Tosatti, E., and Parrinello, M., *J. Phys. Chem. Solids* **56,** 501–505 (1995).
Bernasconi, M., Benoit, M., Parrinello, M., Chiarotti, G. L., Focher, P., and Tosatti, E., *Phys. Scripta* **T66,** 98–101 (1996).
Bernasconi, M., Chiarotti, G. L., Focher, P., Parrinello, M., and Tosatti, E., *Phys. Rev. Lett.* **78,** 2008–2011 (1997).

Bernholc, J., Brabec, C., Nardelli, M. B., Maiti, A., Roland, C., and Yakobson, B. I., *Appl. Phys. A Mater. Sci. Process.* **67**, 39–46 (1998).
Bittner, E. R., and Kosov, D. S., *J. Chem. Phys.* **110**, 6645–6656 (1999).
Blöchl, P. E., *Phys. Rev. B Condensed Matter* **50**, 17953–17979 (1994).
Bockstedte, M., Kley, A., Neugebauer, J., and Scheffler, M., *Comput. Phys. Commun.* **107**, 187–222 (1997).
Boero, M., Parrinello, M., and Terakura, K., *J. Am. Chem. Soc.* **120**, 2746–2752 (1998).
Boese, A. D., Doltsinis, N. L., Handy, N. C., and Sprik, M., *J. Chem. Phys.* **112**, 1670–1678 (2000).
Bornemann, F. A., and Schutte, C., *Numer. Math.* **83**, 179–186 (1999).
Car, R., and Parrinello, M., *Phys. Rev. Lett.* **55**, 2471 (1985).
Carloni, P., and Rothlisberger, U., In *Theoretical Biochemistry—Process and Properties of Biological Systems* (L. Eriksson, ed.), Elsevier Science, Amsterdam, 2001.
Carter, E. A., Ciccotti, G., Hynes, J. T., and Kapral, R., *Chem. Phys. Lett.* **156**, 472 (1989).
Charlier, J. C., DeVita, A., Blase, X., and Car, R., *Science* **275**, 646–649 (1997).
Cohen-Tannoudji, C., Diu, B., and Laloe, F., *Quantum Mechanics,* John Wiley & Sons, New York, 1977.
Curtiss, L. A., Raghavachari, K., Redfern, P. C., and Pople, J. A., *J. Chem. Phys.* **106**, 1063–1079 (1997).
Debernardi, A., Bernasconi, M., Cardona, M., and Parrinello, M., *Appl. Phys. Lett.* **71**, 2692–2694 (1997).
de Wijs, G. A., Kresse, G., and Gillan, M. J., *Phys. Rev. B Condensed Matter.* **57**, 8223–8234 (1998).
Dupuis, M., Marquez, A., and Davidson, E. R., Available from the Quantum Chemistry Program Exchange, Indiana University, Bloomington.
Estreicher, S. K., Hastings, J. L., and Fedders, P. A., *Phys. Rev. Lett.* **82**, 815–818 (1999).
Focher, P., Chiarotti, G. L., Bernasconi, M., Tosatti, E., and Parrinello, M., *Europhys. Lett.* **26**, 345–351 (1994).
Frank, I., Parrinello, M., and Klamt, A., *J. Phys. Chem. A* **102**, 3614–3617 (1998).
Frenkel, D., and Smit, B., *Understanding Molecular Simulations,* Academic Press, San Diego, CA, 1996.
Galli, G., and Parrinello, M., In *Computer Simulations in Materials Science,* Kluwer, Dordrecht, 1991.
Galli, G., and Pasquarello, A., In *Computer Simulation in Chemical Physics,* Kluwer, Dordrecht, 1993.
Gillan, M. J., *Contemp. Phys.* **38**, 115–130 (1997).
Giurumescu, C., and Trout, B. L., in preparation (2001).
Grumbach, M. P., and Martin, R. M., *Phys. Rev. B Condensed Matter* **54**, 15730–15741 (1996).
Haase, F., Sauer, J., and Hutter, J., *Chem. Phys. Lett.* **266**, 397–402 (1997).
Halley, J. W., Mazzolo, A., Zhou, Y., and Price, D., *J. Electroanal. Chem.* **450**, 273–280 (1998).
Hamann, D. R., *Phys. Rev. Lett.* **81**, 3447–3450 (1998).
Hehre, W. J., Radom, L., v. R. Schleyer, P., and Pople, J. A., *Ab Initio Molecular Orbital Theory,* Wiley, New York, 1986.
Hohenberg, P., and Kohn, W., *Phys. Rev.* **136**, B864–B871 (1964).
Hutter, J., Alavi, A., Deutsch, T., Bernasconi, P., Goedecker, S., Marx, D., Tuckerman, M., and Parrinello, M., *CPMD*, version 3.3, MPI für Festkorperforschung and IBM Zurich Research Laboratory, 1995–1999.
Ikeda, T., Sprik, M., Terakura, K., and Parrinello, M., *J. Chem. Phys.* **111**, 1595–1607 (1999).
Janotti, A., Fazzio, A., Piquini, P., and Mota, R., *Phys. Rev. B Condensed Matter* **56**, 13073–13076 (1997).
Kirkpatrick, S., Gelatt, C. D., and Vecchi, M. P., *Science* **220**, 671–680 (1983).

Klesing, A., Labrenz, D., and van Santen, R. A., *J. Chem. Soc. Faraday Trans.* **94**, 3229–3235 (1998).
Kohn, W., and Sham, L. J., *Phys. Rev.* **140**, A1133 (1965).
Kresse, G., and Furthmuller, J., *Phys. Rev. B Condensed Matter* **54**, 11169–11186 (1996).
Laasonen, K. E., and Klein, M. L., *J. Phys. Chem. A* **101**, 98–102 (1997).
Lee, C., Yang, W., and Parr, R. G., *Phys. Rev. B* **37**, 785 (1988).
Lee, C. Y., and Gonze, X., *Phys. Rev. B Condensed Matter* **56**, 7321–7330 (1997).
Levine, I. N., *Quantum Chemistry,* Prentice–Hall, Upper Saddle River, NJ, 2000.
Lippert, G., Hutter, J., and Parrinello, M., *Mol. Phys.* **92**, 477–487 (1997).
Lippert, G., Hutter, J., and Parrinello, M., *Theor. Chem. Acc.* **103**, 124–140 (1999).
Liu, Z. F., Siu, C. K., and Tse, J. S., *Chem. Phys. Lett.* **311**, 93–101 (1999a).
Liu, Z. F., Siu, C. K., and Tse, J. S., *Chem. Phys. Lett.* **314**, 317–325 (1999b).
Marx, D., and Hutter, J., In *Modern Methods and Algorithms of Quantum Chemistry* (J. Grotendorst, ed.), NIC, Forschungszentrum, Julich, 2000, pp. 301–449.
Marx, D., and Parrinello, M., *J. Chem. Phys.* **104**, 4077–4082 (1996).
Marx, D., and Parrinello, M., *Science* **284**, 59 (1999).
Marx, D., Sprik, M., and Parrinello, M., *Chem. Phys. Lett.* **273**, 360–366 (1997).
Marx, D., Tuckerman, M. E., and Martyna, G. J., *Comput. Phys. Commun.* **118**, 166–184 (1999).
Meijer, E. J., and Sprik, M., *J. Am. Chem. Soc.* **120**, 6345–6355 (1998a).
Meijer, E. J., and Sprik, M., *J. Phys. Chem. A* **102**, 2893–2898 (1998b).
Molteni, C., Francis, G. P., Payne, M. C., and Heine, V., *Phys. Rev. Lett.* **76**, 1284–1287 (1996a).
Molteni, C., Francis, G. P., Payne, M. C., and Heine, V., *Mater. Sci. Eng. B Solid State Mater. Adv. Technol.* **37**, 121–126 (1996b).
Papoulias, P., Morgan, C. G., Schick, J. T., Landman, J. I., and Rahhal-Orabi, N., *Defects in Semiconductors, Icds-19, Pts. 1–3* **258**(2), 923–927 (1997).
Parr, R. G., and Yang, W., *Density Functional Theory of Atoms and Molecules,* Oxford University Press, New York, 1989.
Parrinello, M., *Solid State Commun.* **102**, 107–120 (1997).
Pavese, M., Berard, D. R., and Voth, G. A., *Chem. Phys. Lett.* **300**, 93–98 (1999).
Perdew, J. P., and Wang, Y., *Phys. Rev. B* **45**, 13244–13249 (1992).
Perdew, J. P., Burke, K., and Ernzerhof, M., *Phys. Rev. Lett.* **77**, 3865–3868 (1996).
Piana, S., and Carloni, P., *Proteins Struct. Funct. Gen.* **39**, 26–36 (2000).
Pickett, W. E., *Comp. Phys. Rep.* **9**, 115–198 (1989).
Radeke, M. R., and Carter, E. A., *Annu. Rev. Phys. Chem.* **48**, 243–270 (1997).
Ramaniah, L. M., Bernasconi, M., and Parrinello, M., *J. Chem. Phys.* **111**, 1587–1591 (1999).
Raugei, S., Cardini, G., and Schettino, V., *J. Chem. Phys.* **111**, 10887–10894 (1999).
Rothlisberger, U., *J. Mol. Graph.* **16**, 275–276 (1998).
Rousseau, R., Boero, M., Bernasconi, M., Parrinello, M., and Terakura, K., *Phys. Rev. Lett.* **83**, 2218–2221 (1999).
Sandré, E., and Pasturel, A., *Mol. Simul.* **20**, 63–77 (1997).
Silvestrelli, P. L., and Parrinello, M., *J. Chem. Phys.* **111**, 3572–3580 (1999).
Silvestrelli, P. L., Alavi, A., Parrinello, M., and Frenkel, D., *Phys. Rev. B Condensed Matter* **53**, 12750–12760 (1996).
Silvestrelli, P. L., Bernasconi, M., and Parrinello, M., *Chem. Phys. Lett.* **277**, 478–482 (1997).
Soper, A. K., Bruni, F., and Ricci, M. G., *J. Chem. Phys.* **106**, 247 (1997).
Sprik, M., and Ciccotti, G., *J. Chem. Phys.* **109**, 7737–7744 (1998).
Sprik, M., Hutter, J., and Parrinello, M., *J. Chem. Phys.* **105**, 1142–1152 (1996).
Stadler, R., Alfe, D., Kresse, G., de Wijs, G. A., and Gillan, M. J., *J. Non-Cryst. Solids* **250**, 82–90 (1999).
Sugino, O., and Car, R., *Phys. Rev. Lett.* **74**, 1823–1826 (1995).

Szabo, A., and Ostlund, N. S., *Modern Quantum Chemistry,* McGraw–Hill, New York, 1989.
Trout, B. L., and Parrinello, M., *Chem. Phys. Lett.* **288,** 343–347 (1998).
Trout, B. L., and Parrinello, M., *J. Phys. Chem. B* **103,** 7340–7345 (1999).
Tuckerman, M., Laasonen, K., Sprik, M., and Parrinello, M., *J. Chem. Phys.* **103,** 150–161 (1995).
Tuckerman, M. E., Marx, D., Klein, M. L., and Parrinello, M., *J. Chem. Phys.* **104,** 5579–5588 (1996).
Tuckerman, M. E., Marx, D., Klein, M. L., and Parrinello, M., *Science* **275,** 817–820 (1997).
Ursenbach, C. P., Calhoun, A., and Voth, G. A., *J. Chem. Phys.* **106,** 2811–2818 (1997).
Valladares, A., White, J. A., and Sutton, A. P., *Phys. Rev. Lett.* **81,** 4903–4906 (1998).
Wengert, S., Nesper, R., Andreoni, W., and Parrinello, M., *Phys. Rev. Lett.* **77,** 5083–5085 (1996).
Wentzcovitch, R. M., da Silva, C., Chelikowsky, J. R., and Binggeli, N., *Phys. Rev. Lett.* **80,** 2149–2152 (1998).
Woo, T. K., Margl, P. M., Blöchl, P. E., and Ziegler, T., *J. Phys. Chem. B* **101,** 7877–7880 (1997a).
Woo, T. K., Margl, P. M., Ziegler, T., and Blöchl, P. E., *Organometallics* **16,** 3454–3468 (1997b).
Woo, T. K., Margl, P. M., Deng, L., Cavallo, L., and Ziegler, T., *Catal. Today* **50,** 479–500 (1999).
Yamataka, H., Aida, M., and Dupuis, M., *Chem. Phys. Lett.* **300,** 583–587 (1999).

THEORY OF ZEOLITE CATALYSIS

R. A. van Santen* and X. Rozanska

**Schuit Institute of Catalysis, Eindhoven University of Technology
Eindhoven 5600 MB, The Netherlands**

I. Introduction	400
II. The Rate of a Catalytic Reaction	401
III. Zeolites as Solid Acid Catalysts	403
IV. Theoretical Approaches Applied to Zeolite Catalysis	407
A. Simulation of Alkane Adsorption and Diffusion	407
B. Hydrocarbon Activation by Zeolitic Protons	414
C. Kinetics	427
V. Concluding Remarks	432
References	433

The reactivity of acidic zeolites to the activation of hydrocarbons is used to illustrate different modeling approaches applied to catalysis. Quantum-chemical calculations of transition-state and ground-state energies can be used to determine elementary rate constants. But to predict overall kinetics, quantum-mechanical studies have to be complemented with statistical methods to compute adsorption isotherms and diffusion constants as a function of micropore occupation. The relatively low turnover frequencies of zeolite-catalyzed reactions compared to superacid-catalyzed reactions are due mainly to high activation energies of the elementary rate constants of the proton-activated reactions. These high values are counteracted by the significant interaction energies of hydrocarbons with the zeolite micropore wall dominated by van der Waals interactions. © 2001 Academic Press.

*To whom correspondence should be addressed.

I. Introduction

Predictability of activity, selectivity, and stability based on known structures of catalysts can be considered the main aim of the theoretical approaches applied to catalysis. Here, for a particular class of heterogeneous catalysts, namely, acidic zeolites, we present the theoretical approaches that are available to accomplish this goal, which lead to a better understanding of molecular motion within the zeolitic micropores and the reactivity of zeolitic protons. It is not our aim to introduce the methods as such, since introductory treatments on those can be found elsewhere. Rather we focus on their application and use to solve questions on mechanisms and reactivity in zeolites. The discussion is focused on an understanding of the kinetics of zeolite-catalyzed reactions.

We use the activation of linear alkanes and their conversion to isomers and cracked products as the main motive of our discussion. This class of reactions catalyzed by acidic zeolites is an ideal choice to illustrate the state of the art of theoretical molecular heterogeneous catalysis, because the reaction mechanisms, zeolite micropore structure, and structure of the catalytically reactive sites are rather well understood.

Whereas simulation methods are available to model zeolite structures as a function of their composition as well as their topology [1–9], we do not discuss

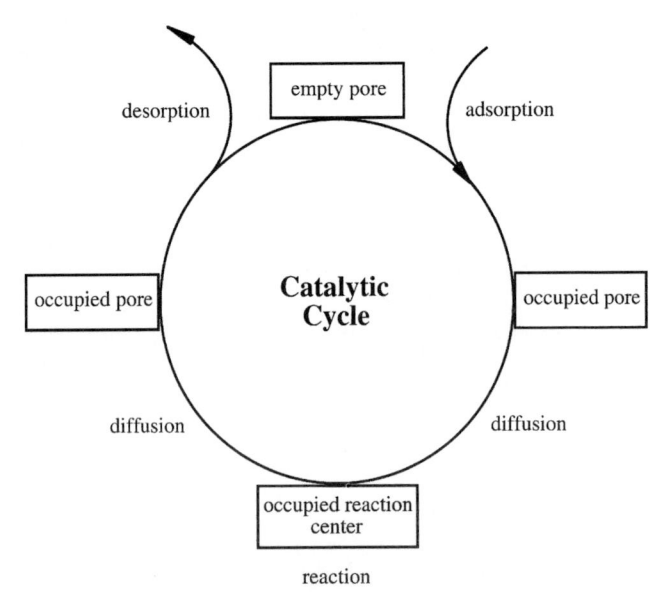

FIG. 1. The catalytic cycle of a zeolite-catalyzed reaction.

those techniques and their results here but refer to some of the results when discussing chemical reactivity and adsorption. Perpetual improvements in both computers and *ab initio* codes allow nowadays calculations on realistic structures [10–12]. Also, we do not present an overview of the deep mechanistic insights that have been obtained recently on a molecular level for many hydrocarbon conversion reactions. Most of the practically important reactions have now been analyzed by quantum-chemical techniques [13–31] (viz., reactions with aromatic species [13–16], olefins and alkanes [17–22], water and methanol [23–26], metallic clusters supported on zeolite or metal exchanged zeolites [27–30], and acetonitrile [31]).

We are concerned with the kinetics of zeolite-catalyzed reactions. Emphasis is put on the use of the results of simulation studies for the prediction of the overall kinetics of a heterogeneous catalytic reaction. As we will see later, whereas for an analysis of reactivity the results of mechanistic quantum-chemical studies are relevant, to study adsorption and diffusion, statistical mechanical techniques that are based on empirical potentials have to be used.

II. The Rate of a Catalytic Reaction

A catalytic reaction is the result of a cyclic process that consists of many elementary reaction steps. The essence of a catalytic reaction is that the catalytic reactive center reappears after each cycle in which reactant molecules are converted into products. Since zeolites are microporous systems, a special feature is the coupling of reaction at the protonic centers with diffusion of the molecules through the micropores to and from the zeolite exterior. The zeolite catalytic cycle is sketched in Fig. 1. To reach the catalytic reactive center, molecules have to adsorb in the mouth of a micropore and diffuse to the catalytic center, where they can react. Product molecules have to diffuse away and, once they reach the micropore mouth, will desorb. Clearly, then, one has to complement quantum-chemical information on reactivity, concerned with the interaction of zeolite protons with reactants, with information on diffusion and on adsorption of reactants and products.

Again, we do not exhaustively discuss molecular theories of diffusion and adsorption in zeolites but refer to other studies [32–34]. However, we highlight some important results significant to the kinetic analysis we are presenting.

A characteristic time scale of a catalytic reaction event such as proton-activated isomerization of an adsorbed alkene molecule is 10^2 s. However, quantum-chemical calculations predict energies of electrons with characteristic time scales of 10^{-16} s. Of use to us are the potential energies that such methods generate that determine the forces that act on the nuclei, which determine the vibrational frequencies. The characteristic time of vibrational motion is 10^{-12} s. In zeolites, diffusional times can be as short as 10^{-8} s and adsorption time scales are typically 10^{-6} s. Dependent on the free energies of adsorption, the time scale of desorption of a molecule is 10^{-4} s or longer. The time scales of the proton-activated elementary reactions are of the order of 10^{-4} s due to their high activation energies. The long time scale of reaction compared to that of vibrational motion implies that thermal energy exchange between reaction molecule and zeolite wall is fast. This is the justification for the use of transition reaction rate theory [35–37]. In its most elementary form, the rate expression,

$$r_{TST} = \frac{kT}{h} e^{(\Delta S\#/R)} e^{-(E_{act}/RT)}, \qquad (1)$$

can be shown to be rigorously valid, if the rate of energy exchange is fast. In Eq. (1), the probability of passing the activation energy barrier by reactants with sufficient energy has been assumed to be equal to one. In this equation, k is Boltzmann's constant, h is Planck's constant, $\Delta S\#$ is the activation entropy, and E_{act} is the transition state barrier height. R is the gas constant.

The major advance of the past decade is that, using quantum-chemical computations, activation energies (E_{act}) as well as activation entropies ($\Delta S\#$) can be predicted *a priori* for systems of catalytic interest. This implies much more reliable use of the transition-state reaction rate expression than before, since no assumption of the transition state-structure is necessary. This transition-state structure can now be predicted. However, the estimated absolute accuracy of computed transition states is approximately of the order of 20–30 kJ/mol. Here, we do not provide an extensive introduction to modern quantum-chemical theory that has led to this state of affairs: excellent introductions can be found elsewhere [38, 39]. Instead, we use the results of these techniques to provide structural and energetic information on catalytic intermediates and transition states.

Because of the size of the reaction centers to be considered, a breakthrough in quantum chemistry had been necessary to make computational studies feasible on systems of catalytic interest. This breakthrough has been provided by density functional theory (DFT). Whereas in Hartree–Fock-based methods, used mainly before the introduction of DFT, electron exchange had to be accounted for by computation of integrals that contain products of four occupied orbitals, in DFT these integrals are replaced by functionals that depend only on the electron density. An exchange-correlation functional can be defined that accounts for exchange as well as correlation energy. The correlation energy is the error made in Hartree–Fock-type theories by the use of the mean-field approximation for electronic motion.

The still unresolved problem is the determination of a rigorously exact functional, as all currently used DFT functionals are approximate.

By computation of the stationary points of the n-dimensional energy diagram of the interacting system, the structure of local energy minima as well as the transition state can be determined. For many systems, these interaction energies have an accuracy of the order of 5–15 kJ/mol. For this reason, as we illustrate later, discrepancies between theoretical calculation and experiment can be often related to shortcomings in model assumptions rather than to quantum-chemical approximations. Around the ground-state energy minimum and the transition-state saddle point, the potential can be expanded as a function of displacement of the normal coordinates with respect to their stationary values. Within the harmonic approximation, the vibrational spectrum and hence the corresponding entropy can be easily computed. The entropy follows from the expression

$$S = -k \ln pf \tag{2a}$$

$$= -k \ln \Pi_i \frac{1}{1 - e^{-(h v_i / kT)}}, \tag{2b}$$

where i sums over the n normal vibration modes of the ground-state system and the $n - 1$ normal modes of the transition state. pf stands for the partition function and v_i is the normal mode frequency.

III. Zeolites as Solid Acid Catalysts

The structure of a zeolite is illustrated in Fig. 2, using mordenite as an example. The zeolitic framework can consist of four valent (Si), three valent (Al, Ga, Fe), or five valent (P) cations, tetrahedrally coordinated with four oxygen atoms. The oxygen atoms bridge two framework cations. When the lattice cation is Si^{+IV}, the framework charge is neutral. When Al^{+III} substitutes for Si^{+IV}, the framework becomes negatively charged. Zeolites of such a composition can exist when extra framework cations are ion exchanged into the zeolite cavities. In the case where this cation is NH_4^+, the heating of the material induces ammonia to desorb. The proton that is left behind will bind to an oxygen atom bridging Al and Si framework cations (see Fig. 3). For an extensive discussion of zeolite crystals of catalytic interest, refer to Refs. 40 and 41. Also, several available reviews discuss the nature of the proton bonded to the bridging oxygen atom [42–45].

Although strongly covalently bonded to the zeolite, hydrogen attached to the bridging oxygen atom reacts as a proton with reactant molecules and

FIG. 2. The structure of mordenite zeolite.

hence induces transformation reactions known also from superacid catalysis [46–48]. However, as will be explained, the detailed mechanism of activation is very different from that known in superacid catalysis. Whereas in superacids low-temperature reactions generate carbonium and carbenium ions as stable but sometimes short-lived intermediates, the nature of carbonium

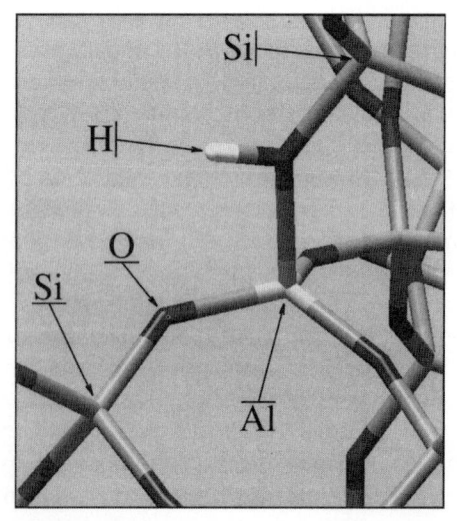

FIG. 3. Hydrogen bonded to a bridging O atom in mordenite.

and carbenium ion intermediates is often quite different in solid acid catalysis. Essentially, carbonium and carbenium cation intermediates are parts of transition states or activated intermediate states through which reactions proceed. The high activation barriers of the elementary reaction steps imply that zeolitic solid acids are weakly acidic compared to superacids. One way to understand this is that, different from superacid media, the dielectric constant of a zeolite is low ($\varepsilon \sim 4$) [49–51] and hence charge separation has a high energy cost [3].

It is now also well established that the zeolite framework has some flexibility. Due to the dominance of directed covalent bonding, the tetrahedra are rather rigid, but the bond-bending potential energy curve for bending of the Si–O–Si angle is rather flat. Local distortions of the framework can be easily accommodated because of the low energy cost of bending of the Si–O–Si angle. This is important because, as we will see later, attachment of a proton to the bridging oxygen atom increases the Si–O and Al–O distances and hence requires a larger volume than a free Si–O–Al unit. Upon deprotonation the Si–O and Al–O bonds decrease and hence the effective volume of the concerned tetrahedron decreases.

For low Al/Si ratio zeolites, differences in acidity of the zeolitic proton relate to slight differences in the local relaxability of the zeolite lattice around the protons [1–9]. For zeolites with a high Al/Si ratio, the zeolite proton interaction increases. Compositional variations may cause the deprotonation energy to vary by $\sim$1–5%.

For a proper understanding of the interaction of hydrocarbons with zeolitic protons, it is important to realize that the oxygen atoms, with a computed charge close to -1, and the Si atoms, with a computed charge close to $+2$, have very different sizes. This is illustrated in Fig. 4. Oxygen atoms are relative large and the framework cations are small. This implies that hydrocarbon molecules, when adsorbed in the zeolite micropores, will experience mainly interactions with the large oxygen atoms, which inhibit direct interactions with the smaller lattice cations.

Bonding within the zeolite framework is mainly covalent, and ionic bonding contributes only $\sim$10% [52–54]. The attractive interaction between the siliceous zeolite framework and hydrocarbons can best be described as a van der Waals dispersive interaction due to the attraction of fluctuating dipole moments of electronic motion on the oxygen atoms and hydrocarbons [55–59]. The van der Waals interaction

$$V_{\text{v.W}} = -C \frac{\alpha_i \alpha_j}{r_{ij}} \qquad (3)$$

is proportional to the polarizabilities α_i of zeolitic oxygen atoms and, for instance, for alkanes the polarizabilities α_j of CH_2 or CH_3 units. The polarizability α_i is approximately proportional to the volume of a molecular unit.

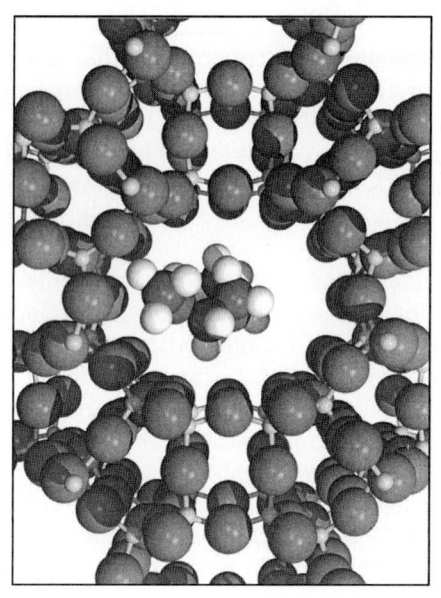

FIG. 4. Hexane occlude in a zeolite micropore.

In Eq. (3), C is a constant, and r_{ij} the distance between the centers of mass of units i and j [55–57]. Density functional theory quantum-chemical codes do not predict this van der Waals dispersive interaction accurately [60, 61].

One has to include the repulsive contribution to obtain the interaction potential. In the Lennard–Jones expression, the attractive and repulsive contributions can be described by

$$V_{\text{v.W}} = -4\varepsilon_{ij}\left(\left(\frac{\sigma_{ij}}{r_{ij}}\right)^6 - \left(\frac{\sigma_{ij}}{r_{ij}}\right)^{12}\right), \tag{4a}$$

where σ_{ij} is the sum of the van der Waals radius of i and j, and ε_{ij} the well depth [62]. However, the repulsive part of the potential is best described by the Born repulsion expression, leading to the Buckingham interaction potential:

$$V_{\text{v.W}} = -C\frac{\alpha_i \alpha_j}{r_{ij}^6} + Ae^{-ar_{ij}}. \tag{4b}$$

The Born repulsion expression stems from the Pauli principle, which results from the prohibition of electrons with the same spin occupying the same wavefunction, a situation that occurs when doubly occupied orbitals interact. Whereas the van der Waals interaction cannot be computed accurately from

DFT quantum-chemical codes, the Born repulsive interaction is accurately computed. Expressions (4a) and (4b) are two-body interaction potentials commonly used in codes that predict energies or geometries based on empirical potentials.

IV. Theoretical Approaches Applied to Zeolite Catalysis

We now report how theoretical methods can be used to provide information on the adsorption, diffusion, and reactivity of hydrocarbons within acidic zeolite catalysts. In Section A, dealing with adsorption, the physical chemistry of molecules adsorbed in zeolites is reviewed. Furthermore, in this section the results of hydrocarbon diffusion as these data are obtained from the use of the same theoretical methods are described. In Section B we summarize the capability of the quantum-chemical approaches. In this section, the contribution of the theoretical approaches to the understanding of physical chemistry of zeolite catalysis is reported. Finally, in Section C, using this information, we study the kinetics of a reaction catalyzed by acidic zeolite. This last section also illustrates the gaps that persist in the theoretical approaches to allow the investigation of a full catalytic cycle.

A. SIMULATION OF ALKANE ADSORPTION AND DIFFUSION

1. Methods and Theory

Configurationally biased Monte Carlo techniques [63–65] have made it possible to compute adsorption isotherms for linear and branched hydrocarbons in the micropores of a siliceous zeolite framework. Apart from Monte Carlo techniques, docking techniques [69] have also been implemented in some available computer codes. Docking techniques are convenient techniques that determine, by simulated annealing and subsequent freezing techniques, local energy minima of adsorbed molecules based on Lennard–Jones- or Buckingham-type interaction potentials.

The interaction energy is determined by potentials as defined in Eq. (4). Bond-bending energy terms within the hydrocarbons are also included. The parameters of the interaction potential with zeolite have to be determined by a fit of experiment with theory. June et al. [66] and Smit and Maesen [67] used slightly different parameters (see Table I). The latter used parameters fitted on experimental adsorption isotherms of hexane in silicalite. Whereas in siliceous materials the dominant interaction term is given by the

TABLE I

VAN DER WAALS INTERACTION POTENTIALS FOR HYDROCARBONS ADSORBED IN ZEOLITES

Parameter set	$\sigma_{CH_3,O} = \sigma_{CH_2,O}$ (Å)	$\varepsilon_{CH_3,O}$ (K)	$\varepsilon_{CH_2,O}$ (K)
June et al. [66]	3.364	83.8	83.8
Smit and Maesen [67]	3.64	87.5	54.4

Buckingham potential, in zeolites with protons or cations additional interaction terms become important. We discuss the interactions with protons later, but here we comment briefly on the effect of the presence of cations in zeolite micropores. Kiselev and Quang Du [68] were among the first to analyze adsorption in the hydrocarbon–zeolite system. Cations other than protons have a charge close to their formal valency. Therefore their bond with the negative charge of the framework is predominantly of an ionic nature. These cations will generate large electrostatic fields. The interactions with apolar hydrocarbons can best be described by the inductive interaction of a polarizable particle i defined by its polarizability α_i and an ion j with charge q_j:

$$V_{\text{ind}} = -q_j^2 \frac{\alpha_i}{2r_{ij}^4}, \tag{5}$$

where r_{ij} is the distance between particle i and particle j [55–57, 68]. Kiselev and co-workers provided empirical potentials that can be used for adsorption in ion-exchanged zeolites.

The equilibrium Monte Carlo method computes thermodynamic averages by simulating equilibrium configurations. Hence it is an adequate technique to compute free energies. In essence, four steps are used.

- Give the adsorbate a random new position and random new configuration.
- Calculate the energy difference between this new configuration and the old configuration.
- Accept this new configuration with a probability P proportional to Boltzman's weight factor at temperature T:

$$P \cong \exp\left(-\frac{E_i}{kT}\right).$$

- Repeat; this generates a chain of configurations Γ_i.

The thermodynamic average is then computed from

$$\langle A \rangle = \frac{1}{m} \sum_i^m A(\Gamma_i), \tag{6}$$

where i sums up the number of configurations.

2. Simulation of Alkane Adsorption Isotherms

One of the most fascinating results of these simulations is the strong dependence of the cavity site on the size and shape of the adsorbing molecule and cavity. For instance, for the zeolite ferrierite, theory and experiment show that whereas propane preferentially adsorbs in the 8-oxygen atom ring pockets, pentane and the longer hydrocarbons preferentially adsorb in the one-dimensional 10-ring channel system [70, 71]. Only at higher occupations do some of the molecules become forced into the eight-ring pocket. In addition, at higher coverages orientational packing of molecules may occur, again dependent on the match of the molecule length and the free micropore length, giving rise to plateaux in the adsorption isotherm [72, 73].

Figure 5 illustrates the dependence of the energy of adsorption for linear alkanes of increasing chain length for different zeolites. An optimum in the

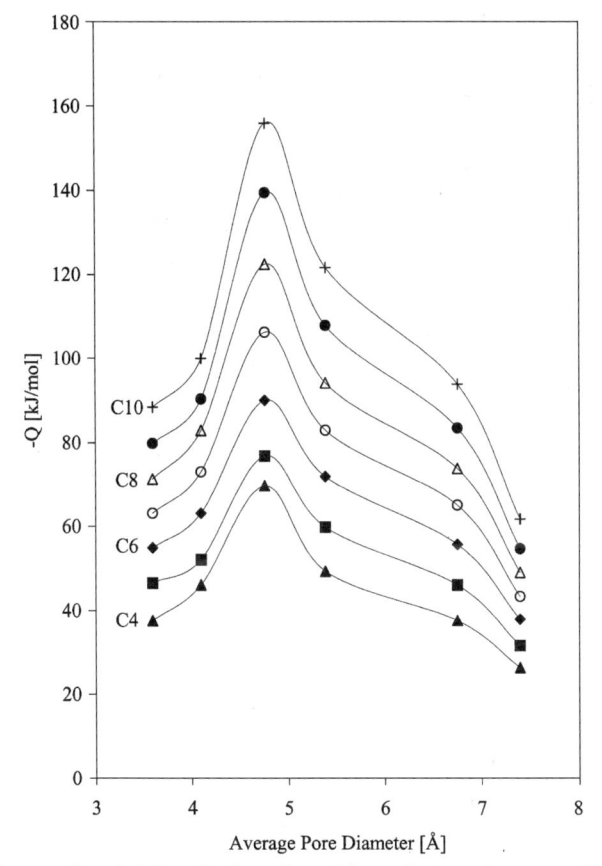

FIG. 5. Computed enthalpies of adsorption of linear hydrocarbons as a function of the average zeolite micropore diameter [70].

adsorption energy is found as a function of the average micropore diameter. Also, an incremental increase in the energy of adsorption with increasing chain length is observed. The basic feature that Fig. 5 illustrates is that the energy of adsorption increases with decreasing microchannel dimension. The optimum arises since there is no access of micropore or cavity when their size is too small. To the left of the optimum, zeolites have a bimodal micropore distribution of very narrow inaccessible pores and wider pores with low interaction energies. The increase in energy of adsorption with decreasing micropore channel is due to the increasing attractive van der Waals interaction with decreasing distance of the alkane CH_2 and CH_3 groups with micropore oxygen atoms.

Table II [74] shows a comparison of computed and measured energies of adsorption for two hydrocarbons and two zeolites. Whereas mordenite (MOR) has a one-dimensional 12-ring channel system, ZSM-22 (TON) has a one-dimensional 10-ring channel system. One notes again the increase in energy of adsorption with decreasing channel dimension. There is also good agreement with experiment, except for zeolites that contain a substantial concentration of protons. In this case there is an increase in energy by 8–15 kJ/mol. Interestingly at $T = 513$ K, a typical reaction temperature, the equilibrium constants of adsorption K_{ads} are smaller for the narrow-micropore zeolite TON than the wider-micropore zeolite MOR. The larger loss in mobility due to the larger confinement in the smaller pore causes a larger loss of entropy and hence counteracts the gain in the energy of adsorption in the smaller micropore. As a consequence, at this temperature the micropore occupation of the medium-sized MOR pore by hydrocarbons is

TABLE II
MEASURED AND SIMULATED HEATS OF ADSORPTION AS WELL AS HENRY ADSORPTION COEFFICIENTS FOR LINEAR HYDROCARBONS IN SILICEOUS ZEOLITES

	$-\Delta H_{ads}$ (kJ/mol)		K_{ads} ($T = 513$ K) (mmol/gPa)	
	Simulation [74]	Literature	Simulation [74]	Literature
n-Pentane/TON	63.6	71 [75]	$4.8 \cdot 10^{-6}$	$4.6 \cdot 10^{-6}$ [75]
		62.1 [76]		$6.53 \cdot 10^{-6}$ [76]
n-Pentane/MOR	61.5	59 [75]	$4.8 \cdot 10^{-5}$	$6.8 \cdot 10^{-5}$ [75]
		55.7 [76]		$8.6 \cdot 10^{-5}$ [76]
n-Hexane/TON	76.3	82 [75]	$1.25 \cdot 10^{-5}$	$8.0 \cdot 10^{-6}$ [75]
		75.0 [76]		$1.99 \cdot 10^{-5}$ [76]
		72.3 [77]		$7.8 \cdot 10^{-6}$ [77]
n-Hexane/MOR	69.5	69 [75]	$1.26 \cdot 10^{-4}$	$1.9 \cdot 10^{-4}$ [75]
		67.1 [76]		$5.5 \cdot 10^{-4}$ [76]
		69 [77]		$1.3 \cdot 10^{-4}$ [77]

substantially higher than that of the smaller-pore zeolite TON. The general consequence of these results is that an optimum zeolite micropore size provides for a maximum pore occupation. When the pore size is too large (e.g., faujasite or a clay surface), the low energy of adsorption gives a low micropore occupation. On the other hand, when the accessible pore size becomes too small, again, a low micropore coverage results because of the increasing loss of entropy.

For the overall rate of a catalytic reaction, this is an important conclusion because the rate of a catalytic reaction is proportional to the site occupancy and the rate constant of molecular activation, when the latter step is rate limiting. For a monomolecular reaction this follows from an elementary expression for the overall rate r:

$$r = V\frac{d[P]}{dt} = k_i \cdot \theta, \tag{7}$$

where k_i is the elementary rate constant of activation of the adsorbed molecule and θ is the coverage by reactant molecules on catalytically active sites. $[P]$ is the concentration of product molecules and V the volume. We discuss in the next section the computation of k_i using quantum-chemical methods. Equation (7) illustrates the proportionality of rate with micropore filling. It can also be used to illustrate that even when k_i does not depend strongly on micropore size (which is often the case), the overall rate may depend strongly on the micropore size. This follows from the micropore size dependence of θ. In the case that θ can be described by Langmuir adsorption and the competitive product or intermediate adsorption is ignored, expression (7) can be rewritten

$$r = k_i \cdot \frac{K_{ads} \cdot [R]}{1 + K_{ads} \cdot [R]} \tag{8}$$

where $[R]$ is the concentration in reactant. From (8), one finds the following expression for the apparent activation energy of the rate r:

$$E_{app} = E_{act}^i + (1-\theta)E_{ads}, \tag{9}$$

where E_{act}^i is the activation energy of the intrinsic rate constant k_i, and E_{ads} the energy of adsorption. Usually, E_{act}^i is only weakly dependent on the micropore size. However, Eq. (9) implies an important dependence of the rate of the overall reaction as a function of the micropore size and shape because of their strong relation to E_{ads}.

Self-diffusivity as a function of chainlength in MOR at 333K

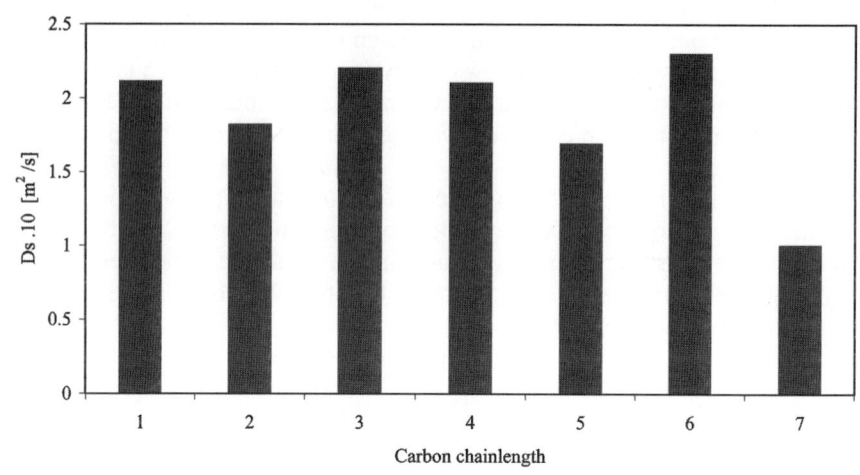

FIG. 6. Simulated self-diffusivity diffusion constant as a function of linear hydrocarbon chain length in mordenite [78].

3. Simulation of Diffusion

Rates of diffusion of linear and branched alkanes by zeolites within medium-sized pores are very fast. This is illustrated for linear alkanes in Figs. 6 and 7. As shown in Fig. 6, the self-diffusion constant in mordenite that has one-dimensional micropores is fast and independent of the length of hydrocarbons [78]. On the other hand, for the wider-pore zeolite faujasite, there is a decrease with chain length. This illustrates that in the cavities of mordenite, diffusion is controlled by the interaction with the zeolite wall. Diffusion within mordenite can be characterized as a creeping motion. In the larger micropores of faujasite, diffusion is more liquid-like [79]. Figure 7 [78] shows a higher initial diffusion rate for the zeolite with a larger micropore. Very similar and still fast rates of self-diffusion are found when the micropores become half-filled.

The diffusion constants are computed by solving Newton's equation of motion, maintaining the average kinetic energy equal to $3/2\ kT$. Whereas generally there is no need to include explicitly the fluctuating motion of lattice atoms, this is critical when diffusion causes molecules to move through windows of size comparable to the lateral dimension of these molecules [85–88]. The activation energies for self-diffusion may differ widely. For highly branched molecules or alkyl-substituted aromatic molecules, it may severely inhibit actual mobility at reaction temperatures [89]. Under such conditions not only conversion may become diffusion limited, but also

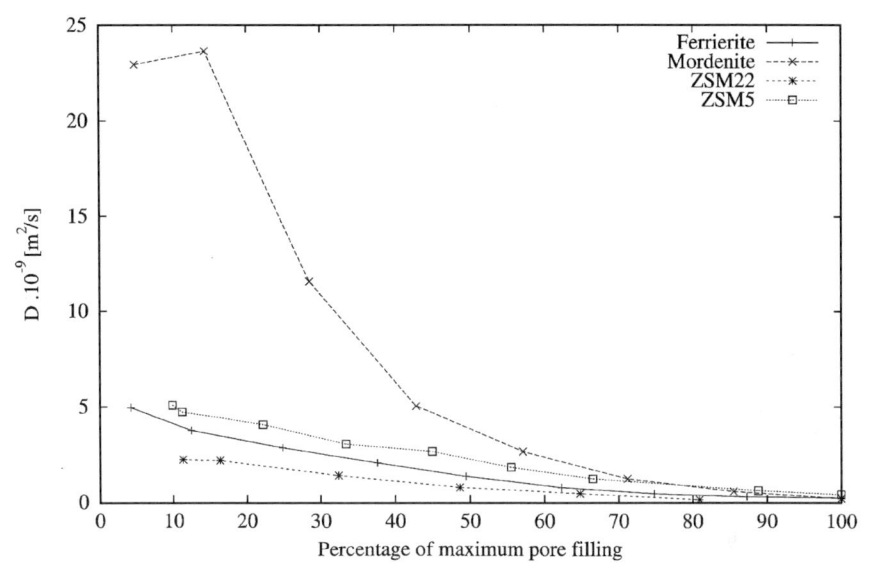

FIG. 7. Simulated n-butane self-diffusivity constant as a function of micropore filling at 333 K [78].

selectivity. For instance, alkylation of toluene into the p-xylene catalyzed by ZSM-5 is highly selective when the rate becomes diffusion limited [90]. Then only p-xylene has a diffusional rate high enough to move out of the zeolite crystallites. To compute the yield of a reaction in such a case, coupling of the reaction rate and diffusion will have to be explicitly considered. We refer to the work of Hinderen and Keil [91], who introduced kinetic Monte Carlo techniques to predict zeolite kinetics. In the next section we discuss the kinetics of zeolite-catalyzed reactions for which self-diffusion is not rate limiting.

A general experimental result is the difference between measured rates of diffusion in macroscopic experiments and measurements of self-diffusion by spectroscopic techniques such as gradient field NMR [80–84]. The difference between the microscopic measurement and the macroscopic experiment is desorption and reentry of molecules in zeolite microcrystallites. In this respect, it is important to remember the Biot condition, which states the condition when the measured rate of diffusion is independent of the rate of desorption:

$$k_{\text{des}} \gg \frac{D}{R^2}. \qquad (10)$$

It shows that when the size of a particle is large enough the time scale of diffusion becomes long compared to the time scale of desorption. When the rate of desorption interferes a strong dependence of the overall rate of diffusion on the particle size should be measured.

B. HYDROCARBON ACTIVATION BY ZEOLITIC PROTONS

The great contribution of quantum chemistry has been the proper understanding of the nature and energetics of carbonium and carbenium ion intermediates, as they occur in zeolites. Carbonium ions, organic cations in which one or more carbon atoms become five coordinated due to proton attachment, and carbenium ions, organic cations in which a carbon atom is three coordinated and has become planar, have been observed as short-lived species using NMR spectroscopy in superacids [92, 93]. Because of the similarity of the type of reactions that are catalyzed, it has been thought that zeolites have a similar superacidity. However, there is now ample experimental and theoretical evidence showing that this is in fact not the case. We summarize the theoretical evidence here.

The elementary reaction steps of the hydrocarbons considered in this section are summarized in Fig. 8. The occurrence of monomolecular reactions with linear hydrocarbons that produce hydrogen and alkane fragments was first demonstrated by Haag and Dessau [94]. For convenience, the zeolite lattice to which the proton is attached is not explicitly shown in the scheme. However, it will become clear later that proton activation cannot be understood properly without explicitly taking into account the interaction of the carbonium and carbenium ion intermediates with the negatively charged zeolite wall.

Russian scientists concluded at the end of the seventies that proton activation and formation of organic cations could occur only when the positively charged protonated intermediates become stabilized by the negative charge left on the zeolite lattice [95–100]. Ten years later quantum-chemical calculations confirmed this conjecture. The finding was that these organic cations cannot exist as stable intermediates unless sterically constrained [101]. They are typically part of transition states. This is illustrated in Figs. 9a and b [102]. For carbenium ion intermediates from alkenes, this was first shown in an ab initio calculation by Kazansky and Senchenya [103] and Kramer et al. [104] for carbonium ion intermediates. Physically, this arises from the high energy cost of splitting the OH bond heterolytically. The corresponding energy cost is $\sim$1250 kJ/mol. Fortunately, when a molecule is protonated or, in other words, activated by a proton, no full cleavage of the OH bond is necessary. Typical activation energies are of the order of 10% of the full OH bond cleavage energy. The energy cost of bond cleavage is compensated for by the proton attachment energy ($\sim$200 kJ/mol) and electrostatic

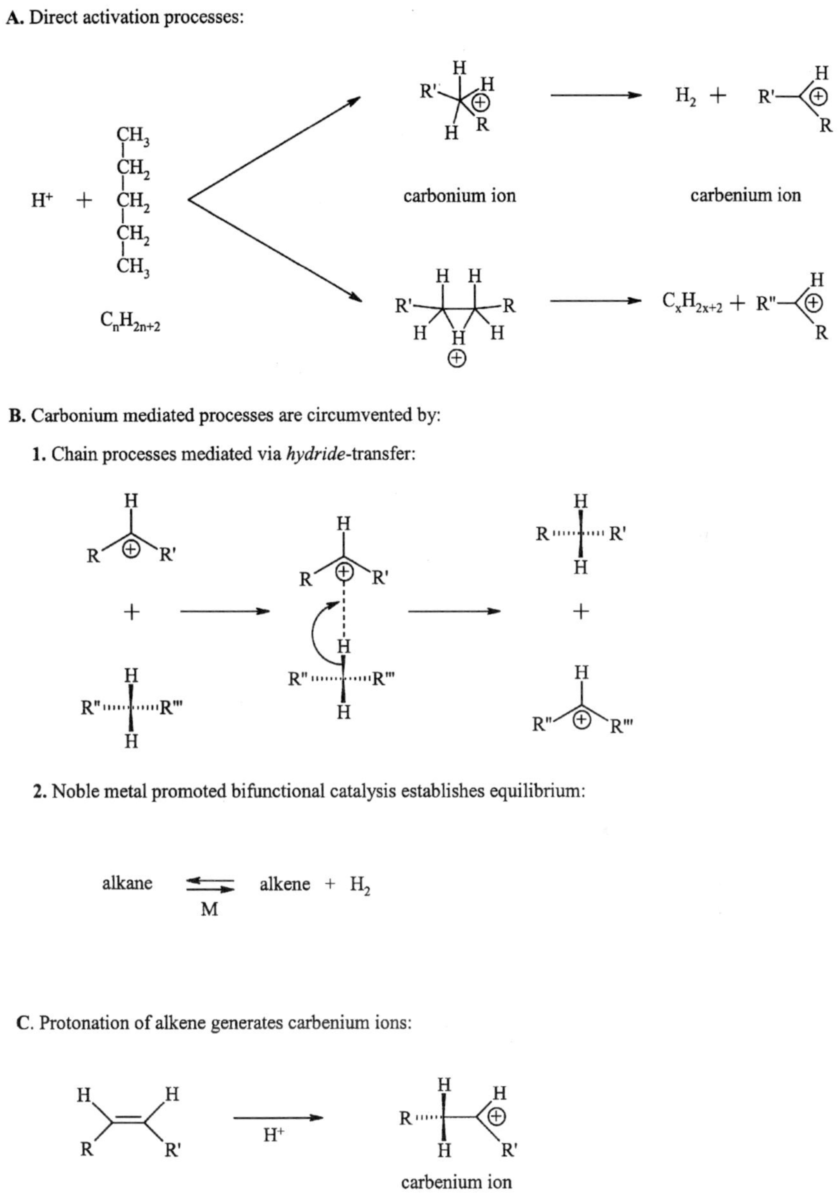

A. Direct activation processes:

$H^+ + C_nH_{2n+2}$ (n-pentane: CH$_3$-CH$_2$-CH$_2$-CH$_2$-CH$_3$)

→ carbonium ion → H_2 + carbenium ion

→ carbonium ion → C_xH_{2x+2} + carbenium ion

B. Carbonium mediated processes are circumvented by:

1. Chain processes mediated via *hydride*-transfer:

2. Noble metal promoted bifunctional catalysis establishes equilibrium:

$$\text{alkane} \underset{M}{\rightleftharpoons} \text{alkene} + H_2$$

C. Protonation of alkene generates carbenium ions:

alkene + H^+ → carbenium ion

FIG. 8. Initial elementary reaction steps of linear alkanes.

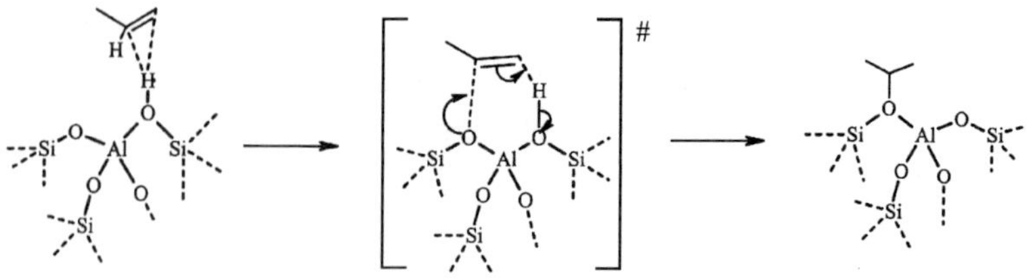

π-adsorption of propene   Transition State   σ-adsorption of propene
isopropyl alcoxy species

FIG. 9a. Schematic representation of the geometries of propene during its protonation attack mechanism by chabazite.

stabilization (~900 kJ/mol) with the negative charge that develops on the zeolite framework when the proton is transferred. The major difference between superacids and zeolites is that in superacids cations and anions become solvated by the polar solvent molecules. No such stabilization occurs in zeolites. The chemistry in zeolites, when applied in hydrocarbon catalysis, is much closer to that in a vacuum (dielectric constant $\varepsilon \sim 4$) than to that in a polar medium such as a polar superacid (dielectric constant $\varepsilon \sim 80$–100).

Until quite recently, quantum-chemical calculations on proton activation by zeolites were possible only using the cluster approximation [13–31]. A representative cluster is shown in Fig. 10 [105]. It initially seems surprising that calculations based on such small clusters can actually predict a property of a periodical system such as a zeolite. Two basic features of bonding in the zeolitic framework are relevant in this respect. First, chemical bonding is largely covalent: 90% of bonding can be described as due to covalent interaction, and only 10% of the bond strength is due to electrostatic interaction ions. Second, the flexible nature of the zeolite lattice can readily accommodate deformations due to the presence of the proton at a relatively low energy cost. Key to the use of the cluster approximation is termination of the cluster by hydride, a hydroxyl ion, so as to make the cluster neutral [1–8, 17]. The activation energies of protons in reactions with hydrocarbons can be computed with an accuracy of ~30 kJ/mol, using programs based on DFT [106–108]. Such quantum-chemical calculations scale much more favorably with the number of electrons in a system ($n^{2.6}$) than Hartree–Fock-based codes ($n^{4.2}$ for the MP2 method).

Due to the covalent nature of chemical bonding, properties of zeolitic protons depend on local electronic structure details. This is illustrated in Fig. 10 [105]. Using a basis set that predicts the NMR quadrupole coupling constants (QCC), accurately measured QCCs in zeolite ZMS-5 for Al tetrahedra in the proton form and deprotonated Al tetrahedra (e.g., by cation exchange) are compared with predicted QCCs from cluster model calculations. One

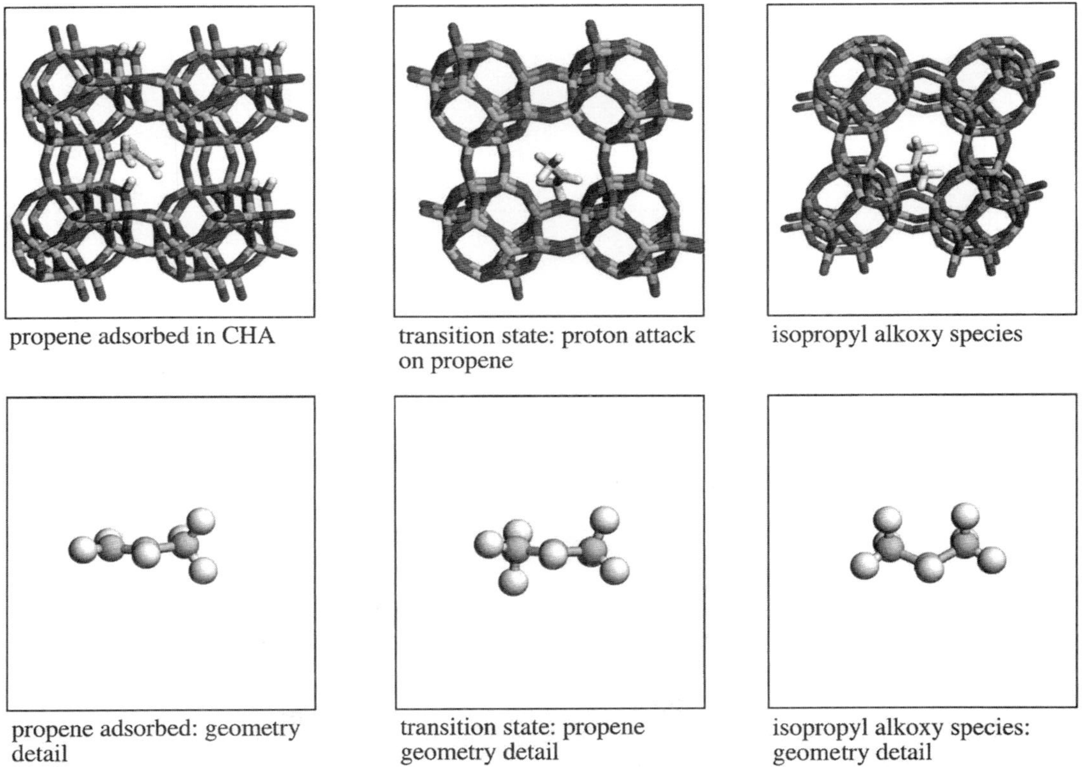

FIG. 9b. Geometries of propene during its protonation attack mechanism by chabazite as obtained from DFT periodical structure optimization calculations. Details of the propene geometry (values in degrees) are shown. The geometry of propene when the transition state occurs is very similar to the geometry of an isopropyl carbenium [102].

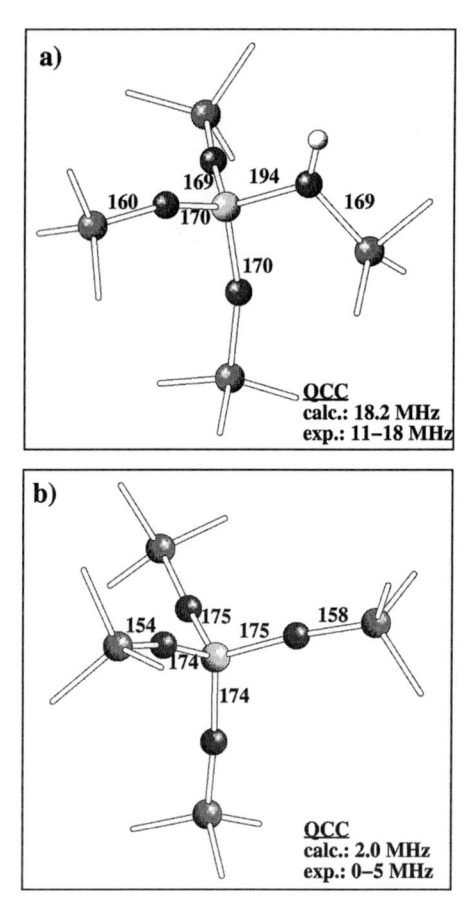

FIG. 10. The geometries of zeolitic site clusters for a protonated cluster (a) and a deprotonated cluster (b) (bond lengths in pm). The values are the quadrupole interaction parameters computed from the cluster models compared with experimental data [105].

notes the agreement between theory and experiment. Such agreement is typical also for other physical properties of the proton and its charge as reflected in the proton chemical shift [109, 110] or vibrational frequency [111, 112]. Note the asymmetric distortions of the protonated cluster in Fig. 10. The computed Al–O bond length of the cluster in its global energy minimum lengthens when a proton becomes attached to the bridging oxygen, which is a behavior indicative of covalent bonding. These results also illustrate the importance of geometry optimization of the computational systems. Only when this is done properly do computed interactions start to converge with experiment. When the protonic site is part of the

TABLE III
CALCULATED ACTIVATION ENERGIES ($E^\#$) FOR SOME ELEMENTARY
REACTION STEPS OF ISOBUTANE COMPARED WITH EXPERIMENTAL NUMBERS
EXTRACTED FROM KINETIC DATA [17, 113]

Reaction	Calculated $E^\#$ (kJ/mol)	Experimental range (kJ/mol)
Protolytic cracking	50.4	43–53
Protolytic dehydrogenation	60.6	40–50
Hybride transfer	36.0	<40–53
To s-C_3	41.0	
To t-C_3		

zeolite lattice, clearly the large distortions of the cluster when reaction with proton transfer occurs have to be accommodated by lattice relaxation for the cluster approximation to be used at all. As discussed before this can occur with very little expense of lattice energy. A major zeolite property that cannot be studied with this very small cluster is the effect of steric matching of the zeolite micropore and the size of the transition state complex. Using periodical calculations based on plane waves, this now becomes directly accessible to calculation, as we discuss later. We analyze transition-state geometry and energy results in clusters first and return to the size issue later.

Table III gives a comparison between computed and experimental activation energies for elementary proton-activated reaction steps of isobutane computed on a small cluster [17, 113]. We recognize the three reactions shown in Fig. 8. The first two reactions are monomolecular reaction steps. The bimolecular hydride transfer reaction is an elementary reaction step that maintains the catalytic reaction cycle: without a carbonium ion an intermediate transition-state formation occurs. As also shown in Fig. 8, an "intermediate" carbenium species (which actually exists as an alkoxy intermediate) can react in two ways to yield a product molecule. Hydride transfer from a reactant molecule produces an alkane, and proton backdonation produces an olefin. In the latter case, the reaction cycle can be continued only when a reactant molecule becomes activated by direct reaction with the proton. As explained before, the relatively high values for the activation energies stem from strong covalent bonding of the zeolite O–H and zeolite O–C bonds of intermediate alkoxy species. The high energy cost is due to the need to stretch these bonds to reach a transition state in which the carbenium ion is more or less free to react. One way of formulating this is that in zeolites reactions occur on top of the potential energy hills, instead of close to the energy valleys as for solvated species.

As has been shown, especially by Zygmunt *et al.* [114] but also by others [113, 115], computed activation energies depend substantially on the size and connectivity of the zeolite model clusters. In general, slow convergence is obtained for the deprotonation energies as a function of cluster size. There is an elegant procedure to obtain rapid convergence [17]. The activation energy (E_{act}) depends on the difference in energy of the cationic transition state (E_{cat}) with a very weak covalent but strong electrostatic interaction and the deprotonation energy in which a strongly covalent bond is broken:

$$E_{act} = E_{cat} - E_{deprotonation}. \quad (11)$$

One expects E_{act} to vary as a function of the cluster size mainly because of variations in the deprotonation energy. Hence a plot of E_{act} versus $E_{deprotonation}$ should result in a straight line (see Fig. 11). For each deprotonation energy the line gives the predicted value of the activation energy. Spectroscopic measurements of spectral shifts by adsorption of weakly interacting basic molecules allow the determination of deprotonation energies for a proton [117–119]. One notes in Table III the general agreement between computed and experimentally determined activation energies. Whereas the orders of magnitudes are correct, one also notes, however, that values tend to be overestimated by 30 or 40 kJ/mol.

Recently, studies have been made on related transition states but in periodical structure calculations [120–122]. The results for cluster and periodical calculations have been compared: significant reductions of the energy have been found for transition-state energies. Also, calculations from large cluster models that contain cavities have led to the same results [50, 111]. Whereas the energies of transition states appear to be influenced strongly by the zeolite topology, this is much less the case for the initial or final steps in the protonation reaction. Very differently from the dominantly covalent bond of the ground-state OH or alkoxy groups, the structures that correspond to transition states are highly polar. Positive charge is distributed over the cationic part of the molecule and negative charge is localized around the Al-containing lattice tetrahedral site. The transition-state complex creates a large dipole in the cavity. The high polarizability of the oxygen atom cavity stabilizes this complex by a screening interaction, E_{scr},

$$E_{scr} \approx -\frac{1}{2}\alpha_c E_\Gamma^2 = -\frac{1}{2}\alpha_c \frac{\mu^2}{R^4}, \quad (12)$$

where α_c is the polarizability of the cavity, $\vec{E}_\Gamma$ the electrostatic field from the transition-state dipole ($\bar{\mu}$), and R the approximate radius of the cavity. Of course this stabilization is expected to be larger for a small cavity than a larger one. Stabilization energies as high as 50–70 kJ/mol have been reported

(a)

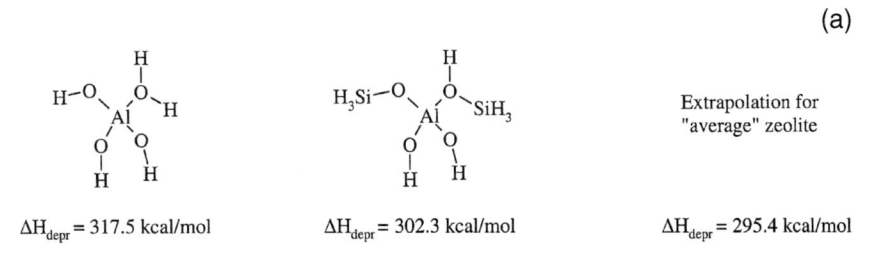

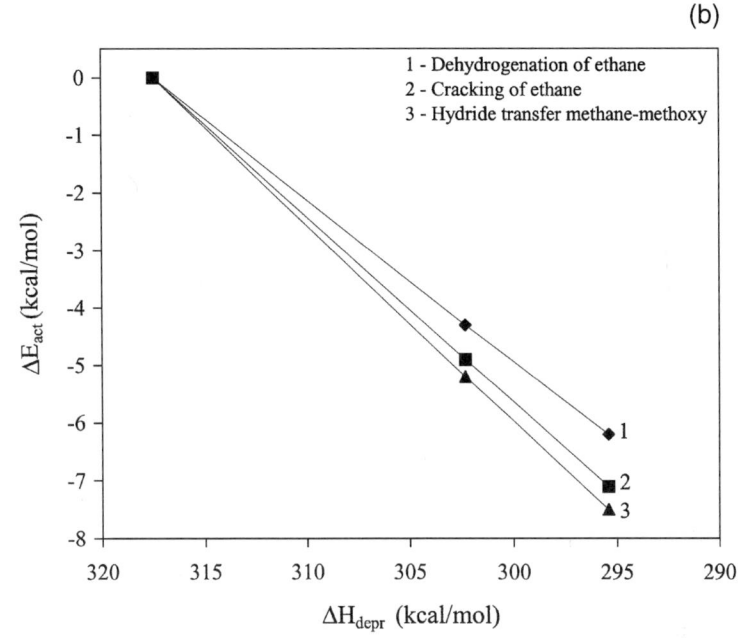

H-O\\Al/O-H with H on top O and H,H on bottom O's   H₃Si-O\\Al/O-SiH₃ with H on top O and H,H on bottom O's   Extrapolation for "average" zeolite

ΔH_{depr} = 317.5 kcal/mol ΔH_{depr} = 302.3 kcal/mol ΔH_{depr} = 295.4 kcal/mol

(b)

1 - Dehydrogenation of ethane
2 - Cracking of ethane
3 - Hydride transfer methane-methoxy

[plot of ΔE_{act} (kcal/mol) vs ΔH_{depr} (kcal/mol)]

FIG. 11. Corrections to the calculated activation energies for the cluster acid strength. Activation energies for protolytic cracking of ethane, protolytic dehydrogenation of ethane, methane–methoxy hybrid transfer, and methane deprotonation energies are computed at the MP2/6-31++G**//HF/6-31G* level with the ZPE corrections [17, 113].

[43, 123]. The van der Waals interaction between cavity and hydrocarbon is expected not to change significantly between the ground and the transition state, because the size of the hydrocarbon part does not change. A comparison of cluster-computed and periodical DFT-computed chemisorption of propylene is shown in Table IV.

Before continuing the analysis of pore size dependence, a remark must be made on the comparison between theoretical and measured data as done

TABLE IV
DFT COMPUTED VALUES OF THE CHEMISORPTION REACTION OF
PROPYLENE ON AN ACIDIC ZEOLITE CATALYST[a]

Energy comparison	Cluster approach (kJ/mol)	Periodical approach (kJ/mol)
E_π	−23	−27
E_{act}	105	47
E_σ	−52	−61

[a] The DFT energies of the reaction on a 4T zeolite cluster are compared with the periodical electronic structure results for chabazite. The computed energies are the π-complex energies with respect to the gas-phase system (E_π), the activation energies with respect to the π-complex (E_{act}), and the energies of σ-complex (E_σ) with respect to the gas phase [122].

in Table III. The experimental kinetic parameters were obtained from a kinetic analysis in which overall experimental data were deconvoluted so as to obtain elementary rate constants. In the next section, we show how this can be done. A prerequisite is knowledge of the reaction mechanism. The reaction mechanism used to deconvolute the experimental data used in Table III has been chosen as consistent with that as theoretically proposed and discussed in this paper.

Apart from the use of periodical or very large cluster quantum-chemical calculations, embedding techniques can be applied to analyze consequences of steric constraints. These embedding methods are quantum mechanics/molecular mechanics-related techniques [11, 50, 112, 123–128]. The quantum-chemical cluster is embedded in a lattice described by classical interactions, such as electrostatic, van der Waals, or repulsive interactions. Care has to be taken to compensate for boundary effects between the quantum-mechanical neutral cluster, terminated with hydride or hydroxyl groups, and the lattice atoms. The lattice atoms and cluster atoms can be imagined as being connected by bonds that give relaxation constraints to the cluster. Sauer and co-workers [11, 50, 112] have developed an embedding approach that enables embedded calculation of ground states and transition states, but separate development of potentials for both states using quantum-mechanical cluster calculations to provide potential energy surfaces is necessary.

An alternative method is the Chem-Shell embedding procedure developed by Greatbanks et al. [124]. This method does not require the development of new force fields as a function of intermediate states [123–128]. Using the Chem-Shell procedure, an interesting result has been obtained for the protonation of isobutene in chabazite. The results are illustrated in Fig. 12 [123]. The quantum-mechanical calculations were done within the

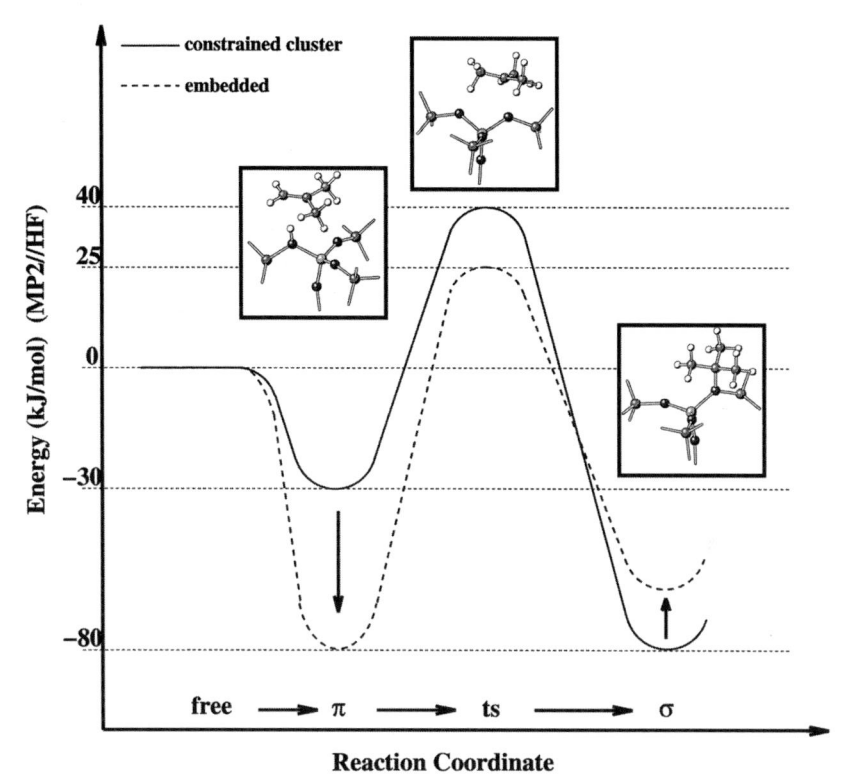

FIG. 12. Effect of embedded environment on the stability of *i*-butene activation at a Brønsted-acid site in chabazite [123].

Hartree–Fock approximation. The interaction of the π bond with a proton leads to different states: the first one is a weakly π-bonded state for which the interaction energy between the proton and the olefin π bond is about 30 kJ/mol, and the second one is a σ-bonded state or alkoxy species in which protonation and chemisorption of isobutene have occurred. The protonation of isobutene gives a $C^+(CH_3)_3$ carbenium ion that binds to a lattice oxygen atom that bridges lattice Al and Si. In the transition state from the π-state to the σ-state, protonation and chemisorption are associated. The curves drawn in Fig. 12 present the relative energies of the π-bonded state, σ-bonded state, and transition state between these two states. One notes the strong stabilization of the σ-bonded protonated state compared to the weakly bonded π-state. Interestingly, the relative energies of the π- and σ-bonded state change completely for the embedded situation. Compared to the π-bonded state, the σ-bonded alkoxy state now appears to be destabilized. This is because of steric repulsive interactions that arise when the alkoxy O–C bond brings the methyl groups of isobutyl close to oxygen atoms in lattice tetrahedra neighboring the Al-containing tetrahedron. The strong repulsive interactions that arise prevent the formation of a strong O–C alkoxy bond. Alkoxy formation is actually prevented and the promoted intermediate is now close to a free carbenium ion, a result recently confirmed by full DFT periodical electronic structure calculations of SAPO-5 [129], which has cavities similar to those of chabazite. For the hydrogen-bonded π-state with a larger O–C distance, a more favorable interaction than the free cluster is found. This is due to the additional stabilizing interaction of the isobutene molecule now possible with the other oxygen atoms from the chabazite cage. The extra stabilization is here due mainly to the attractive van der Waals interaction. These results illustrate the importance of including the attractive van der Waals interaction as well as Born repulsion effects for bulky molecules and intermediates in estimates of protonation of molecules. Both contributions are strongly micropore size and shape dependent. The protonation of the olefin now leads to a state close to a carbenium ion.

We now introduce the two ways in which these pore size-dependent terms can be added as corrections to cluster calculations on small clusters not containing any information on the micropore as shown in Fig. 10. It is important to remember that the attractive van der Waals interaction cannot be properly calculated with currently used quantum-chemical DFT methods. How to correct for the attractive part is illustrated in Fig. 13. The figure illustrates that the protonation energies with respect to the gas phase have to be considered as the summation of two energetic interactions:

$$\Delta E_{\text{sys}} = \Delta E_{\text{ads}} + \Delta E_{\text{sys}}(\text{local}), \tag{13}$$

where ΔE_{ads} is the energy of adsorption of the molecule in the siliceous part of the zeolite. This is the strongly micropore size- and shape-dependent adsorption energy discussed in Section IV. ΔE_{sys} (local) is the quantum-chemical enrgy of the considered system (i.e., π-state, transition state, or σ-state) with respect to its noninteracting state (or gas-phase state). Expression (13) is based on the approximation that in the different states the van der Waals interaction energy does not change, since protonation concerns only a small part of the molecule [130]. The implication of expression (13) is that differences in protonation energies of different zeolites are due mainly to differences in the ΔE_{ads} term of the expression. However, one should not assume that corrected protonation or activation energies obtained from calculations simply reflect the differences in chemical reactivity of the zeolitic proton. Experimentally, the zeolite deprotonation energy of differently located protons varies between 10 and 30 kJ/mol for a zeolite with a Si/Al ratio below 10. Embedded calculations or periodical calculations can be used to estimate this quantum-chemical protonation energy as a function of

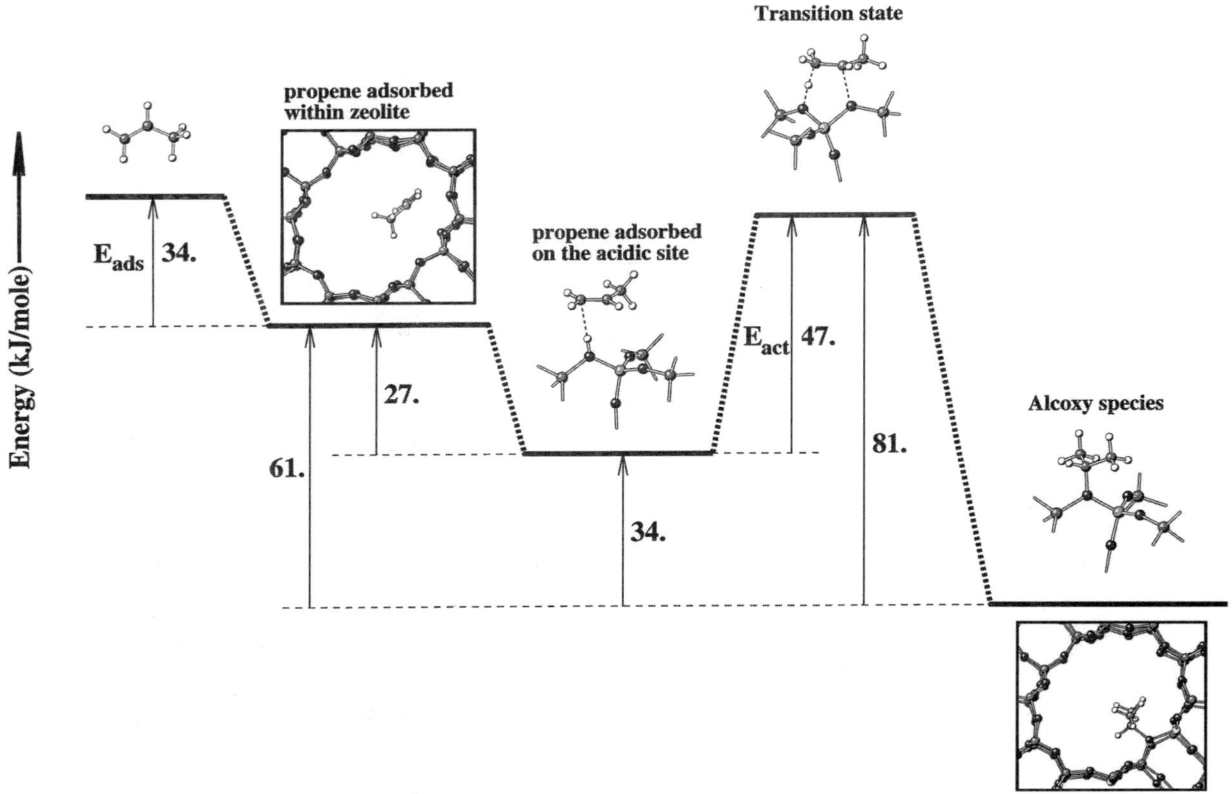

FIG. 13. Changes in energy of adsorption of a protonated molecule with respect to the gas phase for protonation of propylene in chabazite. E_{ads} is the adsorption energy of propene in chabazite.

the proton position in a zeolite. As we will see the additivity concept basic to expression (13) is important in understanding catalytic acidity. It is important to decouple the cavity size-dependent van der Waals contribution, which relates to adsorption, from the chemical reactivity, which relates to the interaction with the proton. The repulsive interactions of bulky molecules with the zeolite micropore are very important for selectivity differences of conversion reactions controlled by steric constraints.

Embedded or periodical calculations are essential to compute these effects properly. A recent study of alkylation of toluene using a DFT periodical electronic structure code demonstrated the feasibility of this approach [131]. We illustrate here how Eq. (13) can be used to compute corrections to the activation energy from the cluster approach by adding the repulsive interaction that results from the formation of bulky intermediates. We use as an example the bimolecular hydride transfer reaction. The activation energy of the corresponding elementary rate constant can then be written

$$\Delta E_{act} = \Delta E_{act}(\text{cluster}) + \Delta E_{steric}, \qquad (14)$$

where ΔE_{steric} is the repulsive term in the activation energy, which depends strongly on the micropore volume and shape.

Transition states of the hydride transfer reaction have been studied quantum chemically using the cluster approach [132, 133]. In the transition state, the hydrogen atom becomes positioned between the two carbon atoms, in between which the hydrogen atom changes its position. In Fig. 14, the schematic

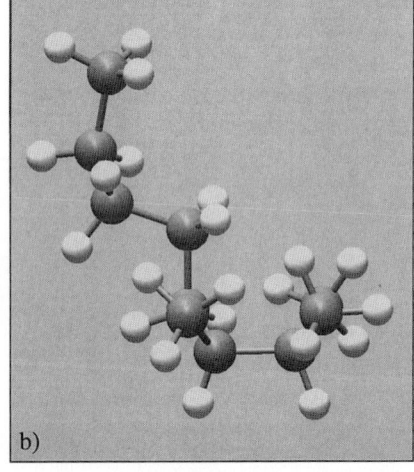

FIG. 14. (a) Schematic geometry of the transition state for hydrogen transfer between heptane and propyl alkoxy. (b) Schematic geometry of the 2-propyl-2 heptane isomer of decane [17].

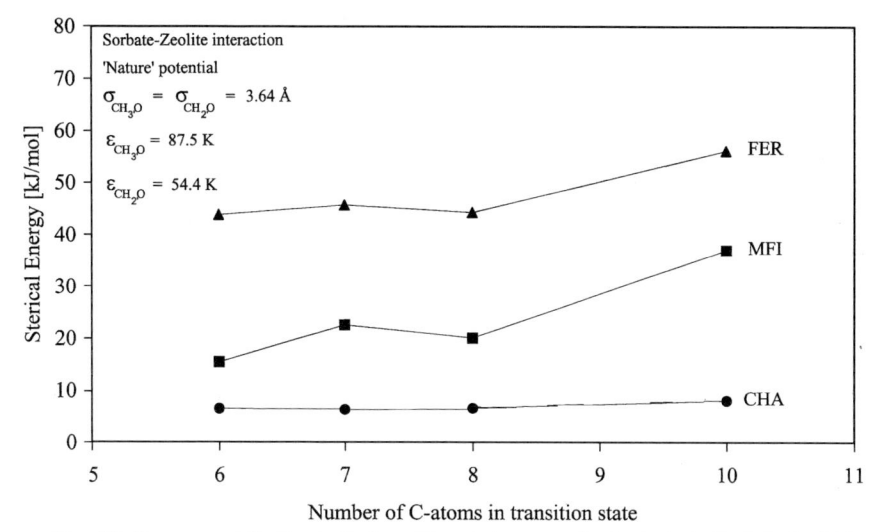

FIG. 15. Energy contribution to the transition-state energy due to sterical effects for hydride transfer reactions in three zeolites [74].

transition state for hydride transfer between heptane and propyl is sketched. As can be noted from Fig. 14b its structure is close to that of the 2-propyl-2-heptane isomer of decane.

Figure 15 gives the computed differences in energy according to a Monte Carlo simulation for a few zeolites [74]:

$$\Delta E_{\text{steric}} = E_{\text{ads}} \text{(intermediate)} - E_{\text{ads}} \text{(reactant 1)} - E_{\text{ads}} \text{(reactant 2)}. \quad (15)$$

One notes the strong dependence of ΔE_{steric} as a function of the type of zeolite and also the large repulsive correction ΔE_{steric} for the zeolite with the narrow one-dimensional pore (ferrierite). We illustrate in the next section how the values of ΔE_{steric} can be used to assist kinetic analysis.

C. Kinetics

The aim of this section is to illustrate how the mechanistic principles as well as the rate constant predictions outlined in the previous section can be used to simulate the overall rate of a reaction. It also provides an opportunity to discuss more extensively the meaning of relation (13). This is done using experimental data on the hydroisomerization of hexane.

The hydroisomerization reaction occurs in an excess of hydrogen (H_2/hexane ~ 30) catalyzed by noble metal (e.g., Pt)-activated acidic zeolites. The mechanism of the reaction is rather well understood and the sequence of

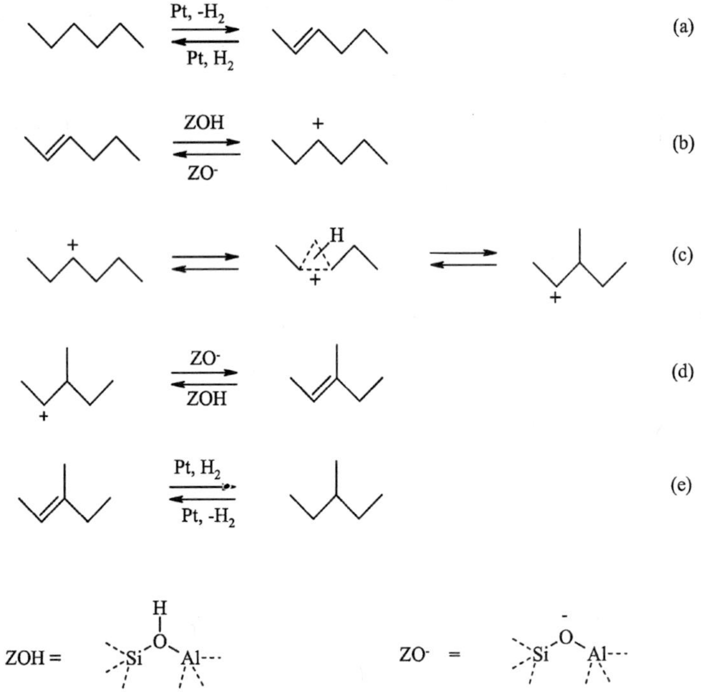

FIG. 16. Reactions involved in the hydroisomerization reaction: (a) equilibration of n-hexane and i-hexane (leading experimentally to a [hexane]/[hexene] ratio of approximately 10^{-6}), (b) protonation of hexene, (c) isomerization of n-alkoxy species to i-alkoxy species, (d) deprotonation of i-alkoxy species, and (e) equilibration of i-hexene and i-hexane.

the mechanisms involved in this reaction is summarized in Fig. 16 [134, 135]. The amount of the noble metal, such as Pt, added is such that hexane/hexene equilibrium is established. Under the reaction conditions applied experimentally (230–250°C), this implies that the hexene/hexane ratio is about 10^{-6}, which suppresses undesirable deactivating oligomerization reactions. Hexene reacts readily with the zeolitic proton. An intermediate alkoxy species is formed, which can isomerize to the isoalkoxy, which upon deprotonation gives isoalkene. Isoalkene is subsequently hydrogenated to hexane.

It has been explained in a previous section that diffusion of hexane and isohexane can be considered to be fast. From a combination of theoretical and experimental data, a reaction energy scheme corresponding to the catalytic reaction cycle that converts n-hexene into i-hexene has been deduced. This is shown in Fig. 17 [136, 137]. As in Section V, adsorption on the siliceous part of the zeolite micropore is considered to be independent of proton activation.

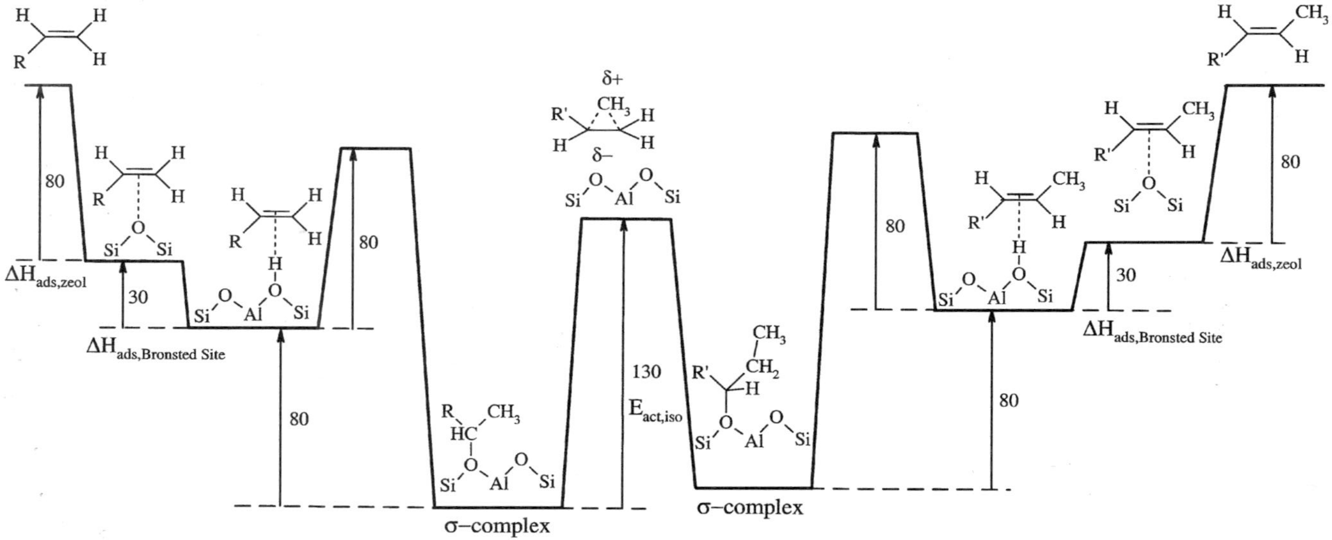

FIG. 17. Reaction energy diagram of the isomerization reaction of hexene (energy values in kJ/mol) [136].

Step 1 in the reaction energy diagram is the strongly micropore-dependent energy of adsorption of reactant hexene. In the diagram, the computed adsorption energy value in siliceous mordenite is used. The olefinic part of the molecule becomes π and, subsequently, σ coordinated in the subsequent steps. The best current estimation of the activation energy of the chemisorption reaction is ~60 kJ/mol (the cluster value, ~100 kJ/mol, corrected for by the screening contribution, ~40 kJ/mol). No accurate theoretical predictions of the activation energy of isomerization are yet available but experiment indicates that this value has to be between 120 and 140 kJ/mol [136–138]. As rate-limiting steps the rates of desorption and isomerization can compete. An estimate of preexponents is necessary to decide whether this is really the case.

We explained in Section IV. B that the positively charged transition states are stabilized by the interaction with the negatively charge zeolite wall. We also discussed the large entropy loss of adsorbed molecules compared to that in the gas phase. Hence, the transition state for the desorbing molecule will have a significantly larger entropy gain than occurs in proton activation. One may expect differences in the preexponent of $\sim 10^4$ [139]. Whereas this implies that for hexane isomerization the elementary rate constant of isomerization will be rate limiting, for larger molecules with higher adsorption energies, the rates of desorption and proton activation may compete. Under conditions where the rate of desorption is rate limiting, dynamic Monte Carlo simulations show that product molecules remain occluded in the one-dimensional channels. The micropore mouths become blocked by incoming molecules. As a consequence, the rate constant per proton decreases (see Fig. 18) [140].

In the absence of the need to include diffusion explicitly because it is fast, the kinetic equations for hexane isomerization reduce to classical steady-state expressions for the rate of isomerization:

$$r_{iso} = k_{iso} \cdot \Theta_{alcoxy} \tag{16a}$$

and

$$k_{ads.alkene} P_{alkene} (1 - \Theta_{alkane} - \Theta_{alkane} - \Theta_{alkoxy}) = k_{des} \Theta_{alkene} + k_{isom} \Theta_{alkoxy}. \tag{16b}$$

To deduce r_{iso}, the elementary rate constant of isomerization, k_{iso}, has been assumed to be rate limiting. The competition of protonic sites adsorption for alkane or alkene has been explicitly included in the kinetic scheme of reactions. Using available data on the adsorption isotherms of alkane and theoretical protonation energies, the elementary rate constant parameters of k_{iso} can be deduced from experiment by measuring the rate of isomerization

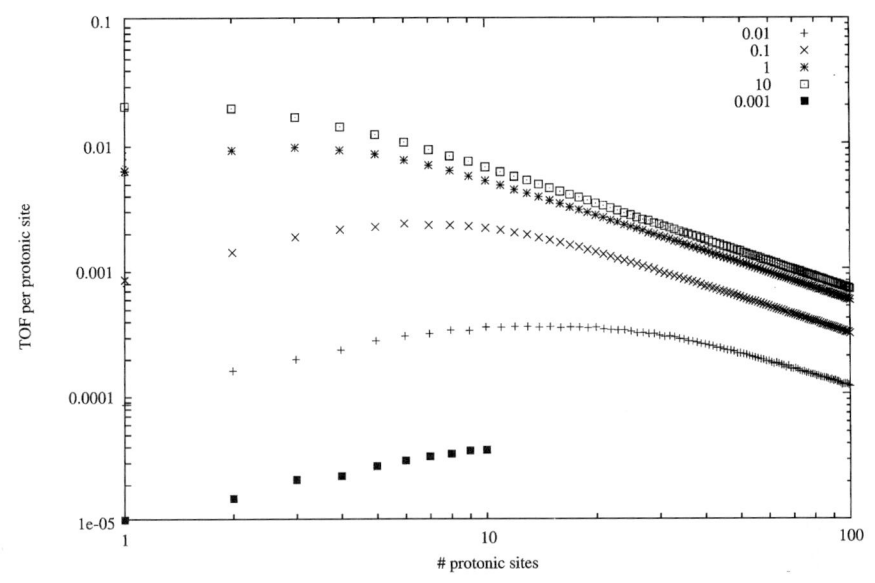

FIG. 18. Simulated turnover frequency as a function of the number of protonic sites. The loading is about 5%, every nine site can become a protonic site, and the rate constant for A → B transformation is 10, 1, 0.1, 0.01, and 0.001 times the rate constant for desorption from the top to the bottom curves [140].

as a function of the partial pressure of hexane [138]:

$$\frac{1}{r_{iso}} = a + \frac{b}{P_{hexane}}$$

$$\begin{cases} a = \dfrac{1}{k_{iso}} + \dfrac{P_{H_2}}{k_{iso} K_{prot}} \dfrac{K_{ads,hexane}}{K_{ads,hexene} K_{dehydr}} \\ b = \dfrac{P_{H_2}}{k_{iso} K_{prot} K_{dehydr} K_{ads,hexene}}. \end{cases}$$

Table V collects values for the activation energies of isomerization and protonation as deduced by De Gauw and van Santen [138] from kinetic measurements. A comparison of the turnover frequency per proton (TOF) and k_{iso} is made in Table V. One notes that the large differences in measured overall TOFs of different zeolites disappear for the elementary rate constant k_{iso}. This implies that the difference in apparent acidity of the zeolite is due mainly to the difference in adsorption isotherms of the different zeolites. One notes the small variation in activation and protonation energy values, which implies a slight dependence of protonation on the micropore channel size and dimension.

TABLE V
MEASURED VALUES OF THE HYDROISOMERIZATION REACTION OF HEXANE CATALYZED BY DIFFERENT ZEOLITES[a]

	ΔH_{prot}	E_{iso}^{act}	k_{iso}^{b}	TOF[b]
H-Beta	−46	145	$1.7 \cdot 10^{-2}$	$5.0 \cdot 10^{-3}$
H-Mordenite	−19	120	$2.7 \cdot 10^{-2}$	$1.1 \cdot 10^{-2}$
H-ZSM-5	−23	105	$2.8 \cdot 10^{-2}$	$4.1 \cdot 10^{-3}$
H-ZSM-22	−35	106	$1.7 \cdot 10^{-2}$	$1.6 \cdot 10^{-3}$

[a] The values are the enthalpies of protonation (ΔH_{prot}; kJ/mol), the activation energies of the isomerization reaction (E_{iso}^{act}; kJ/mol), the rate constants of isomerization (k_{iso}; s^{-1}), and the turnover frequencies (TOF; s^{-1}) [139].
[b] Measured at 240°C.

V. Concluding Remarks

The aim of this paper has been to present an overview of the current status of atomistic simulations in relation to an understanding or prediction of zeolite catalyst performance. In the initial sections, we have described that the attractive van der Waals and the repulsive Born interactions determine micropore size- and shape-dependent differences between different zeolite structures. This affects overall catalytic rates because it controls the degree of micropore filling under catalytic conditions.

The discussion of reactivity focused on the activation of hydrocarbons by zeolitic protons. The deprotonation energy of a proton is weakly dependent on the zeolite crystallographic position but may be strongly zeolite composition dependent, especially at high concentrations of three valent cations (Al, Ga) in the zeolite framework. Nonetheless, the deprotonation energy is a local property of the OH bond, which can be estimated using quantum-chemical calculations by extrapolation from properly terminated cluster calculations.

The elementary rate constant for proton activation is weakly dependent on the micropore size as long as steric constraints do not affect the transition state. Because of the zwitterionic nature of the transition state, dielectric screening by the oxygen atoms of the micropore tends to decrease the cluster-calculated transition state energies to 10 to 30% of the activation energies. Steric constraints on the transition state may substantially increase the cluster-computed activation energies by similar amounts. These steric constraints can be computed from periodical DFT calculations or from transition-state model structures using Monte Carlo adsorbate–zeolite pore interaction calculations.

The overall concept that appears central is that zeolite activity depends on two energy terms: the energy of adsorption, controlled by the micropore size-dependent van der Waals forces; and the protonation energy, which is a local property usually rather independent of the micropore size, unless steric constraints limit optional interactions between proton and substrate.

The activation energy for proton transfer can be viewed as a lattice oxygen Lewis-base and proton Brønsted-acid synergetic event [3]. One generally finds that activation energies of proton-activated reactions are rather high: between 100 and 200 kJ/mol for proton-activated elementary reaction steps in hydrocarbon conversion catalysis. This is the main reason for the relatively low TOF per proton ($\sim 10^2$ s^{-1}) for this type of reaction. Similarly to enzymes [31], the weak van der Waals-type interaction determines the size- and shape-dependent behavior.

REFERENCES

1. Catlow, C. R. A, Cormack, A. N., and Theobald, F., *Acta Crystallogr. B* **40**, 195 (1984).
2. Catlow, C. R. A., Ackermann, L., Bell, R. G., Corà, F., Gay, D. H., Nygren, M. A., Pereira, J. C., Sastre, G., Slater, B., and Sinclair, P. E., *Faraday Discuss.* **106**, 1 (1997).
3. van Santen, R. A., and Kramer, G. J., *Chem. Rev.* **95**, 637 (1995).
4. Sauer, J., in *Modeling of Structure and Reactivity in Zeolites* (Catlow, C. R. A., ed.), Academic Press, London, 1992, p. 183.
5. Sauer, J., in *Cluster Models for Surface and Bulk Phenomena: Proceedings of a NATO Advanced Research Workshop* (Pacchioni, G., Bagus, P. S., and Parmigiani, F., eds.), Plenum Press, London, 1992, p. 533.
6. van Santen, R. A., De Bruyn, D. P., Den Ouden, C. J. J., and Smit, B., in *Introduction to Zeolite Science and Practice* (Van Bekkum, H., Flaningen, E. M., and Jansen, J. C., eds.), Elsevier, Amsterdam, 1991, Vol. 58, p. 317.
7. Corma, A., Llopis, F., Viruela, P., and Zicovich-Wilson, C., *J. Am. Chem. Soc.* **116**, 134 (1994).
8. Sauer, J., Ugliengo, P., Garrone, E., and Saunders, V. R., *Chem. Rev.* **94**, 2095 (1994).
9. Nicholas, J. B., *Topics Catal.* **4**, 157 (1997).
10. Civalleri, B., Zicovich-Wilson, C. M., Ugliengo, P., Saunders, V. R., and Dosevi, R., *Chem. Phys. Lett.* **292**, 394 (1998).
11. Brändel, M., and Sauer, J., *J. Am. Chem. Soc.* **120**, 1556 (1998).
12. Jeanvoine, Y., Ángyán, J. G., Kresse, G., and Hafner, J., *J. Phys. Chem. B* **102**, 5573 (1998).
13. Blaszkowski, S. R., and van Santen, R. A., in *Transition State Modeling for Catalysis* (Truhlar, D. G., and Morokuma, K., eds.), ACS Symposium Series 721, Am. Chem. Soc., 19xx, Chap. 24
14. Beck, L. W., Xu, T., Nicholas, J. B., and Haw, J. F., *J. Am. Chem. Soc.* **117**, 11594 (1995).
15. Corma, A., Sastre, G., and Viruela, P. M., *J. Mol. Catal. A* **100**, 75 (1995).
16. Saintigny, X., van Santen, R. A., Clémendot, S., and Hutschka, F., *J. Catal.* **183**, 107 (1999).
17. Frash, M. V., and van Santen, R. A., *Topics Catal.* **9**, 191 (1999).
18. Kazansky, V. B., *Catal. Today* **51**, 419 (1999).

19. Boronat, M., Viruela, P., and Corma, A., *J. Phys. Chem. A* **102,** 982 (1998).
20. Evleth, E. M., Kassab, E., Jessrich, H., Allavena, M., Montero, L., and Sierra, L. R., *J. Phys. Chem.* **100,** 11368 (1996).
21. Esteves, P. M., Nascimento, M. A. C., and Mota, C. J. A., *J. Phys. Chem. B* **103,** 10417 (1999).
22. Frash, M. V., Kazansky, V. B., Rigby, A. M., and van Santen, R. A., *J. Phys. Chem. B* **101,** 5346 (1997).
23. Sauer, J., Horn, H., Häser, M., and Ahlrichs, R., *Chem. Phys. Lett.* **173,** 26 (1990).
24. Blaszkowski, S. R., and van Santen, R. A., *J. Am. Chem. Soc.* **118,** 5152 (1996).
25. Blaszkowski, S. R., and van Santen, R. A., *J. Am. Chem. Soc.* **119,** 5020 (1997).
26. Sierra, L. R., Kassab, E., and Evleth, E. M., *J. Phys. Chem.* **97,** 641 (1993).
27. Yoshizawa, K., Shiota, Y., Yumura, T., and Yamabe, T., *J. Phys. Chem. B* **104,** 734 (2000).
28. Barbosa, L. A. M. M., and van Santen, R. A., *J. Mol. Catal. A* **166**(1), 101 (2001).
29. Rice, M. J., Chakraborty, A. K., and Bell, A. T., *J. Phys. Chem. A* **102,** 7498 (1998).
30. Fois, E., Gamba, A., and Tabachi, G., *Phys. Chem. Chem. Phys.* **1,** 531 (1999).
31. Barbosa, L. A. M. M., and van Santen, R. A., *J. Catal.* **191,** 200 (2000).
32. Frenkel, D., and Smit, B., in *Understanding Molecular Simulation: From Algorithms to Applications,* Academic Press, San Diego, 1996.
33. Bates, S. P., and van Santen, R. A., *Adv. Catal.* **42,** 1 (1998).
34. Bell, A. T., Maginn, E. J., and Theodorou, D. N., in *Handbook of Heterogeneous Catalysis* (Ertl, G., Knözinger, H., and Weitkamp, J., eds.), Wiley–VCH, Weihheim, Germany, 1997, Vol. 3, p. 1165.
35. Kramers, H. A., *Physica* **7,** 284 (1940).
36. Gilbert, R. G., and Smith, S. C., *Theory of Unimolecular and Recombination Reactions,* Blackwell, Oxford, 1990.
37. van Santen, R. A., and Niemantsverdriet, J. W., *Chemical Kinetics and Catalysis,* Plenum, New York, 1995.
38. Schleyer, P. v. R. (ed.), in *Encyclopedia of Computational Chemistry,* John Wiley & Sons, New York, 1998.
39. Bernardi, F., and Robb, M. A., in *Ab Initio Methods in Quantum Chemistry—I* (Lawley, K. P., ed.), John Wiley & Sons, New York, 1987, p. 155.
40. Venuto, P. B., *Microporous Matter* **2,** 297 (1994).
41. Hölderich, W. F., and Van Bekkum, H., in *Introduction to Zeolite Science and Practice* (Van Bekkum, H., Flaningen, E. M., and Jansen, J. C., eds.), Elsevier, Amsterdam, 1991, Vol. 58, p. 631 .
42. Nicholas, J. B., *Topics Catal.* **4,** 157 (1997).
43. Rigby, A. M., Kramer, G. J., and van Santen, R. A., *J. Catal.* **170,** 1 (1997).
44. Koller, H., Engelhardt, G., and van Santen, R. A., *Topics Catal.* **9,** 163 (1999).
45. van Santen, R. A., *Catal. Today* **38,** 377 (1997).
46. Olah, G. A., and Donovan, D. J., *J. Am. Chem. Soc.* **99,** 5026 (1977).
47. Olah, G. A., Prakash, G. K. S., and Sommer, J., in *Superacids,* Wiley Interscience, New York, 1985.
48. Olah, G. A., Prakash, G. K. S., Williams, R. E., Field, L. D., and Wade, K., in *Hypercarbon Chemistry,* Wiley Interscience, New York, 1987.
49. De Man, A. J. M., Van Beest, B. W. H., Leslie, M., and van Santen, R. A., *J. Phys. Chem.* **94,** 2524 (1990).
50. Schröder, K.-P., and Sauer, J., *J. Phys. Chem.* **100,** 11043 (1996).
51. Levien, L., Previtt, C. T., and Weider, D. J., *Am. Mineral.* **65,** 925 (1980).
52. De Man, A. J. M., and van Santen, R. A., *Zeolites* **12,** 269 (1992).
53. Lee, C., Parrillo, D. J., Gorte, R. J., and Farneth, W. E., *J. Am. Chem. Soc.* **118,** 3262 (1996).

54. Gorte, R. J., *Catal. Lett.* **62,** 1 (1999).
55. Bezuz, A. G., Kiselev, A. G., Lopatkin, A. A., and Quang Du, P., *J. Chem. Soc. Faraday Trans. II* **74,** 367 (1978).
56. Kiselev, A. A., and Quang Du, P., *J. Chem. Soc. Faraday Trans. II* **77,** 17 (1981).
57. Bezuz, A. Z., Kocirik, M, Kiselev, A. V., Lopatkin, A. A., and Vasilyena, E. A., *Zeolites* **6,** 101 (1986).
58. Meinander, N., and Tabisz, G. C., *J. Chem. Phys.* **79,** 416 (1983).
59. Bell, R. G., Lewis, D. W., Voigt, P., Freeman, C. M., Thomas, J. M., and Catlow, C. R. A., in *Proceedings of the 10th International Conference, Garmisch-Partenkirchen, 1994* (Weitkamp, J., Karge, H. G., Pfeifer, H., and Hölderich, W., eds.), Elsevier, Amsterdam, 1994, Part C, p. 2075.
60. Kristyan, S., and Pulay, P., *Chem. Phys. Lett.* **229,** 175 (1994).
61. Pelmenschikov, A., and Leszcynski, J., *J. Phys. Chem. B* **103,** 6886 (1999).
62. Allen, M. P., and Tildesley, D. J., in *Computer Simulation of Liquids,* Oxford Science, Oxford, 1987.
63. Frenkel, D., in *Proceeding of the Euroconference on Computer Simulation in Condensed Matter Physics and Chemistry, Como, 1995* (Binder, K., and Ciccotti, G., eds.), Italian Physical Society, Bolognia, 1995, Chap. 7.
64. Smit, B., and Siepmann, J. I., *J. Chem. Phys.* **98,** 8442 (1994).
65. Smit, B., *Mol. Phys.* **85,** 153 (1995).
66. June, R. L., Bell, A. T., and Theodorou, D. N., *J. Phys. Chem.* **96,** 1051 (1992).
67. Smit, B., and Maesen, T. L. M., *Nature* **374,** 42 (1995).
68. Kiselev, A. V., and Quang Du, P., *J. Chem. Soc. Faraday Trans.* **77,** 1 (1981).
69. Freeman, C. M., Catlow, C. R. A., Thomas, J. M., and Brode, S., *Chem. Phys. Lett.* **186,** 137 (1991).
70. Bates, S. P., Van Well, W. J. M., van Santen, R. A., and Smit, B., *J. Am. Chem. Soc.* **118,** 6753 (1996).
71. Bates, S. P., Van Well, W. J. M., van Santen, R. A., and Smit, B., *J. Phys. Chem.* **100,** 17573 (1996).
72. Smit, B., and Maesen, T. L. M., *Nature* **374,** 42 (1995).
73. Van Well, W. J. M., Cottin, X., and Smit, B., *J. Am. Chem. Soc.* **37,** 1081 (1998).
74. Schumacher, R., and van Santen, R. A., in preparation.
75. Eder, F., Ph.D. thesis, TU, Twente, 1996.
76. Denayer, J. F., Baron, G. V., Martens, J. A., and Jacobs, P. A., *J. Phys. Chem. B* **102,** 3077 (1998).
77. Noordhoek, N. J., Van Ijzendoorn, L. J., Anderson, B. A., De Gauw, F. J., van Santen, R. A., and De Voigt, M. J., *Ind. Eng. Chem. Res.* **37,** 825 (1998).
78. Schuring, D., Jansen, A. P. J., and van Santen, R. A., *J. Phys. Chem. B* **104,** 941 (2000).
79. Clark, L. A., Ye, G. T., Gupta, A., Hall, L. L., and Snurr, R. O., *J. Chem. Phys.* **111,** 1209 (1999).
80. Heink, W., Kärger, J., Pfeifer, H., and Stallmach, F. J. M. M., *J. Am. Chem. Soc.* **112,** 2175 (1990).
81. Kärger, J., and Rutheven, D. M., in *Diffusion in Zeolites and Other Microporous Solids,* Wiley–Interscience, New York, 1992.
82. Callaghan, P. T., in *Principles of Nuderar Magnetic Resonance Microscopy,* Clarendon, Oxford, 1991.
83. Hong, U., Kärger, J., and Pfeifer, H., *J. Am. Chem. Soc.* **113,** 4812 (1991).
84. Kärger, J., and Pfeifer, H., in *NMR Techniques in Catalysis* (Bell, A. T., and Pines, A., eds.), Marcel Dekker, New York, 1994, p. 69.
85. Yashonath, S., and Santikary, P., *Mol. Phys.* **78,** 1 (1993).

86. Yashonath, S., and Santikary, P., *J. Chem. Phys.* **100,** 4013 (1994).
87. Yashonath, S., and Santikary, P., *J. Phys. Chem.* **97,** 3849 (1993).
88. Smit, B., Loyens, L. D. J. C., and Verbist, G. L. M. M., *Faraday Discuss.* **106,** 93 (1997).
89. Karpinski, Z., Gandhi, S. N., and Sachtler, W. M. H., *J. Catal.* **141,** 337 (1993).
90. Fraenkel, D., and Levy, M., *J. Catal.* **118,** 10 (1989).
91. Hinderen, J., and Keil, F., *J. Chem. Eng. Sci.* **51,** 2667 (1996).
92. Olah, G. A., and Schosberg, R. H., *J. Am. Chem. Soc.* **90,** 2726 (1968).
93. (a) Haw, J. F., Richardson, B. R., Oshiro, I. S., Lazo, N. D., and Speed, J. A., *J. Am. Chem. Soc.* **111,** 2052 (1989). (b) Haw, J. F., Nicholas, J. B., Xu, T., Beck, L. W., and Ferguson, D. B., *Acc. Chem. Res.* **29,** 259 (1996).
94. Haag, W. O., and Dessau, R. M., in *Proce. 8th Int. Congr. Catal.*, Berlin, 1984, Vol. 2, p. 305.
95. Mikheikin, I. D., Senchenya, I. N., Lumpov, A. I., Zhidomirov, G. M., and Kazansky, V. B., *Kinet. Katal.* **20,** 496 (1979).
96. Lumpov, A. I., Mikheikin, I. D., Zhidomirov, G. M., and Kazansky, V. B., *Kinet. Katal.* **20,** 1979 (1979).
97. Senchenya, I. N., Mikheikin, I. D., Zhidomirov, G. M., and Kazansky, V. B., *Kinet. Katal.* **22,** 1174 (1980).
98. Senchenya, I. N., Mikheikin, I. D., Zhidomirov, G. M., and Kazansky, V. B., *Kinet. Katal.* **21,** 1184 (1980).
99. Senchenya, I. N., Mikheikin, I. D., Zhidomirov, G. M., and Kazansky, V. B., *Kinet. Katal.* **23,** 591 (1982).
100. McVicker, G. B., Kramer, G. M., and Ziemiak, J. J., *J. Catal.* **83,** 286 (1983).
101. Bordiga, S., Civalleri, B., Spoto, G., Pazè, C., Lamberti, C., Ugliengo, P., and Zecchina, A., *Chem. Soc. Faraday Trans.* **21,** 3893 (1997).
102. Rozanska, X., van Santen, R. A., Hutscka, F., and Hafner, J., unpublished results.
103. Kazansky, V. B., and Senchenya, I. N., *J. Catal.* **119,** 108 (1989).
104. Kramer, G. J., van Santen, R. A., Emeis, C. A., and Nowak, A. K., *Nature* **363,** 529 (1993).
105. Koller, H., Meijer, E. L., and van Santen, R. A., *Solid State Nucl. Magn. Res.* **9,** 165 (1997).
106. Parr, R. G., and Yang, W., in *Density-Functional Theory of Atoms and Molecules*, Oxford University, New York, 1989.
107. Ziegler, T., *Chem. Rev.* **91,** 651 (1991).
108. Labonowski, J. K., and Andzelm, J. W. (eds.), in *Density Functional Methods in Chemistry*, Springer, Berlin, 1991.
109. Krossner, M., and Sauer, J., *J. Phys. Chem.* **100,** 6199 (1996).
110. (a) Haase, F., and Sauer, J., *J. Am. Chem. Soc.* **117,** 3780 (1995). (b) Moravetski, V., Hill, J., Eichler, U., Cheetman A. K., and Sauer, J., *J. Am. Chem. Soc.* **118,** 13015 (1996).
111. Schröder, K.-P., Sauer, J., Leslie, M., Catlow, C. R. A., and Thomas, J. M., *Chem. Phys. Lett.* **188,** 320 (1992).
112. Sauer, J., Eichler, U., Meier, U., Schäfer, A., von Arnim, M., and Ahlrichs, R., *Chem. Phys. Lett.* **308,** 147 (1999).
113. Brand, H. V., Curtiss, L. A., and Iton, L. E., *J. Phys. Chem.* **96,** 7725 (1992).
114. Zygmunt, S. A., Curtiss, L. A., Iton, L. E., and Erhardt, M. K., *J. Phys. Chem.* **100,** 6663 (1996).
115. Brand, H. V., Curtiss, L. A., and Iton, L. E., *J. Phys. Chem.* **97,** 12773 (1993).
116. Ugliengo, P., Ferrari, A. M., Zecchina, A., and Garrone, E., *J. Phys. Chem.* **100,** 3632 (1996).
117. Civalleri, B., Garrone, E., and Ugliengo, P., *J. Phys. Chem. B* **102,** 2373 (1998).
118. Frash, M. V., Makarova, M. A., and Rigby, A. M., *J. Phys. Chem. B* **101,** 2116 (1997).
119. Bonn, M., Bakker, H. J., Domen, K., Hirose, C., Kleyn, A. W., and van Santen, R. A., *Catal. Rev. Sci. Eng.* **40,** 127 (1998).

120. Haase, F., Sauer, J., and Hutter, J., *J. Chem. Phys. Lett.* **266,** 397 (1997).
121. Boronat, M., Zicovich-Wilson, C. M., Corma, A., and Viruela, P., *Phys. Chem. Chem. Phys.* **1,** 537 (1999).
122. Rozanska, X., van Santen, R. A., Hutschka, F., and Hafner, J., *J. Am. Chem. Soc.* **123,** (2001).
123. Sinclair, P. E., De Vries, A., Sherwood, P., Catlow, C. R. A., and van Santen, R. A., *J. Chem. Soc. Faraday Trans.* **94,** 3401 (1998).
124. Greatbanks, S. P., Hillier, I. H., and Sherwood, P., *J. Comp. Chem.* **18,** 562 (1997).
125. Sherwood, P., De Vries, A. H., Collins, S. J., Greatbanks, S. P., Burton, S. A., Vincent, M. A., and Hillier, I. H., *Faraday Discuss.* **106,** 79 (1997).
126. De Vries, A. H., Sherwood, P., Collins, S. J., Rigby, A. M., Rigutto, M., and Kramer, G. J., *J. Phys. Chem. B* **103,** 6133 (1999).
127. Gao, J., in *Reviews in Computational Chemistry* (Lipkowitz, K. B., and Boyd, D. B., eds.), VCH, New York, 1996, Vol. 7.
128. Pisani, C., and Birkenheuer, U., *Int. J. Quant. Chem.* **29,** 221 (1995).
129. Parsons, D., Ángyán, J. G., and Hafner, J., in preparation.
130. Klein, H., Kirschhoch, C., and Fuess, H., *J. Phys. Chem.* **98,** 12345 (1994).
131. Vos, A., Rozanska, X., Schoonheydt, R. A., van Santen, R. A., Hutschka, F., and Hafner, J., *J. Am. Chem. Soc.* **123,** 2799 (2001).
132. Kazansky, V. B., Frash, V. B., and van Santen, R. A., *Catal. Lett.* **48,** 61 (1997).
133. Kazansky, V. B., Frash, V. B., and van Santen, R. A., *Stud. Surf. Sci. Catal.* **105,** 2283 (1997).
134. Weisz, P. B., in *Advances in Catalysis and Related Subjects.* (Eley, D. D., Selwood, P. W., and Weisz, P. B., eds.), Academic Press, London, 1963, Vol. 13, p. 157.
135. Maxwell, I. E., and Stork, W. H. J., in *Introduction to Zeolite Science and Practice* (Van Bekkum, H., Flaningen, E. M., and Jansen, J. C., eds.), Elsevier, Amsterdam, 1991, Vol. 58, p. 571.
136. Van de Runstraat, A., Van Grondelle, J., and van Santen, R. A., *Ind. Eng. Chem. Res.* **36,** 3116 (1997).
137. Van de Runstraat, A., Kamp., J. A., Stobbelaar, P. J., Van Grondelle, J., Krijnen, S., and van Santen, R. A., *J. Catal.* **171,** 77 (1997).
138. De Gauw, F. J. M. M., and van Santen, R. A., in preparation.
139. van Santen, R. A., and Niemantsverdriet, J. W., in *Chemical Kinetics and Catalysis,* Plenum, New York, 1995, p. 159.
140. Jansen, A. P. J., unpublished results.

MORPHOLOGY, FLUCTUATION, METASTABILITY, AND KINETICS IN ORDERED BLOCK COPOLYMERS

Zhen-Gang Wang

Division of Chemistry and Chemical Engineering, California Institute of Technology, Pasadena, California 91125

I. Introduction	439
II. Anisotropic Fluctuations in Ordered Phases	441
III. Kinetic Pathways of Order–Order and Order–Disorder Transitions	445
IV. The Nature and Stability of Some Nonclassical Phases	450
V. Long-Wavelength Fluctuations and Instabilities	452
VI. Morphology and Metastability in ABC Triblock Copolymers	456
VII. Conclusions	460
References	460

I. Introduction

Block copolymers are macromolecules consisting of two or more distinct types of linear polymer chains that are connected by covalent bonds. Normally, the distinct chemical blocks tend not to be miscible, yet the chemical bonds that connect them make macroscopic phase separation impossible. These competing effects lead to the fascinating phenomenon of microphase separation in these systems: the spontaneous formation of ordered microstructures with characteristic periodicities in the range of 10–100 nm (Bates and Fredrickson, 1990, 1999). The morphology of the ordered states depends on the composition, the interaction energies between the distinct blocks, and the particular molecular architecture. For simple AB diblock copolymers, as the composition symmetry decreases, a progression from layered lamellae (LAM), through bicontinous gyroid (G), to hexagonally packed cylinders (HEX), and, finally, to body-center-cubic spheres (BCC) is usually followed. The morphological complexity increases drastically when another chemically different block is added. While block copolymers have been traditionally used as bulk thermoplastic materials, the principle of microphase ordering in block copolymers is being exploited in novel materials applications, such as nanospheres, fibers, and channels (Ding and Liu, 1998; Liu et al., 1999a,b) and nano-organic/inorganic composites (Förster and Antonietti, 1998; Templin et al., 1998; Avgeropoulos et al., 1998; Chan et al., 1999), as well as optoelectric materials (Morkved et al., 1994; Fink et al., 1997).

From a theoretical standpoint, block copolymers are ideal systems for studying many fundamental issues in the thermodynamics and dynamics of self-assembly of soft materials. This is so because of the relatively large length scale of macromolecules, the slow dynamics associated with the relevant structural relaxations, and the ease of control of the molecular characteristics, such as the molecular weight, compositions, and architecture.

The essential physics leading to microphase ordering in simple diblock copolymers is fairly well understood. For most systems, at elevated temperatures, the free energy is dominated by the configurational entropy, which favors a homogeneous, disordered state. As the temperature decreases, the free energy cost due to unfavorable interaction between the two distinct blocks outweighs the entropic gain in the mixed state, leading to separation of the two blocks, constrained to the size scale of the macromolecule by the connectivity of the polymer chain. Theoretical methods have been developed for a quantitative description of the competing interactions leading to the ordering of block copolymers, as well as of transitions between different morphological structures (Helfand, 1975; Helfand and Wasserman, 1976, 1978, 1980; Leibler, 1980; Semenov, 1985). At the mean-field level, the most significant advance in recent years is the self-consistent field theory of Matsen and Schick (1994). The phase diagram predicted by Matsen and Schick is in good agreement with experimental phase diagrams on model diblock copolymers. In particular, their calculation established, for the first time, the thermodynamic stability of the bicontinuous gyroid phase.

Because of the softness of interactions in block copolymers (here we restrict our consideration to flexible molten blocks above the glass transition temperature), thermal fluctuation in these systems is expected to be significant, especially near the order–disorder transition temperatures (Fredrickson and Helfand, 1987). In addition, the long relaxation times, due to the slowness of the motion of polymers, often lead to metastable and other kinetic states. Thus, full understanding of the self-assembly in block copolymers requires understanding of the nature of fluctuation, metastability, and kinetic pathways for various transitions. Most of this article is focused on theoretical studies of these issues in the simpler AB block copolymers. A key concept that emerges from these studies is the concept of anisotropic fluctuations: first, these fluctuations determine the stability limit of an ordered phase; second, they are responsible for the emergence of new structures when the initial structure becomes unstable; and third, they are important for consistent interpretation of small-angle neutron or X-ray scattering data in weakly ordered systems.

While the morphology of AB-type block copolymers is limited to a relatively small number of possibilities, the morphology of ABC triblock copolymers is considerably richer and complex (Bates and Fredrickson, 1999). New, exotic, and often unexpected phases are continually being discovered by experimentalists (Breiner *et al.*, 1998). There is a need for theoretical methodology that is capable of predicting, *a priori*, the possible new morphologies in these more complex block copolymers. We describe one such method in this article.

The organization of the rest of the article is as follows: In Section II, we present the general equilibrium theory of anisotropic order parameter fluctuation in the ordered phases of AB-type block copolymers, focusing on the largest fluctuation modes and the scattering patterns due to the fluctuations. The concept of anisotropic fluctuation is then used in Section III to understand the kinetics of various order–order and order–disorder transitions. In Section IV, we show how the theory of anisotropic fluctuation can be used to understand the nature and stability of some nonclassical phases in diblock copolymers. Section V extends our discussion of fluctuation to long wavelengths, where we describe the continuum elasticity description of ordered block copolymer phases and certain long wavelength instabilities. In Section VI, we discuss some of the novel features in ABC triblock copolymers and an efficient method for discovering new ordered phases in these multiblock systems. Section VII contain some brief concluding remarks. The scope of this review is limited to the bulk behavior of flexible block copolymers under quiescent conditions. Many important topics, most notably thin films, surface effects, and flow-induced structures, are left out. These obviously deserve their own articles by experts in the respective fields.

II. Anisotropic Fluctuations in Ordered Phases

Because of the broken translational and rotational symmery, order-parameter fluctuations in ordered phases are anisotropic. The importance of such anisotropic fluctuations was mentioned in the work of Bates and co-workers, who conjectured that anisotropic fluctuations might play a role in stablizing nonclassical ordered structures (Hamley *et al.*, 1993; Bates *et al.*, 1994a,b). The concept was also invoked by this group in understanding the effect of shear on the HEX-to-DIS transitions in diblock copolymer melts (Almdal *et al.*, 1996).

Fluctuation in the order parameter is determined by the free energy change due to variation of the order parameter from its thermal average. For incompressible diblock block copolymer melts, the order parameter is defined as the local volume fraction of A-monomers relative to its global value, $\psi(\vec{r}) = \phi_A(\vec{r}) - f_A$. Because of the periodic nature of the ordered phases, it is more convenient to use the Fourier transform of $\psi(\vec{r})$, which we denote $\psi(\vec{k})$. We separate $\psi(\vec{k})$ into its average value $\psi_0(\vec{k})$ plus a fluctuation part $\Delta\psi(\vec{k})$. To quadratic order in $\Delta\psi(\vec{k})$, the free energy change takes the form of

$$F[\psi(\vec{k})] = F[\psi_0(\vec{k})] + \frac{1}{2} \sum_{\vec{k}} \sum_{\vec{k}'} \Delta\psi(\vec{k}) \Gamma(\vec{k}, \vec{k}') \Delta\psi(\vec{k}'). \tag{1}$$

For periodic structure, the average order parameter can be written $\psi_0(\vec{k}) = \sum_{\vec{G}} A_{\vec{G}} \delta(\vec{k} - \vec{G})$, where $\vec{G}$ is the set of reciprocal lattice wavevectors of the ordered phase, and $A_{\vec{G}}$ is obtained by minimizing the free energy. Taking advantage of the periodicity of the ordered phase, we can write the free energy as

$$F[\psi(\vec{k})] = F[A_{\vec{G}}] + \frac{1}{2} \sum_{\vec{k}} \sum_{\vec{G}} \sum_{\vec{G}'} \Delta\psi(-\vec{k} - \vec{G}) \Gamma_{\vec{k}}(\vec{G}, \vec{G}') \Delta\psi(\vec{k} + \vec{G}'), \tag{2}$$

where $\vec{k}$ is now confined to the first Brillouin zone of the reciprocal space. The nature of fluctuation and the stability of the ordered phase are determined by the matrix $\Gamma_{\vec{k}}(\vec{G}, \vec{G}')$, which is in general nondiagonal and anisotropic. A stable structure, either locally or globally, is characterized by the positive definiteness of $\Gamma_{\vec{k}}(\vec{G}, \vec{G}')$. The matrix can be diagonalized to yield a set of eigenvectors and eigenvalues. The ordered phase reaches its spinondal when the lowest eigenvalue turns negative, the corresponding eigenvector dictating the potential direction for the spontaneous emergence of a new structure. When the system is stable, the structure factor can be obtained by inverting the matrix $\Gamma_{\vec{k}}(\vec{G}, \vec{G}')$.

A rigorous method for calculating the matrix $\Gamma_{\vec{k}}(\vec{G}, \vec{G}')$ and for addressing anisotropic fluctuations in ordered block copolymer phases at equilibrium was developed by Shi et al. (1996), Yeung et al. (1996), and Laradji et al. (1997a,b) in a series of papers. The theory is based on an expansion of the free energy of an ordered broken symmetry phase around the exact self-consistent field solution of this ordered state. Since the chain conformation in an ordered periodic structure is analogous to the problem of an electron in a periodic potential, the study of order-parameter fluctuations bears an important and useful analogy with the theory of the electronic band structure in crystalline solids. Many of the theoretical concepts and techniques can then be applied to obtain the spectrum of the anisotropic fluctuations.

A simpler though approximate theory for anisotropic fluctuations was developed by Qi and Wang (1997, 1998) based on a Ginzburg–Landau–Leibler free energy approach, valid in the weak segregation regime. In this approach, the anisotropic fluctuation is seen to arise from the coupling between the fluctuation part of the order parameter and the mean-field order parameter. For asymmetric diblock copolymers, the most nontrivial anisotropic fluctuation (near the high-temperature spinodal) is due to the cubic term in the free energy, which contains interaction between the fluctuation and the linear order term of the mean-field order parameter. In the weak segregation limit, both the mean-field order parameter and the fluctuations are dominated by a single wavenumber, k_0. This result, together with the momentum conservation

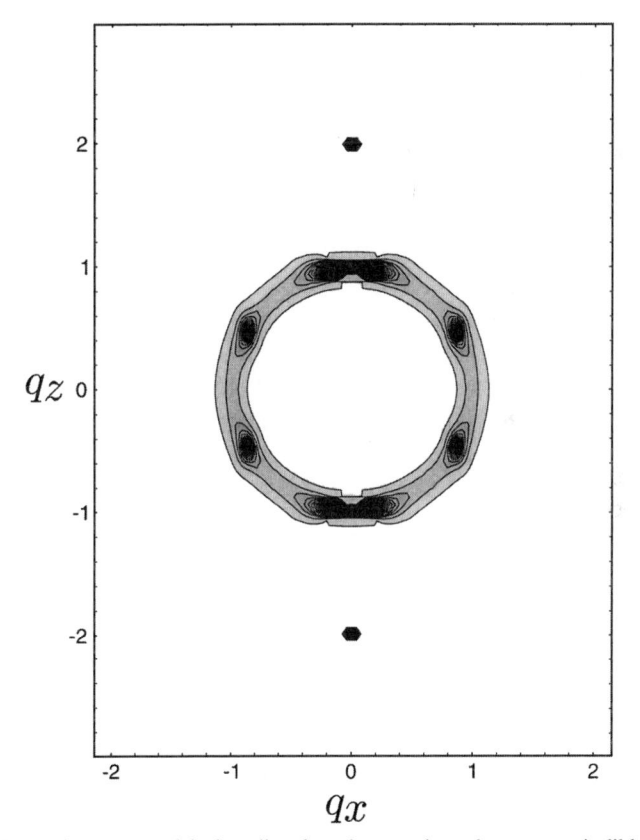

FIG. 1. Scattering pattern of the lamellar phase for a moderately asymmetric diblock copolymer near its spinodal in the $q_y = 0$ scattering plane. The plot shows q_x and q_z in units of k_0. The lamellae are oriented in the z-direction. Courtesy of An-Chang Shi.

that is obeyed in the cubic coupling term, allows one to locate easily the largest fluctuation modes in the reciprocal space.

The scattering function of the lamellar phase near its high-temperature mean-field spinodal is shown in Fig. 1 for the $q_y = 0$ plane. The normal of the lamellar layers is taken to be in the z-direction. In addition to the two strong Bragg peaks at $q_z = k_0$, $q_x = q_y = 0$, and the two higher-order peaks at $q_z = 2k_0$, the most notable new feature is the four peaks at $q_z = k_0/2$, $q_x = (\sqrt{3}/2)k_0$. Because of the rotational symmetry about the z-axis, these four peaks actually correspond to a ring of strong fluctuations at $q_z = k_0/2$. Although this scattering ring lacks any in-layer structure (in the x, y plane), the finding that the dominant fluctuations occur at $q_z = k_0/2$ immediately leads to the conclusion that any structures that form as a result of instability of the LAM phase will have a periodicity of two layers in the z-direction. This conclusion has important implications for the kinetics of the LAM-to-HEX transition as well as for understanding the structural nature of the perforated layer phases, as discussed later.

The fluctuations in the HEX phase exhibit similar nontrivial features. Figure 2a shows the real-space contour plot of the order parameter in the

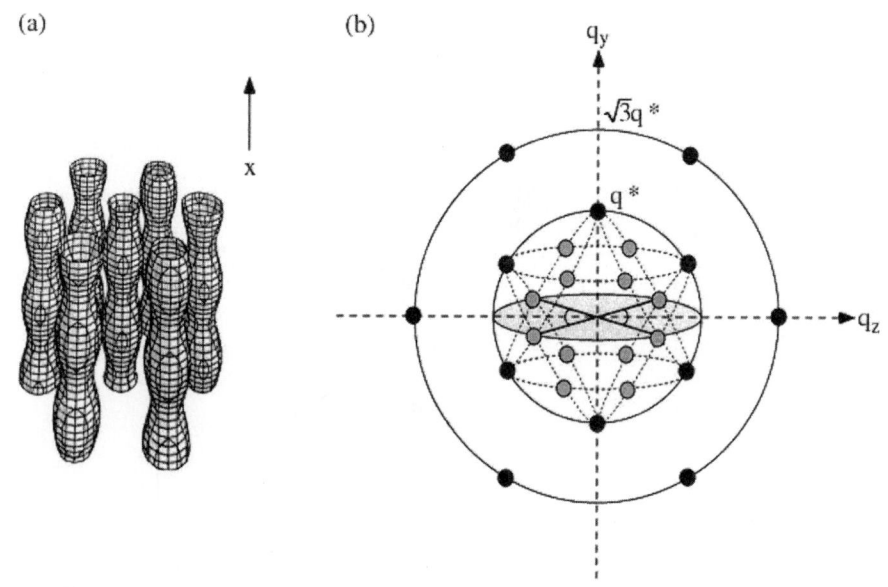

FIG. 2. Dominant fluctuation modes in the HEX phase. (a) Real-space representation according to the work of Laradji et al. (1997). (b) Location of the fluctuation peaks, shown as gray dots, in reciprocal space proposed by Qi and Wang (1998). The black dots are the Bragg peaks of the HEX phase. Reproduced with permission from C. Y. Ryu, M. S. Lee, D. A. Hajduk, and T. P. Lodge. *J. Polym. Sci. B Polym. Phys.*, 1998;81:5345–5357.)

HEX phase with the least stable fluctuation modes, following the work of Laradji et al. (1997b). Figure 2b demonstrates the location of the largest fluctuation modes in the reciprocal space proposed by Qi and Wang (1998). As the HEX phase approaches its high-temperature spinodal, the largest fluctuation mode leads to undulation of the cylinders. When the amplitudes of the fluctuation become sufficiently high, the undulated cylinders break into ellipsoids centered on the bcc lattice whose [111] direction is along the original cylinders. Qi and Wang (1998) point out that the fluctuation spectrum at the high-temperature spinodal of the HEX phase has the symmetry of a twinned bcc structure rather than the simple bcc structure, as can be seen clearly from Fig. 2b. Experiments by Ryu et al. (1997, 1998) on the HEX → BCC transition using both small-angle X-ray scattering (SAXS) and transimission electron microscopy (TEM) clearly show the undulation of cylinders as the transition point is approached. These authors term this undulation pretransitional fluctuation. The scattering pattern obtained from the SAXS as well as from Fourier transform of the TEM image is consistent with the twinned bcc structure.

III. Kinetic Pathways of Order–Order and Order–Disorder Transitions

Since the phase boundaries near the order–disorder transition temperature are strongly dependent on temperature, an interesting question naturally arises as to how a system transforms from one morphology to another after a temperature change. In recent years, several groups have experimentally studied the kinetics during the various order–order (Sakurai et al., 1993a,b; Hajduk et al., 1994a; Schulz et al., 1994; Koppi et al., 1994; Kim et al., 1998; Krishnamoorti et al., 2000a,b; Kimishima et al., 2000) and order–disorder (Hashimoto et al., 1986a,b; Singh et al., 1993; Floudas et al., 1996; Adams et al., 1996; Newstein et al., 1998; Balsara et al, 1998; Kim et al., 1998) transitions in block copolymers. Qi and Wang (1996, 1997, 1998) performed extensive computer simulation studies of the kinetic pathways of order–order and order–disorder transitions in the direction of temperature jumps. These authors showed, that depending on the extent of the temperature jump, these transitions often occur in several stages and can involve nontrivial intermediate states. For example, it was found that transition from the LAM phase to the HEX phase goes through a perforated lamellar (PL) state within a certain temperature range; see Fig. 3. Extensive experimental studies by Hajduk et al. (1997) showed convincingly that the PL, which was previously believed to be a new thermodynamic phase, is a kinetic state *en route* from the LAM to HEX or LAM to G phase, thus providing indirect evidence of the mechanism suggested by the simulation.

(a) t=0 step

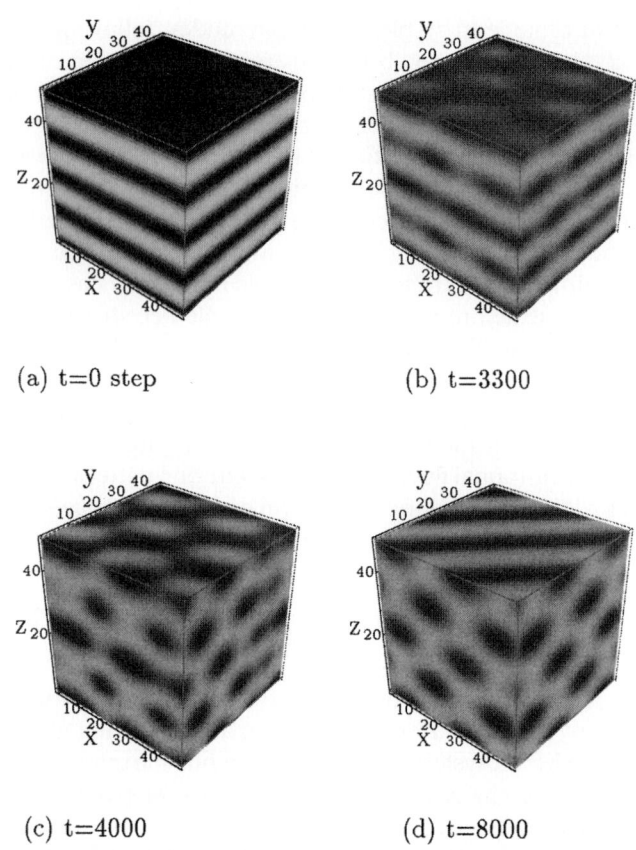

(b) t=3300

(c) t=4000

(d) t=8000

FIG. 3. Computer simulation results using a time-dependent Ginzburg–Landau approach, showing the microstructural evolution after a temperature jump from the lamellar phase to the hexagonal cylinder phase for a moderately asymmetric diblock copolymer. The time units are arbitrary. (Reprinted with permission from *Polymer* 39, S. Y. Qi and Z.-G. Zheng, Weakly segregated block copolymers: Anisotropic fluctuations and kinetics of order-order and order-disorder transitions, 4639–4648, copyright 1998, with permission of Excerpta Medica Inc.)

Qi and Wang also observed a transient undulated HEX state during the melting of the HEX phase to the disordered state. This is consistent with the experimental observation in the shear cessation experiments of Bates and co-workers (Bates *et al.*, 1994a; Almdal *et al.*, 1996). In these experiments, an initially disordered phase of the asymmetric poly(ethylenepropylene-b-ethylethylene) (PEP-PEE) diblock copolymer close to the order–disorder boundary is subjected to a steady shear which induces a transition to the HEX phase. The shear is then suddenly stopped and a transient modulated state is observed as the system relaxes back to the DIS. Insofar as a HEX phase is created and then the condition is changed to favoring the disordred phase (DIS), the shear-cessation experiment can be likened to a temperature jump.

The appearance of nontrivial intermediate state during the order–order and order–disorder transitions can be understood using the concept of anisotropic fluctuations: the most unstable fluctuation modes lead to the emergence of new structures from the initial phase when that phase becomes unstable. For example, the intermediate states in the LAM-to-HEX transition shown in Fig. 3 is a result of the anisotropic fluctuations located on the spherical shell $q = k_0$ with $q_z = k_0/2$.

Laradji et al. (1997a,b) used the most unstable modes of an ordered structure to infer the kinetic pathways in several order–order transitions. Among other things, these authors predict that the LAM-to-HEX and the HEX-to-LAM transitions are not reversible in their kinetic pathways, whereas the transition from LAM to HEX goes through a perforated layer structure, the reverse transition proceeds directly without an intermediate state. Similarly, whereas the HEX-to-BCC transition proceeds directly, the BCC-to-HEX transition is predicted to go through an intermediate modulated layer state. The absence of an intermediate state during the HEX-to-LAM transition was reported in the experiments of Sakurai et al. (1993a). The absence of an intermediate state during the HEX-to-BCC transition is in agreement with the simulation result of Qi and Wang (1996, 1997).

The anisotropic fluctuations discussed in the last section refer to fluctuations at equilibrium. However, usually the various order–order and order–disorder transitions are caused by large deviations from equilibrium conditions, which result in large deterministic driving forces given by nonvanishing first derivatives of the free energy. Qi and Wang (1998) performed a linear stability analysis by separating the time-dependent order parameter into a mean-field part and a fluctuation part. The evolution of the meanfield order parameters changes their magnitudes without leading to new structures. Emergence of new structures is associated with fluctuations that lead to deviation from the mean-field path. Qi and Wang calculated the growth rate of the fluctuations as a function of the instantaneous value of the mean-field order parameter and were able to predict at which point along the meanfield path new structures begin to appear. Their analysis also explains why hexagonal cylinders melt uniformly for large temperature jumps but proceed through a BCC modulated hexagonal cylinder state when the temperature jump is small, as observed in their simulation studies. By projecting the order parameter space to a reduced set containing the order parameter of the initial structure and the largest fluctuation modes, they were able to describe qualitatively the full nonlinear evolution of the microstructures in the various transitions.

The concept of anisotropic fluctuations leads to a simple explanation of the epitaxial relationships observed in the transition from one ordered phase to another (Schulz et al., 1994; Koppi et al., 1994). Because of the broken symmetry in a periodically ordered structure, the dominant fluctuation modes reside at specific locations in the reciprocal space, determined by the symmetry of the reciprocal lattice of the initial ordered structure. This spatial relationship is maintained when the new structure—dictated by the most unstable fluctuation modes—grows out of the initial structure, giving rise to the observed epitaxy.

The epitaxial relationship between ordered phases connected by an order–order transition leads to an intriguing phenomenon—the proliferation of HEX cylinder orientations during repeated heating–cooling cycles across the HEX/BCC phase boundary starting from a well-aligned HEX sample. The epitaxy in the HEX-to-BCC transition dictates that the [111] direction of the BCC coincides with the direction of the cylinder axis of the original HEX phase. However, as discussed in the last section, the symmetry of the fluctuation in the HEX is that of a twinned BCC as opposed to a single BCC (Qi and Wang, 1998) (cf. Fig. 2b); thus transition from HEX to BCC should yield two BCC orientations corresponding to the twins. There are four equivalent [111] directions in each of the BCC twins, but one of the four is shared by both, corresponding to the initial cylinder orientation. Therefore, the twins contribute seven distinct [111]-equivalent directions; see Fig. 4. During the reverse transition, each of these directions can become the cylinder direction of the HEX phase. Thus in one cycle of HEX → BCC → HEX, seven cylinder orientations are generated from a well-aligned HEX with a single cylinder orientation. By the same reasoning, it can be shown that another cycle will give rise to 18 new cylinder directions. Clearly this process can be repeated *ad infinitum,* with each cycle

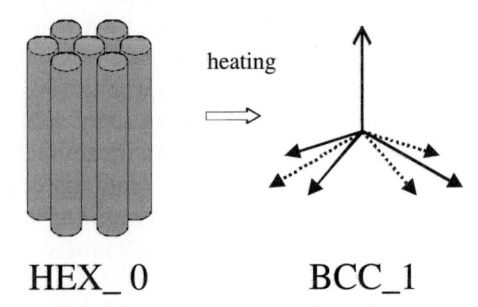

HEX_ 0   **BCC_1**

FIG. 4. Seven distinct [111] directions in the BCC phase are produced from a well-aligned HEX phase. These seven [111] directions become the orientations of the cylinders upon cooling back into the HEX phase.

producing more new cylinder directions in a completely deterministic manner. Preliminary SAXS and birefringence data on poly(styrene-b-isoprene) (PS-PI) diblock copolymers support this mechanism (Lee, 2000).

The deterministic proliferation of HEX cylinder orientations during repeated heating–cooling cycles is a unique phenomenon and is a consequence of the fact that the HEX produces a twinned BCC. Indeed, if the HEX-to-BCC transition yielded only a simple BCC structure, then although the first reverse transition would give four cylinder orientations, repeated heating/cooling would not yield any new cylinder orientations since these four directions form a closed set. In contrast, in the LAM → HEX transition, the epitaxial relationship only constrains the cylinders to be imbedded in the layers of the minority component of the LAM, but the orientation of the HEX is determined by random fluctuations.

While the foregoing discussions focus primarily on spinodal like kinetics, it is worth mentioning studies that deal with other issues in the order–order and order–disorder transitions. Goveas and Milner (1997) studied the LAM–HEX transition in weakly segregated diblock copolymers by addressing the propagation of the stable phase into the metastable phase. Using an approximate free energy based on the Leibler free energy, these authors obtained the interfacial profile at the propagating front as well as the front velocity. Zhang *et al.* (1997) studied the effects of directional quench from the disordered phase in producing ordered phases that are well oriented. The focus on interfaces and growth in this work complements studies by Shi *et al.* and Qi and Wang which focus on global, spinodal-type kinetics.

Another important issue is that of nucleation/growth kinetics in the order–order and order–disorder transitions. When an ordered morphology is taken beyond its spinodal, transition to another morphology should occur immediately throughout the system. However, some recent experiments on the LAM-to-G transition (Hajduk *et al.*, 1998) and on the HEX-to-BCC transition (Ryu *et al.*, 1997) seem to suggest that the transitions occur by nucleation and growth. Nucleation of the LAM phase from the DIS phase was studied by Fredrickson and Binder (1989). Matsen (1998) recently studied the transition between the HEX and the G phases showing a nucleation/growth mechanism. Whether a transition occurs by spinodal or nucleation/growth depends of course on whether the system is within or beyond the spinodal limit or whether a spinodal exists at all. However, in all order–order transitions, except in the very weak segregation limit, there is mismatch in the lattice spacing of the ordered phases. Such a mismatch would lead to a finite strain in the new ordered phase if the new phase appeared uniformly throughout the system; such a strain could erect a free energy barrier in the transition, thus changing an otherwise spinodal mechanism to a nucleation and growth mechanism. It will be interesting to study how fluctuation of the order parameter is coupled to lattice distortions. Such phonon-type distortions are discussed in Section IV.

IV. The Nature and Stability of Some Nonclassical Phases

Besides the three classical phases, namely, the LAM, the HEX and the BCC, several complex morphologies have been obtained in experiments. For diblocks with moderate asymmetries, a bicontinuous double-gyroid (G) phase has been commonly observed (Hajduk *et al.*, 1994b, 1995), in which the minority blocks form domains consisting of two interweaving threefold coordinated lattices. The G phase was previously mistaken as the double diamond (D) structure formed from two fourfold coordinated lattices (Thomas *et al.*, 1986). Another complex morphology is the perforated layer (PL) structure (Thomas *et al.*, 1988) observed in the weak-to-intermediate segregation regime of moderately asymmetric diblock copolymers discovered by Bates and co-workers (Hamley *et al.*, 1993; Bates *et al.*, 1994; Förster *et al.*, 1994b; Khandpur *et al.*, 1995).

The thermodynamic stability of the G phase was demonstrated first by Matsen and Schick (1994) using the reciprocal space self-consistent field (SCF) method they developed. It was shown that the G phase is the lowest free energy state compared to the classical phases and the PL and close-packed spheres in certain parts of the phase diagram in the intermediate segregation regime, ending at a triple point with the LAM and the HEX phase on the high-temperature side. Using a larger number of basis functions, Matsen and Bates (1996) studied the phase behavior at stronger segregations; their calculation shows that the G phase become unstable at values of $N\chi > 60$. Matsen and Bates (1997) subsequently performed SCF calculations aimed at a physical understanding of the complex phase behavior of block copolymers in the intermediate segregation regime. They suggested that the competition between chain packing and interfacial tension used to describe the behavior in the strong segregation regime (Semenov, 1985) can also be used to explain the phase behavior in the intermediate segration regime. Interfacial tension prefers the formation of constant mean curvature surfaces to reduce the interfacial area (Thomas *et al.*, 1988), while chain stretching favors domains of uniform thickness. Whereas the classical phases can satisfy both, the complex phases cannot. Of the complex phases considered, only the G phase is least frustrated and is consequently stable at intermediate degrees of segregation.

The calculations mentioned above draw their conclusion about the thermodynamic stability based on comparing free energy among a few candidate morphologies. They did not address the local stability of the phases. Laradji *et al.* examined the stability of the phases by studying the anisotropic fluctuation. Their calculation shows (Laradji *et al.*, 1997a,b) that the G phase possesses a negative eigenvalue in its fluctuation spectrum, indicating that the G phase may be a saddle point on the free energy surface. This conclusion caused a debate between Matsen (1998) and Laradji *et al.* (1998). No satisfactory resolution of this issue has been reached.

The PL phase discovered by Bates and co-workers was envisioned to consist of alternating minority and majority component layers in which the minority component domains are modulated in thickness or perforated by the majority component with a hexagonal in-plane symmetry. However, the three-dimensional stacking of the layers, i.e., whether they are stacked like abab... or abcabc... was not established by these researchers. Also, several key features in the small-angle neutron scattering (SANS) data, such as the six weak broad peaks in the in-plane scattering pattern (Hamley *et al.*, 1993; Bates *et al.*, 1994b) as well as the nonsixfold in-plane scattering pattern in some samples (Förster *et al.*, 1994; Khandpur *et al.*, 1995), were never satisfactorily explained. The lack of definitive structural models for the PL led several theoretical calculations (Fredrickson, 1991; de la Cruz *et al.*, 1991; Hamley and Bates, 1994) to assume different structures, which resulted in conflicting conclusions with regard to their thermodynamic stability.

The computer simulation result of Qi and Wang (1997) on the LAM–HEX transition suggested that the PL phase might be a kinetic state between the LAM and the HEX or the G phase. Yeung *et al.* (1996) and Laradji *et al.* (1997a,b) suggested that the PL state is a fluctuating lamellar phase without any specific in-layer structures. A unified description of the structural nature, stability, and possible mechanism of formation of the PL structure in diblock copolymer melts was proposed by Qi and Wang (1997). Using a simple Leibler free energy, it is shown that the PL develops from anisotropic fluctuations in the lamellar phase as it approaches its limit of metastability. Two PL models are proposed, one based on a hexagonal close-packed lattice and the other based on a BCC lattice, with nearly degenerate free energies. In the framework of this free energy, it is found that the PL structure is pseudostable (corresponding to a saddle point in the free energy surface) in the weak segregation limit but can be metastable in the intermediate segregation regime. Their proposed structures produce features that are consistent with all known structural data on the ordered PL phases. In particular, the location and intensity of the in-plane scattering peaks can be understood as arising from anisotropic fluctuations in the metastable PL states, rather than from Bragg scattering.

The stability of the PL morphologies was reexamined by Hajduk *et al.* (1997) in a number of block copolymer melts of low to moderate molecular weight. Using SAXS and rheological measurements, they showed rather conclusively that the perforated layer structures, which were initially believed to be new equilibrium phases, are long-lived nonequilibrium states which convert to the G or HEX phase upon isothermal annealing. Thus there seems to be agreement between theory and experiments on the stability of the PL state.

V. Long-Wavelength Fluctuations and Instabilities

The ordered phases formed in block copolymer melts are liquid–crystalline mesophases in that they possess properties intermediate between liquids and solids. This is so because of the different degrees of ordering that can take place at two very different length scales: on the microscopic level, the monomeric units possess quite a bit of fluidity, lacking positional order, and yet on length scales larger than the chain size, the system can exhibit long-range periodic structures. This structural regularity implies that the system is capable of sustaining anisotropic elastic deformation. At long length scales, such anisotropic elastic deformation is described in terms of continuum elasticity. For example, the free energy change due to long wavelength deformation in the lamellar phase can be written

$$\Delta F = \frac{1}{2} \int d\vec{r} \left[B \left(\frac{\partial u}{\partial z} \right)^2 + K \left(\nabla_\perp^2 u \right)^2 \right], \tag{3}$$

where $u(\vec{r})$ is the displacement, defined as the change of the layer position at $\vec{r}$ from its equilibrium position, and $\nabla_\perp$ is the gradient operator in the transverse directions to the layer normal. B and K are termed the compressional and bending moduli, respectively. It can be seen that this free energy has the same form as the free energy of deformation in smectic-A liquid crystals (de Gennes and Prost, 1993). Indeed, the long-wavelength deformation of the ordered block copolymer phase can be described in the same way as the deformation in crystalline solids and liquid crystals. More generally, the elastic deformation free energy of a block copolymer mesophase can be written

$$\Delta F = \frac{1}{2} \int d\vec{r} (B_{ijkl} \partial_i u_k \partial_j u_l + K_{ijklmn} \partial_i \partial_j u_m \partial_k \partial_l u_n), \tag{4}$$

where summation of repeated indices is implied. For a given block copolymer mesophase, symmetry arguments can be used to reduced the number of independent deformation modes and hence the number of terms appearing in the above expression (de Gennes and Prost, 1993).

Long-wavelength deformations/fluctuations in block copolymers are important for several reasons. First, they are responsible for the mechanical properties of a block copolymer mesophase. Second, they determine the degree of long-range order and the effects of defects in the ordered phases. Finally, they are responsible for certain novel long-wavelength instabilities.

The key input in the continuum elasticity descprition of long-wavelength deformations is the set of elastic moduli that appear in Eq. (4). Several authors have calculated the elastic moduli for the lamellar phase of block copolymers. Amundson and Helfand (1993) derived the elastic deformation free energy for a lamellar phase in the weak segregation limit starting from a fluctuation renormalized free energy derived by Fredrickson and Binder (1989). Wang (1994) obtained the compressional and binding moduli for the diblock copolymer phases in both the strong and the weak segregation limit and used the results to examine the thermomechanical behavior and stress-induced melting. Chakraborty and Fredrickson (1994) calculated the compressional modulus for a lamellar phase of AB random copolymers in the strong segregation limit, showing a nontrivial dependence on the sequence distribution. Hamley (1994) derived the elastic free energy of the HEX phase in the weak segregation limit and examined the mechanical instabilities under tension. These studies deal with specific ordered block copolymer mesophases in either the weak or the strong segregation regime; extension to more complex morphologies at arbitrary degrees of segregation is not obvious.

A different route for computing the elastic moduli was taken by Yeung *et al.* (1996), who obtained the elastic moduli for symmetric diblock copolymer lamellar by examing the long-wavelength limit of the anisotropic fluctuation spectrum. This procedure is similar to earlier density functional calculation of elastic moduli for crystalline solids (Lipkin *et al.*, 1985). A similar phase Hamiltonian formulation using an approximate density functional was developed by Kawasaki and Ohta (1986) to compute the elastic moduli of the three classical phases of diblock copolymers in the strong segregation limit. However, a common assumption shared by these works is that of affine deformation, which in this context means that the only change in the density profile is through a change in the coordinates induced by the deformation. While such an approximation may not cause significant errors for hard atomistic solids, adjustment of the profile in systems with soft interactions to imposed deformation could have more pronounced effects. Indeed, Matsen (1999) showed that for the bending modulus of a diblock copolymer film separating the A-rich phase from the B-rich phase in an AB binary blend, the assumption of affine deformation can lead to a change in the sign.

Shi and Wang (unpublished work) have recently developed a rigorous theory for extracting the elastic moduli of ordered block copolymer phases from the anisotropic order-parameter fluctuation spectrum, without invoking the affine deformation assumption. In this new theory, both the phase (displacement) and the amplitudes of the density wave in an ordered block copolymer phase are allowed to vary upon a long-wavelength deformation. A free energy for continuum elasticity is obtained starting from Eq. (2) by integrating the changes in the amplitudes, subject to a given deformation u. In Fourier representation, the resulting elastic free energy takes the form

$$\Delta F = \frac{1}{2} \sum_{\vec{k}} \mathbf{S}^{-1}(\vec{k}) : \vec{u}(\vec{k})\vec{u}(-\vec{k}). \tag{5}$$

The matrix $\mathbf{S}^{-1}$ is a second rank tensor defined as

$$\mathbf{S}^{-1}(\vec{k}) = \sum_{\vec{G}>0}\sum_{\vec{G}'>0} [Q_{\vec{k}}(\vec{G}, \vec{G}') - R_{\vec{k}}(\vec{G}, \vec{G}')] A_{\vec{G}} A_{\vec{G}'} \vec{G}\vec{G}', \tag{6}$$

where the functions $Q_{\vec{k}}(\vec{G}, \vec{G}')$ and $R_{\vec{k}}(\vec{G}, \vec{G}')$ can be obtained from the matrix $\Gamma_{\vec{k}}(\vec{G}, \vec{G}')$. $\mathbf{S}^{-1}(\vec{k})$ can be expanded for small k, the quadratic coefficients yielding B_{ijkl} and the quartic coefficient giving K_{ijklmn} in Eq. (4). The $R_{\vec{k}}(\vec{G}, \vec{G}')$ term reflects the renormalization of the elastic moduli due to the coupling between the amplitude of the density wave and the displacements. Numerical implementation of the method using the exact self-consistent field solutions is quite involved and is now in progress. An interesting and important application of the theory is the study of mechanical properties of the bicontinuous gyroid phase. In addition, the theory can be used to understand the very different degrees of long-range order obtained in the HEX phases on the two sides of the phase diagram in conformational asymmetric diblock copolymers (Gido and Wang, 1997).

The coupling between the displacements and the amplitude of the density wave manifests itself in a long-wavelength instability in ordered phases with soft directions (such as LAM and HEX) upon a temperature quench (Qi and Wang, 1999). This phenomenon is most easily illustrated using the example of the LAM phase in the weak segregation limit. At a temperature T_0 below the order–disorder transition temperature, the density wave can be represented as a simple sinusoidal wave:

$$\psi(\vec{r}) = A_0 \cos\left(\frac{2\pi}{d}z\right), \tag{7}$$

where d is the period of the LAM and we have taken the lamellar normal to be in the z-direction. Upon a decrease in temperature to $T_1 < T_0$ (but still in the weak segregation limit so that the period d remains unchanged), the amplitude will evolve to a new equilibrium value. However, the initial

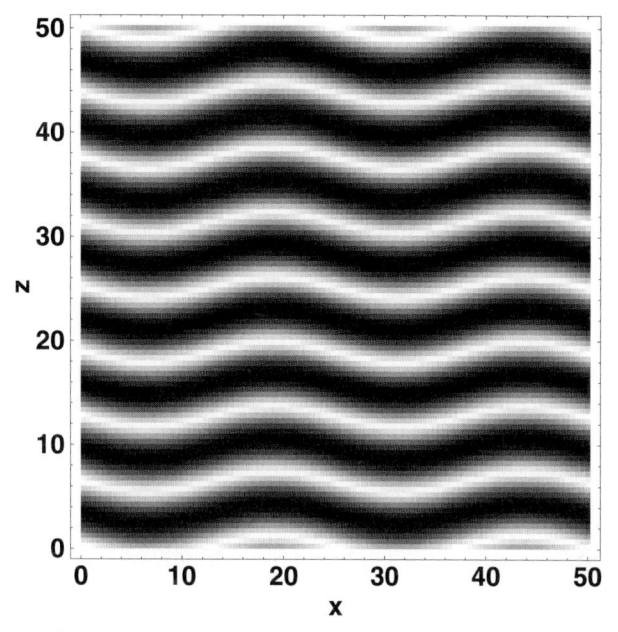

FIG. 5. A two-dimensional illustration of the undulation instability caused by a sudden temperature decrease within the LAM phase.

deviation from equilibrium causes a buckling instability (see Fig. 5), so that the time evolution of the order-parameter takes the form

$$\psi(\vec{r},t) = A(t) \cos\left[\frac{2\pi}{d}(z - u(\vec{r},t))\right] \quad (8)$$

while the relaxation of the amplitude A is exponentially fast, the relaxation of u is only algebraic ($\approx t^{-1/4}$). This would lead to a long-lived transient transverse broadening of the Bragg peaks in the scattering function.

A similar instability exists for the HEX phase, resulting in the undulation of the cylinder axes upon a temperature quench (still in the HEX phase). If we take the special case of $A_0 = 0$, our analysis should apply equally to the ordering kinetics after a temperature quench from the disordered state. Thus the predicted instability might be responsible for the unusual kinetics observed by Balsara *et al.* (1998) in a temperature quench from the disordered state to the HEX state for PS-PI copolymer melts. Using a combination of small-angle X-ray scattering and time-resolved depolarized light scattering, they demonstrate that for a large quench depth, hexagonal order occurs before the development of a coherent structure along the cylinder direction (which they call the liquid direction). They envisage the formation of arrays of hexagonally packed cylinders with a limited persistent length at the initial stage of the transition and subsequent straightening of the cylinders at late stages, with the second process occuring at a much slower rate.

VI. Morphology and Metastability in ABC Triblock Copolymers

In contrast to the simpler AB diblock copolymers, where the morphology is determined by the relative lengths of the two blocks and by a single interaction parameter characterizing the incompatibility of the two blocks (at the mean-field level), the morphology of ABC triblock copolymers depends crucially on the architecture of the triblocks (i.e., A–B–C, C–A–B, or A–C–B, linear vs star) and on the relative strengths of three pair interaction parameters (Gido et al., 1993; Mogi et al., 1993; Sioula et al., 1998a,b). Consequently, the morphology of ABC triblock copolymers is considerably richer and more complex (Auschra and Stadler, 1993; Mogi et al., 1994; Breiner et al., 1998; Brinkman et al., 1998). In the strong segregation limit, several theoretical calculations using the known structures of the ordered block copolymer phases have been carried out (Nakazawa and Ohta, 1993; Stadler et al., 1995; Zheng and Wang, 1995; Phan and Fredrickson, 1998). The calculated phase diagram is in general agreement with experimental results. Werner and Fredrickson (1997) studied the spinodal limit of the disordered state in ABC triblock copolymers and found several cases with reentrant behaviors. Self-consistent field calculations have been performed (Matsen, 1997) to elucidate the nature of the tricontinuous phases in ABC triblock copolymers.

In spite of these successes, and in spite of the high level of sophistication in block copolymer theory, almost all the nontrivial new morphologies were first discovered experimentally. The role of theory has been primarily to compute and compare the free energies for a number of candidate morphologies with presumed symmetries. Therefore, there is a need for theoretical methods capable of discovering or predicting new ordered morphologies. This need becomes particularly acute in view of the large parameter space and the increased number of possible architectures in multiblock copolymers. An important step in this direction was taken recently by Drolet and Fredrickson (1999), who proposed exploring the morphological space by solving the self-consistent field equations in real space and varying the initial conditions. Their method involves discretizing the self-consistent equations on grid points within a fixed volume. To minimize the finite size effects, a large volume (with each dimension several times larger than the typical domain size of the ordered phases) and hence a large number of grid points are required.

Bohbot-Raviv and Wang (2000) developed a similar method with an important improvement over that of Drolet and Fredrickson: instead of minimizing the free energy over a large but fixed volume, the method of Bohbot-Raviv and Wang performs free energy minimization in real space in an arbitrary *unit cell* with respect to the composition profile *and* the dimensions of the unit cell. The use of a unit cell considerably reduces the number of grid points needed for the numerical minimization, while allowing the dimensions of the unit cell to vary, removing any finite size/strain effects that exist in straightforward discretization with a fixed volume. In addition, since the primary focus is on discoverying new structures, rather than accurate evaluation of the free energy, Bohbot-Raviv and Wang constructed a simple, approximate free energy functional where the chain connectivity is retained at the RPA level. This simplification further increases the effciency of their method. Figure 6 shows two complex two-dimensionally ordered new morphologies discovered by this method. Figure 6a is similar to the knitting pattern discovered by Breiner *et al.* (1998). A promising direction is to combine the efficiency of this method with the accuracy of self-consistent field theory, wherein new ordered morphologies identified by the approach of Bohbot-Raviv and Wang serve as input for more elaborate SCF calculations. This combination will provide the long desired

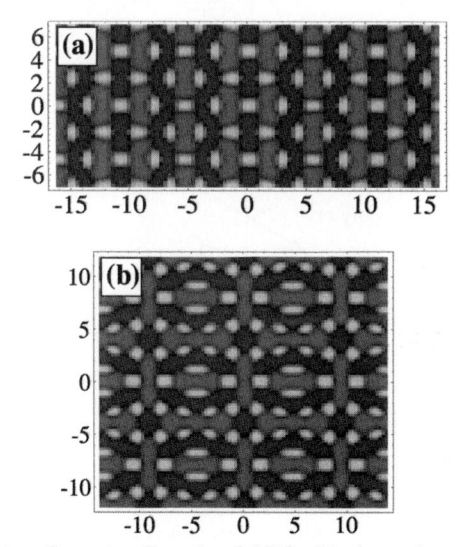

FIG. 6. Complex two-dimensionally ordered ABC triblock copolymer phases produced using the method of Bohbot-Raviv and Wang (2000). The parameters are (a) $f_A = 0.36$, $f_B = 0.31$, $N\chi_{AB} = 30$, $N\chi_{BC} = 32$, and $N\chi_{AC} = 22$; (b) $f_A = 0.36$, $f_B = 0.31$, $N\chi_{AB} = 30$, $N\chi_{BC} = 35$, and $N\chi_{AC} = 22$.

theoretical tool for efficiently and accurately predicting, *a priori,* complex phase diagrams of multiblock copolymer melts.

Because the strengths of the three pair interactions in an ABC triblock are in general unequal and have different temperature dependences, as the temperature is decreased from the disordered (well-mixed) state, segregation will occur first for the most incompatible pair(s). At even lower temperatures, segregation between the other pair(s) of blocks will follow, which is often but not always accompanied by a phase transition, depending on whether a symmetry breaking is involved. This stagewise segregation can give rise to interesting metastable or otherwise kinetically trapped states. An example is given in Fig. 7 for an A–B–C sequenced triblock where the repulsion between the two terminal blocks, A and C, is less than that between the middle block B and the two terminal blocks. Coming from the disordered phase, the first segregation will be between block B and the two terminal blocks, resulting in a two-domain structure (i) with a B domain and a mixed A/C domain. In this domain structure, the chains can assume both loop and bridge conformations, where the bridges can be oriented in either direction. At some lower temperature, the A and C blocks will begin to segregate from each other, eventually resulting in a fully segregated three-domain structure (ii). In such a structure, the triblocks take predominantly

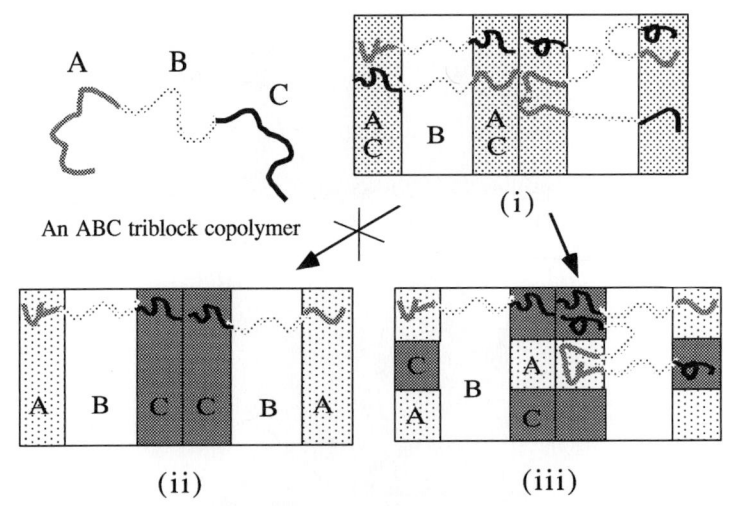

FIG. 7. Transition from a two-domain structure to a three-domain structure in an ABC triblock copolymer lamellar phase. Although state ii is thermodynamically the most stable state, transition to this state from i is hindered because of the high free energy cost in switching the orientation of the bridges and in turning the loops into bridges. Thus, a kinetically more likely process is for the A and C blocks from the bridge conformation to separate laterally, with the loops straddling the interfaces between the A and the C domains.

the bridge conformation. Although state (ii) is thermodynamically the most favorable, the transition from (i) to (ii) is hindered because of the high free energy cost in switching the orientation of the bridges and in turning the loops into bridges. Thus a kinetically more likely process is for the A and C blocks from the bridge conformation to separate laterally, with the loops straddling the interfaces between the A and the C domains (iii). Such a topologically trapped state has been observed by Kornfield's group at Caltech for the S–I–R triblock copolymer (S, polystyrene; I, polyisoprene; R, random styrene–isoprene copolymer).

Figure 8 similarly shows the effects of preordering on the subsequent segregation. The interaction parameters and blocks lengths in both cases are identical to those in Fig. 6b (connected-wheel pattern); the only difference lies in the intial condition. In Fig. 8a, the initial state is that of hexagonally ordered domains of A (gray) in a mixed B/C (dark/light) matrix, while in Fig. 8b a hexagonally ordered C in an A/B matrix is used. Both patterns have retained features of the initial domain structure, though with considerable distortion and modification. Not surprisingly, these kinetically trapped metastable states both have free energies higher than the morphology shown in Fig. 6b. The dependence of these kinetically trapped states on the processing route can be exploited to obtain structures that would otherwise be inaccessible.

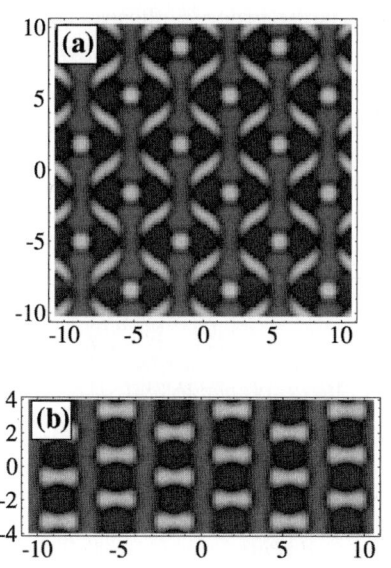

FIG. 8. Metastable states produced from preordered states. The parameters are identical to those used in Fig. 7b.

VII. Conclusions

The microphase ordering in block copolymers is one of the best examples of self-assembly in soft matter and exemplifies many of the principal features of this phenomenon, such as competing interactions and hierachical order (Muthukumar *et al.*, 1997). In contrast to atomic or small molecular systems whose thermodynamic and dynamic properties are governed by strong and detailed interactions at the atomic scale, the effects of such atomic-level interactions on the behaviors of block polymers are usually manifested in an average manner, allowing a coarse-grained description of the systems. As a result, it is often possible to describe accurately the physics of block copolymers in terms of a few course-grained parameters, such as the molecular weight, composition, and Flory–Huggins parameters, which can often be fairly readily controlled experimentally. This, together with the large length scales and slow relaxation times, makes block copolymers ideal model systems for addressing many fundamental issues in the thermodynamic and dynamics in soft self-assembling systems. Our understanding of these issues in block copolymers can often be helpful in understanding related issues in smaller molecular systems, such as in small molecular surfactants (Muller and Schick, 1998; Li and Schick, 2000). On the other hand, topological constraints, and different architectural designs in polymeric systems, lead to unique phenomena such as topologically frustrated states, as well as the formation of complex, exotic ordered phases. These phenomena add new richness to self-assembly and can be utilized in the design of novel materials.

REFERENCES

Adams, J. L., Quiram, D. J., Graessley, W. W., Register, R. A., and Marchand, G. R., *Macromolecules* **29,** 2929–2938 (1996).
Almdal, K., Mortensen, K., Koppi, K. A., Tirrell, M., and Bates, F. S., *J. Phys. II* **6,** 617–637 (1996).
Amundson, K., and Helfand, E., *Macromolecules* **26,** 1324–1332 (1993).
Auschra, C., and Stadler, R., *Macromolecules* **26,** 2171–2174 (1993).
Avgeropoulos, A., Chan, V. Z. H., Lee, V. Y., Ngo, D., Miller, R. D., Hadjichristidis, N., and Thomas, E. L., *Chem. Mater.* **10,** 2109–2115 (1998).
Balsara, N. P., Garetz, B. A., Newstein, M. C., Bauer, B. J., and Prosa, T. J., *Macromolecules* **31,** 7668–7675 (1998).
Bates, F. S., and Fredrickson, G. H., *Annu. Rev. Phys. Chem.* **41,** 525–557 (1990).
Bates, F. S., and Fredrickson, G. H., *Phys. Today* **52,** 32–38 (1999).
Bates, F. S., Koppi, K. A., Tirrell, M., Almdal, K., and Mortensen, K., *Macromolecules* **27,** 5934–5936 (1994a).

Bates, F. S., Schulz, M. F., Khandpur, A. K., Frster, S., and Rosedale, J. H., *Faraday Discuss.* **98**, 7–18 (1994b).
Bohbot-Raviv, Y., and Wang, Z.-G., *Phys. Rev. Lett.* **85**, 3428–3431 (2000).
Breiner, U., Krappe, U., Thomas, E. L., and Stadler, R., *Macromolecules* **31**, 135–141 (1998).
Brinkmann, S., Stadler, R., and Thomas, E. L., *Macromolecules* **31**, 6566–6572 (1998).
Chan, V. Z. H., Hoffman, J., Lee, V. Y., Iatrou, H., Avgeropoulos, A., Hadjichristidis, N., Miller, R. D., and Thomas, E. L., *Science* **286**, 1716–1719 (1999).
Chakraborty, A. K., and Fredrickson, G. H., *Macromolecules* **27**, 7079–7084 (1994).
de Gennes, P.-G., and Prost, J., *The Physics of Liquid Crystals,* Claredon Press, Oxford, 1993.
de la Cruz, M. O., Mayes, A. M., and Swift, B. W., *Macromolecules* **25**, 944–948 (1991).
Ding, J. F., and Liu, G. J., *Chem. Mater.* **10**, 537–542 (1998).
Drolet, F., and Fredrickson, G. H., *Phys. Rev. Lett.* **83**, 4317–4320 (1999).
Fink, Y., Urbas, A. M., Bawendi, M. G., Joannopoulos, J. D., and Thomas, E. L., *J. Lightwave Technol.* **17**, 1963–1969 (1999).
Floudas, G., Pispas, S., Hadjichristidis, N., Pakula, T., and Erukhimovich, I., *Macromolecules* **29**, 4142–4154 (1996).
Förster, S., and Antonietti, M., *Adv. Mater.* **10**, 196–218 (1998).
Förster, S., Khandpur, A. K., Zhao, J., Bates, F. S., Hamley, I. W., Ryan, A. J., and Bras, W., *Macromolecules* **27**, 6922–6935 (1994).
Fredrickson, G. H., *Macromolecules* **24**, 3456–3458 (1991).
Fredrickson, G. H., and Binder, K., *J. Chem. Phys.* **91**, 7265–7275 (1989).
Fredrickson, G. H., and Helfand, E., *J. Chem. Phys.* **87**, 697–705 (1987).
Gido, S. P., and Wang, Z.-G., *Macromolecules* **30**, 6771–6782 (1997).
Gido, S. P., Schwark, D. W., Thomas, E. L., and do Carmo, G. M., *Macromolecules* **26**, 2636–2640 (1993).
Goveas, J. L., and Milner, S. T., *Macromolecules* **30**, 2605–2612 (1997).
Hajduk, D. A., Gruner, S. M., Rangarajan, P., Register, R. A., Fetters, L. J., Honeker, C., Albalak, R. J., and Thomas, E. L., *Macromolecules* **27**, 490–501 (1994a).
Hajduk, D. A., Harper, P. E., Gruner, S. M., Honeker, C. C., Kim, G., Thomas, E. L., and Fetters, L., *Macromolecules* **27**, 4063–4075 (1994b).
Hajduk, D. A., Harper, P. E., Gruner, S. M., Honeker, C. C., Thomas, E. L., and Fetters, L. J., *Macromolecules* **28**, 2570–2573 (1995).
Hajduk, D. A., Takenouchi, H., Hillmyer, M. A., Bates, F. S., Vigild, M. E., and Almdal, K., *Macromolecules* **30**, 3788–3795 (1997).
Hajduk, D. A., Ho, R. M., Hillmyer, M. A., and Bates, F. S., *J. Phys. Chem. B* **102**, 1356–1363 (1998a).
Hajduk, D. A., Tepe, T., Takenouchi, H., Tirrell, M., Bates, F. S., Almdal, K., and Mortensen, K., *J. Chem. Phys.* **108**, 326–333 (1998b).
Hamley, I. W., *Phys. Rev. E* **50**, 2872–2880 (1994).
Hamley, I. W., and Bates, F. S., *J. Chem. Phys.* **100**, 6813–6817 (1994).
Hamley, I. W., Koppi, K., Rosedale, J. H., Bates, F. S., Almdal, K., and Mortensen, K., *Macromolecules* **26**, 5959–5970 (1993).
Hashimoto, T., Kowsaka, K., Shibayama, M., and Suehiro, S., *Macromolecules* **19**, 750–754 (1986a).
Hashimoto, T., Kowsaka, K., Shibayama, M., and Kawai, H., *Macromolecules* **19**, 754–762 (1986b).
Helfand, E., *Macromolecules* **8**, 552–556 (1975).
Helfand, E., and Wasserman, Z. R., *Macromolecules* **9**, 879–888 (1976).
Helfand, E., and Wasserman, Z. R., *Macromolecules* **11**, 960–966 (1978).
Helfand, E., and Wasserman, Z. R., *Macromolecules* **13**, 994–998 (1980).

Kawasaki, K., and Ohta, T., *Physica A* **139,** 223–255 (1986).
Khandpur, A. K., Förster, S., Bates, F. S., Hamley, I. W., Ryan, A., Bras, W., Almdal, K., and Mortensen, K., *Macromolecules* **28,** 8796–8806 (1995).
Kim, J. K., Lee, H. H., Gu, Q. J., Chang, T., and Jeong, Y. H., *Macromolecules* **31,** 4045–4548 (1998).
Kimishima, K., Koga, T., and Hashimoto, T., *Macromolecules* **33,** 968–977 (2000).
Koppi, K. A., Tirrell, M., and Bates, F. S., *Phys. Rev. Lett.* **70,** 1449–1452 (1993).
Koppi, K. A., Tirrell, M., Bates, F. S., Almdal, K., and Mortensen, K., *J. Rheol.* **38,** 999–1027 (1994).
Krishnamoorti, R., Silva, A. S., Modi, M. A., and Hammouda, B., *Macromolecules* **33,** 3803–3809 (2000a).
Krishnamoorti, R., Modi, M. A., Tse, M. F., and Wang, H. C., *Macromolecules* **33,** 3810–3817 (2000b).
Lammertink, R. G. H., Hempenius, M. A., van den Enk, J. E., Chan, V. Z. H., Thomas, E. L., and Vancso, G. J., *Adv. Mater.* **12**(2), 98–103 ((2000).
Laradji, M., Shi, A. C., Desai, R. C., and Noolandi, J., *Phys. Rev. Lett.* **78,** 2577–2580 (1997a).
Laradji, M., Shi, A. C., Noolandi, J., and Desai, R. C., *Macromolecules* **30,** 3242–3255 (1997b).
Laradji, M., Desai, R. C., Shi, A. C., and Noolandi, J., *Phys. Rev. Lett.* **80,** 202–202 (1998).
Lee, H. H., *The Kinetics and Mechanism of Order to Order Transition in Block Copolymers*, Ph.D. thesis, Pohang University of Science and Technology, Pohang, Korea, 2000.
Leibler, L., *Macromolecules* **13,** 1602–1617 (1980).
Li, X. J., and Schick, M., *Biophys. J.* **78,** 34–46 (2000).
Lipkin, M. D., Rice, S. A., and Mohanty, U., *J. Chem. Phys.* **82,** 472–479 (1985).
Liu, G. J., Ding, J. F., Hashimoto, T., Kimishima, K., Winnik, F. M., and Nigam, S., *Chem. Mater.* **11,** 2233–2240 (1999a).
Liu, G. J., Ding, J. F., Qiao, L. J., Guo, A., Dymov, B. P., Gleeson, J. T., Hashimoto, T., and Saijo, K., *Chem.-Eur. J.* **5,** 2740–2749 (1999b).
Matsen, M. W., *Phys. Rev. Lett.* **80,** 201–201 (1998a).
Matsen, M. W., *Phys. Rev. Lett.* **80,** 4470–4473 (1998b).
Matsen, M. W., *J. Chem. Phys.* **108,** 785–796 (1998c).
Matsen, M. W., *J. Chem. Phys.* **110,** 4658–4667 (1999).
Matsen, M. W., and Bates, F. S., *J. Chem. Phys.* **106,** 1–13 (1997).
Matsen, M. W., and Schick, M., *Phys. Rev. Lett.* **72,** 2660–2663 (1994).
Matsen, M. W., and Bates, F. S., *Macromolecules* **29,** 1091–1098 (1996).
Mogi, Y., Mori, K., Kotusji, H., Matsushita, Y., Noda, I., and Han, C. C., *Macromolecules* **26,** 5169–5173 (1993).
Mogi, Y., Nomura, M., Kotusji, H., Ohnishi, K., Matsushita, Y., and Noda, I., *Macromolecules* **27,** 6755–6760 (1994).
Morkved, T. L., Wiltzius, P., Jaeger, H. M., Grier, D. G., and Witten, T. A., *Appl. Phys. Lett.* **64,** 422–424 (1994).
Muller, M., and Schick, M., *Phys. Rev. E* **57,** 6973–6978 (1998).
Muthukumar, M., Ober, C. K., and Thomas, E. L., *Science* **277,** 1225–1232 (1997).
Nakazawa, H., and Ohta, T., *Macromolecules* **26,** 5503–5511 (1993).
Newstein, M. C., Garetz, B. A., Balsara, N. P., Chang, M. Y., and Dai, H. J., *Macromolecules* **31,** 64–76 (1998).
Phan, S., and Fredrickson, G. H., *Macromolecules* **31,** 59–63 (1998).
Qi, S. Y., and Wang, Z.-G., *Phys. Rev. Lett.* **76,** 1679–1682 (1996).
Qi, S. Y., and Wang, Z.-G., *Phys. Rev. E* **55,** 1682–1697 (1997a).
Qi, S. Y., and Wang, Z.-G., *Macromolecules* **30,** 4491–4497 (1997b).
Qi, S. Y., and Wang, Z.-G., *Polymer* **39,** 4639–4648 (1998).

Qi, S. Y., and Wang, Z.-G., *J. Chem. Phys.* **111,** 10681–10688 (1999).
Ryu, C. Y., Lee, M. S., Hajduk, D. A., and Lodge, T. P., *J. Polym. Sci. B Polym. Phys.* **35,** 2811–2823 (1997).
Ryu, C. Y., Vigild, M. E., and Lodge, T. P., *Phys. Rev. Lett.* **81,** 5354–5357 (1998).
Sakurai, S., Kawada, H., and Hashimoto, T., *Macromolecules* **26,** 5796–5802 (1993a).
Sakurai, S., Momii, T., Taie, K., Shibayama, M., Nomura, S., and Hashimoto, T., *Macromolecules* **26,** 485–491 (1993b).
Schulz, M. F., Bates, F. S., Almdal, K., and Mortensen, K., *Phys. Rev. Lett.* **73,** 86–89 (1994).
Semenov, A. N., *Sov. Phys. JETP* **61,** 733–742 (1985).
Shi, A. C., Noolandi, J., and Desai, R. C., *Macromolecules* **29,** 6487–6504 (1996).
Singh, M. A., Harkless, C. R., Nagler, S. E., Shannon, R. F., and Ghosh, S. S., *Phys. Rev. B* **47,** 8425–8435 (1993).
Sioula, S., Hadjichristidis, N., and Thomas, E. L., *Macromolecules* **31,** 5272–5277 (1998a).
Sioula, S., Hadjichristidis, N., and Thomas, E. L., *Macromolecules* **31,** 8429–8432 (1998b).
Stadler, R., Auschra, C., Beckmann, J., Krappe, U., Voigtmartin, I., and Leibler, L., *Macromolecules* **28,** 3080–3097 (1995).
Templin, M., Franck, A., Duchesne, A., Leist, H., Zhang, Y. M., Ulrich, R., Schadler, V., and Wiesner, U., *Science* **278,** 1795–1798 (1997).
Thomas, E. L., Alward, D. B., Kinning, D. J., Martin, D. C., Handlin, D. L., and Fetter, L. J., *Macromolecules* **19,** 2197–2202 (1986).
Thomas, E. L., Anderson, D. M., Henkee, C. S., and Hoffman, D., *Nature* **334,** 598–601 (1988).
Wang, Z.-G., *J. Chem. Phys.* **100,** 2298–2309 (1994).
Werner, A., and Fredrickson, G. H., *J. Polym. Sci. Pol. Phys.* **35,** 849–864 (1997).
Yeung, C., Shi, A. C., Noolandi, J., and Desai, R. C., *Macromol. Theory Simul.* **5,** 291–298 (1996).
Zhang, H. D., Zhang, J. W., Yang, Y. L., and Zhou, X. D., *J. Chem. Phys.* **106,** 784–792 (1997).
Zheng, W., and Wang, Z.-G., *Macromolecules* **28,** 7215–7223 (1995).

INDEX

A

Ab initio quantum chemistry
  average interaction energies, 314
  clusters of strongly interacting molecules, 341–342
  mixtures, 314
  solvent as polarizable continuum, 314
  solving Schrödinger equation, 316
  see also Quantum chemical calculations
Acetonitrile
  correlation plots of pair interaction energies, 319
  critical properties of vapor-liquid equilibria, 321
  isotropic and anisotropic polarizability, 323–324
  phase behavior, 321
  vapor pressure and vapor-liquid coexistence, 320
Acetylene, pressure-induced polymerization, 387, 388
Acid catalysts, solid, *see* Zeolite catalysis theory
Activated carbon, pore size distribution, 220–221
Activation energies, hydrocarbon, 420
Activity coefficients
  predictive methods, 341
  UNIQUAC and Wilson models, 342, 344–345
Adam–Gibbs relation, 59–60
Adsorbates, structure and energetics, 382
Adsorption isotherms
  Lennard–Jones methane in silica xerogel, 215
  mesocarbon microbead (MCMB), 212–213
  simulation of alkane, 409–411
Adsorption process
  models, 205
  see also Disordered structure models; Simple geometric pore structure models

Alkane adsorption and diffusion
  methods and theory, 407–408
  simulation, 407–414
  simulation of adsorption isotherms, 409–411
  simulation of diffusion, 412–414
  see also Zeolite catalysis theory
Alkanes, *see* Hydrocarbon activation
Amino acid substitution
  hierarchical evolution protocol, 105
  Monte Carlo simulation results, 107
  schematic diagram, 106
Amorphous silicon film
  chemically reactive minority species, 286–290
  clusters containing Si atoms, 288–290
  growth mechanism, 280–281
  mechanism of SiH radical attachment, 286–287
  SiH penetration mechanism, 287–288
  surface evolution and structural characterization, 281–283
  see also Silicon film growth
Anisotropic fluctuations, ordered block copolymers, 441–445
Argon, adsorption isotherms and pore size distribution, 230
Atomistic density functional theory, 205

B

Barrett–Joyner–Halenda (BJH) method, Kelvin equation, 240
Base substitutions, DNA copying and amplification, 100
Basin enumeration function
  description, 60–61
  isobaric, 62, 63
  obtaining from experimental data, 61–62
  schematic, 65
Basis set expansion, Car–Parrinello, 362–363
Basis set superposition error (BSSE), 317
Bicontinuous networks, *see* Microemulsions

Bicontinuous phase reactions
  anomalous mean-field (AMF) kinetics, 129
  diffusion equations, 127–128
  dynamics of particle, 130
  effect of thermal fluctuations of, 135–136
  flow equations, 133–134
  Kubo formula, 130
  Langevin equation for particle in potential field, 130
  mean-field analysis, 129–132
  objectives, 128–129
  physical implications of kinetics, 134
  presence of potential, 129–132
  probability distribution, 130–131
  renormalization group theory (RG), 132–134
  result counterintuitive to expectations, 134–135
  self-consistent analysis, 131–132
Bimolecular reactions, droplet phase, 136–137
Binary collision theory, ion-surface scattering, 179–180
Biological systems, Car–Parrinello, 389–392
Blends, *see* Homopolymer solutions and blends
Block copolymers
  anisotropic fluctuations in ordered phases, 441–445
  approximate theory for anisotropic fluctuations, 442–443
  composition, 439–440
  computer simulation by Ginzburg–Landau approach, 446
  dominant fluctuation modes in HEX phase, 444, 445
  fluctuation in order parameter, 442
  hyperparallel tempering Monte Carlo (HPTMC), 17–18
  kinetic paths of order-order/order-disorder transitions, 445–450
  long-wavelength deformations/fluctuations, 452–454
  long-wavelength fluctuations and instabilities, 452–456
  long-wavelength instability in ordered phases, 454–456
  metastable states from preordered states, 459
  method for calculating matrix at equilibrium, 442
  microphase ordering, 440
  microphase separation, 439
  morphology, 441
  morphology and metastability in ABC triblock, 456–459
  nature and stability of nonclassical phases, 450–452
  nucleation/growth kinetics in order-order/order-disorder, 449–450
  phase diagram, 18
  proliferation of HEX cylinder orientations, 448–449
  scattering function of lamellar phase, 443–444
  stability of PL morphologies, 451–452
  thermodynamic stability of G phase, 450–451
  three pair interactions in ABC triblock, 458–459
  transition from two- to three-domain structure in triblock, 458
  two-dimensionally ordered ABC triblock phases, 457
Bonding, zeolite framework, 405–406
Born–Oppenheimer molecular dynamics, Car–Parrinello, 368–370
Born repulsion expression, 406–407
Bulk materials, processes, Car–Parrinello, 376–378

C

Carbon tetrachloride, interaction model, 324
Car–Parrinello methods
  advances in methodology, 392–393
  basis set expansion, 362–363
  Born–Oppenheimer molecular dynamics, 368
  chain termination and branching in olefin polymerization, 38
  choice of model, 363
  classes of problems best applicable, 356–357
  classical molecular dynamics, 357–358
  coupling two methanol at acid site of chabazite, 385–386
  cubic supercell of water molecules, 365

INDEX

density functional theory (DFT), 358–362
  DFT to compute electronic state, 357
  diffusivities of Si, Mg, and Li in conductor, 376–377
  disadvantage of path integral methods, 392–393
  dynamics of proton motion in aspartyl dyad, 390–392
  exposing chemical engineering community, 355–356
  formation of defects in GaAs, 371–372
  gas-phase processes, 372–376
  Grand Challenge Problem, 354
  Hartree–Fock method for single electron equation, 358–359
  Hellman–Feynman theorem, 368
  heterogeneous processes on surfaces, 382–386
  history, 353–354
  infrared spectrum of D_2O, 380, 381
  interpretation, 354–355
  Kohn–Sham approach, 361–362
  many-body electronic system via electronic density, 359
  many-electron system by Hohenberg and Kohn, 359–360
  melting point of solid materials, 387, 389
  methanol adsorption at activity sites of chabazite, 371
  molecular dynamics, 368–370
  new algorithms and methods, 392
  objectives, 356–357
  periodic boundary conditions of plane waves, 363, 367
  phases of electron transfer process, 372–373, 374
  phase transitions, 386–389
  plane waves form, 363
  polymerization reactions on Ziegler–Natta catalysts, 383–385
  pressure-induced acetylene polymerization, 387, 388
  processes in biological systems, 389–392
  processes in bulk materials, 376–378
  propagating orbitals with nuclei without reoptimizing, 369
  properties of liquids, solvation, and reactions in liquids, 378–382
  pseudopotential method, 367
  purposes of simulations, 371
  reactivity of ketones with OH using, 373–375, 376
  slab model of platinum(111) surface, 366
  sliding of grain boundaries in germanium by *ab initio*, 377
  structural and bonding properties of water via supercells, 378–380
  structure and dynamics of aqueous proton transfer, 382
  superposition of oscillations, 369–370
  unit cell of zeolite chabazite, 364
Catalysts, *see* Zeolite catalysis theory
Catalytic reaction, rate, 401–403
Chabazite
  coupling two methanol molecules, 385–386
  energy of adsorption of protonated propylene, 425
  propene during protonation attack, 416, 417
  methanol adsorption, 371
  stability of isobutene activation, 423
  unit cell, 364
Charging free energy
  continuum solvation models, 327–328
  contributions, 337–338
Chemical reactivity, microemulsions, 125
Chem-Shell embedding procedure, 422–424
Chlorine ions, *see* Ion-assisted etching
Clusters
  formation in surfactant solutions, 308
  impact on film defects, 290
  interactions of Si_6H_{13} with a-Si:H film, 289
  neutral and cationic with Si atoms, 288–289
  structure of Si_6H_{13}, 289
Collision cascade, 157
Combinatorial chemistry
  goal, 81–82
  steps of first round, 89
  task of searching composition space, 82
  see also Materials discovery; Protein molecular evolution
Complex fluids, molecular simulation, 1–2
Computational quantum chemistry
  isolated molecule, 313
  see also Quantum chemical calculations
Condensed phases, topographic viewpoint, 34–35
Controlled pore glasses (CPG), disordered structure models, 206–209
Copolymers, *see* Block copolymers

Copolymers, block and random, hyperparallel tempering, 17–18
Corradini site, Ziegler–Natta catalysts, 383–384
Critical micelle concentration (cmc) determination, 306–307
  reduced osmotic pressure vs. surfactant, 306
  versus inverse temperature for systems forming micelles, 307
Crystal nucleation, pure melt, 38

D

Debye equation, rotational motion, 31
Density functional theory (DFT)
  adsorption models, 225–231
  Car–Parrinello simulations, 357, 358–362
  catalytic reaction rate, 402–403
  choice of weighting function, 227–228
  comparing to GEMC model isotherms, 229
  DFT excess isotherm, 228
  energy minimization step, 343
  equilibrium density profile, 228
  model isotherms for various gases, 228–229
  pore filling pressure and pore width, 226
  pore filling pressure correlations for MK-BET and DFT, 243
  pore size distribution results, 231
  radical-surface interactions, 257–258
  test of robustness, 229–230
Deposition
  ion-surface interactions, 180–198
  see also Ion-assisted etching
Desorption isotherms, methane in silica xerogel, 215
Diffusion equations
  bicontinuous phase, 127–128
  droplet phase reactions, 139–141
  simulation, 412–414
Disease organism, evolved resistance, 117
Disilane formation, initial stages of film growth, 278–279
Disordered structure models
  carbon-carbon radial distribution functions (RDFs), 211
  experimental RDFs, 209
  grand conanical Monte Carlo (GCMC), 214, 216

geometric definition of pore size distribution, 208
geometric pore size distribution, 207–208
illustration of porous matrix via templating, 216
microporous carbons, 209–213
model porous glasses via quench molecular dynamics (MD), 207
Percus–Yevick closure, 217
perturbations of GCMC simulation, 213–214
physical properties vs. reverse Monte Carlo (RMC), 212–213
porous glasses, 206–209
replica Ornstein–Zernicke (ROZ) integral equations, 217–218
RMC, 209–211
templated porous materials, 216–218
visualizations from GCMC of confined fluids in xerogels, 215
xerogels, 213–216
see also Simple geometric pore structure models
Dissociative adsorption, Si incorporation, 274–275
DNA, mutation tendencies, 111, 112
DNA base mutation, protein expression, 111, 113
DNA base substitution, local protein space, 115
DNA shuffling
  connection between codon usage and, 113
  discovery, 100–101
  genes and operons evolved by, 101
  hierarchical evolution protocol, 105, 107
  incapable of evolving new protein folds, 116
  Monte Carlo simulation results, 107
  schematic diagram, 106
DNA swapping
  evolution of *E. coli* from *Salmonella*, 115
  penicillin-binding proteins, 115
Dominant precursor
  deposition, 284–286
  SiH_3 for silicon film growth, 285–286
Droplet phase reactions
  analysis of leading order corrections, 145–146
  bimolecular reactions $A + B \zeta 0$, 136
  considering temporal regimes, 141–143

diffusion equation and perturbation
  expansion, 139–141
dynamical fluctuations enhancing rate,
  143
effect of Péclet number, 144–145
effects for future work, 146
fluctuations, 137–138
intermediate times, 143
short time regime, 143–144
steady-state limit of expression, 142
Dubinin adsorption models
  Dubinin and Astakhov (DA) equation,
    236–237
  Dubinin–Radushkevitch (DR) equation,
    236
  Dubinin–Stoechli (DS) method, 237
  pore size distribution (PSD) comparisons,
    238
  semiempirical, 205, 236–239

E

Electrolyte solutions
  binodal curves for asymmetric ionic
    systems, 8–9
  clusters from simulation of $\lambda = 0.1$ system,
    10
  critical parameters as function of size
    asymmetry, 9
  fraction of ions in clusters, 11
  hyperparallel tempering Monte Carlo
    (HPTMC), 7–11
  integral-equation theoretical predictions,
    9–10
  parallel tempering methods, 8
  primitive model, 7–11
Eley–Rideal reaction, hydrogen abstraction,
  277
Embedding techniques
  Chem-Shell, 422–424
  hydrocarbon activation, 422–424
Empirical interatomic interactions, 258–260
Energy landscape
  application of landscape-based ideas,
    34–35
  constant volume and pressure, 38–39
  crystal nucleation form pure melt, 38
  deepest basin in each symmetry sector,
    35–36

identification of inherent structures, 35
insight into stretched liquids and glassy
  state, 67–68
N-body potential energy function, 33–34
partitioning configuration space, 35
potential energy range, 36
potential enthalpy, 39
schematic for many-particle system, 34
single-sector distribution, 37–38
steepest descent mapping, 35
supercooled liquids and glasses, 33–39
temperature variations, 37
thermal equilibrium, 37–38
topographic viewpoint of condenses
  phases, 34
vibrational partition function for each
  basin, 36–37
void geometry and connections to, 40–45
Enthalpy
  potential, 39
  relationship to temperature, 26
Entropy
  configurational, 59–60
  liquid versus crystal, 29, 30
Etching, see Ion-assisted etching
Ethanol in heptane, infinite dilution activity
  coefficient, 333
Ethylene polymerization, Ziegler–Natta,
  382–385
Exchange energy parameters, surfactants,
  300–301

F

Film growth, see Silicon film growth
Flexible polymers
  liquid-liquid phase diagram, 17
  phase diagram, 16
  see also Polymers
Fluctuations
  droplet phase reactions, 137–138
  effect of, of bicontinuous phase on
    kinetics, 135–136
Fluorine ions, see Ion-assisted etching
Frederickson–Andersen kinetic Ising model,
  55
Fredholm integral equations
  adsorption models, 219–220
  circumventing problems by, 220

G

GaAs, formation of detects, 371–372
Gas-phase processes
  Car–Parrinello, 372–376
  formaldehyde radical anion with methyl chloride, 372, 373
  phases of electron-transfer process, 372–373, 374
  reactivity of ketones with OH, 373–375, 376
Gaussian landscapes
  1-propanol and 3-methylpentane, 66
  relations for systems with, 66–67
  thermodynamic properties of macroscopic system with, 66–67
GCSKOW model, *see* Group contribution solvation (GCS) model
Geometric pore size distribution
  advantage of definition, 208
  determination for porous glasses, 207–208
  disordered porous carbon model, 212
  two-dimensional illustration, 208
  see also Pore size distribution (PSD)
Geometry
  algorithm determining volume and surface area, 40–41
  analysis of void space in inherent structures, 43
  average pressure in inherent structures, 42–43
  complexity of hydrogen bonds in aqueous systems, 45
  expanding system to lower densities, 43–44
  morphology of Lennard–Jones inherent structures, 42
  question from Lennard–Jones fluid results, 44
  shape of pressure versus density curve, 44
  shifted-force Lennard–Jones system, 42
  supercooled liquids and glasses, 39–40
  void, 40–45
  Voronoi–Delaunay tessellation, 41
Germanium, sliding of grain boundaries, 377
Gibbs ensemble Monte Carlo (GEMC)
  advantage over grand canonical Monte Carlo (GCMC), 225
  comparing to DFT isotherms, 229
  critical properties of vapor-liquid equilibria, 321

isotherms for nitrogen adsorption, 225, 226
molecular simulation, 204
phase equilibria of confined fluids, 224
pore filling pressure and pore width, 226
potentials for predicted phase behavior, 319–321
simulation adsorption model, 222, 224–225
Glass transition temperature, Tg
  definition, 24
  dependence on cooling rate, 24–25
  transformation range, 25–26
Glasses
  chemical engineering science, 22–23
  commercial processes and applications, 22, 71
  see also Supercooled liquids and glasses
Gram–Schmidt procedure, subspace identification, 90–91
Grand canonical Monte Carlo (GCMC)
  GCMC adsorption simulation method, 214, 216
  GCMC with percolation theory, 222
  isotherms for adsorption of bases, 222
  Lennard–Jones methane in xerogel, 215
  molecular simulation, 204
  perturbations for xerogels, 213
  simulation of xerogels, 213–216
  visualizations of confined fluids in xerogels, 215
Grand Challenge Problem, multiscale modeling, 354
Group contribution methods
  correcting for structure and proximity effects, 336
  failure from distortion of electron distribution, 336
  influence of proximity effects, 339–341
  physical and thermodynamic properties, 335
  shortcomings, 335–336
Group contribution solvation (GCS) model
  derivation for octanol-water partition coefficient (GCSKOW), 332
  GCSKOW predictions using simple group contribution, 337
  GCSKOW predictions with multipole corrections, 337, 338–339
  infinite dilute activity coefficients, 330–332
  scale factor and temperature dependence, 334–335

Group scale factor, functional groups in different solvents, 329

H

Halogen ions for etching, *see* Ion-assisted etching
Hartree–Fock (HF) method
  energy calculations, 344
  energy minimization, 343
  Schrödinger to single-electron equation, 358–359
Hellman–Feynman theorem, 368
Hexane, isomerization, 430–431, 432
Hexene, isomerization reaction, 428–430
Hierarchical approach, radical-surface interactions, 255–257
HIV-1 protease, Car–Parrinello, 390–392
Homopolymer chains
  mole fraction in dilute phase vs. inverse temperature, 303
  phase behavior, 303
  surfactant solution simulation, 302–304
Homopolymer solutions and blends
  coexistence curves, 13–14
  compressibility effects of phase behavior, 14–15
  end-to-end vector autocorrelation function, 12
  hyperparallel tempering Monte Carlo (HPTMC), 11–15
  performance of hyperparallel tempering, 12–13
  phase diagram of asymmetric blends, 15
  phase diagram of long chains, 13
  scaling critical density with chain length, 14
Horvath–Kawazoe (HK) method
  adsorbate-adsorbate interaction, 233
  causes for differences in model results, 234
  dispersion coefficients using Kirkwood–Muller, 233
  illustration of modified HK, 236
  limitation, 235
  modified HK, 235
  pore condensation pressure and slit pore width, 233–234
  pore filling correlation, 234
  pore filling pressure and pore width, 226

  semiempirical adsorption model, 231, 233–234
  semiempirical models, 205
Hydride transfer reaction, transition states, 426–427
Hydrocarbon activation
  activation energies, 420
  activation energies for proton-activated isobutane, 419
  energy of adsorption for protonated propylene, 425
  Chem-Shell embedding procedure, 422–424
  corrections to activation energies for cluster acid strength, 421
  elementary reaction steps, 414, 415
  embedding techniques, 422
  energy differences using Monte Carlo, 427
  formation of organic cations, 414, 416
  geometries of propene protonation, 416, 417
  geometries of zeolitic site clusters, 418
  corrections to cluster calculations, 424, 426
  properties of zeolitic protons, 416, 418–419
  quantum-chemical calculations, 416
  stability of *i*-butene activation, 423
  transition-state energies, 420–421
  transition states of hydride transfer, 426–427
  see also Zeolite catalysis theory
Hydrogen abstraction
  Eley–Rideal reaction, 277
  Langmuir–Hinselwood mechanism, 277–278
  SiH$_3$ abstracting from H-terminated Si surface, 276
  silicon film growth, 276–278
Hydrogen bonding, configurational freedom, 346
Hydroisomerization reaction, kinetics, 427–428
Hyperparallel tempering Monte Carlo
  acceptance rate for trial swap, 19
  advantages, 18
  block and random copolymers, 17–18
  factors affecting performance, 19
  grand canonical ensemble, 3
  homopolymer solutions and blends, 11–15
  Lennard–Jones fluid, 5–7
  methodology, 3–5

Hyperparallel tempering Monte
  Carlo (*continued*)
 partition function of composite
  ensemble, 3
 preweighting function, 4
 primitive model electrolyte solutions, 7–11
 schematic of implementation, 5
 semiflexible polymers and blends with
  flexible, 15–17
 swapping, 4–5
 three types of trial moves, 4
 weighting function, 4

I

Ideal glass, description, 64
Immune system, evasion of, 117
Infinite dilution activity coefficients
 comparing predictions, 331
 continuum solvation models, 326–327
 ethanol in heptane, 333
 group contribution solvation model,
  330–332
Infrared spectrum, deuterated water,
  380, 381
Interatomic interactions, radical-surface,
  258–260
Intermediate time regime, droplet phase, 143
Ion-assisted etching
 Ar from bare Si surface, 177
 arguments about mechanisms, 161–162
 atomic configuration after fluence
  of CF_3^+, 185
 average product molecular weight vs. ion
  energy, 190
 average yield of sputtered species per ion,
  188
 binary collision (BC) theory, 179–180
 bond energy distribution for F/Si top layer,
  166
 C and F content as function of CF_3^+
  fluence, 183
 C, F, and Si depth profiles for CF^+
  bombardment, 196, 197
 cascade energy distribution, 162–163
 CF^+ and CF_2^+ deposition on Si, 190–198
 CF_3^+ etching of silicon, 182–186
 chlorosilyl layer from Cl bombardment,
  173
 coordinate system and angle definition,
  176
 decomposition yield of F and C vs. CF^+
  and CF_2^+, 195
 deposition and etching, 180–198
 depth profiles for Si, F, and C after CF_3^+,
  186
 distributions of total scattering angle, 181
 etch product flux versus energy, 163
 etch yield and fluorocarbosilyl thickness
  vs. CF_3^+, 187
 etch yield vs. ion energy from MD
  simulation, 167
 fluorosilyl layer from F bombardment,
  171
 experimental studies of mechanisms,
  162–164
 fluorine coverage vs. F^+ fluence, 170
 flux probability density, 163
 interfacial $Si_xC_yF_z$ thickness vs. CF^+ and
  CF_2^+, 198
 ion impact and scattering, 174
 ion-surface scattering dynamics, 172,
  174–180
 kinetic energy distribution of sputter
  products, 168
 mechanisms, 161
 molecular dynamics studies of,
  mechanisms, 164–172
 perfluorinated silicon layers with Ar^+, 165,
  166
 polar angle scattering, Cl-covered Si,
  177–178
 product angular distributions, 192
 product distributions in CF_3^+ etching,
  186–190
 product kinetic energy distributions, 191
 reflected energy fraction vs. total scatter
  angle, 179
 role of surface roughness, 174–178
 results of MD, 172
 second stage of simulation, 172
 Si surface content as function of ion
  fluence, 184
 Si surface requiring amount of halogen,
  167–168
 side and top view of surface layers, 175
 silicon etching with F or Cl, 169
 square wave modulated ion bean and flux
  of XeF_2, 163–164

steady-state etching of Si by energetic F or Cl, 168–169, 171
sticking coefficients of F and C vs. CF^+ and CF_2^+, 195
uptake of F and C as function of fluence of CF^+, 193
uptake of F and C as function of fluence of CF_2^+, 194
views of Si–Cl surface layers, 176
yields of major products versus ion energy, 189
see also Molecular dynamics (MD)
Ion-surface interactions
collision cascade, 157
deposition and etching, 180–198
heating from energetic ion impact, 159–160
hierarchical scheme for atomistic simulations, 159
molecular dynamics, 156–161
plasma processing, 155–156
quantum mechanics and Born–Oppenheimer approximation, 158–159
side view of simulated Cu layer with Cu^+, 157
simulating, 157–158
simulating long times between impacts, 160–161
size of cell crucial, 160
sputtering, 157
see also Ion-assisted etching
Isobutane, activation energies for proton-activated, 419
Isobutene, effect of embedded environment on stability of activation, 423
Isomerization
hexane, 430–431, 432
hexene, 428–430

K

Kauzmann curve, 68–69
Kauzmann paradox, violation of Third Law of Thermodynamics, 29–30
Kauzmann temperature
configurational entropy, 59–60
description, 29, 30
Kelvin equation

Barrett–Joyner–Halenda (BJH) method, 240
classical adsorption model, 239
classical thermodynamic models, 205
improved model of statistical adsorbed film thickness, 242, 244
mean pore sizes of adsorbents, 240
model nitrogen adsorption isotherms via modified Kelvin-BET, 243
modified Kelvin (MK) equation, 239–240
pore filling correlations for nitrogen in MCM-41, 241
pore filling pressure correlations for MK-BET and DFT, 243
pore size distribution (PSD) using BJH method, 239
PSD for model porous silica glasses, 241
schematic of vapor-liquid equilibrium in wetted pore, 239
shortcomings spurring efforts to modify, 240, 242
Ketones, reactivity with OH, 373–375
Kinetic Monte Carlo (KMC)
modeling growth process over coarser time scales, 257
radical-surface interactions, 255–257
Kinetics
hexane isomerization, 430–431
hydroisomerization reaction, 427–428
isomerization reaction of hexene, 428–430
Kohlrausch–Williams–Watts (KWW)
relaxation kinetics, 54–55
temporal behavior of response function, 28–29
Kohn–Sham approach, 360–361

L

Landscape dynamics and relaxation
Adam–Gibbs relation, 59–60
basin shapes and transitions, 52–53
basins and kinetic pathways, 58–59
configurational entropy, 59–60
dielectric relaxation time, 60
driving supercooled system from equilibrium, 54
fragile glass formers, 56–57

Landscape dynamics and
    relaxation (*continued*)
  Fredrickson–Andersen kinetic Ising
    model, 55
  isochoric dynamical evolution of
    N-particle system, 50–51
  kinetics of interbasin transitions and
    relaxation, 53–54
  Kohlrausch–Williams–Watt (KWW)
    stretched exponential, 54–55
  long-pathway rearrangement processes for
    fragile materials, 58
  mean relaxation time, 54–55
  mean-squared atomic displacements vs.
    temperature, 52
  mode coupling theory (MCT), 51–52
  models for supercooling and glass
    formation, 55–56
  shear viscosity measures, 56
  Stokes–Einstein relation, 58
  supercooled liquids and glasses, 50–60
  system configuration point, 51
  tiling model, 55
  topographic differences, 57–58
  transitions between neighboring
    basins, 53
  viscosity for molten silica (SiO_2), 56
Langmuir–Hinshelwood mechanism, H
    abstraction, 277–278
Length scales, plasma processing, 152–153
Lennard–Jones fluid
  adsorption and desorption isotherms for
    methane in silica xerogel, 215
  adsorption simulation, 222
  density variation of inherent structure
    pressure, 43
  geometric algorithm exploring
    morphology, 42
  hyperparallel tempering Monte Carlo
    (HPTMC), 5–7
  phase behavior, 5–6
  phase diagram, 7
  phase diagram calculated vs. literature,
    6–7
  question from results, 44
  replica number as function of Monte Carlo
    steps, 6
  shifted-force, 42
Library design and redesign
  closing loop between, 96–97

  materials discovery, 85–87
Liquids, *see* Supercooled liquids and glasses
Lubachevsky–Stillinger protocol,
    preparation method, 49

M

Many-particle system, energy landscape, 34
Materials discovery
  acceptance probability of exchanging
    samples, 89
  analogy with Monte Carlo (MC), 85
  closing loop between library design and
    redesign, 96–97
  combinatorial approach, 82
  effectiveness of MC strategies, 95–96
  figure-of-merit landscape, 87
  further development, 96–97
  Gram–Schmidt procedure, 90–91
  grid method, 95
  grid search on composition and
    noncomposition variables, 85, 86
  library design and redesign, 85–87
  low-discrepancy sequence (LDS)
    method, 95
  material optimization or development, 83
  maximum figure of merit for different
    protocols, 96
  MC library design and redesign strategy, 88
  MC protocols, 94–95
  method for changing variables, 87
  parallel tempering as MC extension, 89
  phase points as function of dimension and
    spacing, 93
  random phase volume model, 92–94
  searching multidimensional space, 84
  searching variable space by MC, 87–89
  significance of sampling, 91–92
  simplex of allowed compositions, 89–91
  space of variables, 84–85
  steps of combinatorial chemistry, 89
  swapping, 87, 88
  synthetic approach, 82
  ways of changing variables, 87–88
  see also Protein molecular evolution
MCM-41
  pore filling correlations for nitrogen
    in, 241
  pore size distribution, 231, 232

INDEX

Mean-field analysis
  anomalous mean-field (AMF) kinetics, 129
  bicontinuous phase, 129–132
  presence of potential, 129–132
  self-consistent analysis, 131–132
Mean-spherical approximation (MSA) theory, 9–10
Melting point, Car–Parrinello, 387, 389
Mesocarbon microbead (MCMB), see Microporous carbons
Metastability, ABC triblock copolymers, 456–459
Methane, activated carbon to, adsorption, 220–221
Methanol
  pair interaction energies, 319
  phase behavior, 321
  vapor-liquid coexistence diagram, 320
  vapor-liquid equilibria, 321
  vapor pressure, 320
Micelles, see Surfactant solutions
Microemulsions
  applications utilizing, 125
  importance of fluctuations, 125
  microstructures of, phase, 126
  phase behavior, 124
  practical utility, 125
  primary interest in reactions A + B ⋜ 0, 125–126
  reaction kinetics at oil–water interface, 126
  reactions in bicontinuous phase, 127–136
  reactions in droplet phase, 136–146
  see also Bicontinuous phase reactions; Droplet phase reactions
Microphase ordering, block copolymers, 440
Microporous carbons
  agreement between experimental and simulated, 212–213
  carbon–carbon radial distribution functions (RDFs), 211
  carbon plate and microstructure from reverse Monte Carlo (RMC), 210
  nitrogen adsorption isotherms, 212–213
  experimental RDFs, 209
  geometric pore size distribution, 212
  morphological models by RMC, 209–211
  RMC procedure, 210–211

Mixing
  hierarchical evolution protocol, 108–109
  Monte Carlo simulation results, 107
Mode coupling theory (MCT), dynamics and relaxation processes, 51–52
Modified Horvath–Kawazoe method
  illustration, 236
  semiempirical adsorption model, 235
Modified Kelvin equation (MK)
  relationship between pore filling pressure and pore width, 226
  see also Kelvin equation
Molecular disorder
  bond-orientational order parameter, 46–47
  crystal-independent measures, 48
  degree of translational order, 47–48
  jammed hard-sphere structures, 50
  Lubachevsky–Stillinger protocol, 49
  non-equilibrium preparation method, 49
  ordering phase diagram, 48–50
  quantifying, 45–50
  random system, 46
  specific bond-orientational order parameter, 47
  structural order parameter, 46
Molecular dynamics (MD)
  Car–Parrinello, 368–370
  Car–Parrinello, classical, 357–358
  classical mechanics, 158
  collision cascade, 157
  heating by energetic ion impact, 159–160
  hierarchical approach for radical-surface interactions, 255–257
  hierarchical scheme for atomistic simulations, 159
  ion-assisted etching mechanisms, 164–172
  ion-surface interactions, 157–158
  plasma processing, 149–152
  quantum mechanics and Born–Oppenheimer approximation, 158–159
  radical-surface interactions, 254–255
  side view of copper layer with Cu^+ bombardment, 157
  simulating long times between ion impacts, 160–161
  simulation procedure, 156–161
  size of cell, crucial factor, 160
  sputtering, 157

476　INDEX

Molecular dynamics (MD) (*continued*)
　see also Ion-assisted etching; Plasma
　　processing
Molecular evolution
　experimental applications of technology,
　　116
　see also Protein molecular evolution
Molecular statics (MS)
　hierarchical approach for radical-surface
　　interactions, 255–257
　radical-surface interactions, 254
Monte Carlo methods
　alkane adsorption and diffusion, 408
　analogy with combinatorial chemistry, 85
　benefit of swapping, 2
　complex fluids, 2
　conventional, 2
　effectiveness in materials discovery,
　　95–96
　first molecular simulations, 315
　grid method, 94–95
　hierarchical approach for radical-surface
　　interactions, 255–257
　library design and redesign, 88
　low-discrepancy sequence (LDS) method,
　　94–95
　methods requiring energy of assembly of
　　molecules, 315
　parallel tempering as extension of, 89
　protocols, 94–95
　radical-surface interactions, 254
　searching variable space by, 87–89
　specifying thermodynamic state of replica,
　　2–3
　see also Grand canonical Monte Carlo
　　(GCMC); Hyperparallel tempering
　　Monto Carlo; Kinetic Monte Carlo
　　(KMC); Reverse Monte Carlo (RMC)
Mordenite
　hydrogen bonded to bridging O atom, 404
　self-diffusivity diffusion constant, 412
　structure, 403, 404
　see also Zeolite catalysis theory
Morphology
　ABC triblock copolymers, 456–459
　block copolymers, 441
Multipool swapping
　hierarchical evolution protocol, 109
　Monte Carlo simulation results, 107
　schematic diagram, 106

N

Nanostructure formation, *see* Surfactant
　solutions
Nitrogen
　adsorption isotherm on mesocarbon
　　microbead (MCMB), 212–213
　adsorption isotherms and pore size
　　distribution, 230
NK model
　energy functions, 103
　interaction between secondary structures,
　　104
　molecular evolution protocols, 102–104
　random energy model, 102

O

Olefin polymerization, chain termination
　and branching, 380
Order-disorder transitions, *see* Block
　copolymers
Order-order transitions, *see* Block
　copolymers

P

Parallel tempering
　basic idea, 2
　schematic of swapping, 108
　see also Hyperparallel tempering Monto
　　Carlo
Parasites, evading chemical control, 116
Péclet number, effect in droplet phase
　reactions, 144–145
Penicillin-binding proteins, 115
Percolation theory
　adsorption and percolation, 223
　grand canonical Monte Carlo
　　(GCMC), 222
Percus–Yevick closure, templated
　structures, 217
Perturbation expansion, droplet phase
　reactions, 139–141
Perturbation methods, computing
　interaction energies, 318
Pest organisms, evading chemical
　control, 116

Phase behavior, predictive methods, 341
Phase diagrams
　amphiphiles not forming micelles, 304
　asymmetric polymer blends, 15
　block and random copolymers, 18
　Car–Parrinello simulations, 386–389
　Lennard–Jones fluid, 7
　liquid-liquid, of semiflexible/flexible polymer mix, 17
　long polymer chains, 13
　semiflexible and flexible polymer, 16
　soft-sphere plus mean-field model, 70
　two-parameter ordering, 49
Phase separation, see Surfactant solutions
Plane-wave basis sets
　Car–Parrinello, 363, 367
　Hellman–Feynman theorem, 368
Plasma-enhanced chemical vapor deposition (PECVD)
　amorphous films on substrates by MD, 262–263
　growth of thin films, 252
　hierarchical approach to plasma-surface interactions, 255–257
　methods of surface preparation, 260–263
　plasma-surface interactions, 252–253
　procedure for a-Si:H/c-Si film/substrate systems, 261
　silane and hydrogen containing discharges, 253
　surface characterization and reaction analysis, 263–264
　see also Radical-surface interactions; Surface chemical reactivity, SiH$_x$ radicals
Plasma processing
　crystalline silicon surface exposed to chlorine plasma, 154
　examples of etch profiles, 153
　ion-surface interactions, 155–156
　length scales, 152–153
　molecular dynamics (MD) simulations, 149–150
　nature of plasma-surface interactions, 153–155
　processing wafers, 151–152
　schematic of typical plasma reactor, 151
　semiconductors, 150–152
　simulation procedure, 156–161
　typical conditions of nonequilibrium plasmas, 151
　see also Ion-assisted etching; Molecular dynamics (MD)
Plasma-surface interactions
　nature, 153–155
　steady-state depth composition profile, 154
Platinum, slab model, 366
Polarizable continuum model (PCM), 328
Polarizable potentials, multibody interactions, 322–323
Polymerization of olefins, 380
Polymers
　block and random copolymers, 17–18
　configuration snapshot of semiflexible systems, 16
　hyperparallel tempering Monte Carlo (HPTMC), 11–15
　liquid-liquid phase diagram for semiflexible/flexible mix, 17
　phase diagram for semiflexible and flexible polymers, 16
　semiflexible, and blends with flexible, 15–17
Pore size distribution (PSD)
　activated carbon to methane adsorption, 220–221
　density functional theory (DFT) results, 231, 232
　Dubinin–Stoeckli (DS), Horvath–Kawazoe (HK) and DFT methods, 238
　grand canonical Monte Carlo (GCMC) for CH_4, CF_4, and SF_6 isotherms, 223
　model silica glasses, 240, 241
　testing robustness for DFT, 229–230
　see also Geometric pore size distribution
Porous glasses
　generation of model, 207
　geometric pore size distribution, 207–208
　interpreting adsorption isotherms, 208–209
　quench molecular dynamics methods, 206–209
　reverse Monte Carlo technique, 209
Porous materials
　classical thermodynamic models, 205
　disordered model microstructure, 204
　modeling pore structure, 204

Porous materials (*continued*)
  molecular simulation calculations, 204
  pore structure characterization, 203–204
  semiempirical models, 205
  simple geometric model, 204
  statistical thermodynamic theories, 205
  templated, 216–218
  thermodynamic models, 204–205
  see also Disordered structure models; Simple geometric pore structure models
Propene, geometries during protonation, 416, 417
Propylene, changes in energy of adsorption of protonated, 425
Protein molecular evolution
  ab initio evolution of protein with specific function, 99
  amino acid substitution, 105
  background on experimental, 100–102
  base substitutions, 100
  challenge of swapping protocols, 110
  connection between codon usage and DNA shuffling, 113
  current state-of-the-art techniques, 101–102
  description, 98–99
  discovery of DNA shuffling, 100–101
  DNA base mutation, 111, 112, 113
  DNA base substitution, 115
  DNA shuffling, 105, 107, 116
  energy functions of NK form, 103
  energy of interaction between secondary structures, 104
  evasion of immune system, 117
  evolution of multiprotein complex, 99
  evolution of new life forms, 99
  evolved resistance, 117
  examples of exploiting codon potentials, 113
  examples of swapping by nature, 114–115
  experimental conditions and constraints, 104–105
  generalized NK model, 102–104
  genes and operons evolved by DNA shuffling, 101
  hierarchical decomposition of protein space, 97–98
  hierarchical decomposition of sequence space, 99
  hierarchical evolution protocols, 105–109
  mixing, 108–109
  Monte Carlo results of evolution protocols, 107
  multipool swapping, 109
  mutation tendencies for DNA, 112
  natural analogs of protocols, 113–115
  parallel tempering, 108–109
  pests and parasites evading chemical control, 116
  possible experimental implementations, 109–111
  random or directed mutagenesis, 100
  searching space, 97
  simulated molecular evolution protocols, 106
  single-pool swapping, 107
  specific energy function as selection criterion, 103
Proximity effect, group contribution models, 336
Pseudopotential method, Car–Parrinello, 367–368

Q

Quantum chemical calculations
  ab initio molecular simulations, 316
  activity coefficients, 341
  application of group contribution methods to large compounds, 341
  application to chemical engineering, 346–347
  availability of computational packages, 346
  charging free energy contributions, 337–338
  charging free energy from continuum solvation models, 327–328
  chosen pair potential and polarization term, 323
  correcting for structure and proximity effects, 336
  dilute activity coefficients, 331
  energies by Hartree–Fock with 6-311++G(3d,2p) basis set, 344
  GCS model for octanol-water partition coefficient (GCSKOW), 332

INDEX 479

GCSKOW model with multipole correction method, 337, 338–339
GCSKOW model with simple group contribution method, 337
Gibbs ensemble Monte Carlo (GEMC) for predicting phase behavior, 319–321
group contribution solvation (GCS) model, 330–332
group scale factor for functional groups, 329
hydrogen bonding, 346
infinite dilute activity coefficient of ethanol in heptane, 333
infinite dilute activity coefficients, 326–327
influence of proximity effects, 339–341
interaction model including polarization, 324
isotropic and anisotropic polarizability, 323–324
MC computer simulation methods, 315
MC simulation of water, 316–317
molecular simulations, 315
multibody interactions, 321–322
pair interaction energies of acetonitrile and methanol, 319
perturbation methods for computing interaction energies, 318
phase behavior, 321
phase equilibria predictions, 341–346
polarizable continuum model (PCM), 328
potential energy surface between two molecules, 317
procedure to determine interaction energies, 342
scale factor in charging free energy calculations, 330
scale factors and temperature dependence, 334–335
sequential use of *ab initio* and molecular simulations, 316
shortcomings of group contribution, 335–336
simulations (GEMC) of vapor-liquid equilibria, 321
small clusters of strongly interacting molecules, 341–342
solvation models for thermodynamic properties, 325–326
solvation process in two steps, 326

supermolecular method, 317–318
temperature effects by continuum solvation, 332–334
thermodynamic property predictions, 335–341
two-body potentials and macroscopic property predictions, 318–319
UNIQUAC and Wilson activity coefficient models, 342, 344–345
using polarizable potentials, 322–323
vapor-liquid coexistence diagrams, 320
vapor pressure, 320
water models, 324–325

R

Radial distribution functions (RDFs), 209, 211
Radical-surface interactions
atomic-scale simulation studying deposition, 255
atomic-scale theoretical studies, 254
atomistic simulation methods, 254–255
characterizing surfaces, reactivity, 264–265
clusters containing Si atoms in plasmas, 288–289
density functional theory (DFT), 257–258
disilane formation, 278–279
dissociative adsorption mechanism, 274–275
dominant deposition precursor, 284–286
Eley–Rideal H abstraction, 277
empirical description of interatomic, 258–260
film growth mechanisms by PECVD, 273
film surface composition, 283–284
hierarchical approach, 255–257
hydrogen abstraction by SiH_3 radical, 276–278
impact of Si_6H_{13} clusters on film defects, 290
kinetic Monte Carlo (KMC) methods, 255
Langmuir–Hinselwood H removal, 277–278
mechanism of amorphous silicon film growth, 280–281
mechanism of attachment for SiH radical, 286–287
mechanism of SiH penetration, 287–288

INDEX

Radical-surface interactions (*continued*)
  nanosecond-time scale MD of film growth, 273
  reactions of SiH$_2$ radical with H:Si surface, 268–269
  role of chemically reactive minority species, 286–290
  SiH$_2$ radical, possible adsorption sites, 268
  SiH$_3$ insertion reactions and dissociative adsorption events, 275
  SiH$_3$ radical attaching to pristine Si surface, 269
  SiH$_x$ radicals with amorphous silicon films, 270–272
  SiH$_x$ radicals with crystalline silicon surfaces, 266–269
  Si$_6$H$_{13}$ cluster with a-Si:H surface, 289
  structure of crystalline and amorphous silicon, 265–266
  surface characterization and reaction analysis, 263–264
  surface chemical reactions during film growth, 274–280
  surface evolution and characterization, 281–283
  surface preparation, 260–263
  see also Surface chemical reactivity, SiH$_x$ radicals
Random copolymers
  hyperparallel tempering Monte Carlo (HPTMC), 17–18
  phase diagram, 18
Random phase volume model
  functional form, 94
  materials discovery, 92–94
  number of phase points, 93
Reactivity of surfaces, *see* Surface chemical reactivity, SiH$_x$ radicals
Reactor design and kinetics, 123–124
Relaxation, viscous liquids, 28
Renormalization group theory
  bicontinuous phase reactions, 132–134
  flow equations, 133–134
  framework for identifying system behavior, 132
  implementing analysis, 133
  phase-space flow diagram, 134
Replica Ornstein–Zernicke (ROZ) integral equations, 217–218
Reverse Monte Carlo (RMC)
  agreement with physical properties, 212–213
  morphological model for microporous carbon, 209–211
  procedure, 210–211
Rotational diffusion, 31–32

S

Sampling, materials discovery, 91–92
Sastry density, description, 68
Sastry point
  superheated liquid and, 68
  theoretical predictions, 69
Self-assembled system, applications, 124
Semiconductors, plasma processing, 150–152
Semiflexible polymers
  configuration snapshot, 16
  liquid-liquid phase diagram, 17
  phase diagram, 16
  see also Polymers
Sequence space, hierarchical decomposition, 99
Short time regime, droplet phase, 143–144
SiH$_x$ radicals, *see* Surface chemical reactivity, SiH$_x$ radicals
Silica (SiO$_2$)
  strong glass former, 56
  topographic illustration, 57
  see also Supercooled liquids and glasses
Silicon
  atomic configuration after CF$_3^+$ fluence, 185
  average yield of major sputtered species, 188
  C and F content as function of CF$_3^+$ fluence, 183
  C, F, and Si depth profiles for CF$^+$, 196, 197
  CF$^+$ and CF$_2^+$ deposition, 190–198
  CF$_3^+$ etching, 182–186
  deposition yield of F and C for CF$^+$ and CF$_2^+$, 195
  depth profiles for Si, F, and C after CF$_3^+$, 186
  etch yield and fluorocarbosilyl layer thickness, 187
  interfacial Si$_x$C$_y$F$_z$ layer thickness for CF$^+$ and CF$_2^+$, 198

product angular distributions as function of angle, 192
product distributions in CF_3^+ etching, 186–190
product kinetic energy distributions, 191
sticking coefficients of F and C for CF^+ and CF_2^+, 195
surface content, 184
uptake of F and C as function of CF^+ fluence, 193
uptake of F and C as function of CF_2^+ fluence, 194
yields of major products, 189
see also Ion-assisted etching
Silicon film growth
clusters with Si atoms in plasmas, 288–289
disilane formation, 278–279
dissociative adsorption, 274–275
Eley–Rideal H abstraction, 277
evolution of surface roughness in MD simulation, 282–283
film surface composition, 283–284
hydrogen abstraction by SiH_3 radicals, 276–278
impact of Si_6H_{13} clusters, 290
Langmuir–Hinshelwood H removal, 277–278
MD simulations of Si_6H_{13} clusters, 289
mechanism of amorphous, 280–281
mechanism of SiH radical attachment, 286–287
mobility of SiH_3 radical on amorphous growth surface, 279–280
plasma-surface interactions, 273
role of chemically reactive minority species, 286–290
role of dominant deposition precursor, SiH_3, 284–286
SiH_3 insertion reactions, 275
SiH penetration mechanism, 287–288
surface chemical reactions during, 274–280
surface evolution and characterization, 281–283
see also Radical-surface interactions; Surface chemical reactivity, SiH_x radicals
Simple geometric pore structure models
adsorption isotherms and pore size distribution (PSD), 230

carbon micropore distributions, 237, 239
classical adsorption models, 239–244
comparing pore filling correlations for nitrogen in MCM-41, 241
adsorption and percolation, 223
density functional theory (DFT)
adsorption models, 225–231
DFT model isotherms for various gases, 228–229
Dubinin adsorption models, 236–239
Dubinin and Astakhov (DA) equation, 236–237
Dubinin–Radushkevitch (DR) equation, 236
Dubinin–Stoeckli (DS) method, 237
excess adsorption by adsorption integral equation, 218–219
Fredholm integral equations, 219–220
Gibbs ensemble Monte Carlo (GEMC), 222, 224–225
grand canonical Monte Carlo (GCMC), 222
Horvath–Kawazoe (HK) method, 231, 233–234
interchange steps, 224
Kelvin equation, 239
Lennard–Jones models, 222
mean pore sizes of adsorbents, 240
model nitrogen adsorption isotherms, 243
modified Horvath–Kawazoe models, 235
modified Kelvin (MK) equation, 239–240
molecular displacements, 224
molecular simulation adsorption models, 222–225
percolation theory with GCMC, 222
pore filling model, 242
pore filling pressure and pore width, 226
pore filling pressure correlations, 243
pore size distribution (PSD) for methane adsorption, 220–221
PSD using GCMC for CH_4, CF_4, and SF_6, 223
PSDs by various methods, 238
PSDs for model porous silica glasses, 241
schematic of GEMC method, 224
semiempirical adsorption models, 231, 233–239
shortcomings spurring efforts to modify Kelvin, 240, 242
volume exchange steps, 225

Simple geometric pore structure models (*continued*)
  weighting function for DFT, 227–228
Single-pool swapping, 107
Slab model, platinum(111) surface, 366
Solid acid catalysts, *see* Zeolite catalysis theory
Solids, melting points by Car–Parrinello, 387, 389
Solvation process, 325, 326
Space of variables, materials discovery, 84–85
Sputtering
  description, 157
  see also Ion-assisted etching
Steepest descent mapping, 35
Stokes–Einstein equation, 31, 58
Structural order parameters
  bond-orientational, 46–47
  crystal-independent measures, 48
  Lubachevsky–Stillinger protocol, 49
  measuring degree of translational order, 47–48
  ordering phase diagram, 48–50
  specific bond-orientational, 47
  terminology, 46
  Voronoi–Delaunay tessellation, 46–47
Superacid catalysis, 403–405
Supercells
  structural and bonding properties of water, 378–380
  water, 365
Supercooled liquids and glasses
  bond-orientational order parameters, 46–47
  characterization of voids, 40
  chemical engineering science, 22–23
  crystal-independent measures, 48
  density dependence of average pressure, 42–43
  energy landscape, 33–39
  entropy for liquids and stable crystals, 29–30
  expanding system to lower densities, 43–44
  geometric algorithm, 40–41
  geometric analysis of void space, 43
  glass transition temperature, 24
  glass transition transformation range, 25–26
  good glass-forming materials, 44–45

  hydrogen bonds in aqueous systems, 45
  isobaric relationship between volume and temperature, 25
  Kauzmann paradox, 30
  Kauzmann temperature, 29–30
  kinetics and thermodynamics, 30–31
  Kohlrausch–Williams–Watts (KWW) function, 28–29
  landscape dynamics and relaxation phenomena, 50–60
  morphology of Lennard–Jones inherent structures, 42
  nonexponential relaxation, 28
  ordering phase diagram, 48–50
  parameter for degree of translational order, 47–48
  quantifying molecular disorder, 45–50
  questions from Lennard–Jones fluid, 44
  relationship between enthalpy and temperature, 26
  relationship between volume and temperature, 24, 25
  rotational diffusion and viscosity, 31–32
  scalar measures for translational order, 47
  shape of pressure versus density curve, 44
  shifted-force Lennard–Jones system, 42
  specific bond-orientational order parameter, 47
  statistical geometry and structure, 39–50
  strong-fragile classification of liquids, 28
  structural order parameter, 46
  substances known to form glasses, 24
  Tg depending on cooling rate, 24–25
  thermodynamics, 60–70
  tools and protocols, 32
  translational and rotational motion, 31–32
  truly random system, 46
  viscosity by Vogel–Tammann–Fulcher (VTF), 26–27
  viscosity of glass-forming liquids, 27, 28
  viscosity of supercooled liquids, 27
  vitrification, 23–29
  void geometry and energy landscape, 40–45
  volume and surface area of packings, 40, 41
  see also Energy landscape; Landscape dynamics and relaxation; Thermodynamics of supercooled liquids and glasses
Superionic conductor, 376–377

Supermolecular method, 317–318
Surface chemical reactivity, SiH$_x$ radicals
  characterizing surface reactivity, 264–265
  crystalline and amorphous silicon surfaces, 265–266
  possible adsorption sites of SiH$_2$ radical, 268
  projections of SiH$_3$ center-of-mass trajectories, 271
  radical impingement on a-Si:H surfaces, 272
  reactions of SiH$_2$ radical with H:Si surface, 268–269
  reactivity maps, 266, 270
  SiH radical, 272
  SiH$_3$ radical attaching to pristine Si surface, 269
  SiH$_x$ radicals and silicon surfaces, 267–269
  SiH$_x$ radicals with amorphous silicon films, 270–272
  SiH$_x$ radicals with crystalline silicon surfaces, 266–269
Surface roughness
  evolution during film growth, 282–283
  ion-surface scattering, 174–178
Surfactant solutions
  aggregation to large, stable clusters, 302
  assembly into nanostructures, 298
  behavior of (I1) and (I2) interaction sets, 304–305
  cmc vs. inverse temperature, 307
  critical micelle concentrations (cmc), 306–307
  dimensionless ratio, 298
  exchange energy parameters, 300–302
  homopolymer chains, 302–304
  macroscopic phase separation to micellar aggregates, 308–309
  methodological issues, 301–302
  micellization processes, 299–300
  models and methods, 300–301
  mole fraction in dilute phase vs. inverse temperature, 303
  normalized cluster probability distributions, 308
  phase behavior of homopolymers, 303
  phase diagrams of amphiphiles, 304
  reduced osmotic pressure vs. volume fraction surfactant, 306
  role of different interaction sets, 304–308
  self-assembly of amphiphilic molecules, 299
  simulation details, 300–302
  snapshot of H$_1$T$_4$(I2), 305
  snapshot of H$_2$T$_4$(I1), 305
  snapshot of H$_2$T$_4$(I2), 306
  surfactant architectures and interaction energy sets, 300
  time scales for micellar aggregate formation, 299
Swapping
  acceptance rate, 19
  benefit, 2
  changing variables, 87–88
  DNA, 115
  examples by nature, 114–115
  hyperparallel tempering Monte Carlo, 4–5
  ITCHY and SCRATCHY techniques, 110–111
  Monte Carlo simulation results, 107
  multipool, 109
  schematic diagram, 106
  single-pool, 107
  technical challenge by protocols, 110
Symmetry-adapted perturbation theory (SAPT), 318

T

Temperature effects, 332–334
Templated porous materials
  illustration of porous matrix by templating, 216
  Monte Carlo simulations, 217–218
  Percus–Yevick closure, 217
  replica Ornstein–Zernicke (ROZ) integral equations, 217–218
Templating, porous matrix, 216–218
Temporal regimes, droplet phase, 141–143
o-Terphenyl (OTP)
  fragile glass former, 56–57
  topographic illustration, 57
Thermodynamics of supercooled liquids and glasses
  basin enumeration function, 60–62
  disordered materials, 69–70
  macroscopic system with Gaussian landscape, 66–67

Thermodynamics of supercooled liquids and glasses (*continued*)
  Gaussian landscape parameters, 66
  Helmholtz energy, 61
  ideal glass, 64
  isobaric basin enumeration functions, 62, 63, 65
  isobaric excitation profiles, 64
  isobaric excitation profiles and isochoric profiles, 65–66
  low-temperature termination of Kauzmann curve, 68–69
  maximum tensile strength of amorphous material at Sastry density, 68
  partition function as one-dimensional integral over basin depth, 61
  phase diagram of soft-sphere plus mean-field model, 70
  relations for systems with Gaussian landscapes, 66–67
  spinodal curve for superheated liquid and Sastry point, 68
  stretched liquids and glassy state, 67–68
  temperature-dependent basin depth, 62–64
  theoretical predictions of Sastry point, 69
  thermodynamic properties and energy landscape, 61
Third Law of Thermodynamics, Kauzmann's paradox, 29–30
Tiling model, supercooling and glass formation, 55
Transition state energies, hydrocarbon activation, 420–421
Triblock copolymers
  complex two-dimensionally ordered ABC phases, 457
  metastable states from preordered states, 459
  morphology and metastability, 456–459
  transition from two- to three-domain structure, 458
  see also Diblock copolymers

U

UNIQUAC model, activity coefficients, 342, 344–345

V

Vapor-liquid coexistence diagram
  acetonitrile, 320
  methanol, 320
Vapor-liquid equilibrium diagram, acetone + water, 345
Vapor pressure
  acetonitrile, 320
  methanol, 320
Variables, space of, materials discovery, 84–85
Viscosity
  glass-forming liquids, 26–27
  glass-forming liquids in Arrhenius fashion, 27, 28
  rotational diffusion and, 31–32
  Vogel–Tammann–Fulcher (VTF) equation, 26–27
Vitrification, *see* Supercooled liquids and glasses
Vogel–Tammann–Fulcher (VTF) equation, viscosity, 26–27
Void space, geometry of pore structure, 40
Voronoi–Delaunay tessellation, 41

W

Wafers processing, 151–152
Water
  cubic supercell, 365
  models based on effective potential, 324–325
  Monte Carlo from *ab initio* calculations, 316–317
  structural and bonding properties, supercells, 378–380
Water, deuterated, infrared spectrum, 380, 381
Wilson model, activity coefficients, 342, 344–345

X

Xerogels
  grand canonical Monte Carlo (GCMC), 213–216
  perturbations, 213

simulated adsorption and desorption
isotherms, 215
visualizations from GCMC, 215

Z

Zeolite catalysis theory
activation energies for isomerization and
protonation, 431, 432
bonding within zeolite framework,
405–406
Born repulsion expression, 406–407
catalytic cycle of zeolite-catalyzed
reaction, 400
density functional theory (DFT), 402–403
dependence of energy of adsorption for
linear alkanes, 409–410
energies of adsorption for hydrocarbons
and zeolites, 410–411
entropy expression, 403
flexibility of zeolite framework, 405
hexane in zeolite micropore, 406
hydrocarbon activation by zeolitic protons,
414–427
hydrogen attached to bridging oxygen
atom, 403–404
hydroisomerization reaction, 427–428
interaction potential, 406
isomerization reaction of hexene, 428–430
kinetics, 427–431
kinetics for hexane isomerization,
430–431
methods and theory of alkane adsorption
and diffusion, 407–408
Monte Carlo computing thermodynamic
averages, 408
overall rate of catalytic reaction, 411
quantum-chemical computations, 402
rate of catalytic reaction, 401–403
simulation of alkane adsorption and
diffusion, 407–414
simulation of alkane adsorption isotherms,
409–411
simulation of diffusion, 412–414
structure of mordenite, 403, 404
superacid catalysis, 403–405
transition rate theory, 402
zeolites as solid acid catalysts, 403–407
see also Hydrocarbon activation
Zeolites
unit cell of chabazite, 364
see also Chabazite
Ziegler–Natta catalysts
Corradini site, 383, 384
polymerization reactions, 382–385

CONTENTS OF VOLUMES IN THIS SERIAL

Volume 1

J. W. Westwater, *Boiling of Liquids*
A. B. Metzner, *Non-Newtonian Technology: Fluid Mechanics, Mixing, and Heat Transfer*
R. Byron Bird, *Theory of Diffusion*
J. B. Opfell and B. H. Sage, *Turbulence in Thermal and Material Transport*
Robert E. Treybal, *Mechanically Aided Liquid Extraction*
Robert W. Schrage, *The Automatic Computer in the Control and Planning of Manufacturing Operations*
Ernest J. Henley and Nathaniel F. Barr, *Ionizing Radiation Applied to Chemical Processes and to Food and Drug Processing*

Volume 2

J. W. Westwater, *Boiling of Liquids*
Ernest F. Johnson, *Automatic Process Control*
Bernard Manowitz, *Treatment and Disposal of Wastes in Nuclear Chemical Technology*
George A. Sofer and Harold C. Weingartner, *High Vacuum Technology*
Theodore Vermeulen, *Separation by Adsorption Methods*
Sherman S. Weidenbaum, *Mixing of Solids*

Volume 3

C. S. Grove, Jr., Robert V. Jelinek, and Herbert M. Schoen, *Crystallization from Solution*
F. Alan Ferguson and Russell C. Phillips, *High Temperature Technology*
Daniel Hyman, *Mixing and Agitation*
John Beck, *Design of Packed Catalytic Reactors*
Douglass J. Wilde, *Optimization Methods*

Volume 4

J. T. Davies, *Mass-Transfer and Interfacial Phenomena*
R. C. Kintner, *Drop Phenomena Affecting Liquid Extraction*
Octave Levenspiel and Kenneth B. Bischoff, *Patterns of Flow in Chemical Process Vessels*
Donald S. Scott, *Properties of Concurrent Gas–Liquid Flow*
D. N. Hanson and G. F. Somerville, *A General Program for Computing Multistage Vapor–Liquid Processes*

Volume 5

J. F. Wehner, *Flame Processes–Theoretical and Experimental*
J. H. Sinfelt, *Bifunctional Catalysts*
S. G. Bankoff, *Heat Conduction or Diffusion with Change of Phase*

George D. Fulford, *The Flow of Liquids in Thin Films*
K. Rietema, *Segregation in Liquid–Liquid Dispersions and Its Effect on Chemical Reactions*

Volume 6

S. G. Bankoff, *Diffusion-Controlled Bubble Growth*
John C. Berg, Andreas Acrivos, and Michel Boudart, *Evaporation Convection*
H. M. Tsuchiya, A. G. Fredrickson, and R. Aris, *Dynamics of Microbial Cell Populations*
Samuel Sideman, *Direct Contact Heat Transfer between Immiscible Liquids*
Howard Brenner, *Hydrodynamic Resistance of Particles at Small Reynolds Numbers*

Volume 7

Robert S. Brown, Ralph Anderson, and Larry J. Shannon, *Ignition and Combustion of Solid Rocket Propellants*
Knud Østergaard, *Gas–Liquid–Particle Operations in Chemical Reaction Engineering*
J. M. Prausnitz, *Thermodynamics of Fluid–Phase Equilibria at High Pressures*
Robert V. Macbeth, *The Burn-Out Phenomenon in Forced-Convection Boiling*
William Resnick and Benjamin Gal-Or, *Gas–Liquid Dispersions*

Volume 8

C. E. Lapple, *Electrostatic Phenomena with Particulates*
J. R. Kittrell, *Mathematical Modeling of Chemical Reactions*
W. P. Ledet and D. M. Himmelblau, *Decomposition Procedures for the Solving of Large Scale Systems*
R. Kumar and N. R. Kuloor, *The Formation of Bubbles and Drops*

Volume 9

Renato G. Bautista, *Hydrometallurgy*
Kishan B. Mathur and Norman Epstein, *Dynamics of Spouted Beds*
W. C. Reynolds, *Recent Advances in the Computation of Turbulent Flows*
R. E. Peck and D. T. Wasan, *Drying of Solid Particles and Sheets*

Volume 10

G. E. O'Connor and T. W. F. Russell, *Heat Transfer in Tubular Fluid–Fluid Systems*
P. C. Kapur, *Balling and Granulation*
Richard S. H. Mah and Mordechai Shacham, *Pipeline Network Design and Synthesis*
J. Robert Selman and Charles W. Tobias, *Mass-Transfer Measurements by the Limiting-Current Technique*

Volume 11

Jean-Claude Charpentier, *Mass-Transfer Rates in Gas–Liquid Absorbers and Reactors*
Dee H. Barker and C. R. Mitra, *The Indian Chemical Industry–Its Development and Needs*
Lawrence L. Tavlarides and Michael Stamatoudis, *The Analysis of Interphase Reactions and Mass Transfer in Liquid–Liquid Dispersions*
Terukatsu Miyauchi, Shintaro Furusaki, Shigeharu Morooka, and Yoneichi Ikeda, *Transport Phenomena and Reaction in Fluidized Catalyst Beds*

Volume 12

C. D. Prater, J. Wei, V. W. Weekman, Jr., and B. Gross, *A Reaction Engineering Case History: Coke Burning in Thermofor Catalytic Cracking Regenerators*

Costel D. Denson, *Stripping Operations in Polymer Processing*

Robert C. Reid, *Rapid Phase Transitions from Liquid to Vapor*

John H. Seinfeld, *Atmospheric Diffusion Theory*

Volume 13

Edward G. Jefferson, *Future Opportunities in Chemical Engineering*

Eli Ruckenstein, *Analysis of Transport Phenomena Using Scaling and Physical Models*

Rohit Khanna and John H. Seinfeld, *Mathematical Modeling of Packed Bed Reactors: Numerical Solutions and Control Model Development*

Michael P. Ramage, Kenneth R. Graziano, Paul H. Schipper, Frederick J. Krambeck, and Byung C. Choi, *KINPTR (Mobil's Kinetic Reforming Model): A Review of Mobil's Industrial Process Modeling Philosophy*

Volume 14

Richard D. Colberg and Manfred Morari, *Analysis and Synthesis of Resilient Heat Exchanger Networks*

Richard J. Quann, Robert A. Ware, Chi-Wen Hung, and James Wei, *Catalytic Hydrometallation of Petroleum*

Kent David, *The Safety Matrix: People Applying Technology to Yield Safe Chemical Plants and Products*

Volume 15

Pierre M. Adler, Ali Nadim, and Howard Brenner, *Rheological Models of Suspensions*

Stanley M. Englund, *Opportunities in the Design of Inherently Safer Chemical Plants*

H. J. Ploehn and W. B. Russel, *Interations between Colloidal Particles and Soluble Polymers*

Volume 16

Perspectives in Chemical Engineering: Research and Education

Clark K. Colton, *Editor*

Historical Perspective and Overview

L. E. Scriven, *On the Emergence and Evolution of Chemical Engineering*

Ralph Landau, *Academic–Industrial Interaction in the Early Development of Chemical Engineering*

James Wei, *Future Directions of Chemical Engineering*

Fluid Mechanics and Transport

L. G. Leal, *Challenges and Opportunities in Fluid Mechanics and Transport Phenomena*

William B. Russel, *Fluid Mechanics and Transport Research in Chemical Engineering*

J. R. A. Pearson, *Fluid Mechanics and Transport Phenomena*

Thermodynamics

Keith E. Gubbins, *Thermodynamics*
J. M. Prausnitz, *Chemical Engineering Thermodynamics: Continuity and Expanding Frontiers*
H. Ted Davis, *Future Opportunities in Thermodynamics*

Kinetics, Catalysis, and Reactor Engineering

Alexis T. Bell, *Reflections on the Current Status and Future Directions of Chemical Reaction Engineering*
James R. Katzer and S. S. Wong, *Frontiers in Chemical Reaction Engineering*
L. Louis Hegedus, *Catalyst Design*

Environmental Protection and Energy

John H. Seinfeld, *Environmental Chemical Engineering*
T. W. F. Russell, *Energy and Environmental Concerns*
Janos M. Beer, Jack B. Howard, John P. Longwell, and Adel F. Sarofim, *The Role of Chemical Engineering in Fuel Manufacture and Use of Fuels*

Polymers

Matthew Tirrell, *Polymer Science in Chemical Engineering*
Richard A. Register and Stuart L. Cooper, *Chemical Engineers in Polymer Science: The Need for an Interdisciplinary Approach*

Microelectronic and Optical Materials

Larry F. Thompson, *Chemical Engineering Research Opportunities in Electronic and Optical Materials Research*
Klavs F. Jensen, *Chemical Engineering in the Processing of Electronic and Optical Materials: A Discussion*

Bioengineering

James E. Bailey, *Bioprocess Engineering*
Arthur E. Humphrey, *Some Unsolved Problems of Biotechnology*
Channing Robertson, *Chemical Engineering: Its Role in the Medical and Health Sciences*

Process Engineering

Arthur W. Westerberg, *Process Engineering*
Manfred Morari, *Process Control Theory: Reflections on the Past Decade and Goals for the Next*
James M. Douglas, *The Paradigm After Next*
George Stephanopoulos, *Symbolic Computing and Artificial Intelligence in Chemical Engineering: A New Challenge*

The Identity of Our Profession

Morton M. Denn, *The Identity of Our Profession*

Volume 17

Y. T. Shah, *Design Parameters for Mechanically Agitated Reactors*
Mooson Kwauk, *Particulate Fluidization: An Overview*

Volume 18

E. James Davis, *Microchemical Engineering: The Physics and Chemistry of the Microparticle*
Selim M. Senkan, *Detailed Chemical Kinetic Modeling: Chemical Reaction Engineering of the Future*
Lorenz T. Biegler, *Optimization Strategies for Complex Process Models*

Volume 19

Robert Langer, *Polymer Systems for Controlled Release of Macromolecules, Immobilized Enzyme Medical Bioreactors, and Tissue Engineering*
J. J. Linderman, P. A. Mahama, K. E. Forsten, and D. A. Lauffenburger, *Diffusion and Probability in Receptor Binding and Signaling*
Rakesh K. Jain, *Transport Phenomena in Tumors*
R. Krishna, *A Systems Approach to Multiphase Reactor Selection*
David T. Allen, *Pollution Prevention: Engineering Design at Macro-, Meso-, and Microscales*
John H. Seinfeld, Jean. M. Andino, Frank M. Bowman, Hali J. L. Forstner, and Spyros Pandis, *Tropospheric Chemistry*

Volume 20

Arthur M. Squires, *Origins of the Fast Fluid Bed*
Yu Zhiqing, *Application Collocation*
Youchu Li, *Hydrodynamics*
Li Jinghai, *Modeling*
Yu Zhiqing and Jin Yong, *Heat and Mass Transfer*
Mooson Kwauk, *Powder Assessment*
Li Hongzhong, *Hardware Development*
Youchu Li and Xuyi Zhang, *Circulating Fluidized Bed Combustion*
Chen Junwu, Cao Hanchang, and Liu Taiji, *Catalyst Regeneration in Fluid Catalytic Cracking*

Volume 21

Christopher J. Nagel, Chonghun Han, and George Stephanopoulos, *Modeling Languages: Declarative and Imperative Descriptions of Chemical Reactions and Processing Systems*
Chonghun Han, George Stephanopoulos, and James M. Douglas, *Automation in Design: The Conceptual Synthesis of Chemical Processing Schemes*
Michael L. Mavrovouniotis, *Symbolic and Quantitative Reasoning: Design of Reaction Pathways through Recursive Satisfaction of Constraints*
Christopher Nagel and George Stephanopoulos, *Inductive and Deductive Reasoning: The Case of Identifying Potential Hazards in Chemical Processes*
Keven G. Joback and George Stephanopoulos, *Searching Spaces of Discrete Solutions: The Design of Molecules Processing Desired Physical Properties*

Volume 22

Chonghun Han, Ramachandran Lakshmanan, Bhavik Bakshi, and George Stephanopoulos, *Nonmonotonic Reasoning: The Synthesis of Operating Procedures in Chemical Plants*
Pedro M. Saraiva, *Inductive and Analogical Learning: Data-Driven Improvement of Process Operations*

Alexandros Koulouris, Bhavik R. Bakshi and George Stephanopoulos, *Empirical Learning through Neural Networks: The Wave-Net Solution*
Bhavik R. Bakshi and George Stephanopoulos, *Reasoning in Time: Modeling, Analysis, and Pattern Recognition of Temporal Process Trends*
Matthew J. Realff, *Intelligence in Numerical Computing: Improving Batch Scheduling Algorithms through Explanation-Based Learning*

Volume 23

Jeffrey J. Siirola, *Industrial Applications of Chemical Process Synthesis*
Arthur W. Westerberg and Oliver Wahnschafft, *The Synthesis of Distillation-Based Separation Systems*
Ignacio E. Grossmann, *Mixed-Integer Optimization Techniques for Algorithmic Process Synthesis*
Subash Balakrishna and Lorenz T. Biegler, *Chemical Reactor Network Targeting and Integration: An Optimization Approach*
Steve Walsh and John Perkins, *Operability and Control in Process Synthesis and Design*

Volume 24

Raffaella Ocone and Gianni Astarita, *Kinetics and Thermodynamics in Multicomponent Mixtures*
Arvind Varma, Alexander S. Rogachev, Alexandra S. Mukasyan, and Stephen Hwang, *Combustion Synthesis of Advanced Materials: Principles and Applications*
J. A. M. Kuipers and W. P. M. van Swaaij, *Computational Fluid Dynamics Applied to Chemical Reaction Engineering*
Ronald E. Schmitt, Howard Klee, Debora M. Sparks, and Mahesh K. Podar, *Using Relative Risk Analysis to Set Priorities for Pollution Prevention at a Petroleum Refinery*

Volume 25

J. F. Davis, M. J. Piovoso, K. A. Hoo, and B. R. Bakshi, *Process Data Analysis and Interpretation*
J. M. Ottino, P. DeRoussel, S. Hansen, and D. V. Khakhar, *Mixing and Dispersion of Viscous Liquids and Powdered Solids*
Peter L. Silveston, Li Chengyue, Yuan Wei-Kang, *Application of Periodic Operation to Sulfur Dioxide Oxidation*

Volume 26

J. B. Joshi, N. S. Deshpande, M. Dinkar, and D. V. Phanikumar, *Hydrodynamic Stability of Multiphase Reactors*
Michael Nikolaou, *Model Predictive Controllers: A Critical Synthesis of Theory and Industrial Needs*

Volume 27

William R. Moser, Josef Find, Sean C. Emerson, and Ivo M. Krausz, *Engineered Synthesis of Nanostructured Materials and Catalysts*

Bruce C. Gates, *Supported Nanostructured Catalysts: Metal Complexes and Metal Clusters*
Ralph T. Yang, *Nanostructured Adsorbents*
Thomas J. Webster, *Nanophase Ceramics: The Future Orthopedic and Dental Implant Material*
Yu-Ming Lin, Mildred S. Dresselhaus, and Jackie Y. Ying, *Fabrication, Structure, and Transport Properties if Nanowires*

Volume 28

Qiliang Yan and Juan J. DePablo, *Hyper-Parallel Tempering Monte Carlo and Its Applications*
Pablo G. Debenedetti, Frank H. Stillinger, Thomas M. Truskett, and Catherine P. Lewis, *Theory of Supercooled Liquids and Glasses: Energy Landscape and Statistical Geometry Perspectives*
Michael W. Deem, *A Statistical Mechanical Approach to Combinatorial Chemistry*
Venkat Ganesan and Glenn H. Fredrickson, *Fluctuation Effects in Microemulsion Reaction Media*
David B. Graves and Cameron F. Abrams, *Molecular Dynamics Simulations of Ion-Surface Interactions with Applications to Plasma Processing*
Christian M. Lastoskie and Keith E. Gubbins, *Characterization of Porous Materials Using Molecular Theory and Simulation*
Dimitrios Maroudas, *Modeling of Radical-Surface Interactions in the Plasma-Enhanced Chemical Vapor Deposition of Silicon Thin Films*
Sanat Kumar, M. Antonio Floriano, and Athanassios Z. Panagiotopoulos, *Nanostructured Formation and Phase Separation in Surfactant Solutions*
Stanley I. Sandler, Amadeu K. Sum, and Shiang-Tai Lin, *Some Chemical Engineering Applications of Quantum Chemical Calculations*
Bernhardt L. Trout, *Car-Parrinello Methods in Chemical Engineering: Their Scope and Potential*
R. A. van Santen and X. Rozanska, *Theory of Zeolite Catalysis*
Zhen-Gang Wang, *Morphology, Fluctuation, Metastability and Kinetics in Ordered Block Copolymers*

ISBN 0-12-008528-3

DATE DUE